Measuring and Modeling the Universe

We have just concluded a remarkable century: the 1917 publication of Einstein's general theory of relativity, Carnegie astronomer Edwin Hubble's 1929 discovery of the expansion of the universe, evidence for the existence of dark matter, and the discovery of a mysterious dark energy, which is causing the universe to speed up its expansion. This comprehensive volume reviews the current theory and measurement of various parameters related to the evolution of the universe. Topics include inflation, string theory, the history of cosmology in the context of current measurements being made of the Hubble constant, the matter density, and dark energy, including observational results from the Sloan Digital Sky Survey, Keck, Magellan, cosmic microwave background experiments, *Hubble Space Telescope*, and *Chandra*. With chapters by leading authorities in the field, this book is a valuable resource for graduate students and professional research astronomers.

WENDY L. FREEDMAN is the Crawford H. Greenewalt Director at the Carnegie Observatories of the Carnegie Institution of Washington in Pasadena, California. She received a Carnegie Fellowship at the Observatories in 1984, and joined the permanent faculty in 1987. Dr. Freedman is a member of the National Academy of Sciences. She received the 2002 American Philosophical Society's Magellanic Prize, and the 1994 Marc Aaronson Prize and Lectureship, both citing her fundamental contributions to cosmology. One of her principal research interests is aimed at measuring an accurate value for the rate at which the universe is expanding, a quantity that yields both the age and size of the universe.

This series of four books celebrates the Centennial of the Carnegie Institution of Washington, and is based on a set of four special symposia held by the Observatories in Pasadena. Each symposium explored an astronomical topic of major historical and current interest at the Observatories, and each resulting book contains a set of comprehensive, authoritative review articles by leading experts in the field.

Series Editor: Luis C. Ho.
Luis Ho received his undergraduate education at Harvard University and his Ph.D. in Astronomy from the University of California at Berkeley. He is currently a Staff Astronomer at the Carnegie Observatories, where he conducts research on black holes, accretion physics in galactic nuclei, and star formation processes.

Carnegie Observatories Astrophysics Series
Volume 2

MEASURING AND MODELING THE UNIVERSE

Edited by

WENDY L. FREEDMAN

CAMBRIDGE
UNIVERSITY PRESS

CAMBRIDGE
UNIVERSITY PRESS

University Printing House, Cambridge CB2 8BS, United Kingdom

Cambridge University Press is part of the University of Cambridge.

It furthers the University's mission by disseminating knowledge in the pursuit of education, learning and research at the highest international levels of excellence.

www.cambridge.org
Information on this title: www.cambridge.org/9780521755764

First published 2004

A catalogue record for this publication is available from the British Library

ISBN 978-0-521-75576-4 Hardback

Contents

Introduction

The Carnegie Institution of Washington celebrated its centennial in the year 2002. Following on a suggestion of staff astronomer Luis Ho, the Carnegie staff unanimously agreed that a fitting way to celebrate the Centennial of the Observatories was to have series of symposia devoted to topics where Carnegie astronomers have been actively engaged. And cosmology, with its longstanding tradition going back to Edwin Hubble near the turn of the last century, is certainly an ideal topic with which to mark a centennial celebration. Hence, Symposium 2 of the series was devoted to "Measuring and Modeling the Universe." This symposium was held 17–22 November, 2002, in Pasadena, California, with approximately 100 people in attendance. The meeting focused on current theory and measurement of various parameters related to the evolution of the Universe. This volume presents a set of refereed review articles, written mostly by invited review speakers.

Astronomy in Pasadena, and indeed for modern cosmology, was catalyzed at the turn of the last century by George Ellery Hale. It seems fitting at the occasion of the Carnegie Centennial to recall some of the history that led to the development of Mount Wilson, and the establishment of the Observatory. In 1902, Hale read in the *Chicago Tribune* that the steel magnate, Andrew Carnegie had donated a gift of $10,000,000, founding the Carnegie Institution of Washington "to encourage investigation, research, and discovery in the broadest and most liberal manner, and the application of knowledge to the improvement of mankind." Hale secured funding from the Carnegie Institution for the construction a giant, 60-inch reflecting telescope to be erected on Mount Wilson. He was convinced that the way to make progress in optical astronomy was to build large, reflecting telescopes. Remarkably, in 1906, two years before the 60-inch had even begun its operations, Hale was already devising a means for the construction of a 100-inch reflector. This telescope came into operation in 1917.

Discoveries followed rapidly. First, our Sun was displaced from the center of the Universe with Harlow Shapley's study of the globular cluster system in the Milky Way. Soon thereafter followed Edwin Hubble's discovery of Cepheids in NGC 6822, M31, and M33, establishing the existence of galaxies other than the Milky Way. By 1929, Hubble had demonstrated the correlation of distance and recession velocity. In 1915, Albert Einstein had formulated his theory of general relativity, and recognized that a stationary universe would not survive for very long — that is, it would tend to either contract or expand. General relativity provided a framework for understanding the unexpected motions of galaxies observed by Hubble, and led to our current picture of an expanding Universe, the backbone of modern cosmology.

Hale's journey did not end with the Mount Wilson 100-inch telescope. His undergraduate days at the Massachusetts Institute of Technology in Boston motivated him to create a similar intellectual center in Pasadena. In 1910, the Throop Polytechnic Institute was a small elementary and high school with a total student body of 31. Elected to the Board of Throop trustees in 1907, Hale ultimately convinced the chemist Arthur Noyes (an old college friend and, at the time, the acting president of M.I.T.) and in 1921 the physicist Robert Millikan (soon to win a Nobel Prize for his experimental work in determining the mass of the electron) to join him in establishing the California Institute of Technology.

In 1921 (preceding Hubble's discovery of galaxies and the expansion of the Universe), Hale returned to his quest for building ever-larger telescopes. He began to rally Mount Wilson astronomer, Francis Pease, to design a 300-inch telescope. In 1928, he successfully persuaded Wickliffe Rose of the Rockefeller Foundation to give a gift of $6,000,000 toward the construction of a 200-inch telescope. The story is often told that Rockefeller emphatically balked at the notion of giving money to an institution of Andrew Carnegie's (it being like bringing coals to Newcastle), and Wickliffe insisted that the money be given to Caltech rather than the Carnegie Institution. Hale had no intention of waiting for another donor (the strong preference of then Carnegie President John C. Merriam). According to his biographer, Helen Wright, Hale wrote: "The general interests of scientific research are far more important than the reputation of any single institution ... The accomplishment of the work is the main thing, the other question after all is an incidental one."

In 1926, an agreement was signed between Carnegie and Caltech. The impact of World War II slowed construction, but in 1948 the 200-inch telescope on Mount Palomar became operational. In the same year an astronomy department was founded at Caltech, and Carnegie/Mount Wilson astronomers were cross-appointed to the Caltech faculty. The two institutions continued to run Palomar jointly until about 1980, as the Hale Observatory. For about two-thirds of the 20th century, it is no exaggeration to say that the telescopes and institutions connected with Mount Wilson and Mount Palomar dominated the world of astronomy. Big telescopes made possible the cosmological discoveries of the 20th century.

Most recently, Carnegie, with its partners, Harvard University, M.I.T., and the Universities of Michigan and Arizona, have built two twin 6.5-meter telescopes at Las Campanas in Chile: The Baade and Clay Telescopes are another very fitting tribute to the Carnegie Centennial. In the past year, three major new instruments have been commissioned on the Magellan telescopes, opening a new era in Carnegie astronomy. And ambitious new plans are now underway for the construction of a 20-meter telescope named the Giant Magellan Telescope. The next century is beginning in as exciting manner as the previous one, and we look forward to equally interesting new results!

The topics in this volume include inflation, string theory, the history of cosmology in the context of current measurements being made of the Hubble constant, the matter density, and dark energy, including observational results from the Sloan Digital Sky Survey, Keck, Magellan, cosmic microwave background experiments, *Hubble Space Telescope* and *Chandra*. Unlike most symposium proceedings, all of the papers in this volume have been refereed by experts in the field. We hope that the proceedings will serve as a valuable resource for graduate students as well as professionals in astronomy and related fields.

This symposium could not have taken place without the expert assistance of a number of individuals. I would like to thank Steve Wilson, Silvia Hutchison, Becky Lynn, Scott Rubel, and Greg Ortiz for organizing the meeting and taking care of the logistics. I thank also the many individuals who served as referees of the papers. To Luis Ho, a very special thank you. Luis conceived of the Carnegie Centennial Symposia, persuaded others of us at the Observatories to become organizers of the various symposia, gently prodded and reminded us to send out notices for the meetings at the appropriate time, served as the super-editor of all four of the volumes, and after I became Director in March, 2003, he assumed most of the workload for this volume, helping to ensure that the papers were submitted by very busy authors, and that they were edited with care. And finally, I thank all of the authors who contributed to the volume.

Wendy L. Freedman
Crawford H. Greenewalt Chair
Director, Carnegie Observatories
February 2004

List of Participants

Agol, Eric	Caltech, USA
Albrecht, Andreas	U. C. Davis, USA
Amblard, Alexandre	U. C. Berkeley, USA
Annis, James	Fermilab, USA
Bahcall, Neta	Princeton University, USA
Bernardi, Mariangela	Carnegie Mellon University, USA
Blandford, Roger	Caltech, USA
Blanton, Michael	New York University, USA
Bloom, Elliott	SLAC/Stanford University, USA
Bond, J. Richard	University of Toronto, Canada
Carroll, Sean	University of Chicago, USA
Chapman, Scott	Caltech, USA
Chartas, George	Penn State University, USA
Chave, Robert	Robert Chave Applied Physics, USA
Christianson, Gale	Indiana State University, USA
Church, Sarah	Stanford University, USA
Clancy, Dominic	University of Crete, Greece
Colless, Matthew	Australian N. University, Australia
Cooray, Asantha	Caltech, USA
Dekel, Avishai	Hebrew University of Jerusalem, Israel
Dressler, Alan	Carnegie Observatories, USA
Eisenstein, Daniel	University of Arizona, USA
Ellis, Richard	Caltech, USA
Faber, Sandra	U. C. Santa Cruz, USA
Farrar, Glennys	New York University, USA
Filippenko, Alex	U. C. Berkeley, USA
Flanagan, Eanna	Cornell University, USA
Freedman, Wendy	Carnegie Observatories, USA
Gingerich, Owen	Harvard University, USA
Gladders, Mike	Carnegie Observatories, USA
Guth, Alan	MIT, USA
Hamuy, Mario	Carnegie Observatories, USA

Hivon, Eric	Caltech, USA
Ho, Luis	Carnegie Observatories, USA
Hoekstra, Henk	CITA, Canada
Holz, Daniel	U. C. Santa Barbara, USA
Impey, Chris	University of Arizona, USA
Jain, Bhuvnesh	University of Pennsylvania, USA
Jensen, Joseph	Gemini Observatory, USA
Kamionkowski, Marc	Caltech, USA
Katz, Neal	University of Massachusetts, USA
Kim, Jihn E.	Seoul National University, Korea
Kochanek, Christopher	Harvard-Smithsonian Center for Astrophysics, USA
Koo, David	UCO/Lick Observatory, USA
Lange, Andrew	Caltech, USA
Lauer, Tod	NOAO, USA
Lazzarini, Albert	Caltech, USA
Lee, Jounghun	University of Tokyo, Japan
Livio, Mario	Space Telescope Science Institute, USA
Longair, Malcolm	Cambridge University, UK
Maller, Ariyeh	University of Massachusetts, USA
Marble, Andy	University of Arizona, USA
McClure, Megan	University of Toronto, Canada
Meylan, Georges	Space Telescope Science Institute, USA
Mould, Jeremy	NOAO, USA
Nishioka, Hiroaki	Hiroshima University, Japan
Novikov, Igor	Theoretical Astrophysics Center, Denmark
Outram, Phil	University of Durham, UK
Page, Lyman	Princeton University, USA
Perlmutter, Saul	U. C. Berkeley, USA
Peterson, Jeff	
Petry, Cathy	Steward Observatory, USA
Phillips, Mark	Carnegie Observatories, Chile
Phinney, E. Sterl	Caltech, USA
Randall, Lisa	Princeton University, USA
Readhead, Anthony	Caltech, USA
Reese, Erik	U. C. Berkeley, USA
Rich, R. Michael	UCLA, USA
Rubin, Vera	DTM, USA
Rusholme, Ben	Stanford University, USA
Sadoulet, Bernard	U. C. Berkeley, USA
Scargle, Jeffrey	NASA, USA
Schwarz, John	Caltech, USA
Seljak, Uros	Princeton University, USA
Shectman, Stephen	Carnegie Observatories, USA
Silk, Joseph	University of Oxford, UK
Steigman, Gary	Ohio State University, USA
Treu, Tommaso	Caltech, USA
Trimble, Virginia	University of Maryland, USA
Turner, Edwin	Princeton University, USA

Turner, Michael	University of Chicago, USA
Tytler, David	U. C. San Diego, USA
Weiner, Ben	U. C. Santa Cruz, USA
Wilson, Brian	University of Toronto, Canada
Wright, Ned	UCLA, USA
Zaldarriaga, Matias	New York University, USA

1

A brief history of cosmology

MALCOLM S. LONGAIR
Cavendish Laboratory, Cambridge, UK

Abstract

Some highlights of the history of modern cosmology and the lessons to be learned from the successes and blind alleys of the past are described. This heritage forms the background to the lectures and discussions at this Second Carnegie Centennial Symposium, which celebrates the remarkable contributions of the Carnegie Institution in the support of astronomical and cosmological research.

1.1   Introduction

It is a great honor to be invited to give this introductory address at the Second Carnegie Centennial Symposium to celebrate the outstanding achievements of the Observatories of the Carnegie Institution of Washington. I assume that the point of opening this meeting with a survey of the history of cosmology is not only to celebrate the remarkable achievements of modern observational and theoretical cosmology, but also to provide lessons for our time, which may enable us all to avoid some of the errors that we now recognize were made in the past. I am bound to say that I am not at all optimistic that this second aim will be achieved. I recall that, when I gave a similar talk many years ago with the same intention, Giancarlo Setti made the percipient remark:

Cosmology is like love; everyone likes to make their own mistakes.

By its very nature, the subject involves the confrontation of theoretical speculation with cosmological observations, the scepticism of the hardened observer about taking anything a theorist says seriously, the problems of pushing observations to the very limits of technological capability, and sometimes beyond these, resulting in dubious data, and so on. These confrontations have happened many times in the past. My suspicion is that the community of astronomers and cosmologists is now sufficiently large for false dogma and insecure observations to have only limited shelf-lives, but we must remain vigilant. Nonetheless, it is intriguing to survey the present state of cosmology, with its extraordinary successes and challenges, and recognize the many similarities to those that faced the great scientists of the past. I leave it to readers to draw their own preferred analogies.

The history of cosmology is a vast and fascinating subject, and I will only touch on some of the highlights of that story. I have given a more detailed account of that history elsewhere (Longair 1995), and it is a subject that repays careful study. To my regret, there will be little space to do justice to the technological achievements that have made modern cosmology a rather exact science (see Longair 2001). Without these developments, none of us would be celebrating the achievements of modern cosmology at this symposium.

Before getting down to the history, let me contribute a personal appreciation of Andrew Carnegie's philanthropy. He gave away \$350M of his fortune of \$400M to charitable causes. Among the more remarkable of these was the founding of about 3000 libraries worldwide, including five in my home town of Dundee in Scotland, only about 30 miles from Dunfermline, Carnegie's birthplace. Here is a quotation from the Carnegie Libraries of Scotland web-site:

> *When the library was officially opened on 22 October 1908, [Charles] Barrie [a former Lord Provost] was asked to perform the opening ceremony ... There was a banquet afterwards in the Victoria Art Galleries hosted by Lord Provost Longair.*

Lord Provost Longair was my great, great uncle. I remember as a small boy going regularly to the Dundee Public Libraries to learn about rockets and space flight. Little did I realise then that, more than 50 years later, I would be participating in the celebrations of the centenary of the founding of the Carnegie Institution.

1.2   Observational Cosmology to 1929

The earliest cosmologies of the modern era were speculative cosmologies. The "island universe" model of Descartes, published in *The World* of 1636, involved an interlocking jigsaw puzzle of solar systems. Wright's *An Original Theory of the Universe* of 1750 involved spheres of stars and solar systems, while Kant in 1755 and Lambert in 1761 developed the first hierarchical, or fractal, pictures of the Universe (see Harrison 2001). The problem with these early cosmologies was that they lacked observational validation. When these ideas were put forward, the only star whose distance was known was the Sun. The first parallax measurements of stars were only made in the 1830s by Friedrich Bessel, Friedrich Georg Wilhelm Struve and Thomas Henderson.

The first quantitative estimates of the scale and structure of the Universe were made by William Herschel in the late 18th century. Herschel's model of the large-scale structure of the Universe was based upon star counts and provided the first quantitative evidence for the "island universe" picture of Wright, Kant, Swedenborg and Laplace. In deriving his famous model for our Galaxy, Herschel assumed that all stars have the same absolute luminosities. The importance of interstellar extinction in restricting the number counts of stars to a relatively local region of our Galaxy was only fully appreciated in the early 20th century.

John Michell had already warned Herschel that the assumption that the stars have a fixed luminosity was incorrect. This is the same John Michell who was Woodwardian Professor of Geology at Queen's College, Cambridge, before becoming the rector of Thornhill in Yorkshire in 1767. He designed and built what we now know as the Cavendish experiment to measure the mean density of the Earth. Nowadays, he is rightly remembered as the inventor of black holes. In 1767, he showed that there must be a dispersion in the absolute luminosities of the stars from observations of bright star clusters. Despite this warning, Herschel ignored the problem and proceeded to produce a number of different versions of the structure of our Galaxy. In 1802, Herschel measured the magnitudes of visual binary stars and was forced to agreed with Michell's conclusion. Equally troubling was the fact that observations with his magnificent 40-foot telescope showed there was no edge to the

Galaxy. He continued to find stars the fainter he looked—evidently, the stellar system was unbounded. Eventually, Herschel lost faith in his model of the Galaxy.

The desire to observe the Universe with telescopes of greater and greater aperture continued throughout the nineteenth century. The largest reflecting telescope constructed during that century was the great 72-inch reflector at Birr Castle in Ireland by William Parsons, the 3rd Earl of Rosse. This "Great Leviathan" was moved by ropes and astronomical objects could be tracked by moving the barrel of the telescope between the two large walls, which also accommodated a movable observing platform at the Newtonian focus of the telescope. Observations were made by eye and so the "length of the exposure" was limited to about a tenth of a second. Despite the difficulties of making observations and the inclement weather in central Ireland, Lord Rosse was able to resolve nebulae into stars and, perhaps most important of all, discovered the spiral structure of galaxies, the most famous drawing being his sketch of M51.

The revolutions that led to the discipline of extragalactic astronomy as we know it today were the use of photography to record astronomical images and the shift from refracting to reflecting telescope designs. The Yerkes 40-inch refractor was the end of the line so far as refracting telescopes were concerned. The much more compact reflecting design had the advantage of greater collecting area, but was much more sensitive to tracking and guiding errors. Many key technologies were developed during the latter half of the nineteenth century, thanks to pioneers such as Lewis Morris Rutherfurd, John Draper, Andrew Common and George Carver. These pioneers solved the problems of the tracking and pointing of reflecting telescopes, an invention of particular importance being the adjustable plate holder, which enables the observer to maintain the pointing of the telescope with high precision.

The resulting technical advances contributed to the remarkable achievement of James Keeler and his colleagues at the Lick Observatory in enhancing the performance of the 91-cm Crossley reflector to become the premier instrument for astronomical imaging. During the commissioning of the Crossley reflector in 1900, Keeler obtained spectacular images of spiral nebulae, including his famous image of M51. Not only were the details of its spiral structure observed in unprecedented detail, but there were also large numbers of fainter spiral nebulae of smaller angular size. If these were objects similar to the Andromeda Nebula M31, they must lie at very great distances from our Solar System. Tragically, just as this new era of astronomy was dawning, Keeler died of a stroke later that same year at the early age of only 42 (Osterbrock 1984).

George Ellery Hale plays a central role in the celebrations of the centenary of the Carnegie Institution. He is rightly regarded as the most successful astronomical entrepreneur of the modern era. He maintained an unswerving determination to construct successively larger and larger telescopes from the time of his directorship of the Yerkes Observation in the 1890s through the period when he became Director of the Mount Wilson Observatory in 1903 until his death in 1938. In 1895, he had persuaded his father to buy the 1.5 meter blank for a 60-inch reflecting telescope. The design was to be an enlarged version of the Calver-Common design for the 91-cm Crossley reflector at the Lick Observatory. Before the 60-inch telescope was completed, however, he persuaded J. D. Hooker to fund an even bigger telescope, the 100-inch telescope to be built on Mount Wilson. The technological challenges were proportionally greater, the mass of the telescope being 100 tons, but the basic Calver-Common design was retained. The optics were the responsibility of George Ritchey, an optical designer of genius, who was to come up with the ingenious optical configuration known as the Ritchey-Chrétien design, which enabled excellent imaging to be achieved over a wide field of view. The 60- and 100-inch telescopes became the prime telescopes for the study of the spiral nebulae, but these accomplishments were not achieved without an enormous effort on Hale's part.

The story of Hale's construction of these telescopes is a heroic tale. Equally impressive is Andrew Carnegie's generosity in enabling Hale to realise his vision. Carnegie's fateful visit to the Mount Wilson Observatory in 1906 has been recorded in the volume *The Legacy of George Ellery Hale* (Wright, Warnow, & Weiner 1972). Carnegie was clearly impressed by what he saw during his visit. As recorded in the local Pasadena newspaper, he remarked:

"We do not know what may be discovered here," he said. Franklin had little idea what would be the result of flying his kite. But we do know that this will mean the increase of our knowledge in regard to this great system of which we are part.

Mr. Hale has discovered here 1600 worlds about one of the stars which were not known before. We have found helium in the Sun, and after finding it there, we find it in the Earth. It all goes to show that all things are of a common origin.

To anyone who has had the fortune to be responsible for the operation of a large observatory facility, these remarks have heartening resonances. Carnegie did not quite get the science right so far as the stars were concerned, but he got it absolutely right so far as helium is concerned. Helium was discovered astronomically long before it was identified in the laboratory and is but one of many examples of how astronomical observations can provide key insights into the behavior of matter under circumstances which are only later reproduced in the laboratory. Plainly, Hale had carried out a very successful campaign in enthusing Carnegie about the importance of progress in astronomy.

Following his visit, Carnegie pledged an additional $10M to the endowment of the Carnegie Institution, specifically requesting that the benefaction be used to enable the work of the Observatory to proceed as rapidly as possible. This is the purist music to the ears of any Observatory Director, who knows that, while it is usually possible in the end to find the capital resources for ambitious projects, these cannot succeed without matching funds for operations in the long term. Carnegie's vision and understanding are models for benefactors of astronomy.

The construction of the 60- and 100-inch telescopes were stressful and Hale suffered a nervous breakdown in 1910. It is touching to read Carnegie's letter to Hale of 1911 (with the original spelling), urging him to take care of his health:

November 27, 1911

Mr dear Frend,-

Delited to read your long note this morning; not too long—every word tells, but pray show your good sense by keeping in check your passion for work, so that you may be spared to put the capstone upon your career, which should be one of the most remarkable ever lived.

Ever yours,

Andrew Carnegie.

Carnegie's benefaction was crucial for the completion of the 100-inch Hooker Telescope at Mount Wilson. The telescope was by far the largest in the world and incorporated all the lessons learned from the works of earlier telescope builders. Completed in 1918, this instrument was to dominate observational cosmology for the next 30 years until the commissioning of the Palomar 200-inch telescope in 1948.

The Hooker 100-inch telescope played a central role in the resolution of what became known as the "Great Debate," which concerned the related issues of the size of our own Galaxy and the nature of the white, or spiral, nebulae. This confrontation between Harlow Shapley and Heber D. Curtis is too well known to need much amplification here (see Christianson, this volume). In 1899, Scheiner had obtained a spectrogram of M31 and stated that the spectrum suggested a "cluster of Sun-like stars." In 1922, Öpik estimated the distance of M31 by comparing the mass-to-light ratio of the central region of M31 with that of our own Galaxy and found a distance of 440 kpc, suggesting that it lay well outside the confines of our Galaxy. The discovery of variable stars in spiral nebulae by Duncan in 1922 led to a flurry of activity and Hubble's famous discovery of Cepheid variables in M31.

Central to Hubble's use of Cepheid variable stars in M31 to measure its distance was the discovery of the period-luminosity relation for Cepheids in the Magellanic Clouds by Henrietta Leavitt (Leavitt 1912). Leavitt, like Annie Cannon, was profoundly deaf. While she is best remembered for her work on the Cepheid variables, her main work was the establishment of the North Polar Sequence, the accurate determination of the magnitude scale for stars in a region of sky which would always be accessible to observers in the Northern Hemisphere. By the time of her death in 1921, she had extended the North Polar Sequence from 2.7 to 21st magnitude, with errors less than 0.1 magnitudes. To achieve this, she used observations from 13 telescopes ranging from 0.5 to 60 inches in diameter and compared her scale using 5 different photographic photometric techniques. Without this fundamental work, the magnitude scale for galaxies could not have been established.

It is intriguing that by far the most stubborn pieces of observational evidence against what might be termed the long distance scale were van Maanen's measurements of the proper motions of spiral arms. It is now well understood how difficult it is to measure tiny displacements of any diffuse object—van Maanen's evidence was only refuted in 1933 by Edwin Hubble after a considerable observational effort.

Hubble's paper of 1925 establishing the extragalactic nature of the spiral nebulae is impressive enough (Hubble 1925), but to my mind his paper of the following year entitled *Extragalactic Nebulae* is even more compelling (Hubble 1926). In this paper, he provided the first more or less complete description of galaxies as extragalactic systems. The paper includes a morphological classification of galaxies into the classic Hubble types, estimates of the relative numbers of different types, estimates of mass-to-luminosity ratios for different types of galaxies and their average number densities. Finally, the mean mass density in galaxies in the Universe as a whole was derived for the first time. Adopting Einstein's static model for the Universe, the radius of curvature of the spherical geometry was $\mathcal{R} = 27,000$ Mpc and the total number of galaxies 3.5×10^{15}. Thus, by 1926, the first application of the ideas of relativistic cosmology to the Universe of galaxies had been made. Hubble concluded that the observations already extended to about 1/300 of the radius of the closed Einstein universe. The prophetic last sentence of his great paper reads

... with reasonable increases in the speed of the plates and size of telescopes, it may become possible to observe an appreciable fraction of the Einstein Universe.

It is no surprise that Hale began his campaign to raise funds for the 200-inch telescope in

1928. By the end of the year, he had received the promise of a grant of $6M from the Rockefeller foundation.

Hubble's important insights were soon followed by an even more remarkable discovery. In 1917, Vesto M. Slipher had published his heroic pioneering spectroscopic observations of 25 spiral nebulae (Slipher 1917). He realised that, for the spectroscopy of low surface brightness objects such as the spiral nebulae, the crucial factor was the f-ratio of the spectrograph camera, not the size of the telescope. The observations involved very long integrations of 20, 40 or even 80 hours with small telescopes.

The velocities of the galaxies inferred from the Doppler shifts of their absorption lines were typically about 570 km s^{-1}, far in excess of the velocity of any known object in our Galaxy. Furthermore, most of the velocities corresponded to the galaxies moving away from the Solar System, that is, the lines were *redshifted* to longer (red) wavelengths. In 1921, Carl Wilhelm Wirtz concluded that, when the data were averaged in a suitable way, "an approximate linear dependence of velocity upon apparent magnitude is visible" (Wirtz 1921). By 1929, Hubble had assembled approximate distances of 24 galaxies for which velocities had been measured, mostly by Slipher, all within 2 Mpc of our Galaxy. I have always been impressed that Hubble was able to find the law which bears his name from the very crude distance indicators which he had available. The first seven objects within 500 kpc had Cepheid distances; the distances of the next 13 were found assuming the brightest stars all had the same absolute magnitude; the last four, in the Virgo cluster, were estimated on the basis of the mean luminosities of nebulae in the cluster. From these meager data, Hubble derived his famous redshift-distance relation (Hubble 1929). If the redshifts z are interpreted as the Doppler shifts of galaxies due to their recession velocities v, the relation can be written $v = H_0 r$, where H_0 is *Hubble's constant*.

Milton Humason had by then mastered the use of the 100-inch telescope for obtaining the spectra of faint galaxies and by 1934 Humason and Hubble had extended the velocity-distance relation to 7% of the speed of light (Humason & Hubble 1934). Furthermore, Hubble realised that he could test for the isotropy and homogeneity of the Universe by counting the numbers of faint galaxies. Hubble established that the numbers of galaxies increased with increasing apparent magnitude in almost exactly the fashion expected if they were uniformly distributed in space.

Even before 1929, however, it was appreciated that Hubble's law was expected according to world models based upon the general theory of relativity.

1.3   Theoretical Cosmology to 1939

Let us turn to theoretical cosmology and the history of Einstein's static model for the Universe. Working independently, Lobachevsky in Kazan in Russia and Bolyai in Hungary solved the problem of the existence of geometries that violated Euclid's fifth axiom in 1825. These were the first self-consistent hyperbolic (non-Euclidean) geometries. In his great text *On the Principles of Geometry* (1825), Lobachevsky worked out the minimum parallax of any star in hyperbolic geometry

$$\theta = \arctan\left(\frac{a}{\mathcal{R}}\right) \tag{1.1}$$

where a is the radius of the Earth's orbit and $\mathcal{R}$ the radius of curvature of the geometry. In his textbook, he found a minimum value of $\mathcal{R} \geq 1.66 \times 10^5$ AU. What is intriguing is that this estimate was made 8 years before Bessell's announcement of the first successful parallax measurement of 61 Cygni. In making his estimate, Lobachevsky used the observational upper limit to the parallax of bright stars. In a statement which will warm the heart of observational astronomers, and which is particularly apposite in the light of what we will hear at this meeting, he remarked

There is no means other than astronomical observations for judging the exactness which attaches to the calculations of ordinary geometry.

The pioneering works of Lobachevsky and Bolyai led to Riemann's introduction of quadratic differential forms, his generalization of their results to non-Euclidean geometries, and his discovery of spaces of positive curvature—that is, spherical non-Euclidean geometries.

Unlike his other great discoveries, Einstein's route to general relativity was long and tortuous. Four ideas were important in his search for a self-consistent relativistic theory of gravity:

- The influence of gravity on light
- The principle of equivalence
- Riemannian spacetime
- The principle of equivalence

Toward the end of 1912, he realised that what was needed was non-Euclidean geometry. Einstein consulted his old school friend, Marcel Grossmann, about the most general forms of transformation between frames of reference for metrics of the form

$$ds^2 = g_{\mu\nu}\, dx^\mu dx^\nu. \tag{1.2}$$

Grossmann soon came back with the answer that the most general transformation formulae were the Riemannian geometries, but that they had the "bad feature" that they are nonlinear. Einstein instantly recognized that, on the contrary, this was a great advantage since any satisfactory theory of relativistic gravity must be nonlinear.

After further years of struggle, during which he and Grossmann were ploughing very much a lone furrow, general relativity was formulated in its definitive form in 1915 (Einstein 1915). In 1916, Willem de Sitter and Paul Ehrenfest suggested in correspondence that a spherical 4-dimensional spacetime would eliminate the problems of the boundary conditions at infinity, which pose insuperable problems for Newtonian cosmological models. In 1917, Einstein realised that, in general relativity, he had for the first time a theory which could be used to construct fully self-consistent models for the Universe as a whole (Einstein 1917). At that time, the expansion of the Universe had not been discovered.

One of objectives of Einstein's program was to incorporate into the structure of general relativity what he called *Mach's Principle*, meaning that the local inertial frame of reference should be determined by the large-scale distribution of matter in the Universe. There was, however, a further problem, first noted by Newton, that static model universes are unstable under gravity. Einstein proposed to solve both problems by introducing an additional term into the field equations, the *cosmological constant* Λ. In Newtonian terms, the cosmological constant corresponds to a repulsive force $\vec{f}$ acting on a test particle at distance $\vec{r}$, $\vec{f} = \frac{1}{3}\Lambda\vec{r}$. Unlike gravity, this force is independent of the density of matter. The Λ-term has negligible influence on the scale of the Solar System and is only appreciable on cosmological scales.

The equation that describes the expansion becomes

$$\frac{d^2 R}{dt^2} = -\frac{4\pi G \rho_0}{3R^2} + \frac{1}{3}\Lambda R \tag{1.3}$$

The first term on the right-hand side describes the deceleration due to gravity and the second what Zel'dovich referred to as the "repulsive effect of the vacuum" (Zel'dovich 1968). At that time, the physical significance of the Λ term was not understood.

Einstein believed that he had incorporated Mach's Principle into general relativity. In his words,

> *The inertial structure of spacetime was to be "exhaustively conditioned and determined" by the distribution of material throughout the Universe.*

Further, he stated that the extension of the field equations was "not justified by our actual knowledge of gravitation," but was "logically consistent." Furthermore, the cosmological term was "necessary only for the purpose of making possible a quasi-static distribution of matter, as required by the fact of the small velocities of stars."

From Einstein's field equations of general relativity, it followed that the geometry of Einstein's static universe is closed and the radius of curvature of the geometrical sections is $\mathcal{R} = c/(4\pi G\rho_0)^{1/2}$, where ρ_0 is the mean density of the static Universe. The value of Λ was directly related to the mean density of the Universe, $\Lambda = 4\pi G\rho_0$. Einstein believed that he had incorporated Mach's Principle into general relativity, in that static solutions of the field equations did not exist in the absence of matter.

Almost immediately, de Sitter (1917) showed that one of Einstein's objectives had not been achieved. He found solutions of Einstein's field equations in the absence of matter, $\rho = p = 0$. The metric he derived had the form

$$ds^2 = dr^2 - R^2 \sin\left(\frac{r}{R}\right)\left(d\phi^2 + \cos^2\phi\, d\theta^2\right) + \cos^2\left(\frac{r}{R}\right)c^2\, dt^2 \tag{1.4}$$

Although there is no matter present in the Universe, a test particle still moves along a perfectly well-defined path through spacetime. As de Sitter remarked, "If no matter exists apart from the test body, has this inertia?" One prediction of de Sitter's paper was the fact that distant galaxies would be observed with a redshift, although in his solution the metric was stationary—this phenomenon became known as the *de Sitter effect*.

In 1922, Kornel Lanczos showed that by a simple change of coordinates, the de Sitter solution could be interpreted as an expansion of the system of coordinates in hyperbolic space (Lanczos 1922).

$$ds^2 = -dt^2 + \cosh^2 t \left[d\phi^2 + \cos^2\phi\left(d\psi^2 + \cos^2\psi\, d\chi^2\right)\right] \tag{1.5}$$

Lanczos wrote that:

> *It is interesting to observe how one and the same geometry can appear with quite different physical interpretations according to the interpretations placed upon the particular coordinates.*

At almost exactly the same time, the Soviet meteorologist and theoretical physicist Alexander Alexandrovich Friedman published the first of his two classic papers on relativistic cosmology (Friedman 1922, 1924). His key realization was that isotropic world models had to have isotropic curvature everywhere. In the paper of 1922, Friedman found solutions for expanding world models with closed spatial geometries, including those that expand to a maximum radius and then collapse to a singularity. In the paper of 1924, he showed that there exist expanding solutions that are unbounded with hyperbolic geometry. The differential equations that he derived were:

$$(1922) \qquad \left(\frac{\dot{R}}{R}\right)^2 + \left(\frac{2R\ddot{R}}{R^2}\right) + \frac{c^2}{R^2} - \Lambda = 0 \qquad (1.6)$$

$$(1924) \qquad \left(\frac{\dot{R}}{R}\right)^2 + \left(\frac{2R\ddot{R}}{R^2}\right) - \frac{c^2}{R^2} - \Lambda = 0 \qquad (1.7)$$

In both cases,

$$\frac{3\dot{R}^2}{R^2} + \frac{3c^2}{R^2} - \Lambda = \kappa c^2 \rho \qquad (1.8)$$

The solutions of these equations correspond exactly to the standard world models of general relativity and are appropriately known as the *Friedman world models*. The history of general relativity in the Soviet Union is a remarkable story and Friedman's role in introducing Soviet scientists to the theory and the subsequent difficult development of these studies in the USSR needs to be better known. It has been carefully described by Zelmanov (1967) in a review that has not been translated into English.

It has always been considered somewhat surprising that it was some years before Friedman's important papers were given the recognition they deserve. In 1923, Einstein believed he had found an error in the first of Friedman's papers and published his concern in *Zeitschrift für Physik*. Friedman showed that Einstein was incorrect and Einstein subsequently published his withdrawal of his objection in the same journal. My guess is that Einstein's concern was remembered, but not his acknowledgment of his error.

In 1927, Georges Lemaître independently discovered the Friedman solutions and only then became aware of Friedman's pioneering contributions (Lemaître 1927). Both Lemaître and Howard P. Robertson (1928) were aware of the fact that the Friedman solutions result locally in a velocity-distance relation. Lemaître derived what he termed the "apparent Doppler effect," in which "the receding velocities of extragalactic nebulae are a cosmical effect of the expansion of the Universe" with $v \propto r$. Robertson found a similar result stating that "we should expect ... a correlation $v \approx (cl/R)$," where l is distance and v the recession velocity. From nearby galaxies, he found a value for Hubble's constant of 500 km s^{-1} Mpc^{-1}.

The discovery of the velocity-distance relation for galaxies was interpreted as evidence for the expansion of the Universe as a whole. There remained problems of interpretation of the notions of time and distance in cosmology because the field equations could be set up in any frame of reference. By 1935, the problem had been solved independently by Robertson and George Walker (Robertson 1935; Walker 1935). For isotropic, homogeneous world models, they showed that the metric of spacetime had to have the form

$$ds^2 = dt^2 - \frac{R^2(t)}{c^2}\left[\frac{dr^2}{(1+\kappa r^2)} + r^2(d\theta^2 + \sin^2\theta d\phi^2)\right] \qquad (1.9)$$

where κ is the curvature of space at the present epoch, r is a comoving radial distance coordinate and $R(t)$ is the scale factor which describes how the distance between any two world lines change with cosmic time t. The *Robertson-Walker metric* contains all the geometries consistent with the assumptions of isotropy and homogeneity of the Universe; the curvature $\kappa = \mathcal{R}^{-2}$, where $\mathcal{R}$, the radius of curvature of the spatial sections of the isotropic curved space, can be positive, negative or zero. The physics of the expansion is absorbed into the scale factor $R(t)$.

With the discovery of the velocity-distance relation, Einstein regretted the inclusion of the cosmological constant into the field equations. According to George Gamow, Einstein stated that the introduction of the cosmological constant was "the biggest blunder of my life" (Gamow 1970). In 1932, Einstein and de Sitter showed that there is one special solution of the equations with $\Lambda = 0$ and $\kappa = 0$, corresponding to Euclidean space sections (Einstein & de Sitter 1932). This *Einstein-de Sitter model* has density at the present epoch $\rho_0 = 3H_0^2/8\pi G$. This density is often referred to as the *critical density* and the Einstein-de Sitter model as the *critical model*, because it separates the ever-expanding models with open, hyperbolic geometries from those that will eventually collapse to a singularity and that have closed, spherical geometry. When Einstein and de Sitter inserted $H_0 = 500$ km s^{-1} Mpc^{-1} into the expression for ρ_0, they found $\rho_0 = 4 \times 10^{-25}$ kg m^{-3}. Although this value was somewhat greater than the mean density in galaxies derived by Hubble, they argued that it was of the correct order of magnitude and that there might well be a considerable amounts of "dark matter" present in the Universe.

1.4   Astrophysical Cosmology up to 1939

1.4.1   Dark Matter

Astrophysical evidence for dark matter was not long in coming. In 1933, Fritz Zwicky made the first dynamical studies of rich clusters of galaxies, in particular, of the Coma cluster (Zwicky 1933, 1937). The method Zwicky used to estimate the total mass of the cluster involved the *virial theorem*, which had been derived by Arthur Eddington in 1916 to estimate the masses of star clusters (Eddington 1916). The theorem relates the total internal kinetic energy $T = \frac{1}{2}M\langle v^2 \rangle$ of the galaxies in a cluster to its gravitational potential energy, $|U| = GM^2/2R_{cl}$ in statistical equilibrium under gravity. Eddington showed that $T = \frac{1}{2}|U|$ and so the mass of the cluster can be found, $M \approx 2R_{cl}\langle v^2 \rangle/G$.

Zwicky measured the velocity dispersion of the galaxies in the Coma cluster and found that there was much more mass in the cluster than could be attributed to the visible parts of galaxies. In solar units, the ratio of mass to optical luminosity of a galaxy such as our own is about 3, whereas for the Coma cluster the ratio was found to be about 500—there must be about 100 times more dark or hidden matter as compared with visible matter in the cluster. Zwicky's pioneering studies have been confirmed by all subsequent studies of rich clusters of galaxies.

1.4.2   The Age of the Universe and Eddington-Lemaître Models

Despite Einstein's renunciation of the cosmological constant Λ, there remained a very grave problem for those models in which Λ is set equal to zero. In all world models with $\Lambda = 0$, the age of the Universe is less than H_0^{-1}. Using Hubble's estimate of $H_0 = 500$ km s^{-1} Mpc^{-1}, the age of the Universe must be less than 2×10^9 years old, a figure in conflict with the age of the Earth derived from studies of the ratios of abundances of long-lived radioactive species, which gave ages significantly greater than this value.

Eddington and Lemaître recognized that this problem could be eliminated if Λ were positive (Eddington 1930; Lemaître 1931a). The effect of a positive cosmological constant is to counteract the attractive force of gravity when the Universe has grown to a large enough size. Among the solutions of Einstein's equations, there are special cases equivalent to the Einstein static Universe, but at some earlier epoch. These models remained in the static Einstein state for an arbitrarily long period and then expanded away from that state under the influence of the cosmological term. In these *Eddington-Lemaître* models, the age of the Universe could be arbitrarily long. As Eddington expressed it, the Universe would have a "logarithmic eternity" to fall back on, and so resolve the conflict between estimates of Hubble's constant and the age of the Earth.

1.4.3   The Origin of the Chemical Elements

In the 1930s, there were two reasons why the synthesis of the chemical elements during the early stages of the Friedman world models was taken seriously. Firstly, the chemical abundances of the elements in stars seemed to be remarkably uniform. Secondly, it appeared that the interiors of stars were not hot enough for nucleosynthesis of the chemical elements to take place in their interiors. A starting point for a cosmological solution to this problem was to work out the equilibrium abundances of the elements at some very high temperature and assume that, if the density and temperature decreased sufficiently rapidly, these abundances would remain "frozen."

In 1931, Lemaître proposed that the initial state of the Friedman models consisted of what he termed a "primaeval atom" (Lemaître 1931b). Following the discovery of the neutron in the following year, this state could be thought of as a sea of neutrons closely packed together. The primaeval neutrons were supposed to decay into protons and the chemical elements, as well as the cosmic rays, form in the subsequent nuclear interactions. These ideas inspired George Gamow's attack upon the problem of the origin of the chemical elements. In 1946, he extrapolated the Freidman models back to epochs when the densities and temperatures were high enough for nucleosynthesis to take place and found that the time scale of the Universe during these early stages was too short to establish an equilibrium distribution of the elements (Gamow 1946).

1.5   The Cosmological Problem in 1939

By the end of the 1930s, there was a common view that the solution of the cosmological problem lay in the determination of the parameters which define the Friedman world models. This became one of the great goals of the programs of observation to be carried out by the Palomar 200-inch telescope (Sandage 1961) and the subsequent generation of 4-meter class telescopes. The challenge was to measure precisely the parameters that characterize the Universe: Hubble's constant, $H_0 = \dot{R}/R$; the deceleration parameter, $q_0 = -\ddot{R}/R^2$; the curvature of space $\kappa = \mathcal{R}^{-2}$; the mean density of matter in the Universe ρ and, in particular, whether or not it attains the critical density ρ_0; the age of the Universe, T_0; and the cosmological constant Λ. These are not independent. According to general relativity,

$$\kappa = \mathcal{R}^{-2} = \frac{(\Omega-1)+\frac{1}{3}(\Lambda/H_0^2)}{(c/H_0)^2} \qquad q_0 = \frac{\Omega}{2} - \frac{1}{3}\frac{\Lambda}{H_0^2},$$

where $\Omega = \rho/\rho_0$ is known as the *density parameter*, where ρ_0 is the critical density. The determination of these parameters turned out to be among the most difficult observational challenges in astronomy, and progress by the traditional techniques of optical astronomy proved to be much more difficult than the optimists of the 1930s must have hoped. The Palomar 200-inch telescope was commissioned in 1948, and much effort was devoted to the determination of cosmological parameters, particularly by Allan Sandage, who has published a splendid review of this heroic endeavor (Sandage 1994).

In compensation, completely new vistas were to open up after the Second World War as the whole of the electromagnetic spectrum became available for astronomical observation and completely new approaches to the determination of cosmological parameters and the origin of structure in the Universe became possible.

1.6   Post-War Cosmology to 1970

1.6.1   *Gamow and the Big Bang*

The first detailed calculations of the expected abundances of the elements according to Lemaître's concept of the "primaeval atom" were carried out in 1942 by Chandrasekhar & Henrick (1942). They confirmed the expectation of equilibrium theory that, if the elements are in equilibrium at temperatures of about 10^{10} K and densities of about 10^9 kg m^{-3}, their abundances should be inversely correlated with their binding energies. There were, however, several gross discrepancies with the observed abundances. The light elements, lithium, beryllium, and boron, were vastly overproduced and iron and the elements with mass numbers greater than about 70 underproduced. They suggested that some non-equilibrium process was required.

In 1946, George Gamow found that the time scale of early expansion of the Universe was indeed too short to establish an equilibrium abundance of the elements (Gamow 1946). Neutron capture cross sections became available in 1946 as a by-product of the nuclear physics programs carried out during the Second World War, and these showed that there is an inverse correlation between these cross sections and the relative abundances of the elements. In the first calculations carried out by Gamow and Ralph Alpher, the computations assumed a sea of free neutrons and that nucleosynthesis only took place after the temperature had fallen below $kT = 0.1$ MeV—the Universe was assumed to be static. This theory was published in 1948 by Alpher, Bethe, & Gamow (1948) and they found reasonable agreement with the observed abundances of the elements. The paper drew attention to the necessity of a hot, dense phase in the early Universe if the elements were to be synthesized cosmologically.

In the same year, Alpher and Robert Herman (1948) carried out improved calculations, including the cosmic expansion into their calculations. They realised that, at such very high temperatures at early epochs, the Universe was radiation rather than matter-dominated and they solved the problem of the subsequent temperature history of the Universe. They came to the far-reaching conclusion that the cooled remnant of the hot early phases should be present in the Universe today and estimated that the temperature of this thermal background should be about 5 K.

There was, however, a major problem with this picture—there are no stable nuclei with mass numbers 5 and 8. Fermi and Turkevich calculated the evolution of the nuclear abundances of the light elements including 28 nuclear reactions for elements up to mass number 7 in a radiation-dominated, expanding Universe and their results were published by Alpher & Herman (1950). These calculations showed that only about one part in 10^7 of the initial mass was converted into elements heavier than helium.

In 1950, Hayashi pointed out that, in the early Universe, at temperatures only ten times greater than that at which the nucleosynthesis took place, the neutrons and protons were brought into thermal equilibrium by the weak interactions:

$$e^+ + n \leftrightarrow p + \bar{\nu}_e \qquad \nu_e + n \leftrightarrow p + e^-$$

(Hayashi 1950). At about the same temperature, electron-positron pair production ensures a plentiful supply of positrons and electrons. Thus, rather than assume arbitrarily that the initial conditions consisted of a sea of neutrons, the equilibrium abundances of protons, neutrons, electrons and all the other constituents of the early Universe could be calculated exactly. In 1953, Alpher, Follin, & Herman (1953) determined the evolution of the proton-neutron ratio as the Universe expanded and obtained an answer remarkably similar to modern calculations. These ideas were, however, overtaken by the discovery of the physics of nucleosynthesis of the chemical elements in stars.

1.6.2   Steady-state Cosmology

Immediately after the War, many new ideas were in the air. Milne had developed his theory of kinematic relativity, in which there are two times, one associated with dynamical phenomena and another with electromagnetic phenomena (Milne 1948). Dirac had been impressed by coincidences between the very large numbers in physics and the properties of the Universe—for example, the square of the ratio of the strengths of electromagnetic and gravitational forces is roughly equal to the numbers of protons in the Universe (Dirac 1937). A consequence of his identification of these large numbers was the idea that the gravitational constant should change with time. Eddington had developed his *Fundamental Theory*, in which the cosmological constant appeared as a fundamental constant of nature (Eddington 1946).

Steady-state cosmology was invented by Hermann Bondi, Thomas Gold and Fred Hoyle in 1948 (Bondi & Gold 1948; Hoyle 1948). They extended the cosmological principle to what they termed the *perfect cosmological principle* according to which the Universe presents the same large-scale picture to all fundamental observers *at all times*. Hence, Hubble's constant becomes a fundamental constant of nature. The perfect cosmological principle led to a unique metric for the dynamics of the Universe with zero spatial curvature. Because of the expansion of the Universe, matter has to be continuously created in order to replace the dispersing matter, the rate of creation amounting to only one particle m^{-3} every 300,000 years. A consequence of the theory was that the Universe was infinite in age, but the age of typical objects observed in the local Universe is only $\frac{1}{3}H_0^{-1}$. It was during a radio program on cosmology in the late 1940s that Hoyle introduced the somewhat pejorative term "Big Bang" to describe the Friedman models with singular origin, which is eliminated in the steady-state picture.

Hoyle set about finding an alternative means of understanding the formation of the chemical elements by nucleosynthesis in stars and these considerations led to his remarkable prediction of the carbon resonance (Hoyle 1953) for the formation of carbon in stars and the important paper on the processes of stellar nucleosynthesis by Burbidge, Burbidge, Fowler, & Hoyle (1957). With these new insights, the abundances of the chemical elements disappeared as evidence for a hot initial phase of the Universe.

In the 1950s, two important results were reported of central importance for cosmology. The first concerned the value of Hubble's constant. At the meeting of the International Astronomical Union in Rome in 1952, Walter Baade announced that the distance to the Andromeda Nebula (M31) had been underestimated by a factor of 2 (Baade 1952). Hubble's constant was therefore reduced to 250 km s^{-1} Mpc^{-1} and H_0^{-1} increased to 4×10^9 years. In 1956, Humason, Mayall, & Sandage (1956) revised Hubble's constant downward again to 180 km s^{-1} Mpc^{-1}. These revisions eliminated the discrepancy between the age of the Earth and the age of the Universe according to the Friedman models with $\Lambda = 0$. By the 1970s, the value was reduced further to between 50 and 100 km s^{-1} Mpc^{-1}. The precise value became a subject of considerable controversy, but this has now been resolved, largely thanks to the leadership of the Hubble Key Program by Dr. Wendy Freedman, the chair of this Symposium (Freedman et al 2001).

The second concerned the number counts of radio sources that showed that there was an excess of faint radio sources. Martin Ryle concluded that the only reasonable interpretation was that there was a much greater comoving number density of extragalactic radio sources at large distances, and hence at earlier cosmic epochs, than nearby. As Ryle expressed it in his Halley Lecture in 1955, "there seems no way in which the observations can be explained in terms of a Steady-State theory" (Ryle 1955). This led to a somewhat bitter controversy, both within the radio astronomical community and with the proponents of the steady-state theory. By the 1960s, it was established that Ryle's conclusion was correct, but the effect was not nearly as large as had been believed in the 1950s, because the importance of source confusion had not been appreciated.

Although steady-state theory is nowadays considered to be largely of historical and scientific sociological interest, there are some features of the theory that have a resonance with contemporary cosmological theories. In the steady-state picture,

- The density of the Universe is a constant
- The spatial geometry is flat
- The scale factor varies as $\exp\{H_0(t_0 - t)\}$

These features have a rather familiar ring about them nowadays and correspond rather precisely to the present best-buy picture of the Universe, in which we are entering a phase when its dynamics are to be dominated by the dark energy, equivalent to the presence of a significant cosmological constant Λ. In Hoyle's version of steady-state theory, these properties are attributed to the action of the creation field C. It is amusing to note that on the occasion of Fred's 80th birthday celebrations in 1995, he gave a splendid lecture to the Cavendish Physical Society in which he stated that, if only he had called the creation field ψ, rather than C, he would now would be remembered as the originator of the inflationary Universe.

William McCrea had, however, already had this deep insight in 1951 (McCrea 1951). McCrea realised that there was a quite different interpretation of what Hoyle had done, which bears a much closer resonance with contemporary cosmology. To quote McCrea,

The single admission that the zero of absolute stress may be set elsewhere than is currently assumed on somewhat arbitrary grounds permits all of Hoyle's results to be derived within the system of general relativity theory. Also, this derivation gives the results an intellectual physical coherence.

McCrea wrote the physics of the steady-state picture in terms of a negative energy equation of state $p = -\rho c^2$ and recovered the three features of the theory listed above. It is intriguing that McCrea had realised that there is nothing intrinsically implausible about a negative energy equation of state. Indeed, this is what we believe dominates the dynamics of the Universe from now on.

1.7 The Helium Problem and the Microwave Background Radiation

In 1961 Osterbrock & Rogerson (1961) showed that the fractional abundance of helium seemed to be remarkably uniform wherever it could be observed and corresponded to about 25% by mass. In 1964 O'Dell, Peimbert, & Kinman (1964) found that the helium abundance in a planetary nebula in the old globular cluster M15 also had helium abundance about 25%, despite the fact that the heavy elements were very significantly depleted relative to their cosmic abundances.

It is not often that a fundamental aspect of contemporary cosmology is developed in the course of a post-graduate lecture course, but this in fact occurred in the solution of the helium problem. In the Lent term of 1964, my first year as a research student in Cambridge, Fred Hoyle gave a post-graduate lecture course entitled *Extragalactic Astrophysics and Cosmology*. It was given twice a week to a remarkable group of research students, many of whom went on to become leaders of astronomy. Fred would turn up with a few notes scribbled on what looked like the traditional envelope and run through what indeed turned out to be many of the key problems of astrophysics during the subsequent decades.

Toward the end of the course, he tackled the problem of the origin of helium in the cosmos, reviewing the early work of Alpher, Gamow and Hermann. Roger Tayler and John Falconer were in the audience, and they realised that they could use the EDSAC-2 computer to carry out predictions of the cosmic helium abundance for a wide range of different cosmological models. In the course of the following two lectures, they unraveled in some detail the implications of these calculations, and the result was the famous *Nature* paper by Hoyle & Tayler (1964), which revived interest in the primordial synthesis of the light elements. They found that about 25% helium by mass is synthesised in the Big Bang, in remarkable agreement with observation, and that this result is essentially independent of the present baryonic density. Hoyle and Tayler did not mention that the cooled remnant of the hot early Universe should be detectable at centimeter wavelengths. Alpher and Herman's prediction had been more or less forgotten when Gamow's theory of primordial nucleosynthesis had failed to account for the creation of the elements.

In the very next year 1965, the microwave background radiation was discovered, more or less by accident, by Arno Penzias and Robert Wilson (1965). During the commissioning of a 20-foot horn antenna designed for telecommunications, they found an excess noise temperature of about 3.5 ± 1 K wherever they pointed their telescope at the sky. Robert Dicke's group in Princeton was attempting exactly this experiment to detect the cooled remnant of the Big Bang—it was very quickly realised that Penzias and Wilson had discovered the signal sought by the Princeton physicists. Within a few months, the Princeton group had measured a background temperature of 3.0 ± 0.5 K at a wavelength of 3.2 cm, confirming the black body nature of the background spectrum (Roll & Wilkinson 1966).

1.8 Conclusions

This seems an appropriate point at which to conclude this brief review. By the late 1960s and early 1970s, the observational evidence strongly favored what has become the standard Big Bang framework for contemporary geometrical and astrophysical cosmology. The 1960s and 1970s were also decades during which the whole face of astronomy, astrophysics and cosmology were revitalized through the opening up of the complete electromagnetic spectrum for astronomical observation. Some highlights of these observational advances would include:

- the discovery of quasars
- the strong cosmological evolution of all classes of active galaxies
- the discovery of deuterium in the interstellar medium
- the discovery of gravitational lensing
- the discovery of the diffuse X-ray emission from clusters of galaxies
- the observation of the Sunyaev-Zel'dovich effect
- *COBE* observations of the spectrum and fluctuations in the cosmic microwave background radiation
- the discovery of the Type Ia supernova technique for estimating cosmological parameters
- precise measurements of the distances of nearby galaxies
- the determination of the mean mass density of the Universe on large scales

A corresponding list could be drawn up for theoretical advances:

- the determination of the detailed thermal history of the Universe
- the theory of the development of fluctuations in standard Big Bang picture and the resulting prediction of the spectrum of fluctuations in cosmic microwave background radiation
- the mathematical analysis of the large-scale structure of the Universe
- constraints on the number of neutrino species and the early dynamics of the Universe from primordial nucleosynthesis
- the cold dark matter scenario for the origin of structure
- the inflationary scenario for the very early Universe
- the use of massive parallel computing to simulate the origin of large-scale structures in the Universe

All these topics are now the bread and butter of modern cosmology and will be discussed *in extenso* during this centennial meeting. It is startling to realise just how far we have come in the matter of only 30 years. Few of us who began our research careers in the early 1960s could have predicted the enormous advances in observational and theoretical cosmology, far less the extraordinary fact that there seems to be a concordance between the many different approaches to geometrical and astrophysical cosmology. None of this could have come about without the pioneering efforts of many great astronomers whose endeavors have been the theme of this survey. In turn, these astronomers could not have made their discoveries without the tools provided by generous benefactors of astronomy. Among these, Andrew Carnegie's name will always be remembered as the founder of the Observatories of the Carnegie Institution of Washington, which has been, and continues to be, at the forefront of the best of contemporary astronomy and cosmology.

References

General References

Many important topics are mentioned only briefly in this short survey. I have found the following books particularly helpful in preparing this paper.

Bernstein, J., & Feinberg, G. 1986, Cosmological Constants: Papers in Modern Cosmology (New York: Columbia Univ. Press)
Bertotti, B., Balbinot, R., Bergia, S., & Messina, A., ed. 1990, Modern Cosmology in Retrospect (Cambridge: Cambridge Univ. Press)
Bondi, H. 1960, Cosmology, 2nd edition (Cambridge: Cambridge Univ. Press)
Gillespie, C. C., ed. 1981, Dictionary of Scientific Biography (New York: Charles Scribner's Sons)
Gingerich, O., ed. 1984, The General History of Astronomy, Vol. 4., Astrophysics and Twentieth-Century Astronomy to 1950: Part A (Cambridge: Cambridge Univ. Press)
Harrison, E. 2001, Cosmology: The Science of the Universe (Cambridge: Cambridge Univ. Press)
Hearnshaw, J. B. 1986, The Analysis of Starlight: One Hundred and Fifty Years of Astronomical Spectroscopy (Cambridge: Cambridge Univ. Press)
——. 1996, The Measurement of Starlight: Two Centuries of Astronomical Photometry (Cambridge: Cambridge Univ. Press)
Lang, K. R., & Gingerich, O., ed. 1979, A Source Book in Astronomy and Astrophysics, 1900–1975 (Cambridge, Mass.: Harvard Univ. Press)
Learner, R. 1981, Astronomy Through the Telescope (London: Evans Brothers Limited)
Leverington, D. 1996, A History of Astronomy from 1890 to the Present (Berlin: Springer-Verlag)
Longair, M. S. 1995, Astrophysics and Cosmology, in 20th Century Physics, 3, ed. L. M. Brown, A. Pais, & A. B. Pippard (Bristol: Institute of Physics Publications: New York: American Institute of Physics), 1691–1821. A much expanded version of this text will be published by Cambridge University Press in 2003.
Martínez, V. J., Trimble, V. & Pons-Bordería, M. J., ed. 2001, Historical Development of Modern Cosmology (San Francisco: ASP), 252
North, J. D. 1965, The Measure of the Universe (Oxford: Clarendon Press)

Literature References

Alpher, R. A., Bethe, H., & Gamow, G. 1948, Phys. Rev., 73, 803
Alpher, R. A., Follin, J. W., & Herman, R. C. 1953, Phys. Rev., 92, 1347
Alpher, R. A., & Herman, R. C. 1948, Nature, 162, 774
——. 1950, Rev. Mod. Phys., 22, 153
Baade, W. 1952, Trans. IAU, 8, 397
Bondi, H., & Gold, T. 1948, MNRAS, 108, 252
Burbidge, E. M., Burbidge, G. R., Fowler, W. A., & Hoyle, F. 1957, Rev. Mod. Phys., 29, 547
Chandrasekhar, S., & Henrick, L. R. 1942, ApJ, 95, 288
de Sitter, W. 1917, MNRAS, 78, 3
Dirac, P. A. M. 1937, Nature, 139, 323
Eddington, A. S. 1916, MNRAS, 76, 525
——. 1930, MNRAS, 90, 669
——. 1946, Fundamental Theory, ed. E. Whittaker (Cambridge: Cambridge Univ. Press)
Einstein, A. 1915, Sitzungsberichte Preuss. Akad.Wissenschaften, 844
——. 1917, Sitzungsberichte Berl. Akad., 1, 142
Einstein, A., & de Sitter, W. 1932, Proc. Natl. Acad. Sciences, 18, 213
Freedman, W. L., et al. 2001, ApJ, 533, 47
Friedman, A. A. 1922, Zeitschrift für Physik, 10, 377
——. 1924, Zeitschrift für Physik, 21, 326
Gamow, G. 1946, Phys. Rev., 70, 572
——. 1970, My World Line (New York: Viking Press), 44
Hayashi, C. 1950, Prog. Theor. Phys. (Japan), 5, 224
Hoyle, F. 1948, MNRAS, 108, 372
——. 1953, ApJS, 1, 121
Hoyle, F., & Tayler, R. J. 1964, Nature, 203, 1108

Hubble, E. P. 1925, Publ. Amer. Astron. Soc., 5, 261

——. 1926, ApJ, 64, 321

——. 1929, Proc. Natl. Acad. Sciences, 15, 168

Hubble, E. P., & Humason, M. 1934, ApJ, 74, 43

Humason, M. L., Mayall, N. U., & Sandage, A. R. 1956, ApJ, 61, 97

Lanczos, C. 1922, Phys. Zeitschrift, 23, 539

Leavitt, H. S. 1912, Harvard College Observatory Circular, No. 173, 1

Lemâitre, G. 1927, Annales Société Scientifique de Bruxelles, A47, 49

——. 1931a, MNRAS, 91, 483

——. 1931b, Nature, 127, 706

Lobachevsky, N. I. 1829–30, On the Principles of Geometry (Kazan Bulletin)

Longair, M. S. 2001, in Historical Development of Modern Cosmology, ed. V. J. Martínez, V. Trimble, &
    M. J.Pons-Bordeía, (San Francisco: ASP), 55

McCrea, W. H. 1951, Proc. Roy. Soc., 206, 562

Milne, E. 1948, Kinematic Relativity (Oxford: Clarendon Press)

O'Dell, C. R., Peimbert, M., & Kinman, T. D. 1964, ApJ, 140, 119

Osterbrock, D. E. 1984, James E. Keeler: Pioneer American Astrophysicist (Cambridge: Cambridge Univ. Press)

Osterbrock, D. E., & Rogerson, J. B. 1961, PASP, 73, 129

Penzias, A. A., & Wilson, R. W. 1965, ApJ, 142, 419

Robertson, H. P. 1928. Phil. Mag., 5, 835

——. 1935, ApJ, 82, 284

Roll, P. G., & Wilkinson, D. T. 1966, Phys. Rev. Lett., 16, 405

Ryle, M. 1955, Observatory, 75, 137

Sandage, A. R. 1961, ApJ, 133, 355

——. 1994, in The Deep Universe, ed. B. Binggeli & R. Buser (Berlin: Springer-Verlag), 1

Slipher, V. M. 1917, Proc. Amer. Phil. Soc., 56, 403

Walker, A. G. 1936, Proc. Lond. Math. Soc., Ser. 2., 42, 90

Wirtz, C. W. 1921, Astronomische Nachrichten, 215, 349

Wright, H., Warnow, J. N., & Weiner, C., ed. 1972, The Legacy of George Ellery Hale (Cambridge, Mass: MIT
    Press)

Zel'dovich, Ya. B. 1968, Uspekhi Fizicheskikh Nauk, 95, 209

Zelmanov, A. 1967, in Development of Astronomy in the USSR, ed V. A. Ambartsumian et al. (Moscow: Nauka),
    320

Zwicky, F. 1933, Helv. Physica Acta, 6, 110

——. 1937, ApJ, 86, 217

2

Edwin Hubble: a biographical retrospective

GALE E. CHRISTIANSON
Indiana State University

A minor but intriguing drama was playing itself out on both sides of the Atlantic in the summer of 1919. Comfortably settled into rooms near the Great Gate of Trinity College, in Cambridge, Major Edwin Hubble was the recipient of a series of increasingly agitated letters postmarked Pasadena, California. Their author was the great solar astronomer and scientific entrepreneur George Ellery Hale, who beseeched Hubble to sail to America posthaste and to exchange his military uniform for the business suit of an astronomer. "Please come as soon as possible," a nervous Hale wrote. "We expect to get the 100-inch telescope into commission very soon, and there should be abundant opportunity for work by the time you arrive."[1]

The Missouri-born Hubble, who had worked mightily to rid himself of a telltale accent, was thirty years old and scarcely bursting with promise. Aside from his military service, he had never held a regular job save for a year spent teaching Spanish and mathematics at New Albany High School in Indiana. Yet he comported himself with an aristocratic air, making it appear that Hale was somehow his inferior. This impression went well beyond their correspondence. Hubble had entered the good graces of the wealthy English astronomer H. F. Newall whose home, Madingley Rise, was located near Cambridge Observatory. Newell had proposed Hubble for membership in the prestigious Royal Astronomical Society, whose outcome was a foregone conclusion. When a delegation of visiting American astronomers was wined and dined by members of the Society two months later, Hubble was seated near the head of the table next to Frank Dyson, the Astronomer Royal. The astronomers Walter Adams, Charles St. John, and Frederick H. Sears were the first from Mount Wilson to take the measure of the tall, crisply dressed major whose prominence was hardly commensurate with any scientific achievements. Galling it must have been to encounter a fellow American who affected an English accent during the period of inflated nationalism following a world war.

Despite his Christian upbringing, Edwin Hubble was a man who believed in destiny—particularly his own—though it was not something he communicated to others in so many words. The closest he came was in two letters to his mother, Virginia James Hubble, written in 1910 during his Rhodes Scholar days at Oxford, when he had donned plus-fours, a Norfolk jacket and cape. "I sometimes feel that there is within me to do what the average man would not do if only I find some principle for whose sake I could leave everything else and devote my life."[2] As one who read the classics in his spare time, he realized that the path traced out by the gods would be anything but easy: "Work, to be pleasant, must be toward some great end; an end so great that dreams of it, anticipation of it, overcomes all aversion to labour."[3]

Having taken his own good time, Hubble finally boarded a ship for New York some four months after Hale's initial plea that he come home. On reaching the East Coast, he took a train to San Francisco, where he received his formal discharge at the Presidio.

Reluctant to forfeit the cachet of his uniform, Hubble was still in full dress when he reached Pasadena in early September. Towering above the valley floor, and accessible only by a zigzagging dirt road, stood 5,714-foot Mount Wilson. Hale had spent his first night on the peak in the summer of 1903, and remembered falling asleep watching the stars pass over a gaping hole in the roof of an abandoned cabin. Having worked his will on the Wisconsin skies above Williams Bay, where he had conceived and overseen the construction of the great Yerkes 40-inch refractor, Hale subsequently raided the observatory of its best astronomers and technicians, then headed for the West Coast and the pristine firmament above the Pacific.

Hale, a consummate salesman, persuaded John D. Hooker, a Los Angeles businessman enamored of astronomy, to underwrite the purchase of a 100-inch glass disk for the most powerful telescope in the world. In return, Hooker would receive the honor of having the magnificent instrument christened his namesake. This promising beginning was followed by a second and far greater coup in early 1910, when Hale persuaded his friend Andrew Carnegie to come visit the mountain. Looking more like a sawed-off Ernest Hemingway than a titan of industry, the Scotsman posed for a photograph with Hale in front of the newly installed 60-inch telescope, the world's largest at the time. Aware of Carnegie's sensitivity about his diminutive stature, the diplomatic Hale made certain that his friend was standing up slope before the shutter clicked. When Carnegie was interviewed about his experience by the press, he spoke of 60,000 new stars already discovered on the mountain, causing the University of Chicago astronomer Forest Ray Moulton to quip that one might as well claim to have "discovered 60,000 new gallons of water in Lake Michigan."[4] No matter, the check for $10 million was in the mail.

Construction of the 100-inch Hooker reflector was nearing completion even as Europe was engulfed in war. The glassworks at St. Gobain in Paris had produced a giant disk weighing some 5 tons and measuring 101 inches across by 13 inches thick. The nerve-racking task of grinding its 7,800 square inches of surface had taken nearly five years to complete, yet so fine was the craftsmanship that every square inch produced the same focal length to within one part in some 90,000.[5] Upon its surface would fall not 60,000 points of light but an estimated 3 billion. The English poet Alfred Noyes would soon commemorate Hale's feat in his epic *Watchers of the Skies*:

> *Where was the gambler that would stake so much —*
> *Time, patience, treasure, on a single throw?*[6]

Back at Yerkes Observatory another young gambler was champing at the bit. Only months before the 100-inch mirror was moved safely to the mountaintop, Congress had declared war on Germany. Hubble, twenty-seven and an inveterate Anglophile, was itching to get into the fray. Abetted by the rising tide of patriotism, he pressured his dissertation advisor and director of Yerkes, Edwin Brant Frost, into moving up the date of his oral examinations. The aging astronomer was uneasy to say the least, but he yielded when Hubble informed him that he was applying for a commission in the Officers Reserve Corps. The document in question, titled "Photographic Investigations of Faint Nebulae," contained all of 17 pages, forcing Hubble himself to admit that it "seems so skimpy."

The work was little based on observations undertaken with the main telescope, for "seeing time" on the 40-inch was largely reserved for astronomers higher in the pecking order than Hubble. Fortunately, the brilliant but mentally unstable George Ritchey, who Hale had been forced to replace during the meticulous grinding of the 100-inch mirror, had designed and built a 24-inch reflector before departing for Mount Wilson. It was on this generally neglected instrument that Hubble cut his teeth as an astronomer.

Hubble took the first of hundreds of glass plates in the autumn of 1915. He described the object, known to astronomers as NGC (*New General Catalogue*) 2261, as the "finest example of a cometary nebula in the northern skies."[7] It, together with some 17,000 similar formations, had already been catalogued. At least 130,000 more were calculated to be within telescopic range. What most intrigued Hubble, who compared his plate with others taken years earlier, was that the nebula had since bowed to display a larger degree of convexity than before. He could only conclude that the object was quite near in astronomical terms, or he could never have observed its subtle change in shape. Displaying an ingrained caution that would one day become legendary, he wrote: "No attempt is here made to explain the phenomenon."[8] Nor did Hubble try to classify this and other nebulae according to type. It appeared that some of them are within our stellar system while others, such as the giant spirals with their enormous radial velocities and insensible proper motions, seemed to lie outside it. More distant still were numberless whiffs of light whose appearance on the photographic plates looked like spattered porridge. For the present there was little further to be done. All would have to await, Hubble concluded, the advent of "instruments more powerful than those we now possess."[9]

Two years later found Hubble in command of the most powerful telescope ever conceived. It was Christmas Eve 1919, and the mountain could be a lonely place, especially during the holidays. The astronomer Wendell Hoge, a victim of that loneliness, had poignantly made the following entry in the 1912 log of the 60-inch: "Merry Christmas to all the Universe."[10] Yet for Hubble, who preferred his own company to that of most others, such isolation was to be savored. What greater gift than to be sitting atop a mountain detached from a small planet circling a middling star, his fingers poised on the control paddles of the massive yet gentle giant?

To Hubble the process was every bit as much aesthetic as scientific. With the celestial map firmly fixed in his head, he waited patiently for the dark matter called night to steal in and fill the dome. The night assistant soon became invisible while the only trace left of Hubble was the Cheshire glow of his signature briar. Sounding like rolling thunder, the dome circled until its opening approached the field he was planning to photograph. Just as it was coming into focus he would suddenly shout the command to clamp the telescope, often catching the assistant off guard. The plate was then exposed as the astronomer settled in to wait and to wonder, periodically adjusting the instrument's position in an effort to capture light as old as creation itself. Few living astronomers can lay claim to this transforming experience, which is fast becoming part of the world we have lost.

Less enamored of the mountain than Hubble, though a brilliant astronomer in his own right, was Harlow Shapley, who could frequently be heard complaining about the numbing cold and lack of sleep. Shapley had preceded Hubble's arrival on Mount Wilson by five years and had virtually laid claim to the Milky Way galaxy, which he equated with the universe itself. However, Shapley's daunting and meticulous study of double stars or binaries, whose distances he had plotted by calculating light curves and rough stellar masses, caused him to reject the generally accepted belief that the Milky Way was only some 30 million light-years in diameter, the figure championed by the Dutch astronomer Jacobus Cornelius Kapteyn. It was Shapley's conclusion that Earth was far removed from the Milky Way's cyclopean eye, and that the distances to the binaries were "pretty darned big."

More recently, Shapley had turned his attention to the study of Cepheid variables, pulsating stars whose surfaces, he postulated, ebb and flow in great cauldron-like waves. The secrets of these intriguing objects were first plumbed by Henrietta Leavitt, a research assistant at Harvard College Observatory. Leavitt had proven that Cepheids can be treated as celestial timepieces, which alternately brighten and dim like clockwork. Employing photographic plates from Harvard's observatory in Peru, she further determined that the longer a Cepheid's period, the brighter its image on the glass exposure, a discovery of profound implications. Henceforth, Cepheids of similar periods could be looked upon as virtual twins, no matter their apparent brightness or location in the sky. As "standard candles" they were nothing less than beacons for calculating distances across the void.

Working on the 60-inch reflector while awaiting the completion of the more powerful Hooker, Shapley scoured the mottled globular clusters for the glimmer of yet unseen variables. By 1918 he had succeed in plotting the period-luminosity relation of 230 pulsating Cepheids, whose cycles ranged from 5 hours to 100 days. Putting pen to paper in a blizzard of publication after four grueling years on the mountain, he created a new and audacious galactic model—one that would do for the Milky Way what Copernicus had done for the solar system nearly four centuries earlier.

Having discerned that more than 40% of the 100 or so globular clusters then known were concentrated in 3% of the sky's area, Shapley reasoned that this massive gathering marked the center of the galaxy, which, in turn, could mean only one thing. Our planet is but a tiny part of a solar system located on the very fringes of the Milky Way, whose diameter had inflated to ten-fold that of the puny Kapteyn universe. In a word employed by Copernicus the cosmos was *"immensum,"* a staggering 300,000 light-years across.

More than satisfied, Shapley came down from the mountain in March of 1921 to become director of Harvard College Observatory, a seeming just reward for what he considered his greatest scientific achievement. Yet his grandiose model of the universe would survive all of five years.

The German scientist and metaphysician Immanuel Kant imagined a system whereby countless stars are gathered together in a common plane or thin disk, like those of the Milky Way, yet so far removed from Earth that the individual components are indistinguishable with a telescope. They would appear as a feebly illuminated spot—circular if its plane is perpendicular to the line of sight; elliptical if viewed obliquely. Thus to Kant the nebulae were nothing less than countless systems of countless suns so distant that they are the weakest candles in the heavens, albeit galaxies not unlike our own. Embracing the principle of the uniformity of nature, the harness maker's son is credited with postulating the theory of island universes, a term coined by the explorer Alexander von Humboldt in the mid-nineteenth century.

Kant's contemporary, the great William Herschel, dictated a description of every chalk-colored web and pinwheel to his gifted sister Caroline as the objects drifted by the lens of his fixed telescope. Embracing Galileo's belief that faintness means distance, Herschel thought it likely that the nebulae could be massive aggregates of stars no different than the Milky Way, some of which "may well outvie [it] in grandeur."[11] Still, who could say for certain?

Shapley's most outspoken critic was the congenial and bespectacled Heber Curtis of

Mount Hamilton's Lick Observatory. Curtis was dubious about Shapley's construct, not least because of the contribution made to it by Shapely's colleague and friend, Adriaan van Maanen. Also of Mount Wilson, van Maanen claimed to have detected rotational movements in several of the most photographed spiral nebulae, including M33, M51, M81, and M101. He calculated the rate of motion at about one hundred-thousandth of a revolution per year. Though this figure may seem small, indeed infinitesimal, when projected on a cosmic scale, it seemed to sound the death knell of Kant's island universe theory. The stellar rim of a spiral only 500,000 light-years from Earth would have to rotate at an incredible 30,000 miles per second. Spirals at far greater distances would not only approach but exceed the speed of light.

Try as he might, Curtis could find no confirmation of any such rotations in plates from van Maanen's arsenal—and Shapley's. After debating this and other issues before a meeting of the National Academy of Sciences in April 1920—the so-called Great Debate, which failed to live up to its billing—both Curtis and Shapley came away more convinced than ever of the validity of their respective positions.

Or so Shapley claimed for his part. Yet not long after the exchange he handed some plates of the great Andromeda nebula, M31, to Milton Humason, the former mule skinner and janitor whom Shapley had helped train to use the telescopes on Mount Wilson. He instructed Humason to examine them on the stereocomparator. As he was blinking the plates, the night assistant marked the locations in ink of images never before discerned, and which Humason believed to be Cepheid variables located beyond the Milky Way. As Humason told the story to astronomer Allan Sandage years later, Shapley was having none of it. He drew a handkerchief from his pocket and wiped the plates clean of Humason's telltale marks, all the while lecturing him on the points he had scored in the Great Debate.[12]

Hubble, who had impressed his fellow Rhodes Scholars by reading Kant in German, had lately embarked on an ambitious scheme encompassing the nature, form, and classification of the nebulae without tipping his hand regarding the Great Debate. On the night of October 4, 1923, he was on his ninth run of the year with the 100-inch telescope, having homed in on a spiral arm of M31. The seeing was rated at less than 1, the worst possible without closing the dome, the kind of conditions that would frequently cause him to dismiss the night assistant, who lost no time making for a boiling cup of coffee and a warm bed. Despite the poor conditions, the 40-minute exposure yielded a "suspected" nova, which Hubble intended to confirm the next night. The viewing had markedly improved, and he increased the exposure time by five minutes, just to be sure. Plate H335H confirmed his suspicions. What was more, a careful examination of the plate revealed two more stars, both of which the astronomer concluded were novae as well.

His run over, the knickers-wearing father of triplets caught the truck bound for the observatory offices on Pasadena's Santa Barbara Street, where he began searching the files for previous photographs of the "novae." He was startled to find that one of the objects on plate H335H—destined to become the most famous ever taken on Mount Wilson—was not a nova at all but a Cepheid variable. He first plotted the light curve of the object, determining that it had a period of 31.415 days. Then, by exploiting Shapley's distance-measuring techniques, he found that it was at least 300,000 parsecs from Earth, the equivalent of one million light-years, easily more than three times the diameter of Shapley's universe. Hubble took out his marking pen and, near the top of the mind-altering plate, crossed out "N" for nova, and in bold capital letters printed "VAR!" for variable, followed by an exclamation point.

Hubble waited four months, during which he discovered and calibrated the distance of a second Cepheid and a half dozen more novae, before writing to Shapley of his celestial tour de force. There was never any love lost between the two Missourians, who were as different as night and day: Shapley, the studied hick who despised Hubble's highhanded airs, English garb, and posturing, not to mention his spoken "Oxford"; Hubble, the patrician who hated Shapley's "Missouri tongue" and thought him a coward for having opposed the Great War. Rubbing it in, Hubble proved that his written Oxford could be as annoying as the spoken variety: "I have a feeling that more variables would be found by a careful examination of long exposures. Altogether next season should be a merry one and will be met with due form and ceremony."[13] Hubble's tone was reminiscent of another letter he had written to his mother from Oxford: "It is always great fun to look a man in the eyes and best him by sheer self-possession."[14]

Hubble's boorishness aside, Shapley well knew who was king of the mountain, as would everyone else soon enough. The Andromeda nebula was nothing less than a separate galaxy composed of stars by the millions, if not more. Graduate student Cecilia Payne happened to be in Shapley's Harvard Observatory office when he opened and read Hubble's letter. He then passed it across his desk to Payne, remarking with the flair of a tragedian: "Here is the letter that has destroyed my universe."[15]

Shapley's capitulation was further signaled in an October 1924 letter to Hubble. Now that Hubble had proven the existence of independent star systems, would it not be more accurate to rename them "galactic nebulae" or perhaps "galaxies," the term Shapley himself preferred?[16] But there was no budging Hubble, who clung to his own nomenclature, calling them "nongalactic nebulae." Indeed, they would remain so on the mountain until his death in 1953.

Hubble's mastery of the 100-inch had recently enabled him to complete another observational program of the first order—this one an act of synthesis some four years in the making. Astronomers from William Herschel to Heber Curtis had long labored to establish a classification scheme of the various galactic nebulae, as well as the types of stars with which they are associated. By April 1921 Hubble had come up with a tentative scheme of his own and was sufficiently emboldened to forward a copy to Lick Observatory astronomer William H. Wright. In the cover letter he admitted to "a few anomalies, but on the whole the progression is surprisingly definite."[17] At the same time, he made it clear that he was by no means proposing a theory of galactic evolution.

Another year passed during which Hubble reinforced his model with new photographs taken with the 60- and 100-inch reflectors. Then, in 1922, he gained a seat on the International Astronomical Union's fourteen-member Commission on Nebulae and Star Clusters. Here was the opportunity he had been waiting for, but when he presented his model to the commission it was relegated to the status of an unpublished report. Others, including certain members of the commission itself, had put forth their own classification schemes. Only Vesto Melvin Slipher, the acting director of Lowell Observatory, thought Hubble's model worthy of being printed.

At this point a frustrated Hubble decided to go it alone. After carefully redrafting his paper, he sent it off to the *Astrophysical Journal*, which published it under the title: "A General Study of Diffuse Galactic Nebulae." So far, so good, yet neither Hubble himself nor his fellow astronomers as yet looked upon his scheme as definitive. Then came word that the commission had decided to compile a new catalogue on the nebulae employing photographic

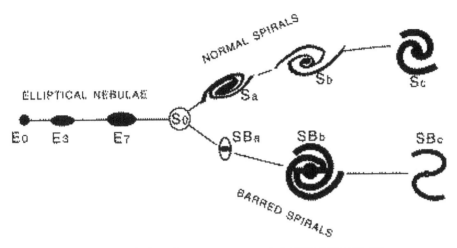

HUBBLE'S CLASSIFICATION OF THE NEBULAE

plates as opposed to traditional visual observations. Still, the project could not advance short of a international consensus on a classification scheme.

Once again Hubble turned to Slipher, who was now chairing the commission. He typed up his extensive notes in July 1923, including revisions of his initial scheme, added a series of photographic plates, and sent the material off to Slipher in Arizona. "The men at Mount Wilson," Hubble wrote, together with a visiting Henry Norris Russell from Princeton, "have looked over the notes and have expressed their approval."[18]

The diagram of Hubble's most recent scheme, which some likened to a "galactic tuning fork," had undergone marked changes from the one published in 1922. While the galactic nebulae remained the same, Hubble now separated the nongalactic nebulae into only two classes: what he had previously termed spindles, ovates, and globulars were subsumed under the categories of ellipticals and spirals. Amorphous in appearance, the ellipticals resembled blobs, which Hubble further divided according to their flatness or lack of same. Those whose shape was perfectly circular were termed E0 nebulae. Several steps removed were the E7 ellipticals, the flattest of their kind and resembling a lens or, to the sports-minded, a football.

More elegant to the astronomer's eye were the spirals whose double arms form cosmic pinwheels. These Hubble separated into two categories, normal spirals and the less common but distinctive barred spirals, so called because their arms originate from the end of a "bar" traversing their nuclei. Both types of spirals were then separated into subclasses based on the tightness of their arms. A normal spiral with tightly wound arms was given an Sa designation while SBa became the symbol for its counterpart in the barred category. Sc and SBc spirals are the most loosely wound of all.

Yet as every astronomer is quick to learn, all is not quite right in the heavens. Hubble was nagged by the anomalous nebulae that could not be easily fit into any of his three categories, forcing him to create the catchall term "irregular nebulae" for the "very few" clusters that otherwise eluded his synthesizing powers.

Although several of Hubble's peers in the IAU were impressed by his revised classification system, they ultimately voted against its adoption. His frustration mounting, Hubble chose to do what he had done before. "Extra-Galactic Nebulae," a seminal paper, appeared in volume 64 of the *Astrophysical Journal*. Thirty years later the gifted astronomer Walter Baade, who was Hubble's colleague at Mount Wilson, told a rapt audience at Harvard that the "systems that really present difficulties to Hubble's classification [are] so small [in number] that I can count them on the fingers of my hand."[19] So it would be until the deep field opened up by the *Hubble Space Telescope*—reaching back in time and across space some 13 billion years—revealed nascent galaxies in the throes of creation, like scattered snippets of cosmic DNA.

Percival Lowell's belief that the surface of the planet Mars was etched by giant canals had caused the astronomer to hire Vesto Melvin Slipher in 1901 for the purpose of verifying the waterways' existence. For years a deeply skeptical Slipher had held Lowell off with one hand while reaching for the stars with the other. Hubble and Slipher had first crossed paths in the autumn of 1914 on the campus of Northwestern University during the annual meeting of the American Astronomical Society. Hubble was heading for Williams Bay to begin his graduate work in astronomy and was in the audience when Slipher announced that the massive Andromeda spiral was blueshifted, barreling toward the Sun at the astonishing speed of 300 kilometers per second. Spectrograms and consequent radial velocities obtained for more than forty other nebulae and star clusters were even more confounding. Unlike Andromeda, most nebulae were redshifted, strong evidence that they were hurtling outward from the Sun at speeds as great as 1,100 kilometers per second.[20] When Slipher finished his paper the audience rose as one and accorded the astronomer a standing ovation, welcome praise indeed for one who was convinced that the velocities of the spirals were too great, their distances too vast, for them to be gravitationally bound to the Milky Way.

Others were considerably more skeptical. Among them was George Ritchey, whose work with the 60-inch reflector on Mount Wilson had produced a series of plates containing spiral nebulae, none of which seemed to possess sufficient matter to be classified as independent galaxies. Then, at some point in 1926, Slipher simply ran out of telescope, his 24-inch refractor a casualty of the increasingly smaller and dimmer objects he was attempting to photograph. Nor had he succeeded in obtaining credible distances to the spiral nebulae, thus leaving the conundrum of expansion unresolved.

Hubble's recent work on Cepheids had enabled him to accumulate a wealth of distance measurements, but he was anything but content. Indeed, he wrote of his work to date as mere "reconnaissance." Elsewhere, I have characterized him as a great mariner in the making, a galactic voyager ready to leave behind familiar shoals and coasts and to strike out for the fabled Indies, supposedly somewhere far off in the murky distance.[21] At his side was the former bellboy, mule driver, failed orchardist, and janitor Milton Humason, whose grammar school education would brand him a pariah in every major observatory today. Yet such were Humason's skills that he commanded Hubble's absolute confidence, which would waver but once during the long, difficult voyage ahead.

From inside the dark and creaking dome the pair slipped into the vast ocean sea of space, or what Hubble would later call "the realm of the nebulae," to begin charting the period-luminosity relation for Cepheids and other stellar bodies. If, as he suspected, a nebula's speed of recession is truly an index of its distance, then the distances of nebulae far across the universe could be inferred by simply measuring their redshifts.

Of all the many plates they would take, the first was destined to be the most memorable. Humason was at the controls of the Hooker and spent two frigid nights photographing through a yellow prism, which blocked the ultraviolet light. The object was a faint nebula chosen because its redshift had eluded Slipher owing to its distance from Earth. After he developed the plate while still on the mountain, Humason scanned it with the aid of a magnifier for the H and K lines generated by calcium atoms in the nebula. They soon appeared, and though the spectrum was faint the telltale vertical marks were shifted to the right or red end, as expected. Humason phoned an anxious Hubble, who was waiting for him at Santa Barbara Street when the assistant astronomer finished his run. Hubble gathered the plate and, after confirming the redshift, calculated its speed of recession at 3,000 kilometers per second, some 1,800 kilometers per second greater than Slipher's largest value. When Humason was asked about his feelings at the moment of discovery years later, he claimed Byron as his model, though he seems not to have had the slightest clue as to who Byron was or what he had done.[22]

As Humanson soon learned, such Faustian arrogance came with a price. The freezing astronomer, his face grotesquely lit by red dark-vision lamps, perched like a Lilliputian on the small Cassegrain platform five stories above the observatory floor, coaxing and prodding the recalcitrant behemoth through moonless nights punctuated by staccato winds and the incessant ticking of the weight-driven clock. When the machinery balked, as it often did, he kept the image steady by forcing his shoulder against more than 100 tons of metal. If all else failed he literally climbed on to the instrument's giant frame, bending his body at painfully awkward angles for the sake of a plate steeping in light from nebulae time out of mind. "You had to stretch out into nothing," he reminisced about his efforts to obtain the long exposures.[23] A deceptively nonchalant Hubble, who was choreographing Humason's every move while putting in his own time on the mountain, dismissed his assistant's acrobatics as Milt's "adventures among the clusters."[24]

Mutiny was in the air. When Hubble pressed Humason to do even more, he recoiled at the prospect of additional pain and suffering. It took the intervention of Hale, who had been succeeded as director of Mount Wilson by Walter Adams, to keep Humason from jumping ship. Hale promised him a new spectrograph and camera, which would shorten the exposure time of plates from nights to hours. Furthermore, Hale was the person who had taken a chance on Humason, elevating him from obscurity to a trusted and highly valued colleague.

So they continued on, the former Rhodes Scholar and the grammar school dropout, forging ever deeper into uncharted waters. Giant spirals like M31, M33, M51, and M101—the very heat of Messier's catalogue—confirmed the redshifts. Galaxies in all directions appeared to be moving away from Earth, or Earth from them. Based on his calibration of Cepheids, Hubble established the first linear relation between the degree of spectral displacement and the calculated distance to the observed object: the greater the redshift, the more remote the source of light. Out of these efforts emerged the revolutionary paper of March 1929, which a cautious Hubble had held back for over a year. He titled it "A Relation Between Distance and Radial Velocity Among Extra-Galactic Nebulae."[25]

In a scant six pages Hubble worked a change in humankind's conception of the cosmos no less profound than that advanced by Copernicus, in 1543. Gone forever was the static universe; in its place was an expanding one, the rate of the mutual recession of its parts increasing with their relative distance.

The coming years would produce more papers and further confirmation of what Hubble had wrought. Slipher's pioneering displacements would seem puny when compared to the new velocities Hubble and Humason were adding at the rate of ten per month. Past Virgo, past Pegasus, past Pisces, Cancer, Perseus, Coma, and Leo, careening into the blackness of space at an incredible 19,700 kilometers per second—the astronomer and his assistant slowly but surely nibbling away at the speed of light. Humason uncorked a bottle of his famous Depression-era "Panther Juice" to mark their success. And later Hubble was overhead speaking to him on the telephone: "Now you are beginning to use the 100-inch the way it should be used."[26] Soon the unlikely duo would publish the jointly authored paper marking the highlight of their collaboration: "The Velocity-Distance Relation Among Extra-Galactic Nebulae."[27] Still, Hubble had good reason to believe that this sudden expansion of the cosmos was more the beginning of the journey than the end. Along the way, he would join other scientific immortals by having a law named after him, *Hubble's law* ($V = Hd$), a measure of the rate of expansion of the universe.

At night, when the waning moon is but a sliver in the distance, the great telescope stands ready to do its master's bidding. Like a birthing ghost, he emerges from the gloaming, tall, slender, confident in knickers and high-topped boots, hands thrust deep into the pockets of his trench coat, sparks rising from his briar into the cavernous dome. His face, much like his thoughts, borders on the opaque, as though glimpsed through a glass darkly.

He pauses for a moment, as if to contemplate what lies ahead, before ascending the steps and iron ladder leading to the observing platform. The dome is open to the chill air, the tops of the tall pines dimly visible against the darkening sky. The master mariner issues orders to the waiting night assistant seated at the console below. So many hours or so many degrees. There follows the metallic whining of the traverse, a series of loud clicks, a final heavy clanging of the Victorian machinery as the 100-inch is clamped. He withdraws a small magnifying glass from his pocket with which to examine the field at the eyepiece. Satisfied that all is well, he eases back into the lone bentwood chair and deliberately fills his pipe. The last traces of daylight have vanished, and the remaining lights are turned out, making way for the soft glow of the stars. Leaning over, he slides the cover from a photographic plate, slips it into its holding frame, and calls out the exposure time to the assistant. Then he tells him: "You can go if you like."[28] Taking control of the telescope himself, he is suddenly alone in the universe with his private thoughts and dreams.

Such solitude had been experienced years earlier by a young and obscure employee in the Swiss patent office at Berne. In the early 1930s, Albert Einstein, by then a Nobel Laureate, had come to Caltech to lecture and to take the measure of the American astronomer who had cast doubt on his belief that the universe is static. It did not take long for Einstein to correct this major blunder, which had provided no explanation for Hubble's redshifts. He modified his calculations on relativity to make theory conform to fact. Then, one afternoon as Hubble's wife Grace was driving the gnomish physicist around Pasadena, Einstein turned to her and issued a compliment for the ages: "Your husband's work," he said, "is beautiful."[29]

References

[1] G.E.H. to E.P.H., Jun. 9, 1919, G.E.H., MWODF: 1904-1923, Box 159, f. E.P.H.
[2] E.P.H. to V.H., c. Dec. 1910 (Oxford correspondence courtesy of J.F.L.)

[3] Ibid.

[4] Quoted in W.S.A. "Early Days at Mount Wilson," 1947, PASP, 59, 296.

[5] W.S.A. Papers, MWOA, Supplement, Box 5, f. 5.154, unp.

[6] Alfred Noyes, *Watchers of the Skies* (New York: Frederick H. Stokes, 1922), 3-4.

[7] E.P.H., "Changes in the Form of the Nebula NGC 2261," 1916. PNAS, 2, 230.

[8] E.P.H., "Recent Changes in Variable Nebula NGC 2261," 1917, ApJ, 45, 352.

[9] E.P.H., "Photographic Investigations of Faint Nebulae," 1920, Pub. Yerkes. Obs., 4, 69.

[10] 60-inch Telescope Log Book, no. 1, Dec. 25, 1912.

[11] William Herschel, "On the Construction of the Heavens," 1785, Philosophical Transactions, 75, 260.

[12] Author's interview with A.S., Jun. 11, 1991.

[13] E.P.H. to H.S., Feb. 19, 1924, HUB, Box 15, f. 611.

[14] E.P.H. to V.H., May 20, 1913.

[15] *Cecilia Payne-Gaposchkin: An Autobiography and Other Recollections*, ed. K. Haramundanis (Cambridge, England: Cambridge Univ. Press, 1984), 209.

[16] H.S. to E.P.H., Oct. 8, 1924, HCOA, 1921-1930, HEP-I, Box 9, UAV 630.22, f. 71.

[17] E.P.H. to W.H.W., Apr. 11, 1921, SALO, Hubble, Edwin P., 1921-1949, f. 741.

[18] E.P.H. to V.M.S., Jul. 24, 1923, HUB, Box 15, f. 620.

[19] Walter Baade, *Evolution of the Stars and Galaxies* (Cambridge, MA: Harvard Univ. Press, 1963), 18.

[20] Vesto Melvin Slipher, "Spectrographic Observations of Nebulae," 1915, Popular Astron., 23, 21.

[21] Gale E. Christianson, *Edwin Hubble: Mariner of the Nebulae* (New York: Farrar, Straus and Giroux, 1995). See, especially, chapters five through eight.

[22] Interview with M.L.H. by B.S., c. 1965, AIP., 1-2.

[23] Ibid., 3-6.

[24] Ibid., 2.

[25] E.P.H., "A Relation Between Distance and Radial Velocity Among Extra-Galactic Nebulae," 1929, PNAS, 15, 173.

[26] Quoted in G.B.H., "E.P.H.: The Astronomer," HUB 82(7), Box 7, 5.

[27] E.P.H. and M.L.H., "The Velocity-Distance Relation Among Extra-Galactic Nebulae," 1931, ApJ, 74, 43.

[28] G.B.H., "E.P.H.: The Astronomer," 2-3.

[29] G.B.H., "E.P.H.: Some People,?"HUB 82(17), Box 8, f. 2.

Abbreviations

Manuscript Collections

AIP — American Institute of Physics, College Park, Maryland

HCOA — Director's Correspondence, Harvard College Observatory Archives, Harvard University, Cambridge, Massachusetts

HUB — Edwin Hubble Manuscript Collection, Henry Huntington Library, San Marino, California

MWODF — Mount Wilson Observatory Archives Director's Files, Henry Huntington Library, San Marino, California

SALO — Mary Lea Shane Archives of the Lick Observatory, University Library, University of California, Santa Cruz

YOA — Director's Papers, Yerkes Observatory Archives, Yerkes Observatory, Williams Bay, Wisconsin

Individuals

A.S. — Allan Sandage

B.S. — Bert Shapiro

E.B.F. — Edwin B. Frost

E.P.H. — Edwin P. Hubble
G.B.H. — Grace Burke Hubble
G.E.H. — George Ellery Hale
H.S. — Harlow Shapley
J.F.L. — John F. Lane
M.L.H. — Milton L. Humason
V.H. — Virginia Hubble
V.M.S. — Vesto M. Slipher
W.H.W. — William H. Wright
W.S.A. — Walter S. Adams

3

Inflation

ALAN H. GUTH
Massachusetts Institute of Technology

Abstract

The basic workings of inflationary models are summarized, along with the arguments that strongly suggest that our universe is the product of inflation. I describe the quantum origin of density perturbations, giving a heuristic derivation of the scale invariance of the spectrum and the leading corrections to scale invariance. The mechanisms that lead to eternal inflation in both new and chaotic models are described. Although the infinity of pocket universes produced by eternal inflation are unobservable, it is argued that eternal inflation has real consequences in terms of the way that predictions are extracted from theoretical models. Although inflation is generically eternal into the future, it is not eternal into the past: it can be proven under reasonable assumptions that the inflating region must be incomplete in past directions, so some physics other than inflation is needed to describe the past boundary of the inflating region. The ambiguities in defining probabilities in eternally inflating spacetimes are reviewed, with emphasis on the youngness paradox that results from a synchronous gauge regularization technique.

3.1 Introduction

I will begin by summarizing the basics of inflation, including a discussion of how inflation works, and why many of us believe that our universe almost certainly evolved through some form of inflation. This material is mostly not new, although the observational evidence in support of inflation has recently become much stronger. Since observations of the cosmic microwave background (CMB) power spectrum have become so important, I will elaborate a bit on how it is determined by inflationary models. Then I will move on to discuss eternal inflation, showing how once inflation starts, it generically continues forever, creating an infinite number of "pocket" universes. If inflation is eternal into the future, it is natural to ask if it can also be eternal into the past. I will describe a theorem by Borde, Vilenkin, and me (Borde, Guth, & Vilenkin 2003), which shows under mild assumptions that inflation cannot be eternal into the past, and thus some new physics will be necessary to explain the ultimate origin of the universe.

3.2 How Does Inflation Work?

The key property of the laws of physics that makes inflation possible is the existence of states with negative pressure. The effects of negative pressure can be seen clearly in the

Friedmann equations,

$$\ddot{a}(t) = -\frac{4\pi}{3}G(\rho+3p)a \tag{3.1a}$$

$$H^2 = \frac{8\pi}{3}G\rho - \frac{k}{a^2} \tag{3.1b}$$

and

$$\dot{\rho} = -3H(\rho+p), \tag{3.1c}$$

where

$$H = \frac{\dot{a}}{a}. \tag{3.2}$$

Here ρ is the energy density, p is the pressure, G is Newton's constant, an overdot denotes a derivative with respect to the time t, and throughout this paper I will use units for which $\hbar = c = 1$. The metric is given by the Robertson-Walker form,

$$ds^2 = -dt^2 + a^2(t)\left\{\frac{dr^2}{1-kr^2} + r^2(d\theta^2 + \sin^2\theta\, d\phi^2)\right\}, \tag{3.3}$$

where k is a constant that, by rescaling a, can always be taken to be 0 or ± 1.

Equation (3.1a) clearly shows that a positive pressure contributes to the deceleration of the universe, but a negative pressure can cause acceleration. Thus, a negative pressure produces a repulsive form of gravity.

Furthermore, the physics of scalar fields makes it easy to construct states of negative pressure, since the energy-momentum tensor of a scalar field $\phi(x)$ is given by

$$T^{\mu\nu} = \partial^\mu\phi\partial^\nu\phi - g^{\mu\nu}\left[\tfrac{1}{2}\partial_\lambda\phi\partial^\lambda\phi + V(\phi)\right], \tag{3.4}$$

where $g^{\mu\nu}$ is the metric, with signature $(-1,1,1,1)$, and $V(\phi)$ is the potential energy density. The energy density and pressure are then given by

$$\rho = T^{00} = \tfrac{1}{2}\dot{\phi}^2 + \tfrac{1}{2}(\nabla_i\phi)^2 + V(\phi), \tag{3.5}$$

$$p = \tfrac{1}{3}\sum_{i=1}^{3}T_{ii} = \tfrac{1}{2}\dot{\phi}^2 - \tfrac{1}{6}(\nabla_i\phi)^2 - V(\phi). \tag{3.6}$$

Thus, any state that is dominated by the potential energy of a scalar field will have negative pressure.

Alternatively, one can show that any state that has an energy density that cannot be easily lowered must have a negative pressure. Consider, for example, a state for which the energy density is approximately equal to a constant value ρ_f. Then, if a region filled with this state of matter expanded by an amount dV, its energy would have to increase by

$$dU = \rho_f\, dV. \tag{3.7}$$

This energy must be supplied by whatever force is causing the expansion, which means that the force must be pulling against a negative pressure. The work done by the force is given by

$$dW = -p_f\, dV, \tag{3.8}$$

where p_f is the pressure inside the expanding region. Equating the work with the change in energy, one finds

$$p_f = -\rho_f , \qquad (3.9)$$

which is exactly what Equations (3.5) and (3.6) imply for states in which the energy density is dominated by the potential energy of a scalar field. [One can derive the same result from Eq. (3.1c), by considering the case for which $\dot{\rho} = 0$.]

In most inflationary models the energy density ρ is approximately constant, leading to exponential expansion of the scale factor. By inserting Equation (3.9) into (3.1a), one obtains a second-order equation for $a(t)$ for which the late-time asymptotic behavior is given by

$$a(t) \propto e^{\chi t} , \text{ where } \chi = \sqrt{\frac{8\pi}{3}G\rho_f} . \qquad (3.10)$$

In the original version of the inflationary theory (Guth 1981), the state that drove the inflation involved a scalar field in a local (but not global) minimum of its potential energy function. A similar proposal was advanced slightly earlier by Starobinsky (1979, 1980) as an (unsuccessful) attempt to solve the initial singularity problem, using curved space quantum field theory corrections to the energy-momentum tensor to generate the negative pressure. The scalar field state employed in the original version of inflation is called a *false vacuum*, since the state temporarily acts as if it were the state of lowest possible energy density. Classically this state would be completely stable, because there would be no energy available to allow the scalar field to cross the potential energy barrier that separates it from states of lower energy. Quantum mechanically, however, the state would decay by tunneling (Coleman 1977; Callan & Coleman 1977; Coleman & De Luccia 1980). Initially it was hoped that this tunneling process could successfully end inflation, but it was soon found that the randomness of the bubble formation when the false vacuum decayed would produce disastrously large inhomogeneities. Early work on this problem by Guth and Weinberg was summarized in Guth (1981), and described more fully in Guth & Weinberg (1983). Hawking, Moss, & Stewart (1982) reached similar conclusions from a different point of view.

This "graceful exit" problem was solved by the invention of the new inflationary universe model by Linde (1982a) and by Albrecht & Steinhardt (1982). New inflation achieved all the successes that had been hoped for in the context of the original version. In this theory inflation is driven by a scalar field perched on a plateau of the potential energy diagram, as shown in Figure 3.1. Such a scalar field is generically called the *inflaton*. If the plateau is flat enough, such a state can be stable enough for successful inflation. Soon afterwards, Linde (1983a, 1983b) showed that the inflaton potential need not have either a local minimum or a gentle plateau: in the scenario he dubbed *chaotic inflation*, the inflaton potential can be as simple as

$$V(\phi) = \frac{1}{2}m^2\phi^2, \qquad (3.11)$$

provided that ϕ begins at a large enough value so that inflation can occur as it relaxes. A graph of this potential energy function is shown as Figure 3.2. The evolution of the scalar field in a Robertson-Walker universe is described by the general relativistic version of the Klein-Gordon equation,

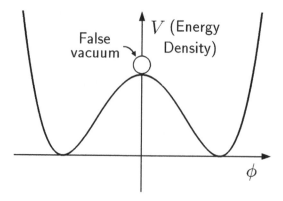

Fig. 3.1. Generic form of the potential for the new inflationary scenario.

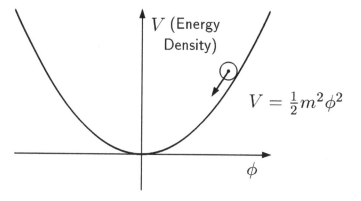

Fig. 3.2. Generic form of the potential for the chaotic inflationary scenario.

$$\ddot{\phi}+3H\dot{\phi}-\frac{1}{a^2(t)}\vec{\nabla}^2\phi=-\frac{\partial V}{\partial\phi}\ . \tag{3.12}$$

For late times the $\vec{\nabla}^2\phi$ term becomes negligible, and the evolution of the scalar field at any point in space is similar to the motion of a point mass evolving in the potential $V(x)$ in the presence of a damping force described by the $3H\dot{\phi}$ term.

For simplicity of language, I will stretch the meaning of the phrase "false vacuum" to include all of these cases; that is, I will use the phrase to denote any state with a large negative pressure.

Many versions of inflation have been proposed. In particular, versions of inflation that make use of two scalar fields [i.e., hybrid inflation (Linde 1991, 1994; Liddle & Lyth 1993; Copeland et al. 1994; Stewart 1995) and supernatural inflation (Randall, Soljačić, & Guth 1996)] appear to be quite plausible. Nonetheless, in this article I will discuss only the basic scenarios of new and chaotic inflation, which are sufficient to illustrate the physical effects that I want to discuss.

The basic inflationary scenario begins by assuming that at least some patch of the early universe was in this peculiar false vacuum state. To begin inflation, the patch must be approximately homogeneous on the scale of χ^{-1}, as defined by Equation (3.10). In the original papers (Guth 1981; Linde 1982a; Albrecht & Steinhardt 1982) this initial condition was motivated by the fact that, in many quantum field theories, the false vacuum resulted naturally from the supercooling of an initially hot state in thermal equilibrium. It was soon found, however, that quantum fluctuations in the rolling inflaton field give rise to density perturbations in the universe, and that these density perturbations would be much larger than observed unless the inflaton field is very weakly coupled (Starobinsky 1982; Guth & Pi 1982; Hawking 1982; Bardeen, Steinhardt, & Turner 1983). For such weak coupling there would be no time for an initially nonthermal state to reach thermal equilibrium. Nonetheless, since thermal equilibrium describes a probability distribution in which all states of a given energy are weighted equally, the fact that thermal equilibrium leads to a false vacuum implies that there are many ways of reaching a false vacuum. Thus, even in the absence of thermal equilibrium—even if the universe started in a highly chaotic initial state—it seems reasonable to assume that some small patches of the early universe settled into the false vacuum state, as was suggested, for example, by Guth (1982). Linde (1983b) pointed out that even highly improbable initial patches could be important if they inflated, since the exponential expansion could still cause such patches to dominate the volume of the universe. If inflation is eternal, as I will discuss in § 3.5, then the inflating volume increases without limit, and will presumably dominate the universe as long as the probability of inflation starting is not exactly zero.

Once a region of false vacuum materializes, the physics of the subsequent evolution is rather straightforward. The gravitational repulsion caused by the negative pressure will drive the region into a period of exponential expansion. If the energy density of the false vacuum is at the grand unified theory scale [$\rho_f \approx (2 \times 10^{16} \text{ GeV})^4$], Equation (3.10) shows that the time constant χ^{-1} of the exponential expansion would be about 10^{-38} s, and that the corresponding Hubble length would be about 10^{-28} cm. For inflation to achieve its goals, this patch has to expand exponentially for at least 65 e-foldings, but the amount of inflation could be much larger than this. The exponential expansion dilutes away any particles that are present at the start of inflation, and also smooths out the metric. The expanding region approaches a smooth de Sitter space, independent of the details of how it began (Jensen & Stein-Schabes 1987). Eventually, however, the inflaton field at any given location will roll off the hill, ending inflation. When it does, the energy density that has been locked in the inflaton field is released. Because of the coupling of the inflaton to other fields, that energy becomes thermalized to produce a hot soup of particles, which is exactly what had always been taken as the starting point of the standard Big Bang theory before inflation was introduced. From here on the scenario joins the standard Big Bang description. The role of inflation is to establish dynamically the initial conditions that otherwise would have to be postulated.

The inflationary mechanism produces an entire universe starting from essentially nothing, so one would naturally want to ask where the energy for this universe comes from. The answer is that it comes from the gravitational field. The universe did not begin with this colossal energy stored in the gravitational field, but rather the gravitational field can supply the energy because its energy can become negative without bound. As more and more positive energy materializes in the form of an ever-growing region filled with a high-energy scalar field, more and more negative energy materializes in the form of an expanding region filled with a gravitational field. The total energy remains constant at some very small value, and could in fact be exactly zero. There is nothing known that places any limit on the amount of inflation that can occur while the total energy remains exactly zero.[*]

Note that while inflation was originally developed in the context of grand unified theories, the only real requirements on the particle physics are the existence of a false vacuum state, and the possibility of creating the net baryon number of the universe after inflation.

3.3   Evidence for Inflation

Inflation is not really a theory, but instead it is a paradigm, or a class of theories. As such, it does not make specific predictions in the same sense that the standard model of particle physics makes predictions. Each specific model of inflation makes definite predictions, but the class of models as a whole can be tested only by looking for generic features that are common to most of the models. Nonetheless, there are a number of features of the universe that seem to be characteristic consequences of inflation. In my opinion, the evidence that our universe is the result of some form of inflation is very solid. Since the term *inflation* encompasses a wide range of detailed theories, it is hard to imagine any reasonable alternative.[*]

The basic arguments for inflation are as follows:

(1) *The universe is big*

First of all, we know that the universe is incredibly large: the visible part of the universe contains about 10^{90} particles. Since we have all grown up in a large universe, it is easy to take this fact for granted: of course the universe is big, it is the whole universe! In "standard" Friedmann-Robertson-Walker cosmology, without inflation, one simply postulates that about 10^{90} or more particles were here from the start. Many of us hope, however, that even the creation of the universe can be described in scientific terms. Thus, we are led to at least think about a theory that might explain how the universe got to be so big. Whatever that theory is, it has to somehow explain the number of particles, 10^{90} or more. One simple way to get such a huge number, with only modest numbers as input, is for the calculation to involve an exponential. The exponential expansion of inflation reduces the problem of explaining 10^{90} particles to the problem of explaining 60 or 70 *e*-foldings of inflation. In fact, it is easy to construct underlying particle theories that will give far more than 70 *e*-foldings of inflation. Inflationary cosmology therefore suggests that, even though the observed universe is incredibly large, it is only an infinitesimal fraction of the entire universe.

(2) *The Hubble expansion*

The Hubble expansion is also easy to take for granted, since we have all known about it from our earliest readings in cosmology. In standard Friedmann-Robertson-Walker cosmology, the Hubble expansion is part of the list of postulates that define the initial conditions. But inflation actually offers the possibility of explaining how the Hubble expansion began. The repulsive gravity associated with the false vacuum is

---

[*] In Newtonian mechanics the energy density of a gravitational field is unambiguously negative; it can be derived by the same methods used for the Coulomb field, but the force law has the opposite sign. In general relativity there is no coordinate-invariant way of expressing the energy in a space that is not asymptotically flat, so many experts prefer to say that the total energy is undefined. Either way, there is agreement that inflation is consistent with the general relativistic description of energy conservation.

[*] The cyclic-ekpyrotic model (Steinhardt & Turok 2002) is touted by its authors as a rival to inflation, but in fact it incorporates inflation and uses it to explain why the universe is so large, homogeneous, isotropic, and flat.

just what Hubble ordered. It is exactly the kind of force needed to propel the universe into a pattern of motion in which each pair of particles is moving apart with a velocity proportional to their separation.

(3) *Homogeneity and isotropy*

The degree of uniformity in the universe is startling. The intensity of the cosmic background radiation is the same in all directions, after it is corrected for the motion of the Earth, to the incredible precision of one part in 100,000. To get some feeling for how high this precision is, we can imagine a marble that is spherical to one part in 100,000. The surface of the marble would have to be shaped to an accuracy of about 1,000 Å, a quarter of the wavelength of light. Although modern technology makes it possible to grind lenses to quarter-wavelength accuracy, we would nonetheless be shocked if we unearthed a stone, produced by natural processes, that was round to an accuracy of 1,000 Å.

The cosmic background radiation was released about 400,000 years after the Big Bang, after the universe cooled enough so that the opaque plasma neutralized into a transparent gas. The cosmic background radiation photons have mostly been traveling on straight lines since then, so they provide an image of what the universe looked like at 400,000 years after the Big Bang. The observed uniformity of the radiation therefore implies that the observed universe had become uniform in temperature by that time. In standard Friedmann-Robertson-Walker cosmology, a simple calculation shows that the uniformity could be established so quickly only if signals could propagate at about 100 times the speed of light, a proposition clearly contradicting the known laws of physics.

In inflationary cosmology, however, the uniformity is easily explained. It is created initially on microscopic scales, by normal thermal equilibrium processes, and then inflation takes over and stretches the regions of uniformity to become large enough to encompass the observed universe and more.

(4) *The flatness problem*

I find the flatness problem particularly impressive, because of the extraordinary numbers that it involves. The problem concerns the value of the ratio

$$\Omega_{\text{tot}} \equiv \frac{\rho_{\text{tot}}}{\rho_c} , \tag{3.13}$$

where ρ_{tot} is the average total mass density of the universe and $\rho_c = 3H^2/8\pi G$ is the critical density, the density that would make the universe spatially flat. (In the definition of "total mass density," I am including the vacuum energy $\rho_{\text{vac}} = \Lambda/8\pi G$ associated with the cosmological constant Λ, if it is nonzero.)

For the past several decades there has been general agreement that Ω_{tot} lies in the range

$$0.1 \lesssim \Omega_0 \lesssim 2 , \tag{3.14}$$

but for most of this period it was very hard to pinpoint the value with more precision. Despite the breadth of this range, the value of Ω at early times is highly constrained, since $\Omega = 1$ is an unstable equilibrium point of the standard model evolution. Thus, if Ω was ever *exactly* equal to one, it would remain exactly one forever. However, if Ω differed slightly from one in the early universe, that difference—whether positive or negative—would be amplified with time. In particular, it can be shown that $\Omega - 1$ grows as

$$\Omega - 1 \propto \begin{cases} t & \text{(during the radiation-dominated era)} \\ t^{2/3} & \text{(during the matter-dominated era)} \end{cases} \tag{3.15}$$

At $t = 1$ s, for example, when the processes of Big Bang nucleosynthesis were just beginning, Dicke & Peebles (1979) pointed out that Ω must have equaled one to an accuracy of one part in 10^{15}. Classical cosmology provides no explanation for this fact—it is simply assumed as part of the initial conditions. In the context of modern particle theory, where we try to push things all the way back to the Planck time, 10^{-43} s, the problem becomes even more extreme. If one specifies the value of Ω at the Planck time, it has to equal one to 58 decimal places in order to be anywhere in the range of Equation (3.14) today.

While this extraordinary flatness of the early universe has no explanation in classical Friedmann-Robertson-Walker cosmology, it is a natural prediction for inflationary cosmology. During the inflationary period, instead of Ω being driven away from one as described by Equation (3.15), Ω is driven toward one, with exponential swiftness:

$$\Omega - 1 \propto e^{-2H_{\text{inf}}t}, \tag{3.16}$$

where H_{inf} is the Hubble parameter during inflation. Thus, as long as there is a sufficient period of inflation, Ω can start at almost any value, and it will be driven to unity by the exponential expansion. Since this mechanism is highly effective, almost all inflationary models predict that Ω_0 should be equal to one (to within about 1 part in 10^4). Until the past few years this prediction was thought to be at odds with observation, but with the addition of dark energy the observationally favored value of Ω_0 is now essentially equal to one. According to the latest *WMAP* results (Bennett et al. 2003), $\Omega_0 = 1.02 \pm 0.02$, in beautiful agreement with inflationary predictions.

(5) *Absence of magnetic monopoles*

All grand unified theories predict that there should be, in the spectrum of possible particles, extremely massive particles carrying a net magnetic charge. By combining grand unified theories with classical cosmology without inflation, Preskill (1979) found that magnetic monopoles would be produced so copiously that they would outweigh everything else in the universe by a factor of about 10^{12}. A mass density this large would cause the inferred age of the universe to drop to about 30,000 years! Inflation is certainly the simplest known mechanism to eliminate monopoles from the visible universe, even though they are still in the spectrum of possible particles. The monopoles are eliminated simply by arranging the parameters so that inflation takes place after (or during) monopole production, so the monopole density is diluted to a completely negligible level.

(6) *Anisotropy of the cosmic microwave background (CMB) radiation*

The process of inflation smooths the universe essentially completely, but density fluctuations are generated as inflation ends by the quantum fluctuations of the inflaton field. Several papers emerging from the Nuffield Workshop in Cambridge, UK, 1982, showed that these fluctuations are generically adiabatic, Gaussian, and nearly scale-invariant (Starobinsky 1982; Guth & Pi 1982; Hawking 1982; Bardeen et al. 1983).*

---

* The concept that quantum fluctuations might provide the seeds for cosmological density perturbations, which

When my colleagues and I were trying to calculate the spectrum of density perturbations from inflation in 1982, I never believed for a moment that it would be measured in my lifetime. Perhaps the few lowest moments would be measured, but certainly not enough to determine a spectrum. But I was wrong. The fluctuations in the CMB have now been measured to exquisite detail, and even better measurements are in the offing. So far everything looks consistent with the predictions of the simplest, generic inflationary models. Figure 3.3 shows the temperature power spectrum and the temperature-polarization cross-correlation, based on the first year of data of the *WMAP* experiment (Bennett et al. 2003). The curve shows the best-fit "running-index" ΛCDM model. The gray band indicates one standard deviation of uncertainty due to cosmic variance (the limitation imposed by being able to sample only one sky). The underlying primordial spectrum is modeled as a power law k^{n_s}, where $n_s = 1$ corresponds to scale-invariance. The best fit to *WMAP* alone gives $n_s = 0.99 \pm 0.04$. When *WMAP* data is combined with data on smaller scales from other observations there is some evidence that n grows with scale, but this is not conclusive. As mentioned above, the fit gives $\Omega_0 = 1.02 \pm 0.02$. The addition of isocurvature modes does not improve the fit, so the expectation of adiabatic perturbations is confirmed, and various tests for non-Gaussianity have found no signs of it.

3.4   The Inflationary Power Spectrum

A complete derivation of the density perturbation spectrum arising from inflation is a very technical subject, so the interested reader should refer to the Mukhanov et al. (1992) or Liddle & Lyth (1993) review articles. However, in this section I will describe the basics of the subject, for single field slow roll inflation, in a simple and qualitative way.

For a flat universe ($k = 0$) the metric of Equation (3.3) reduces to

$$ds^2 = -dt^2 + a^2(t)d\vec{x}^2 . \tag{3.17}$$

The perturbations are described in terms of linear perturbation theory, so it is natural to describe the perturbations in terms of a Fourier expansion in the comoving coordinates $\vec{x}$. Each

goes back at least to Sakharov (1965), was pursued in the early 1980s by Lukash (1980a, 1980b), Press (1980, 1981), and Mukhanov & Chibisov (1981, 1982). Mukhanov & Chibisov's papers are of particular interest, since they considered such quantum fluctuations in the context of the Starobinsky (1979, 1980) model, now recognized as a version of inflation. There is some controversy and ongoing discussion concerning the historical role of the Mukhanov & Chibisov papers, so I include a few comments that the reader can pursue if interested. Mukhanov & Chibisov first discovered that quantum fluctuations prevent the Starobinsky model from solving the initial singularity problem. They then considered the possibility that the quantum fluctuations are relevant for density perturbations, and found a nearly scale-invariant spectrum during the de Sitter phase. Without any derivation that I can presently discern, the 1981 paper gives a nearly scale-invariant formula for the *final* density perturbations after the end of inflation, which is similar but not identical to the result that was later described in detail in Mukhanov, Feldman, & Brandenberger (1992). In a recent preprint, Mukhanov (2003) refers to Mukhanov & Chibisov (1981) as "the first paper where the spectrum of inflationary perturbations was calculated." But controversies surrounding this statement remain unresolved. Why, for example, were the authors never explicit about the subtle question of how they calculated the evolution of the (conformally flat) density perturbations in the de Sitter phase into the (conformally Newtonian) perturbations after reheating? This gap seems particularly evident in the longer 1982 paper. And could the Starobinsky model properly be considered an inflationary model in 1981 or 1982, since at the time there was no recognition in the literature that the model could be used to explain the homogeneity, isotropy, or flatness of the universe? It was not until later that Whitt (1984) and Mijic, Morris, & Suen (1986) established the equivalence between the Starobinsky model and standard inflation. After the 1981 and 1982 Mukhanov & Chibisov papers, the topic of density perturbations in the Starobinsky model was revisited by a number of authors, starting with Starobinsky (1983).

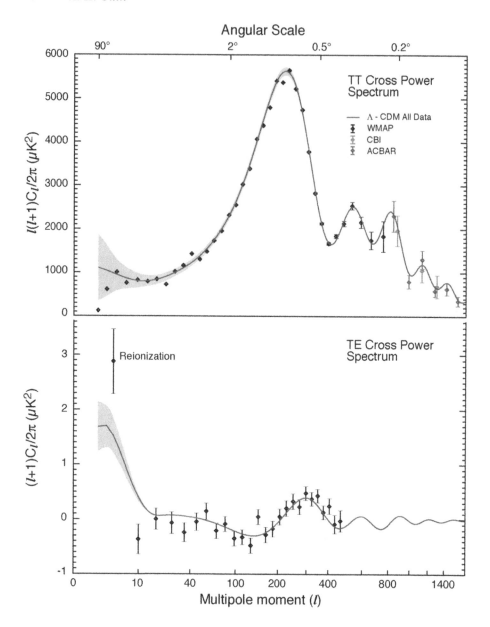

Fig. 3.3. Power spectra of the cosmic background radiation as measured by *WMAP* (Bennett et al. 2003, courtesy of the NASA/*WMAP* Science Team). The top panel shows the temperature anisotropies, and the bottom panel shows the correlation between temperature fluctuations and *E*-mode polarization fluctuations. The solid line is a fit consistent with simple inflationary models.

mode will evolve independently of all the other modes. During the inflationary era the physical wavelength of any given mode will grow with the scale factor $a(t)$, and hence will grow exponentially. The Hubble length H^{-1}, however, is approximately constant during inflation.

The modes of interest will start at wavelengths far less than H^{-1}, and will grow during inflation to be perhaps 20 orders of magnitude larger than H^{-1}. For each mode, we will let t_1 ("first Hubble crossing") denote the time at which the wavelength is equal to the Hubble length during the inflationary era. When inflation is over the wavelength will continue to grow as the scale factor, but the scale factor will slow down to behave as $a(t) \propto t^{1/2}$ during the radiation-dominated era, and $a(t) \propto t^{2/3}$ during the matter-dominated era. The Hubble length $H^{-1} = a/\dot{a}$ will grow linearly with t, so eventually the Hubble length will overtake the wavelength, and the wave will come back inside the Hubble length. We will let t_2 ("second Hubble crossing") denote the time for each mode when the wavelength is again equal to the Hubble length. This pattern of evolution is important to our understanding, because complicated physics can happen only when the wavelength is smaller than or comparable to the Hubble length. When the wavelength is large compared to the Hubble length, the distance that light can travel in a Hubble time becomes small compared to the wavelength, and hence all motion is very slow and the pattern is essentially frozen in.

Inflation ends when a scalar field rolls down a hill in a potential energy diagram, such as Figure 3.1 or 3.2. Since the scalar field undergoes quantum fluctuations, however, the field will not roll homogeneously, but instead will get a little ahead in some places and a little behind in others. Hence inflation will not end everywhere simultaneously, but instead the ending time will be a function of position:

$$t_{\text{end}}(\vec{x}) = t_{\text{end,average}} + \delta t(\vec{x}) \,. \tag{3.18}$$

Since some regions will undergo more inflation than others, we have a natural source of inhomogeneities.

Next, we need to define a statistical quantity that characterizes the perturbations. Letting $\frac{\delta\rho}{\rho}(\vec{x},t)$ describe the fractional perturbation in the total energy density ρ, useful Fourier space quantities can be defined by

$$\left[\frac{\delta\tilde{\rho}}{\rho}(\vec{k},t)\right]^2 \equiv \frac{k^3}{(2\pi)^3} \int d^3x\, e^{i\vec{k}\cdot\vec{x}} \left\langle \frac{\delta\rho}{\rho}(\vec{x},t)\frac{\delta\rho}{\rho}(\vec{0},t) \right\rangle \,, \tag{3.19a}$$

$$\left[\delta\tilde{t}(\vec{k})\right]^2 \equiv \frac{k^3}{(2\pi)^3} \int d^3x\, e^{i\vec{k}\cdot\vec{x}} \left\langle \delta t(\vec{x})\delta t(\vec{0}) \right\rangle \,, \tag{3.19b}$$

where the brackets denote an expectation value.

Since the wave pattern is frozen when the wavelength is large compared to the Hubble length, for any given mode $\vec{k}$ the pattern is frozen between $t_1(\vec{k})$ and $t_2(\vec{k})$. We therefore expect a simple relationship between the amplitude of the perturbation at times t_1 and t_2, where the perturbation at time t_1 is described by a time offset δt in the evolution of the scalar field, and at t_2 it is described by $\delta\rho/\rho$. Since we are approximating the problem with first-order perturbation theory, the relationship must be linear. By dimensional analysis, the relationship must have the form

$$\frac{\delta\tilde{\rho}}{\rho}(\vec{k},t_2(k)) = C_1 H(t_1)\delta\tilde{t}(\vec{k}) \,, \tag{3.20}$$

where C_1 is a dimensionless constant and H is the only quantity with units of inverse time that seems to have relevance. Of course, deriving Equation (3.20) and determining the value of C_1 is a lot of work.

To estimate $\delta\tilde{t}(\vec{k})$, note that we expect its value to become frozen at about time $t_1(k)$. If the classical, homogeneous rolling of the scalar field down the hill is described by $\phi_0(t)$, then the offset in time δt is equivalent to an offset of the value of the scalar field,

$$\delta\phi = -\dot{\phi}_0\,\delta t \ . \tag{3.21}$$

The sign is not very important, but it is negative because inflation will end earliest ($\delta t < 0$) in regions where the scalar field has advanced the most ($\delta\phi > 0$, assuming $\dot{\phi} > 0$). $\dot{\phi}_0$ is in principle calculable by solving Equation (3.12), omitting the spatial Laplacian term. Although it is a second-order equation, for "slow roll" inflation one assumes that the $\ddot{\phi}$ term is negligible, so

$$\dot{\phi} = -\frac{1}{3H}\frac{\partial V}{\partial\phi} \ , \qquad \text{where} \quad H^2 = \frac{8\pi}{3M_p^2}V \ , \tag{3.22}$$

where $M_p \equiv 1/\sqrt{G} = 1.22 \times 10^{19}$ GeV is the Planck mass. $\delta\phi$ can be estimated by defining the quantity $\delta\tilde{\phi}(\vec{k},t)$ in analogy to Equations (3.19), but the quantity on the right-hand side is just the scalar field propagator of quantum field theory. One can approximate $\delta\phi(\vec{x},t)$ as a free massless quantum field evolving in de Sitter space (see, for example, Birrell & Davies 1982). We want to evaluate $\delta\tilde{\phi}(\vec{k},t)$ for $|\vec{k}_{\text{physical}}| \approx H$. Again we can rely on dimensional analysis, since ϕ has the units of mass, and the only relevant quantity with dimensions of mass is H. Thus, $\delta\tilde{\phi} \approx H$, and Equations (3.20)–(3.22) can be combined to give

$$\frac{\delta\tilde{\rho}}{\rho}\big(\vec{k},t_2(k)\big) = C_2\left.\frac{H^2}{\dot{\phi}_0}\right|_{t_1(\vec{k})} = C_3\left.\frac{V^{3/2}}{M_p^3 V'}\right|_{t_1(\vec{k})} \ , \tag{3.23}$$

where C_2 and C_3 are dimensionless constants, and $V' \equiv \partial V/\partial\phi$. The entire quantity on the right-hand side is evaluated at $t_1(\vec{k})$, since it is at this time that the amplitude of the mode is frozen.

Equation (3.23) is the key result. It describes density perturbations which are nearly scale invariant, meaning that $\delta\tilde{\rho}(\vec{k},t_2(k))/\rho$ is approximately independent of k, because typically $V(\phi)$ and $V'(\phi)$ are nearly constant during the period when perturbations of observable wavelengths are passing through the Hubble length during inflation. Since $\delta\tilde{\rho}/\rho$ is measurable and C_3 is calculable, one can use Equation (3.23) to determine the value of $V^{3/2}/(M_p^3 V')$. Using *COBE* data, Liddle & Lyth (1993) found

$$\frac{V^{3/2}}{M_p^3 V'} \approx 3.6 \times 10^{-6} \ . \tag{3.24}$$

While Equation (3.23) describes density perturbations that are nearly scale invariant, it also allows us to express the departure from scale invariance in terms of derivatives of the potential $V(\phi)$. One defines the scalar index n_s by

$$\left[\frac{\delta\tilde{\rho}}{\rho}\big(\vec{k},t_2(k)\big)\right]^2 \propto k^{n_s-1} \ , \tag{3.25}$$

so

$$n_s - 1 = \frac{d\ln\left[\frac{\delta\tilde{\rho}}{\rho}\big(\vec{k},t_2(k)\big)\right]^2}{d\ln k} \ . \tag{3.26}$$

To carry out the differentiation, note that k is related to t_1 by $H = k/(2\pi a(t_1))$. Treating H as a constant, since it varies much more slowly than a, differentiation gives $dk/dt_1 = Hk$. Using Equation (3.22) for $d\phi_0/dt$, one has (Liddle & Lyth 1992)

$$
\begin{aligned}
n_s &= 1 + k \frac{dt_1}{dk} \frac{d\phi_0}{dt_1} \frac{d}{d\phi} \ln \left[C_3 \frac{V^{3/2}}{M_p^3 V'} \right]^2 \\
&= 1 + 2\eta - 6\epsilon ,
\end{aligned}
\tag{3.27}
$$

where

$$
\epsilon = \frac{M_p^2}{16\pi} \left(\frac{V'}{V} \right)^2 , \qquad \eta = \frac{M_p^2}{8\pi} \left(\frac{V''}{V} \right) .
\tag{3.28}
$$

ϵ and η are the now well-known slow-roll parameters that quantify departures from scale invariance. (But the reader should beware that some authors use slightly different definitions.) Alternatively, Equations (3.22) can be used to express ϵ and η in terms of time derivatives of H:

$$
\epsilon = -\frac{\dot{H}}{H^2} , \qquad \eta = -\frac{\dot{H}}{H^2} - \frac{\ddot{H}}{2H\dot{H}} ,
\tag{3.29}
$$

so

$$
n_s = 1 + 4\frac{\dot{H}}{H^2} - \frac{\ddot{H}}{H\dot{H}} .
\tag{3.30}
$$

The above equation can be used to motivate a generic estimate of how much n_s is likely to deviate from 1. Since inflation needs to end at roughly 60 e-folds from the time $t_1(k)$ when the right-hand side of Equation (3.23) is evaluated, we can take $60H^{-1}$ as the typical time scale for the variation of physical quantities. For any quantity X, we can generically estimate that $|\dot{X}| \sim HX/60$, so $n_s - 1 \sim \pm\frac{4}{60} \pm \frac{1}{60}$. We can conclude that typically n_s will deviate from 1 by an amount of order 0.1. Of course, any detailed model will make a precise prediction for the value of n_s.

3.5 Eternal Inflation: Mechanisms

The remainder of this article will discuss eternal inflation—the questions that it can answer, and the questions that it raises. In this section I discuss the mechanisms that make eternal inflation possible, leaving the other issues for the following sections. I will discuss eternal inflation first in the context of new inflation, and then in the context of chaotic inflation, where it is more subtle.

In the case of new inflation, the exponential expansion occurs as the scalar field rolls from the false vacuum state at the peak of the potential energy diagram (see Fig. 3.1) toward the trough. The eternal aspect occurs while the scalar field is hovering around the peak. The first model of this type was constructed by Steinhardt (1983), and later that year Vilenkin (1983) showed that new inflationary models are generically eternal. The key point is that, even though classically the field would roll off the hill, quantum-mechanically there is always an amplitude, a tail of the wave function, for it to remain at the top. If you ask how fast does this tail of the wave function fall off with time, the answer in almost any model is that it falls off exponentially with time, just like the decay of most metastable states (Guth & Pi 1985). The time scale for the decay of the false vacuum is controlled by

$$m^2 = -\frac{\partial^2 V}{\partial \phi^2}\bigg|_{\phi=0}, \tag{3.31}$$

the negative mass-squared of the scalar field when it is at the top of the hill in the potential diagram. This is an adjustable parameter as far as our use of the model is concerned, but m has to be small compared to the Hubble constant or else the model does not lead to enough inflation. So, for parameters that are chosen to make the inflationary model work, the exponential decay of the false vacuum is slower than the exponential expansion. Even though the false vacuum is decaying, the expansion outruns the decay and the total volume of false vacuum actually increases with time rather than decreases. Thus, inflation does not end everywhere at once, but instead inflation ends in localized patches, in a succession that continues *ad infinitum*. Each patch is essentially a whole universe—at least its residents will consider it a whole universe—and so inflation can be said to produce not just one universe, but an infinite number of universes. These universes are sometimes called bubble universes, but I prefer to use the phrase "pocket universe," to avoid the implication that they are approximately round. [While bubbles formed in first-order phase transitions are round (Coleman & De Luccia 1980), the local universes formed in eternal new inflation are generally very irregular, as can be seen for example in the two-dimensional simulation in Figure 2 of Vanchurin, Vilenkin, & Winitzki (2000).]

In the context of chaotic inflationary models the situation is slightly more subtle. Andrei Linde (1986a, 1986b, 1990) showed that these models are eternal in 1986. In this case inflation occurs as the scalar field rolls down a hill of the potential energy diagram, as in Figure 3.2, starting high on the hill. As the field rolls down the hill, quantum fluctuations will be superimposed on top of the classical motion. The best way to think about this is to ask what happens during one time interval of duration $\Delta t = H^{-1}$ (one Hubble time), in a region of one Hubble volume H^{-3}. Suppose that ϕ_0 is the average value of ϕ in this region, at the start of the time interval. By the definition of a Hubble time, we know how much expansion is going to occur during the time interval: exactly a factor of e. (This is the only exact number in this paper, so I wanted to emphasize the point.) That means the volume will expand by a factor of e^3. One of the deep truths that one learns by working on inflation is that e^3 is about equal to 20, so the volume will expand by a factor of 20. Since correlations typically extend over about a Hubble length, by the end of one Hubble time, the initial Hubble-sized region grows and breaks up into 20 independent Hubble-sized regions.

As the scalar field is classically rolling down the hill, the classical change in the field $\Delta\phi_{\rm cl}$ during the time interval Δt is going to be modified by quantum fluctuations $\Delta\phi_{\rm qu}$, which can drive the field upward or downward relative to the classical trajectory. For any one of the 20 regions at the end of the time interval, we can describe the change in ϕ during the interval by

$$\Delta\phi = \Delta\phi_{\rm cl} + \Delta\phi_{\rm qu}. \tag{3.32}$$

In lowest-order perturbation theory the fluctuation is treated as a free quantum field, which implies that $\Delta\phi_{\rm qu}$, the quantum fluctuation averaged over one of the 20 Hubble volumes at the end, will have a Gaussian probability distribution, with a width of order $H/2\pi$ (Vilenkin & Ford 1982; Linde 1982b; Starobinsky 1982, 1986). There is then always some probability that the sum of the two terms on the right-hand side will be positive—that the scalar field will fluctuate up and not down. As long as that probability is bigger than 1 in 20, then the number of inflating regions with $\phi \geq \phi_0$ will be larger at the end of the time interval Δt than it was at the beginning. This process will then go on forever, so inflation will never end.

Thus, the criterion for eternal inflation is that the probability for the scalar field to go up must be bigger than $1/e^3 \approx 1/20$. For a Gaussian probability distribution, this condition will be met provided that the standard deviation for $\Delta\phi_{qu}$ is bigger than $0.61|\Delta\phi_{cl}|$. Using $\Delta\phi_{cl} \approx \dot{\phi}_{cl}H^{-1}$, the criterion becomes

$$\Delta\phi_{qu} \approx \frac{H}{2\pi} > 0.61 |\dot{\phi}_{cl}| H^{-1} \iff \frac{H^2}{|\dot{\phi}_{cl}|} > 3.8 . \tag{3.33}$$

Comparing with Equation (3.23), we see that the condition for eternal inflation is equivalent to the condition that $\delta\rho/\rho$ on ultra-long length scales is bigger than a number of order unity.

The probability that $\Delta\phi$ is positive tends to increase as one considers larger and larger values of ϕ, so sooner or later one reaches the point at which inflation becomes eternal. If one takes, for example, a scalar field with a potential

$$V(\phi) = \frac{1}{4}\lambda\phi^4 , \tag{3.34}$$

then the de Sitter space equation of motion in flat Robertson-Walker coordinates (Eq. 3.17) takes the form

$$\ddot{\phi} + 3H\dot{\phi} = -\lambda\phi^3 , \tag{3.35}$$

where spatial derivatives have been neglected. In the "slow-roll" approximation one also neglects the $\ddot{\phi}$ term, so $\dot{\phi} \approx -\lambda\phi^3/(3H)$, where the Hubble constant H is related to the energy density by

$$H^2 = \frac{8\pi}{3}G\rho = \frac{2\pi}{3}\frac{\lambda\phi^4}{M_p^2} . \tag{3.36}$$

Putting these relations together, one finds that the criterion for eternal inflation, Equation (3.33), becomes

$$\phi > 0.75\,\lambda^{-1/6} M_p . \tag{3.37}$$

Since λ must be taken very small, on the order of 10^{-12}, for the density perturbations to have the right magnitude, this value for the field is generally well above the Planck scale. The corresponding energy density, however, is given by

$$V(\phi) = \frac{1}{4}\lambda\phi^4 = 0.079\lambda^{1/3}M_p^4 , \tag{3.38}$$

which is actually far below the Planck scale.

So for these reasons we think inflation is almost always eternal. I think the inevitability of eternal inflation in the context of new inflation is really unassailable—I do not see how it could possibly be avoided, assuming that the rolling of the scalar field off the top of the hill is slow enough to allow inflation to be successful. The argument in the case of chaotic inflation is less rigorous, but I still feel confident that it is essentially correct. For eternal inflation to set in, all one needs is that the probability for the field to increase in a given Hubble-sized volume during a Hubble time interval is larger than $1/20$.

Thus, once inflation happens, it produces not just one universe, but an infinite number of universes.

3.6   Eternal Inflation: Implications

In spite of the fact that the other universes created by eternal inflation are too remote to imagine observing directly, I nonetheless claim that eternal inflation has real consequences in terms of the way we extract predictions from theoretical models. Specifically, there are three consequences of eternal inflation that I will discuss.

First, eternal inflation implies that all hypotheses about the ultimate initial conditions for the universe—such as the Hartle & Hawking (1983) no boundary proposal, the tunneling proposals by Vilenkin (1984, 1986, 1999) or Linde (1984, 1998), or the more recent Hawking & Turok (1998) instanton—become totally divorced from observation. That is, one would expect that if inflation is to continue arbitrarily far into the future with the production of an infinite number of pocket universes, then the statistical properties of the inflating region should approach a steady state that is independent of the initial conditions. Unfortunately, attempts to quantitatively study this steady state are severely limited by several factors. First, there are ambiguities in defining probabilities, which will be discussed later. In addition, the steady state properties seem to depend strongly on super-Planckian physics, which we do not understand. That is, the same quantum fluctuations that make eternal chaotic inflation possible tend to drive the scalar field further and further up the potential energy curve, so attempts to quantify the steady state probability distribution (Linde, Linde, & Mezhlumian 1994; Garcia-Bellido & Linde 1995) require the imposition of some kind of a boundary condition at large ϕ. Although these problems remain unsolved, I still believe that it is reasonable to assume that in the course of its unending evolution, an eternally inflating universe would lose all memory of the state in which it started.

Even if the universe forgets the details of its genesis, however, I would not assume that the question of how the universe began would lose its interest. While eternally inflating universes continue forever once they start, they are apparently not eternal into the past. (The word *eternal* is therefore not technically correct—it would be more precise to call this scenario *semi-eternal* or *future-eternal*.) The possibility of a quantum origin of the universe is very attractive, and will no doubt be a subject of interest for some time. Eternal inflation, however, seems to imply that the entire study will have to be conducted with literally no input from observation.

A second consequence of eternal inflation is that the probability of the onset of inflation becomes totally irrelevant, provided that the probability is not identically zero. Various authors in the past have argued that one type of inflation is more plausible than another, because the initial conditions that it requires appear more likely to have occurred. In the context of eternal inflation, however, such arguments have no significance. Any nonzero probability of onset will produce an infinite spacetime volume. If one wants to compare two types of inflation, the expectation is that the one with the faster exponential time constant will always win.

A corollary to this argument is that new inflation is not dead. While the initial conditions necessary for new inflation cannot be justified on the basis of thermal equilibrium, as proposed in the original papers (Linde 1982a; Albrecht & Steinhardt 1982), in the context of eternal inflation it is sufficient to conclude that the probability for the required initial conditions is nonzero. Since the resulting scenario does not depend on the words that are used to justify the initial state, the standard treatment of new inflation remains valid.

A third consequence of eternal inflation is the possibility that it offers to rescue the predictive power of theoretical physics. Here I have in mind the status of string theory, or the theory known as M theory, into which string theory has evolved. The theory itself has an elegant uniqueness, but nonetheless it appears that the vacuum is far from unique (Bousso & Polchinski 2000; Susskind 2003). Since predictions will ultimately depend on the properties of the vacuum, the predictive power of string/M theory may be limited. Eternal inflation, however, provides a possible mechanism to remedy this problem. Even if many types of vacua are equally stable, it may turn out that there is one unique metastable state that leads to a maximal rate of inflation. If so, then this metastable state will dominate the eternally inflating region, even if its expansion rate is only infinitesimally larger than the other possibilities. One would still need to follow the decay of this metastable state as inflation ends. It may very well branch into a number of final low-energy vacua, but the number that are significantly populated could hopefully be much smaller than the total number of vacua. All of this is pure speculation at this point, because no one knows how to calculate these things. Nonetheless, it is possible that eternal inflation might help to constrain the vacuum state of the real universe, perhaps significantly enhancing the predictive power of M theory.

3.7 Does Inflation Need a Beginning?

If the universe can be eternal into the future, is it possible that it is also eternal into the past? Here I will describe a recent theorem (Borde et al. 2003) that shows, under plausible assumptions, that the answer to this question is no.*

The theorem is based on the well-known fact that the momentum of an object traveling on a geodesic through an expanding universe is redshifted, just as the momentum of a photon is redshifted. Suppose, therefore, we consider a timelike or null geodesic extended backwards, into the past. In an expanding universe such a geodesic will be blueshifted. The theorem shows that under some circumstances the blueshift reaches infinite rapidity (i.e., the speed of light) in a finite amount of proper time (or affine parameter) along the trajectory, showing that such a trajectory is (geodesically) incomplete.

To describe the theorem in detail, we need to quantify what we mean by an expanding universe. We imagine an observer whom we follow backwards in time along a timelike or null geodesic. The goal is to define a local Hubble parameter along this geodesic, which must be well defined even if the spacetime is neither homogeneous nor isotropic. Call the velocity of the geodesic observer $v^\mu(\tau)$, where τ is the proper time in the case of a timelike observer, or an affine parameter in the case of a null observer. (Although we are imagining that we are following the trajectory backwards in time, τ is defined to increase in the future timelike direction, as usual.) To define H, we must imagine that the vicinity of the observer is filled with "comoving test particles," so that there is a test particle velocity $u^\mu(\tau)$ assigned to each point τ along the geodesic trajectory, as shown in Figure 3.4. These particles need not be real—all that will be necessary is that the worldlines can be defined, and that each worldline should have zero proper acceleration at the instant it intercepts the geodesic observer.

To define the Hubble parameter that the observer measures at time τ, the observer focuses on two particles, one that he passes at time τ, and one at $\tau + \Delta\tau$, where in the end he takes the limit $\Delta\tau \to 0$. The Hubble parameter is defined by

---

* There were also earlier theorems about this issue by Borde & Vilenkin (1994, 1996) and Borde (1994), but these theorems relied on the weak energy condition, which for a perfect fluid is equivalent to the condition $\rho + p \geq 0$. This condition holds classically for forms of matter that are known or commonly discussed as theoretical proposals. It can, however, be violated by quantum fluctuations (Borde & Vilenkin 1997), and so the reliability of these theorems is questionable.

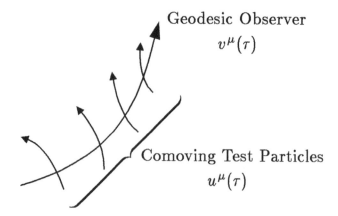

Fig. 3.4. An observer measures the velocity of passing test particles to infer the Hubble parameter.

$$H \equiv \frac{\Delta v_{\text{radial}}}{\Delta r} , \qquad (3.39)$$

where Δv_{radial} is the radial component of the relative velocity between the two particles, and Δr is their distance, where both quantities are computed in the rest frame of one of the test particles, not in the rest frame of the observer. Note that this definition reduces to the usual one if it is applied to a homogeneous isotropic universe.

The relative velocity between the observer and the test particles can be measured by the invariant dot product,

$$\gamma \equiv u_\mu v^\mu , \qquad (3.40)$$

which for the case of a timelike observer is equal to the usual special relativity Lorentz factor

$$\gamma = \frac{1}{\sqrt{1 - v_{\text{rel}}^2}} . \qquad (3.41)$$

If H is positive we would expect γ to decrease with τ, since we expect the observer's momentum relative to the test particles to redshift. It turns out, however, that the relationship between H and changes in γ can be made precise. If one defines

$$F(\gamma) \equiv \begin{cases} 1/\gamma & \text{for null observers} \\ \text{arctanh}(1/\gamma) & \text{for timelike observers} , \end{cases} \qquad (3.42)$$

then

$$H = \frac{dF(\gamma)}{d\tau} . \qquad (3.43)$$

I like to call $F(\gamma)$ the "slowness" of the geodesic observer, because it increases as the observer slows down, relative to the test particles. The slowness decreases as we follow the geodesic backwards in time, but it is positive definite, and therefore cannot decrease below zero. $F(\gamma) = 0$ corresponds to $\gamma = \infty$, or a relative velocity equal to that of light. This bound allows us to place a rigorous limit on the integral of Equation (3.43). For timelike geodesics,

$$\int^{\tau_f} H d\tau \le \operatorname{arctanh}\left(\frac{1}{\gamma_f}\right) = \operatorname{arctanh}\left(\sqrt{1-v_{\text{rel}}^2}\right), \tag{3.44}$$

where γ_f is the value of γ at the final time $\tau = \tau_f$. For null observers, if we normalize the affine parameter τ by $d\tau/dt = 1$ at the final time τ_f, then

$$\int^{\tau_f} H d\tau \le 1. \tag{3.45}$$

Thus, if we assume an *averaged expansion condition*, i.e., that the average value of the Hubble parameter H_{av} along the geodesic is positive, then the proper length (or affine length for null trajectories) of the backwards-going geodesic is bounded. Thus, the region for which $H_{\text{av}} > 0$ is past-incomplete.

It is difficult to apply this theorem to general inflationary models, since there is no accepted definition of what exactly defines this class. However, in standard eternally inflating models, the future of any point in the inflating region can be described by a stochastic model (Goncharov, Linde, & Mukhanov 1987) for inflaton evolution, valid until the end of inflation. Except for extremely rare large quantum fluctuations, $H \gtrsim \sqrt{(8\pi/3)G\rho_f}$, where ρ_f is the energy density of the false vacuum driving the inflation. The past for an arbitrary model is less certain, but we consider eternal models for which the past is like the future. In that case H would be positive almost everywhere in the past inflating region. If, however, $H_{\text{av}} > 0$ when averaged over a past-directed geodesic, our theorem implies that the geodesic is incomplete.

There is, of course, no conclusion that an eternally inflating model must have a unique beginning, and no conclusion that there is an upper bound on the length of all backwards-going geodesics from a given point. There may be models with regions of contraction embedded within the expanding region that could evade our theorem. Aguirre & Gratton (2002, 2003) have proposed a model that evades our theorem, in which the arrow of time reverses at the $t = -\infty$ hypersurface, so the universe "expands" in both halves of the full de Sitter space.

The theorem does show, however, that an eternally inflating model of the type usually assumed, which would lead to $H_{\text{av}} > 0$ for past-directed geodesics, cannot be complete. Some new physics (i.e., not inflation) would be needed to describe the past boundary of the inflating region. One possibility would be some kind of quantum creation event.

One particular application of the theory is the cyclic ekpyrotic model of Steinhardt & Turok (2002). This model has $H_{\text{av}} > 0$ for null geodesics for a single cycle, and since every cycle is identical, $H_{\text{av}} > 0$ when averaged over all cycles. The cyclic model is therefore past-incomplete and requires a boundary condition in the past.

3.8 Calculation of Probabilities in Eternally Inflating Universes

In an eternally inflating universe, anything that can happen will happen; in fact, it will happen an infinite number of times. Thus, the question of what is possible becomes trivial—anything is possible, unless it violates some absolute conservation law. To extract predictions from the theory, we must therefore learn to distinguish the probable from the improbable.

However, as soon as one attempts to define probabilities in an eternally inflating spacetime, one discovers ambiguities. The problem is that the sample space is infinite, in that an eternally inflating universe produces an infinite number of pocket universes. The fraction of universes with any particular property is therefore equal to infinity divided by infinity—a meaningless ratio. To obtain a well-defined answer, one needs to invoke some method of regularization. In eternally inflating universes, however, the answers that one gets depend on how one chooses the method of regularization.

To understand the nature of the problem, it is useful to think about the integers as a model system with an infinite number of entities. We can ask, for example, what fraction of the integers are odd. With the usual ordering of the integers, 1, 2, 3, ..., it seems obvious that the answer is 1/2. However, the same set of integers can be ordered by writing two odd integers followed by one even integer, as in 1, 3, 2, 5, 7, 4, 9, 11, 6 , ... Taken in this order, it looks like 2/3 of the integers are odd.

One simple method of regularization is a cut-off at equal-time surfaces in a synchronous gauge coordinate system. Specifically, suppose that one constructs a Robertson-Walker coordinate system while the model universe is still in the false vacuum (de Sitter) phase, before any pocket universes have formed. One can then propagate this coordinate system forward with a synchronous gauge condition,* and one can define probabilities by truncating the spacetime volume at a fixed value t_f of the synchronous time coordinate t. I will refer to probabilities defined in this way as synchronous gauge probabilities.

An important peculiarity of synchronous gauge probabilities is that they lead to what I call the "youngness paradox." The problem is that the volume of false vacuum is growing exponentially with time with an extraordinarily small time constant, in the vicinity of 10^{-37} s. Since the rate at which pocket universes form is proportional to the volume of false vacuum, this rate is increasing exponentially with the same time constant. This means that for every universe in the sample of age t, there are approximately $\exp\{10^{37}\}$ universes with age $t-(1$ s). The population of pocket universes is therefore an incredibly youth-dominated society, in which the mature universes are vastly outnumbered by universes that have just barely begun to evolve.

Probability calculations in this youth-dominated ensemble lead to peculiar results, as was first discussed by Linde, Linde, & Mezhlumian (1995). Since mature universes are incredibly rare, it becomes likely that our universe is actually much younger than we think, with our part of the universe having reached its apparent maturity through an unlikely set of quantum jumps. These authors considered the expected behavior of the mass density in our vicinity, concluding that we should find ourselves very near the center of a spherical low-density region.

Since the probability measure depends on the method used to truncate the infinite spacetime of eternal inflation, we are not forced to accept the consequences of the synchronous gauge probabilities. A method of calculating probabilities that gives acceptable answers has been formulated by Vilenkin (1998) and his collaborators (Vanchurin et al. 2000; Garriga & Vilenkin 2001). However, we still do not have a compelling argument from first principles that determines how probabilities should be calculated.

3.9　Conclusion

In this paper I have summarized the workings of inflation, and the arguments that strongly suggest that our universe is the product of inflation. I argued that inflation can explain the size, the Hubble expansion, the homogeneity, the isotropy, and the flatness of our

---

* By a synchronous gauge condition, I mean that each equal-time hypersurface is obtained by propagating every point on the previous hypersurface by a fixed infinitesimal time interval Δt in the direction normal to the hypersurface.

universe, as well as the absence of magnetic monopoles, and even the characteristics of the nonuniformities. The detailed observations of the cosmic background radiation anisotropies continue to fall in line with inflationary expectations, and the evidence for an accelerating universe fits beautifully with the inflationary preference for a flat universe. Our current picture of the universe seems strange, with 95% of the energy in forms of matter that we do not understand, but nonetheless the picture fits together very well.

Next I turned to the question of eternal inflation, claiming that essentially all inflationary models are eternal. In my opinion this makes inflation very robust: if it starts anywhere, at any time in all of eternity, it produces an infinite number of pocket universes. Eternal inflation has the very attractive feature, from my point of view, that it offers the possibility of allowing unique (or possibly only constrained) predictions even if the underlying string theory does not have a unique vacuum. I discussed the past of eternally inflating models, concluding that under mild assumptions the inflating region must have a past boundary, and that new physics (other than inflation) is needed to describe what happens at this boundary. I have also described, however, that our picture of eternal inflation is not complete. In particular, we still do not understand how to define probabilities in an eternally inflating spacetime.

The bottom line, however, is that observations in the past few years have vastly improved our knowledge of the early universe, and that these new observations have been generally consistent with the simplest inflationary models.

Acknowledgements. This work is supported in part by funds provided by the U.S. Department of Energy (D.O.E.) under cooperative research agreement #DF-FC02-94ER40818.

References

Aguirre, A., & Gratton, S. 2002, Phys. Rev. D, 65, 083507

——. 2003, Phys. Rev. D, 67, 083515

Albrecht, A., & Steinhardt, P. J. 1982, Phys. Rev. Lett., 48, 1220

Bardeen, J. M., Steinhardt, P. J., & Turner, M. S. 1983, Phys. Rev. D, 28, 679

Bennett, C. L., et al. 2003, ApJS, 148, 1

Birrell, N. D., & Davies, P. C. W. 1982, Quantum Fields in Curved Space (Cambridge: Cambridge Univ. Press)

Borde, A. 1994, Phys. Rev. D, 50, 3692

Borde, A., Guth, A. H., & Vilenkin, A. 2003, Phys. Rev. Lett., 90, 151301

Borde, A., & Vilenkin, A. 1994, Phys. Rev. Lett., 72, 3305

——. 1996, Int. J. Mod. Phys., D5, 813

——. 1997, Phys. Rev. D, 56, 717

Bousso, R., & Polchinski, J. 2000, JHEP, 0006, 006

Callan, C. G., & Coleman, S. 1977, Phys. Rev. D, 16, 1762

Coleman, S. 1977 Phys. Rev. D, 15, 2929 (erratum 1977, 16, 1248)

Coleman, S., & De Luccia, F. 1980, Phys. Rev. D, 21, 3305

Copeland, E. J., Liddle, A. R., Lyth, D. H., Stewart, E. D., & Wands, D. 1994, Phys. Rev. D, 49, 6410

Dicke, R. H., & Peebles, P. J. E. 1979, in General Relativity: An Einstein Centenary Survey, ed. S. W. Hawking & W. Israel (Cambridge: Cambridge Univ. Press), 504

Garcia-Bellido, J., & Linde, A. D. 1995, Phys. Rev. D, 51, 429

Garriga, J., & Vilenkin, A. 2001, Phys. Rev. D, 64, 023507

Goncharov, A. S., Linde, A. D., & Mukhanov, V. F. 1987, Int. J. Mod. Phys., A2, 561

Guth, A. H. 1981, Phys. Rev. D, 23, 347

——. 1982, Phil. Trans. R. Soc. Lond. A, 307, 141

Guth, A. H., & Pi, S.-Y. 1982, Phys. Rev. Lett., 49, 1110

——. 1985, Phys. Rev. D, 32, 1899

Guth, A. H., & Weinberg, E. J. 1983, Nucl. Phys., B212, 321

Hartle, J. B., & Hawking, S. W. 1983, Phys. Rev. D, 28, 2960

Hawking, S. W. 1982, Phys. Lett., 115B, 295

Hawking, S. W., Moss, I. G., & Stewart, J. M. 1982, Phys. Rev. D, 26, 2681

Hawking, S. W., & Turok, N. G. 1998, Phys. Lett., 425B, 25

Jensen, L. G., & Stein-Schabes, J. A. 1987, Phys. Rev. D, 35, 1146

Liddle, A. R., & Lyth, D. H. 1992, Phys. Lett., 291B, 391

———. 1993, Phys. Rep., 231, 1

Linde, A. D. 1982a, Phys. Lett., 108B, 389

———. 1982b, Phys. Lett., 116B, 335

———. 1983a, Pis'ma Zh. Eksp. Teor. Fiz., 38, 149 (JETP Lett., 38, 176)

———. 1983b, Phys. Lett., 129B, 177

———. 1984, Nuovo Cim., 39, 401

———. 1986a, Mod. Phys. Lett., A1, 81

———. 1986b, Phys. Lett., 175B, 395

———. 1990, Particle Physics and Inflationary Cosmology (Chur: Harwood Academic Publishers), Secs. 1.7–1.8

———. 1991, Phys. Lett., 259B, 38

———. 1994, Phys. Rev. D, 49, 748

———. 1998, Phys. Rev. D, 58, 083514

Linde, A. D., Linde, D., & Mezhlumian, A. 1994, Phys. Rev. D, 49, 1783

———. 1995, Phys. Lett., 345B, 203

Lukash, V. N. 1980a, Pis'ma Zh. Eksp. Teor. Fiz., 31, 631

———. 1980b, Zh. Eksp. Teor. Fiz., 79, 1601

Mijic, M. B., Morris, M. S., & Suen, W.-M. 1986, Phys. Rev. D, 34, 2934

Mukhanov, V. F. 2003, astro-ph/0303077

Mukhanov, V. F. & Chibisov, G. V. 1981, Pis'ma Zh. Eksp. Teor. Fiz., 33, 549 (JETP Lett., 33, 532)

———. 1982, Zh. Eksp. Teor. Fiz.83, 475 (JETP, 56, 258)

Mukhanov, V. F., Feldman, H. A., & Brandenberger, R. H. 1992, Phys. Rep., 215, 203

Preskill, J. P. 1979, Phys. Rev. Lett., 43, 1365

Press, W. H. 1980, Physica Scripta, 21, 702

———. 1981, in Cosmology and Particles: Proceedings of the Sixteenth Moriond Astrophysics Meeting, ed. J. Audouze et al. (Gif-sur-Yvette: Editions Frontières), 137

Randall, L., Soljačić, M., & Guth, A. H. 1996, Nucl. Phys., B472, 377

Sakharov, A. D. 1965, Zh. Eksp. Teor. Fiz., 49, 345 (1966, JETP, 22, 241)

Starobinsky, A. A. 1979, Pis'ma Zh. Eksp. Teor. Fiz., 30, 719 (JETP Lett., 30, 682)

———. 1980, Phys. Lett., 91B, 99

———. 1982, Phys. Lett., 117B, 175

———. 1983, Pis'ma Astron. Zh.9, 579 (Sov. Astron. Lett., 9, 302)

———. 1986, in Field Theory, Quantum Gravity and Strings, Lecture Notes in Physics 246, ed. H. J. de Vega & N. Sánchez (Heidelberg: Springer Verlag), 107

Steinhardt, P. J. 1983, in The Very Early Universe: Proceedings of the Nuffield Workshop, Cambridge, 21 June – 9 July, 1982, ed. G. W. Gibbons, S. W. Hawking, & S. T. C. Siklos (Cambridge: Cambridge Univ. Press), 251

Steinhardt, P. J., & Turok, N. G. 2002, Phys. Rev. D, 65, 126003

Stewart, E. 1995, Phys. Lett., 345B, 414

Susskind, L. 2003, hep-th/0302219

Vanchurin, V., Vilenkin, A., & Winitzki, S. 2000, Phys. Rev. D, 61, 083507

Vilenkin, A. 1983, Phys. Rev. D, 27, 2848

———. 1984, Phys. Rev. D, 30, 509

———. 1986, Phys. Rev. D, 33, 3560

———. 1998, Phys. Rev. Lett., 81, 5501

———. 1999, in International Workshop on Particle and the Early Universe (COSMO 98), ed. D. O. Caldwell (New York: AIP), 23

Vilenkin, A., & Ford, L. H. 1982, Phys. Rev. D, 26, 1231

Whitt, B. 1984, Phys. Lett., 145B, 176

4

Update on string theory

JOHN H. SCHWARZ
California Institute of Technology

Abstract
The first part of this report gives a very quick sketch of how string theory concepts originated and evolved during its first 25 years (1968-93). The second part presents a somewhat more detailed discussion of the highlights of the past decade. The final part discusses some of the major problems that remain to be solved.

4.1 Introduction

There are two primary goals in fundamental physics. The first is to construct a unified theory, incorporating relativity and quantum theory, that describes all fundamental forces, as well as all of the elementary particles. The second goal is to understand the origin and evolution of the Universe. Needless to say, these are both incredibly ambitious objectives, representing the extremes of microphysics and macrophysics. The only reason that we can even pose them with a straight face is the fact that so much progress toward each of these objectives has already been achieved.

Superstring theory (also known as M-theory) is a promising candidate for the fundamental underlying theory. It used to be believed that superstring theory is a collection of theories, but as we will discuss, it now appears that (in a certain precise sense) there is a unique theory with no adjustable parameters. But this theory is a work in progress—not yet fully formulated—and it is unclear how far its elucidation will take us toward realizing these lofty goals. Even in the most optimistic scenario, it will certainly take a long time.

This talk will take a somewhat historical tack in describing the subject. The first part will give a very quick sketch of how string theory concepts originated and evolved during its first 25 years (1968-93). In the second part a somewhat more detailed discussion of the highlights of the past decade will be presented. The final section will list and comment on some of the major problems that remain to be solved. The reader who wants a more thorough treatment is referred to the standard texts (Green, Schwarz, & Witten 1987; Polchinski 1998)

4.2 1968–1993

String theory arose in the late 1960s in an attempt to construct a theory of the strong nuclear force. Experiments in the 1960s revealed a rich spectrum of strongly interacting particles (hadrons) lying on nearly straight and parallel Regge trajectories. This means that there were families of particles identified whose spin increased linearly with the square of their mass. This point of view was developed as part of the S matrix theory/bootstrap program that was fashionable in the 1960s. The linear Regge trajectories were interpreted in terms of poles of the analytic S matrix in the angular momentum plane. This picture successfully accounted for certain asymptotic properties of scattering amplitudes at high energy. This led Veneziano (1968) to propose a very simple mathematical function (basically the Euler beta function) as an approximate expression for a scattering amplitude that realized these properties. Over the next two years a small community of theorists generalized this to formulas for N-body scattering amplitudes with various interrelated consistency properties. The fact that this could be done suggested the possibility that Veneziano's formula was not just a phenomenological amplitude (as originally intended) but actually part of a full-fledged theory.

Soon thereafter it was recognized independently by Nambu (1970), Susskind (1970), and Nielsen (1970) that the theory that was being developed could be understood physically as one based on one-dimensional structures (called strings), rather than pointlike particles. It then became clear how such a theory could account for various qualitative features of hadrons and their interactions. The basic idea is that specific particles correspond to specific oscillation modes of the string. S matrix theory, the bootstrap program, and Regge poles were abandoned (for good reasons) a long time ago. However, one cannot deny that they have left a lasting legacy. String theory probably could have been discovered by another route, but it might have taken many more years for that to happen. (Witten used to say that string theory is 21st century physics that happened to be discovered in the 20th century.) Be that as it may, the attempts to construct a string theory of hadrons were not fully successful. Moreover, in the early 1970s a quantum field theory description of the strong force—namely quantum chromodynamics (QCD)—was developed. It was universally accepted and string theory fell out of favor.

4.2.1   Problems with String Theory

Even though string theory had many attractive qualitative features, it began to unravel when one demanded that it should provide a full-fledged self-consistent theory of the hadrons. The original version of string theory turned out to have several fatal flaws. One was that consistency of perturbation theory, beyond the tree approximation, requires 25 spatial dimensions and one time dimension. A second major shortcoming is that this theory does not contain any fermions. Moreover the perturbative spectrum contains tachyons and massless particles. The former imply an unstable vacuum, and the latter are not part of the hadron spectrum.

4.2.2   Supersymmetry

In an attempt to do better, a string theory that does contain fermions was introduced (Neveu & Schwarz 1971; Ramond 1971). It requires only nine spatial dimensions—not the desired answer, but a step in the right direction. The original version of this theory also had a tachyon in its spectrum, but it was later realized that there is a consistent truncation that eliminates it. Very importantly, the study of this theory led to the modern understanding of supersymmetry and superstrings. In the original version (with the tachyon) supersymmetry was present only on the two-dimensional world-sheet, a fact that was first pointed out by Gervais & Sakita (1971). The pioneering work of Wess & Zumino (1974) in the construction of supersymmetric quantum field theories was motivated by the search for four-dimensional analogs of this two-dimensional symmetry. A few years later, Gliozzi, Scherk, & Olive

(1976) realized that the truncation that eliminates the tachyon also results in 10-dimensional spacetime supersymmetry.

Supersymmetry is an important idea for several reasons. For one thing, given very weak assumptions, it is the unique possibility for a nontrivial extension of the usual (Poincaré group) symmetries of space and time. Moreover, it is a symmetry that relates bosons and fermions. That is, they belong to common irreducible representations. Most importantly, there is a good chance that the new particles required by supersymmetry will be discovered at accelerators in this decade. Basically every known elementary particle that is a fermion (i.e., the quarks and leptons) should have bosonic partners and every known elementary particle that is a boson (the gauge bosons and Higgs) should have fermionic partners.

There are several different reasons to expect that the masses of the superpartners should be roughly at the TeV scale. Finding them (perhaps at the LHC) may provide the best experimental support for superstring theory in the foreseeable future. There is a good chance that the lightest super particle (the LSP) is absolutely stable. It is a leading candidate for a cold dark matter WIMP. So it is possible that supersymmetry will be discovered first in dark matter searches.

4.2.3 *Gravity and Unification*

One of the major obstacles to using these string theories to describe hadron physics was the occurrence of massless particles in the string spectrum. This conflicts with the fact that all hadrons are massive. We spent several years unsuccessfully trying to construct string theories that look more like realistic hadron theories. That did not work, yet string theory (especially the 10-dimensional one) was too beautiful to abandon. So we finally decided to try to understand it on its own terms.

Eventually, it was realized (Scherk & Schwarz 1974; Yoneya 1974) that one of the massless particles in the closed string spectrum is spin 2 and interacts at low energy in exactly the right way to be identified as a graviton—the quantum of gravitation. Therefore Scherk and I (1974) proposed using string theory to describe gravity and unification rather than just the strong force. This requires the typical size of a string to be about 10^{-32} cm (the Planck scale) rather than 10^{-13} cm (the typical size of a hadron), as was previously assumed.

This proposal had two immediate benefits: (1) Previous approaches to quantum gravity give unacceptable infinities. String theory does not. (2) Extra dimensions can be acceptable in a gravitational theory, where spacetime geometry is determined dynamically. For these reasons, as well as the fact that string theory requires gravity, Scherk and I were convinced that superstring unification was an important idea. However, for the next 10 years only a few brave souls shared our enthusiasm.

4.2.4 *The First Superstring Revolution*

One of those brave souls was Michael Green. In 1984 my five-year collaboration with Green culminated in an explanation of how superstring theory can be compatible with parity violation (Green & Schwarz 1984), which is an important feature of the standard model. (This was previously considered to be impossible.) Following that other superstring theories were found (Gross et al. 1985), and there were specific proposals for the geometry of the extra six dimensions (Candelas et al. 1985), which come surprisingly close to explaining the standard model.

By the time the dust settled we had five consistent superstring theories:

I, IIA, IIB, HE, HO,

each of which requires 10 dimensions. The Type I and HO theories each have an SO(32) gauge group, whereas the HE theory has an $E_8 \times E_8$ gauge group. Each of these theories has $\mathcal{N} = 1$ supersymmetry in the 10-dimensional sense. This is the same amount of supersymmetry as what is called $\mathcal{N} = 4$ in four dimensions (16 conserved supercharges). The two Type II theories have twice as much supersymmetry, and do not exhibit any nonabelian gauge symmetry in 10 dimensions. It took another 10 years to figure out how to achieve nonabelian gauge symmetry (using D-branes) in these theories.

The most realistic schemes (in those days) involved the HE theory with six dimensions forming a Calabi-Yau space. A Calabi-Yau space is a Kähler manifold of SU(3) holonomy. Such manifolds have been much studied, since their relevance to string theory was recognized, and by now there is a great deal known about them. They have two important topological integers associated with them, called h_{11} and h_{21}. h_{11} is the dimension of the space of Kähler forms, and h_{21} is the dimension of the space of complex structure deformations. In the context of the compactification of the HE theory, one ends up at low energy with a four-dimensional gauge theory with $\mathcal{N} = 1$ supersymmetry and gauge group E_6, though if the Calabi-Yau space is not simply connected this can be broken through the addition of Wilson lines to a gauge group very close to that of the standard model, possibly with additional U(1) factors. The number of generations of quarks and leptons is given by $|h_{11} - h_{21}|/2$. There are a number of known Calabi-Yau spaces that give the desired answer (three), but there are thousands of others that give different answers. There is no particular mathematical reason to prefer any of the three generation models. In any case, the analysis is carried out for weak coupling, and it is not clear that this is a good approximation.

One also needs to analyze the couplings of the gauge theory. Some relevant information, in the case of models with $\mathcal{N} = 1$ supersymmetry, is encoded in a holomorphic function called the superpotential. One wants to know the superpotential not only in the classical approximation but also for the quantum effective action. Over the years one has learned how to derive the effective superpotential, at least in certain cases, and to read off whether interesting phenomena such as confinement or dynamical supersymmetry breaking are implied by its structure.

With these discoveries string theory became a very active (though somewhat controversial) branch of theoretical high-energy physics. Over the subsequent 18 years it has become even more active (but less controversial).

4.2.5   *T Duality*

String theory exhibits many strange and surprising properties. In fact, one could argue that it represents a conceptual revolution comparable to that associated with quantum theory. One surprising property that was discovered in the late 1980s is called T duality. (The letter T has no particular significance. It was the symbol used by some authors for one of the low-energy fields.) A good review has been written by Giveon, Porrati, & Rabinovici (1994). When it is relevant, T duality implies that two different geometries for the extra dimensions are physically equivalent! For example, a circle of radius R can be equivalent to a circle of radius ℓ^2/R, where ℓ is the fundamental string length scale. (String theorists often use $\alpha' = \ell^2$, which is the Regge slope parameter.)

Let me sketch an argument that should make this duality plausible, since at first sight it is highly counterintuitive. When there is a circular extra dimension, the momentum along that direction is quantized: $p = n/R$, where n is an integer. Using the relativistic energy formula $E^2 = M^2 + \sum_i (p_i)^2$ (in units with $c = 1$), one sees that the momentum along the circular dimension can be interpreted as contributing an amount $(n/R)^2$ to the mass squared as measured by an observer in the noncompact dimensions. This is true whether one is considering point particles, strings, or any other kinds of objects. However, in the special case of closed strings, there is a second kind of excitation that can also contribute to the mass squared. Namely, the string can be wound around the circle, so that it is caught up on the topology of the space. The string tension is (now also setting $\hbar = 1$) given by the string scale as $1/(2\pi\ell^2)$. The contribution to the mass squared is the square of this tension times the length of wrapped string, which is $2\pi Rm$, if it wraps m times. Multiplying, the contribution to the mass squared is $(Rm/\ell^2)^2$. Now we can make the key observation: under T duality the role of momentum excitations and winding-mode excitations are interchanged. Note that the contributions to the mass squared match if one interchanges m and n at the same time that $R \to \ell^2/R$.

Usually T duality relates two different theories. Two particularly important examples are

$$\text{IIA} \leftrightarrow \text{IIB} \quad \text{and} \quad \text{HE} \leftrightarrow \text{HO}.$$

Therefore IIA & IIB (also HE & HO) should be regarded as a single theory. More precisely, they represent opposite ends of a continuum of geometries as one varies the radius of a circular dimension. This radius is not a parameter of the underlying theory. Rather, it arises as the vacuum expectation value of a scalar field. Thus, in principle it is determined dynamically, though in the most symmetrical examples there is a flat potential so that any value is possible. (Such scalar fields that do not appear in the potential are called "moduli." Moduli probably should not be part of a realistic solution, since they tend to give a scalar component to long-range gravitational strength forces.)

These T duality identifications reduce the list from five to three superstring theories. When other equivalences (such as S duality discussed below) are also taken into account, we conclude that there is actually a unique theory! What could be better? We are led to a unique underlying theory, free from arbitrary parameters, as the only possibility for a consistent quantum theory containing gravity. Well, there are a few details that still need to be fleshed out. (See the final section of this report.)

There are also fancier examples of T-duality-like equivalences, such as the physical equivalence of Type IIA superstring theory compactified on a Calabi-Yau space and Type IIB compactified on the "mirror" Calabi-Yau space. This mirror pairing of topologically distinct Calabi-Yau spaces was a striking mathematical discovery made by physicists (Greene & Plesser 1990), which has subsequently been explored by mathematicians. The two Hodge numbers h_{11} and h_{21}, discussed above, are interchanged in the mirror transformation.

T duality suggests a possible way for a big crunch to turn into a big bang (Brandenberger & Vafa 1989). The heuristic idea is that a contracting space when it becomes smaller than the string scale can be reinterpreted as an expanding space that is larger than the string scale, without the need for any exotic forces to halt the contraction. However, to be perfectly honest, we do not yet have the tools to analyze such time-dependent scenarios reliably.

T duality implies that usual geometric concepts break down at the string scale. Another manifestation of this is "noncommutative geometry," which arises when certain fields are turned on (Connes, Douglas, & Schwarz 1998; Seiberg & Witten 1999). It results in an "uncertainty relation" of the form $\Delta x \Delta y \geq \theta$, analogous to the more familiar $\Delta x \Delta p \geq h$, limiting one's ability to localize a particle in two orthogonal directions at once.

4.3   1994–Present

For its first 25 years string theory was studied entirely in terms of perturbation expansions in a string coupling constant g, which characterizes the strength of interaction. The Feynman diagrams of string theory are given by two-dimensional surfaces that represent the string world sheet. The story is especially simple in the case of oriented closed strings, since then there is a single Feynman diagram at each order of the perturbation expansion, in striking contrast to quantum field theory. The classification is given in terms of closed Riemann surfaces, with the genus corresponding to the number of loops. Perturbation theory is the way one often studies quantum field theory, as well. As is known from the example of QED, this can work very well when the coupling is small. However, as illustrated by QCD at low energy, for example, such perturbation expansions are not the whole story. New phenomena (such as confinement and chiral symmetry breaking in the case of QCD) arise when g is not small.

In string theory g is not a free parameter. Rather, it is determined dynamically as the value of a certain string field (the dilaton). When perturbation theory makes sense the dilaton is one of the moduli. In a realistic model it probably should not be a modulus, and then other methods of calculation may be required.

4.3.1   S Duality

Another kind of duality—called S duality—was discovered as part of the "second superstring revolution" in the mid 1990s. S duality allows us to go beyond perturbation theory. It relates g to $1/g$ in the same way that T duality relates R to $1/R$. The two basic examples (Hull & Townsend 1995; Witten 1995) are

$$\text{I} \leftrightarrow \text{HO} \quad \text{and} \quad \text{IIB} \leftrightarrow \text{IIB}.$$

Thus we learn how these three theories behave when $g \gg 1$. For example, strongly coupled Type I theory is equivalent to the weakly coupled SO(32) heterotic theory.

The transformation g to $1/g$ (or, more precisely, the corresponding transformation of the dilaton field) is a symmetry of the Type IIB theory. In fact, this is a subgroup of an infinite discrete symmetry group SL(2, Z). If some of the extra dimensions are compactified to give a torus, even larger discrete symmetry groups that combine the S duality and T duality groups arise (Hull & Townsend 1995), in which case one sometimes speaks of U duality. This story has been reviewed by Obers & Pioline (1999).

4.3.2   M Theory

S duality tells us how three of the five original superstring theories behave at strong coupling. This raises the following question: what happens to the other two superstring theories—IIA and HE—when g is large? The answer is quite remarkable: They grow an 11th dimension of size $g\ell$. This new dimension is a circle in the IIA case (Townsend 1995; Witten 1995) and a line interval in the HE case (Horava & Witten 1995). It is not visible in perturbation theory, since that involves expansions about $g = 0$.

When the 11th dimension is large, one is outside the regime of perturbative string theory, and new techniques are required. At low energies, the 11-dimensional theory can be approximated by 11-dimensional supergravity (Cremmer, Julia, & Scherk 1978). However, that is only a classical theory. What we need is a full-fledged quantum theory, for which Witten has proposed the name M theory. Since it is unclear how best to formulate it, he imagined that M stands for "mysterious" or "magical." Others have suggested "membrane" and "mother."

One more or less obvious idea is to try to construct a realistic four-dimensional theory by starting in 11 dimensions and choosing a suitable 7-manifold for the extra dimensions. This raises the question: what 7-manifolds are suitable? It is generally assumed that one wants to end up with a low-energy theory that looks more or less like the minimal supersymmetric extension of the standard model, which has $\mathcal{N} = 1$ supersymmetry. That is what one achieved by Calabi-Yau compactification of the HE theory.

Starting from M theory, one can prove that the way to get $\mathcal{N} = 1$ supersymmetry is to require that the 7-manifold have G_2 holonomy. G_2 is one of the exceptional Lie groups—the only one that can occur as a holonomy group. The study of G_2 manifolds is more difficult and less well understood than that of Calabi-Yau manifolds. Relatively few examples are known. One basic fact is that if the G_2 manifold is smooth, the resulting four-dimensional theory cannot have nonabelian gauge symmetry or chiral fermions. Therefore to get interesting models, it is necessary to consider G_2 manifolds with particular kinds of singularities. This has been an active area of study in the past few years, and some progress has been made, but there is still a long way to go. It is possible that some models constructed this way will turn out to be dual to ones constructed by Calabi-Yau compactification of the HE theory. That would be interesting, because each description would be better suited for exploring certain regimes. For example, the M theory picture should allow one to understand phenomena that are nonperturbative in the heterotic picture.

4.3.3 *p-branes*

Superstring theory requires new objects, called *p*-branes, in addition to the fundamental strings. (*p* is the number of spatial dimensions; e.g., a string is a 1-brane.) All *p*-branes, other than the fundamental string, become infinitely heavy as $g \to 0$, and therefore they do not appear in perturbation theory. On the other hand, at strong coupling this distinction no longer applies, and they are just as important as the fundamental strings.

Superstring theory contains a number of higher rank analogs of gauge fields with antisymmetrized indices, which are interpreted geometrically as differential forms. They can couple to higher dimensional objects, much like U(1) gauge fields can couple to charged point particles. Specifically, a $(p+1)$-form gauge field can couple electrically to a *p*-brane or magnetically to a $(d - p - 3)$-brane, where d is the number of spacetime dimensions. Because of supersymmetry, the energy density of such a charged brane is bounded from below, and when the (BPS) bound is saturated, a certain amount of the supersymmetry remains unbroken, and the brane is stable.

4.3.4 *D-branes*

An important class of *p*-branes, called D-branes, has the (defining) property that fundamental strings can end on them (Polchinski 1995). This implies that quantum field theories are associated with D-branes. These field theories are of the Yang-Mills type, like the standard model (Witten 1996). D-branes have a tension (or energy density) that scales with the string coupling like $1/g$. This is to be contrasted with the more characteristic solitonic $1/g^2$ behavior, exhibited by the NS5-branes, which are the magnetic duals of the fundamental strings in the Type II and heterotic theories.

An interesting possibility is that we experience four dimensions, because we are confined to live on D3-branes, which are embedded in a spacetime with six additional spatial directions. To be compatible with the observed $1/r^2$ force law for gravity, the extra dimensions would need to either have a size $\ll 1$ mm (so as not to have been detected in Cavendish-type experiments) or else be very "warped"—which means that the 4d geometry depends on the position in the other six dimensions.

The stable D-branes in the IIA theory have an even number of spatial dimensions, whereas those in the IIB theory have an odd number of spatial dimensions. Thus, as a particular example, the IIA theory contains D0-branes, which are pointlike objects. These are like extremal black holes that carry just one unit of a certain conserved $U(1)$ gauge symmetry charge. Recall that the IIA theory actually has a circular 11th dimension. The $U(1)$ charge is nothing but the integer that characterizes the momentum along this circle. Thus a D0-brane actually has one unit of momentum along the circle. As we have argued earlier, the mass of such a particle (if there is no other contribution to its mass) should just be $1/R$. However, we have said that the radius is $g\ell$. Therefore, one deduces that the mass of a D0-brane must be $1/(g\ell)$, which is in fact the correct value.

As we indicated earlier, 11-dimensional M theory requires a precise quantum definition, which is not provided by 11-dimensional supergravity. A specific proposal, called matrix theory, was put forward by Banks et al. (1997). In this proposal one builds the theory out of N D0-branes and considers a limit in which $N \to \infty$. This corresponds to going to the infinite momentum frame along the compactified circle. In the limit one can also decompactify the circle, so as to end up with a description of M theory in a flat 11-dimensional spacetime. This proposal has passed a number of nontrivial tests, and it is probably correct. However, it is awkward to work with. In particular, 11-dimensional Lorentz invariance seems very mysterious in this setup.

4.3.5   *Black Hole Entropy*

The gravitational field of the D-branes causes warpage of the spacetime geometry and creates event horizons, like those associated with black holes. In fact, studies of D-branes have led to a much deeper understanding of black hole thermodynamics in terms of string theory microphysics. In special cases, starting with a five-dimensional example analyzed by Strominger & Vafa (1996), one can count the microstates associated with D-brane excitations and compare then with the area of the corresponding black hole's event horizon, confirming the existence of statistical physics underpinnings for the Bekenstein-Hawking entropy formula. Although many examples have been studied and no discrepancies have been found, we are not yet able to establish this correspondence in full generality. The problem is that one needs to extrapolate from the weakly coupled D-brane picture to the strongly coupled black hole one, and mathematical control of this extrapolation is only straightforward when there is a generous measure of unbroken supersymmetry.

4.3.6   *AdS/CFT*

In a remarkable development, Maldacena (1997) conjectured that the quantum field theory that lives on a collection of D3-branes (in the IIB theory) is actually equivalent to Type IIB string theory in the geometry that the gravitational field of the D3-branes creates. This proposal was further elucidated by Witten (1998) and by Gubser, Klebanov, & Polyakov (1998). An excellent review followed a couple years later (Aharony et al. 2000).

In the IIB/D3-brane example a certain (conformally invariant) four-dimensional quantum field theory (CFT), called $\mathcal{N} = 4$ super Yang-Mills theory (Brink, Scherk, & Schwarz 1977; Gliozzi, Scherk, & Olive 1977), is precisely equivalent to Type IIB string theory in a 10-dimensional spacetime that is a product of a five-dimensional anti de Sitter (AdS) spacetime and a five-dimensional sphere (Schwarz 1983). Maldacena also proposed several other analogous dualities. This astonishing proposal has been extended and generalized in a couple thousand subsequent papers.

While I cannot hope to convince you here that these AdS/CFT dualities are sensible, I can point out that the first check is that the symmetries match. An important ingredient in this matching is the fact that the isometry group of anti de Sitter space in $D + 1$ dimensions is SO(D,2), the same as the conformal group in D dimensions. The fact that the four-dimensional gauge theory with $\mathcal{N} = 4$ supersymmetry is conformal at the quantum level is itself a highly nontrivial fact discovered around 1980. It requires that all the UV divergences cancel and the coupling constant does not vary with energy (vanishing beta function) so that renormalization does not introduce a scale.

The study of AdS/CFT duality and its generalizations is serving as a theoretical laboratory for exploring many deep truths about the inner works of superstring theory and its relation to more conventional quantum field theories. As one example of the type of insight that is emerging, let me note that by breaking the conformal symmetry and deforming the AdS geometry one can construct examples in which renormalization group flow in the four-dimensional quantum field theory corresponds to radial motion in the higher dimensional string theory spacetime.

This correspondence raises the hope that (among other things) we can close the circle of ideas—and solve our original problem, namely to find a string theory description of QCD, the theory of the strong nuclear force! That problem is now viewed as a low-energy analog of the unification problem. As yet, the 10-dimensional geometry that gives the string theory dual of QCD has not been identified. There are a number of technical hurdles that need to be overcome that I will not go into here. Presumably, none of them is insuperable.

On the other hand, the duality (discussed above) between supersymmetrical Yang-Mills theory with $\mathcal{N} = 4$ supersymmetry and SU(N) gauge symmetry, and Type IIB superstring theory in an $AdS_5 \times S^5$ spacetime background with N units of five-form (Ramond-Ramond) flux, are very beautiful and surely correct. However, our inability to carry out concrete string theory calculations in this background has been a source of frustration. Even though this geometry has almost as much symmetry as flat spacetime, calculations are much more difficult. As a result, most studies use a low-energy supergravity approximation to the string theory, which corresponds to restricting the dual gauge theory to a certain corner of its parameter space, where gauge theory calculations are difficult. Despite a number of interesting suggestions, I think it is fair to say that no practical scheme for doing string theory calculations in this background is known. While this problem has not been solved, there is an interesting limit in which it can be sidestepped.

4.3.7 The Plane-wave Limit

Recently, using a method due to Penrose (1976), Blau et al. (2002) constructed a plane-wave limit of the Type IIB superstring in the $AdS_5 \times S^5$ spacetime background. The limiting plane-wave background is also maximally supersymmetric. Following that, Metsaev (2002) showed that Type IIB superstrings in this plane-wave background are described in the Green-Schwarz light-cone gauge formalism by free massive bosons and fermions, so that explicit string calculations are tractable. The essential difference from flat space is the addition of mass terms in the world sheet action, something that would have been regarded as very peculiar without this motivation. Then Berenstein, Maldacena, & Nastase (2002) identified the corresponding limit of the gauge theory and carried out some checks of the duality. I will describe this subject in a little more detail than any of the preceding ones, because it is my current research interest.

Let me first describe how the Penrose limit works in this case. The AdS space (in global coordinates) is described by the metric

$$ds^2(AdS_5) = R^2(-\cosh^2 \rho \, dt^2 + d\rho^2 + \sinh^2 \rho \, d\Omega_3^2),$$

where the scale R has been factored out. Similarly the five-sphere is described by the metric

$$ds^2(S^5) = R^2(\cos^2 \theta \, d\phi^2 + d\theta^2 + \sin^2 \theta \, d\tilde{\Omega}_3^2).$$

In order to describe the desired limit, it is convenient to make the changes of variables

$$r = R \sinh \rho, \quad y = R \sin \theta$$

$$x^+ = t/\mu, \quad x^- = \mu R^2 (\phi - t).$$

Here μ is an arbitrary mass scale. The coordinate x^- has period $2\pi \mu R^2$, since ϕ is an angle. Therefore the conjugate (angular) momentum is

$$P_- = J/\mu R^2,$$

where J is an integer.

In terms of the new coordinates, the $AdS_5 \times S^5$ metric becomes

$$ds^2 = 2\left(1 - y^2/R^2\right) dx^+ dx^- - \mu^2(r^2 + y^2)(dx^+)^2$$

$$+ \frac{1}{\mu^2 R^2}\left(1 - y^2/R^2\right)(dx^-)^2 + ds_\perp^2,$$

where

$$ds_\perp^2 = r^2 d\Omega_3^2 + \frac{R^2}{R^2 + r^2} dr^2 + y^2 d\tilde{\Omega}_3^2 + \frac{R^2}{R^2 - y^2} dy^2$$

and

$$y^2 = \sum_{I=1}^{4}(x^I)^2 \text{ and } r^2 = \sum_{I=5}^{8}(x^I)^2.$$

The limit $R \to \infty$ gives the desired plane-wave geometry:

$$ds^2 = 2dx^+ dx^- - \mu^2(x^I)^2(dx^+)^2 + dx^I dx^I.$$

This differs from flat 10-dimensional Minkowski spacetime only by the presence of the mass term, which acts rather like a confining harmonic oscillator potential in the eight transverse dimensions.

This subject has aroused a great deal of interest over the past year, because for the first time we have a setup in which tractable gauge theory calculations can be compared to tractable string theory calculations. However, neither of these is easy, and there are any number of subtle issues to be sorted out. So this keeps a lot of clever people busy.

Let me sketch some of the essential features of the duality. Calling the coupling constant of the gauge theory g_{YM}, $\lambda = g_{YM}^2 N$ is a combination that 't Hooft identified long ago as important in the large-N limit. In the AdS/CFT duality g_{YM}^2 corresponds to the string coupling g, and λ corresponds to $(R/\ell)^4$ in the string theory. The plane-wave limit introduces some additional identifications. For one thing the gauge theory has a global SU(4) symmetry. Picking an arbitrary U(1) subgroup and calling the associated charge J, one identifies J with $\mu R^2 P_-$ in the plane-wave geometry. (We argued earlier that this should be an integer.) Furthermore the limit $R \to \infty$ corresponds in the gauge theory to letting $J, N \to \infty$ while holding fixed the combination $\lambda' = g_{YM}^2 N / J^2$, which corresponds to $(\ell^2 \mu P_-)^{-2}$. The gauge theory expansion parameter λ' for correlation functions of the Berenstein et al. (BMN) operators replaces the usual $\lambda = g_{YM}^2 N$. This is possible because the BMN operators are almost supersymmetric in the large-N limit. This may be a lot to swallow if you have not seen it before. But hopefully, you get the general idea.

To explore this duality in detail, it is necessary to establish a precise dictionary between single-trace gauge-invariant BMN operators and string states. This is fairly straightforward to leading order in the string coupling, but at higher order the correct matching requires mixing of single-trace and double-trace operators in the gauge theory. For the BMN sector of the gauge theory, the perturbation expansion can be organized as a double expansion in λ' (instead of λ) and $g_2 = J^2 / N$. The order in the latter parameter corresponds to the genus of the Feynman diagram (when expressed in double line notation following 't Hooft). In the string side of the duality, the perturbation expansion corresponds to the expansion in g_2, but each order contains the complete λ' dependence, both perturbative and nonperturbative. Therefore computations of the first couple of orders in the string side provide powerful predictions for the gauge theory.

The leading order (in g_2) string prediction was given in the original BMN paper. It has been verified to all orders in λ' in the dual gauge theory (Santambrogio & Zanon 2002). The information at this order consists of a comparison of anomalous dimensions of BMN operators in the gauge theory and light-cone energies in the string theory. The next order in g_2 involves comparing three-string couplings with more complicated correlators in the gauge theory, and is much more challenging.

The only way known to formulate the order g_2 interactions in the string theory is in terms of light-cone gauge string field theory in the plane-wave geometry. The formulas in flat space were worked out long ago (Green & Schwarz 1983; Green, Schwarz, & Brink 1983). However, in flat space there are more efficient formalisms, so that the light-cone gauge string field theory approach was largely ignored until this year when the need arose to generalize it to the plane-wave geometry. That has been achieved by Spradlin & Volovich (2002, 2003) and by Pankiewicz & Stefanski (2003). However, the formulas involve inverses of infinite matrices, which need to be computed before one can make explicit gauge theory predictions to all orders in λ'. I am pleased to report that this has been achieved within the last couple of weeks (He et al. 2002).

4.4   Some Remaining Problems

Let me conclude by listing, and briefly commenting upon, some of the issues that still need to be resolved, if we are to achieve the lofty goals indicated at the beginning of this report. As will be evident, most of these are quite daunting, and the solution of any one of them would be an important achievement. Let me begin with the list that a particle theorist might make.

- *Find a complete and optimal formulation of the theory.* Although we have techniques for identifying large classes of consistent quantum vacua, we do not have a succinct and compelling formulation of the underlying theory of which they are vacua. Many things that we take for granted, such as the existence of a spacetime manifold, should probably be emergent properties of specific vacua rather than identifiable features of the underlying theory. If this is correct, then we clearly need something that is quite unlike all theories with which we are familiar. It is possible that when the proper formulation is found the name "string theory" will become obsolete.

- *Understand why the cosmological constant (the energy density of empty space) vanishes.* The exact value may be nonzero on cosmological scales, but a Lorentz invariant Minkowski spacetime, which requires a vanishing vacuum energy, is surely an excellent approximation to the real world for particle physics purposes. We can achieve an exact cancellation between the contributions of bosons and fermions when there is unbroken supersymmetry. There does not seem to be a good reason for such a cancellation when supersymmetry is broken, however. Many imaginative proposals have been made to solve this problem, and I have not studied each and every one of them. Still, I think it is fair to say that none of them has gained a wide following. My suspicion is that the right idea is yet to be found.

- *Find all static solutions (or quantum vacua) of the theory.* This is a tall order. Very many families of consistent supersymmetric vacua, often with a large number of moduli, have been found. The analysis becomes more difficult as the amount of unbroken supersymmetry decreases and the moduli (including the dilaton) are eliminated or made massive. Vacua without supersymmetry are a real problem. In addition to the issue of the cosmological constant, one must also address the issue of quantum stability. Stable nonsupersymmetric classical solutions are often destabilized by quantum corrections. As far as I am aware, there are no known examples for which this has been proved not to happen.

- *Determine which quantum vacuum describes all of particle physics and understand whether it is special or just an environmental accident.* Presumably, if one had a complete answer to the preceding item, one of the quantum vacua would be an excellent approximation to the microscopic world of particle physics. Obviously, It would be great to know the right solution, but we would also like to understand why it is the right solution. Is it picked out by some special mathematical property or is it just an accident of our particular corner of the Universe? The way this plays out will be important in determining the extent to which the observed world of particle physics can be deduced from first principles.

Next let me turn to the list that a cosmologist might make.

- *Understand why the cosmological constant (the energy density of empty space) is incredibly small, but not zero.* As you know, current observational evidence suggest that about 70% of the closure density is provided by negative pressure "dark energy." The most straightforward candidate is a small cosmological constant, though other possibilities are being considered. Whether or not a cosmological constant is the right answer, string theorists would certainly be pleased if they could give a compelling reason why it should vanish. (See the second item on the particle physics list.) We would then be in a better position to study possible sources of tiny deviations from zero.

- *Understand how string theory prevents quantum information from being destroyed by black holes.* Long ago, Hawking (1976) suggested that when matter falls into black holes and eventually comes back out as thermal radiation, quantum coherence is lost. In short, an initial pure state can evolve into a mixed state, in violation of the basic tenets of quantum mechanics. I am convinced that string theory is a unitary quantum theory in which this can never happen, and so Hawking must be wrong. Still, as far as I know, nobody has formulated a complete explanation of how string theory keeps track of quantum phase relations as black holes come and go.

- *Understand when and how string theory resolves spacetime singularities.* Singularities are a generic feature of nontrivial solutions to general relativity. Not only are they places where the theory breaks down, but, even worse, they undermine the Cauchy problem—the ability to deduce the future from initial data. The situation in string theory is surely better. Strings sense spacetime differently than point particles do. Certain classes of spacelike singularities, which would not be sensible in general relativity, are known to be entirely harmless in string theory. However, there are other important types of singularities that are not spacelike, and where current string theory technology is unable to say what happens. My guess is that some of them are acceptable and others are forbidden. But it remains to be explained which is which and how this works.

- *Understand and classify time-varying solutions.* Only within the past couple of years have people built up the courage to try to construct and analyze time-dependent solutions to string theory. To start with, one goal is to construct examples that can be analyzed in detail, and that do not lead to pathologies. This seems to be very hard to achieve. This subject seems to be badly in need of a breakthrough.

- *Figure out which time-varying solution describes the evolution of our Universe and understand whether it is special or just an environmental accident.* If we had a complete list of consistent time-dependent solutions, then we would face the same sort of question we asked earlier in the particle physics context. What is the principle by which a particular one is selected? How much of the observed large-scale structure of the Universe can be deduced from first principles? Was there a pre-big-bang era and how did the Universe begin?

There is one last item that is of a somewhat different character from the preceding ones, but certainly deserves to be included. We need to

- *Develop the mathematical tools and concepts required to solve all of the preceding problems.* String theory is up against the frontiers of most major branches of mathematics. Given our experience to date, there is little doubt that future developments in string theory will utilize many mathematical tools and concepts that do not currently exist. The need for cutting edge mathematics is promoting a very healthy relationship between large segments of the string theory and mathematics communities. Such relationships were sadly lacking throughout a large part of the twentieth century, and it is pleasing to see them blossoming now.

Acknowledgements. This work was supported in part by the U.S. Dept. of Energy under Grant No. DE-FG03-92-ER40701.

References

Aharony, O., Gubser, S. S., Maldacena, J., Ooguri, H., & Oz, Y. 2000, Phys. Rep. 323, 183

Banks, T., Fischler, W., Shenker, S. H., & Susskind, L. 1997, Phys. Rev. D, 55, 5112

Berenstein, D., Maldacena, J., & Nastase, H. 2002, JHEP, 0204, 013

Blau, M., Figueroa-O'Farrill, J., Hull, C., & Papadopoulos, G. 2002, JHEP, 0201, 047

Brandenberger, R. H., & Vafa, C. 1989, Nucl. Phys. B, 316, 391

Brink, L., Schwarz, J. H., & Scherk, J. 1977, Nucl. Phys. B, 121, 77

Candelas, P., Horowitz, G. T., Strominger, A., & Witten, E. 1985, Nucl. Phys. B, 258, 46

Connes, A., Douglas, M. R., & Schwarz, A. 1998, JHEP, 9802, 003

Cremmer, E., Julia, B., & Scherk, J. 1978, Phys. Lett. B, 76, 409

Gervais, J. L., & Sakita, B. 1971, Nucl. Phys. B, 34, 632

Giveon, A., Porrati, M., & Rabinovici, E. 1994, Phys. Rep., 244, 77

Gliozzi, F., Scherk, J., & Olive, D. 1976, Phys. Lett. B, 65, 282

——. 1977, Nucl. Phys. B, 122, 253

Green, M. B., & Schwarz, J. H. 1983, Nucl. Phys. B, 218, 43

——. 1984, Phys. Lett. B, 149, 117

Green, M. B., Schwarz, J. H., & Brink, L. 1983, Nucl. Phys. B, 219, 437

Green, M. B., Schwarz, J. H., & Witten, E. 1987, Superstring Theory (Cambridge: Cambridge Univ. Press)

Greene, B. R., & Plesser, M. R. 1990, Nucl. Phys. B, 338, 15

Gross, D. J., Harvey, J. A., Martinec, E. J., & Rohm, R. 1985, Phys. Rev. Lett., 54, 502

Gubser, S. S., Klebanov, I. R., & Polyakov, A. M. 1998, Phys. Lett. B, 428, 105

Hawking, S. W. 1976, Phys. Rev. D, 14, 2460

He, Y. H., Schwarz, J. H., Spradlin, M., & Volovich, A. 2002, preprint (hep-th/0211198)

Horava, P., & Witten, E. 1996, Nucl. Phys. B, 460, 506

Hull, C. M., & Townsend, P. K. 1995, Nucl. Phys. B, 438, 109

Maldacena, J. M. 1998, Adv. Theor. Math. Phys., 2, 231

Metsaev, R. R. 2002, Nucl. Phys. B, 625, 70

Nambu, Y. 1970, in Proc. Intern. Conf. on Symmetries and Quark Models, ed. R. Chand (New York: Gordon and Breach), 269

Neveu, A., & Schwarz, J. H. 1971, Nucl. Phys. B, 31, 86

Nielsen, H. B. 1970, unpublished

Obers, N. A., & Pioline, B. 1999, Phys. Rep., 318, 113

Pankiewicz, A., & Stefanski, B. 2003, Nucl. Phys. B, 657, 79

Penrose, R. 1976, in Differential Geometry and Relativity (Dordrecht: Reidel), 271

Polchinski, J. 1995, Phys. Rev. Lett., 75, 4724

——. 1998, String Theory (Cambridge: Cambridge Univ. Press)

Ramond, P. 1971, Phys. Rev. D, 3, 2415

Santambrogio, A., & Zanon, D. 2002, Phys. Lett. B, 545, 425

Scherk, J., & Schwarz, J. H. 1974, Nucl. Phys. B, 81, 118

Schwarz, J. H. 1983, Nucl. Phys. B, 226, 269

Seiberg, N., & Witten, E. 1999, JHEP, 9909, 032

Spradlin, M., & Volovich, A. 2002, Phys. Rev. D, 66, 086004

——. 2003, JHEP, 0301, 036

Strominger, A., & Vafa, C. 1996, Phys. Lett. B, 379, 99

Susskind, L. 1970, Nuovo Cim., 69A, 457

Townsend, P. K. 1995, Phys. Lett. B, 350, 184

Veneziano, G. 1968, Nuovo Cim., 57A, 190

Wess, J., & Zumino, B. 1974, Nucl. Phys. B, 70, 39

Witten, E. 1995, Nucl. Phys. B, 443, 85

——. 1996, Nucl. Phys. B, 460, 335

——. 1998, Adv. Theor. Math. Phys., 2, 253

Yoneya, T. 1974, Prog. Theor. Phys., 51, 1907

5

Dark matter theory

JOSEPH SILK

Department of Physics, University of Oxford

Abstract

I evaluate the dark matter budget and describe baryonic dark matter candidates and their detectability. Dark matter issues in galaxy formation theory are discussed, and I review the prospects for detecting nonbaryonic dark matter. Indirect detection via halo annihilations of the favored dark matter candidate, the SUSY LSP, provides a potential signal. The relic density of dark matter particles specifies the annihilation cross-section within model uncertainties, and indirect detection provides our optimal strategy for confirming the dark matter candidate. Galaxy formation simulations suggest that the predicted clumpiness of the dark halo will facilitate our imaging the dark matter in gamma rays, with cosmic ray signatures providing invaluable confirmation. Similarly, the supermassive black hole in the center of the Milky Way should present a unique signal amplifier by which we can view the dark matter in neutrinos as well as in gamma rays. The astrophysical uncertainties are so large that one has no alternative but to look.

5.1   Introduction

There are two types of dark matter: baryonic and nonbaryonic. Several candidates for the former exist, but the mass fraction is unknown. Conversely, while we have not yet detected the leading nonbaryonic matter (cold dark matter; CDM) candidate, the neutralino, we can calculate its mass fraction and interaction cross-section, within model uncertainties.

Dark matter dominates the Universe, amounting to $\sim 90\%$ of the matter density. Baryonic dark matter, as yet unambiguously identified, comprises up to a third of the baryons, although there are compelling candidates. Elucidating the nature of all of the dark matter is one of the outstanding problems in astrophysics to be addressed over the next decade. As will be discussed in this review, research at the interface of dark matter with galaxy formation has been particularly active in recent years, but has also raised challenges that are leading some to question the entire dark matter edifice.

Dark matter has a venerable history. In the solar system, anomalies in the orbit of Uranus pointed to dark matter in the form of a new planet, and this led to the discovery of the planet Neptune, following the predictions of Adams and Leverrier. The advance of the perihelion of Mercury's orbit also stimulated searches for dark matter in the form of a new planet interior to the orbit of Mercury, dubbed Vulcan. This turned out to be a red herring: the orbital anomalies eventually led Einstein to propose a new theory of gravitation, general relativity.

Similarly, there are parallels that may be drawn today. The modified Newtonian dynamics (MOND) theory seeks to modify the law of gravity in order to dispense with the need for dark matter. There is little in the way of compelling theory or data to support such a position, but, at the same time, MOND is tenaciously difficult to kill (e.g., Sanders & McGaugh 2002). It is certainly worth bearing in mind that general relativity has been thoroughly tested only in the weak-field limit, and on relatively small scales, now extended up to a Mpc or so by gravitational lensing studies.

The modern dark matter problem was first described in 1933 by Fritz Zwicky (Zwicky 1933), who noted that in the Coma cluster of galaxies, the ratio of mass to light as measured by the virial theorem is about 400 $M_\odot/L_\odot$. Since individual galaxies are found to have about 10 $M_\odot/L_\odot$, there is a serious shortfall of luminous matter. When observations of clusters were extended to X-ray frequencies, a significant component of mass was found that was not detected optically. X-ray observations have revealed the presence of a substantial amount of hot intracluster diffuse gas, contributing about 15% of the cluster mass. However, 80% of the cluster mass is not accounted for in any known form. Moreover, this unknown mass component must be nonbaryonic, as the primordial nucleosynthesis measure of global baryon abundance, combined with the mean matter density, is consistent with the observed cluster baryon content in gas and stars.

On larger scales, one can probe dark matter via both Hubble-flow perturbations studied via deep redshift surveys and shear maps from weak gravitational lensing. The dark matter content dominates the total matter content. The general consensus is that $\Omega_m \approx 0.3$, with about 10% of this being in baryons. Most of the baryons must also be dark, although there are persuasive arguments about their nature.

I now describe in more detail the dark matter budget, describe baryonic dark matter candidates and their detectability, and then turn to CDM issues in galaxy formation theory. I conclude with a review of the prospects for detecting nonbaryonic dark matter.

5.2   Global Baryon Inventory

Primordial nucleosynthesis of the light elements demonstrates that the baryon fraction is $\Omega_b = 0.04 \pm 0.004$. In effect, this is measured at $z \approx 10^9$, when the light-element abundances freeze out. There are two other independent measures of the total baryon content of the Universe. The heights of the cosmic microwave background (CMB) acoustic peaks, in particular the odd peaks that correspond to wave compressions and rarefactions on the last-scattering surface scale and at half of this scale, are controlled primarily by the baryon density at $z \approx 1000$. The Lyα forest indirectly measures the baryon density at $z \approx 3$, once a correction is made for the predominant ionized fraction, whose density is inferred from ionization balance by invoking the ionizing radiation field from quasars. All three measures of the baryon density converge on $\Omega_b \approx 0.04$.

At the present epoch, the baryon inventory is rather different, however. Stars account for a modest fraction of all the baryons present, about 0.0026 in spheroid stars and 0.0015 in disk stars and cold gas, in units of the Einstein-de Sitter density (Fukugita, Hogan, & Peebles 1998). These numbers are in approximate agreement with a recent determination, including all cold interstellar matter and stars in low-mass as well as in massive galaxies in the local Universe, which yields 0.0024 $\pm$ 0.001, corresponding to 8% $\pm$ 4% of Ω_b (Bell et al. 2003). Intracluster gas in rich clusters, mapped in X-rays, accounts for another 0.0026 of the closure density, consistent with the conjecture that, outside of the great clusters, most of the hot gas is able to cool and form galaxy disks. In total, only some 20% of the baryons is actually observed at low redshift.

There are strong theoretical indications, however, that most of the baryons, an additional 0.01 to 0.015 in units of the closure density, or a fraction 25% to 40% of Ω_b, are in the form of a warm/hot intergalactic medium (WHIM) in low-density environments outside the rich clusters. A quantitative estimate of the WHIM fraction comes from large-scale numerical simulations of the intergalactic medium (IGM; Davé et al. 2001). The heating is gravitational, due to accretion shocks in filaments and sheets, and generally occurs on the peripheries of galaxies, galaxy groups, and galaxy clusters. The WHIM has recently been detected via excess soft X-ray emission toward clusters of galaxies (Sołtan, Freyberg, & Hasinger 2002) and also via absorption in O VI toward quasars (Simcoe, Sargent, & Rauch 2002), although it is not yet possible to quantify the observed WHIM mass fraction.

The IGM simulations also suggest that there should be some cold intergalactic gas, visible as local, metal-poor, Lyα forest lines, although quantitative predictions are unreliable because most of the cold gas will presumably have formed stars, or at least fallen into galaxies. However, observations of the local Lyα-absorbing intergalactic clouds suggest that they could contribute no more than 0.008 of the closure density, or 20% of Ω_b, toward the baryon budget (Penton, Shull, & Stocke 2000). The net effect is that some 80% of the globally distributed baryons are plausibly accounted for, despite the fact that no more than 20% have hitherto been directly and quantitatively observed. The uncertainties in the gaseous baryon fraction, especially for the warm gas, are such that one can infer that there is no serious global dark baryon problem. However, the situation may be quite different on galactic scales.

5.3   Galactic Baryon Inventory

Rich clusters contain a reservoir of gas that reflects the initial baryon content of the Universe on 10 Mpc scales, and that, for the most part, consists of gas that has been frustrated from forming disks. Processes such as galaxy collisions and tidal harassment inhibit an accumulation of cooling gas in galaxy halos. This frustrates the presumed means by which disks ordinarily accrete cold gas. It is the maintenance of a long-lived supply of cold gas that enables star formation to continue over a Hubble time in low-density regions of the Universe.

The cluster gas fraction is approximately 15%. This plausibly reflects the initial baryon fraction in protodisks. Indeed, simulations of disk formation generally require an initial gas fraction of 15% to 20%. This is necessary for sufficient cooling to have occurred to be able to form the disks. Semi-analytic galaxy formation predicts that halos currently contain large amounts of gas, about as much as eventually forms the disk.

Cold clouds in the halo would be easily observable. The gas, if in the halo, must be at a temperature of several million degrees. However, there is equally a problem for diffuse gas in the halo: the resulting X-ray emission would exceed that observed by up to an order of magnitude (Benson et al. 2000), unless relatively large supernova feedback is implemented (Toft et al. 2002) to eject the gas. If this gas were to end up in the disk, the mass of the disk would presumably exceed observed limits from the rotation curve. Only a small fraction of the disk mass can be present in the halo as hot gas.

The Milky Way stellar disk mass is well constrained by a combination of gravitational microlensing and infrared observations (Klypin, Zhao, & Somerville 2002). The mass of the disk and bulge is $6 \times 10^{11} M_\odot$, whereas the total mass, predominantly in the halo as measured by the rotation curve out to the virial radius, is $10^{12} M_\odot$. If the dark halo profile is NFW-like (Navarro, Frenk, & White 1996), there is no room for any further baryonic component that might be in a noncompact form and not already measured by the powerful combination of bulge microlensing and the Galactic rotation curve. It has been argued that the dark halo must actually have a softer, more isothermal, core than the power-law NFW profile; otherwise, the contribution of the dark matter to the rotation curve in the inner galaxy would be excessive (Binney & Evans 2001).

This leaves us in a quandary: where have the baryons gone, which once were present? There are two possibilities. The baryons may still be present, but hidden. For example, they may be in dense, cold clumps in the outer halo, which are too large, by exceeding the Einstein radius, to give a strong microlensing signal. Or they could be in the form of compact objects, such as massive compact halo objects (MACHOs), since the observational limit on halo MACHOs obtained by microlensing of stars in the Large Magellanic Cloud allows a halo fraction in MACHOs of up to 20%, in compact clump masses below a few solar masses. The positive signal reported by one experiment indeed favors a MACHO mass of about $0.5 M_\odot$. In fact, one only needs to hide a halo mass fraction of about 6%, and possibly as large as 10%, to arrive at an initial baryon mass fraction of $\sim 15\%$. An alternative possibility is that the unaccounted baryons were ejected from the protogalaxy in an early wind.

I now consider these various possibilities in turn. Cold dense clumps of H_2, of Jupiter-like mass and sizes of order a few astronomical units, have been invoked to account for extreme scattering events (Walker & Wardle 1998) and for unidentified SCUBA submillimeter sources (Lawrence 2001). As such halo clumps orbit the Galaxy, they cross the Galactic plane up to 100 times over the age of the disk. These traversals result in the gas clumps acquiring on the order of 1 magnitude of extinction as they sweep up interstellar dust (Kerins, Binney, & Silk 2002). In order to avoid an excessive rate of collisions and yet keep the clumps large enough to be Jeans stable at a given mass and temperature, the clump covering factor must be on the order of 0.0001. Such clumps are potentially detectable via gaseous microlensing events (Rafikov & Draine 2001), as well as by occultations of background stars in a MACHO-type experiment that monitors millions of stars several times per night.

Of course, such clouds would cool to a few degrees Kelvin and collapse, unless heated, and the only plausible proposed heat source is cosmic rays. It is not clear if the clouds could maintain a stable equilibrium (Gerhard & Silk 1996; Wardle & Walker 1999), although the reemitted radiation is consistent with measurements of the far-infrared background (Sciama 2000). If the clumps collapse, the observational motivation for invoking them is removed.

The MACHO option most likely centers on halo white dwarfs. There have been claims and counterclaims of detections of old, high-velocity white dwarfs. The few detected have generally been attributed to the thick disk. If even a few percent of the halo mass was in the form of white dwarfs, a most unusual stellar initial mass function (IMF) would be required for the precursor stars, peaked in the intermediate-mass range. With a solar neighborhood IMF, one could hide only of order 0.1% of the stellar mass in the Milky Way disk in our halo. If the precursor stars had primordial abundances ($\lesssim 10^{-4}$ solar), one could plausibly suppress helium flashes and associated C or N dredge-up, thereby avoiding the obvious concern about stellar ejecta overpolluting the interstellar medium and protogalaxy by a factor of up to 1000 in the elements primarily produced by intermediate-mass stars. One would still perhaps have to tolerate a greatly enhanced rate of Type Ia supernovae (SNIa). However, the likelihood that most of the SNIa ejecta escape from the galaxy via winds, combined with the unknown scenario for SNIa formation, which itself invariably involves old white dwarfs in binary systems, renders such estimates highly uncertain.

5.4   Outflows

An alternative view is that galaxies such as our Milky Way have undergone massive outflows in the past. There is strong evidence that many galaxies presently undergoing starbursts are driving superwinds, with outflows on the order of the current star formation rate. The high-redshift evidence is especially compelling. At $z \approx 3$, the Lyman-break galaxies, many of which have high star formation rates, occasionally reveal inverse P Cygni profiles and, more generally, broad line widths, with inferred outflow velocities of the order of the escape velocity from the galaxy (Pettini et al. 2002; Shapley et al. 2003). Studies of the Lyα forest in the vicinity of the Lyman-break galaxies, via absorption in the spectra of background quasars, reveal Mpc-sized holes. These are seen both in the H I forest and in C IV absorption, centered on the Lyman-break galaxies (Adelberger et al. 2003). A vigorous wind could excavate such holes, although one cannot exclude a photoionization source.

It may be that spirals have expelled a mass in gas comparable to that remaining in stars, whereas, on the basis of gravitational lensing of background quasars, the time-delay evidence suggests that many massive ellipticals have not lost a significant fraction of their initial baryon content. Indeed, ellipticals need to have conserved the primordial baryon fraction, and possibly even accreted more cold baryons, in order to retain consistency both with the Hubble constant and a NFW-like profile (Kochanek 2003).

This would also be in accordance with the following conjecture, namely that the dark halos in massive ellipticals have pristine dark matter profiles, as generated by major mergers and predicted by N-body simulations, whereas the dark matter concentrations in the halos of many spirals may have been modified by the astrophysics of disk formation. Such modifications could plausibly occur as a consequence of the dynamical heating and associated early outflows, linked to the formation of massive, transient protogalactic gaseous bars (Weinberg & Katz 2002). An alternative possibility appeals to the formation of supermassive central black holes and associated early AGN activity that would have resulted in massive winds (Binney, Gerhard, & Silk 2001).

The desired modifications in halos of spirals, both massive and dwarf-like, consist of the following. The amount of substructure must be suppressed, both to be consistent with that viewed directly by counting visible dwarfs, and inferred indirectly by the near-conservation of the initial angular momentum acquired via tidal torques with neighboring protogalaxies that is necessary to account for the observed sizes of galactic disks. The concentration of dark matter within a couple of disk scale lengths may need to be reduced from the 50% or so by mass predicted by the simulations to the 10% or less inferred from microlensing for the Milky Way Galaxy and from dynamical friction of bars acting on the dark halos for spirals with rapidly rotating central bars (Debattista & Sellwood 2000).

Actually, the latter argument is controversial, as it is subject to ignoring destruction and reformation of bars that could occur several times over the disk age, if gas accretion occurs over a similar time scale and at about the same rate as the disk star formation rate (Bournaud & Combes 2002). This might allow bars to be relatively short-lived, decaying via dynamical friction against the dark matter if the halo profile is NFW-like, or decaying on a similar time scale due to the effects of destabilization via central gas accretion and bulge formation.

Provided the disk that reformed via cold gas accretion was sufficiently cold, it would be unstable to bar formation. Presumably the stellar ages of bars would contain clues about the star formation history that could perhaps be unravelled by detailed spectrophotometry.

The dark halos of dwarf spirals generically seem to be mostly cuspless, in some cases in possible contradiction with the extent of a central cusp that is predicted by CDM simulations. The dark matter profiles can be fit either by CDM profiles with anomalously low concentrations or by dark matter profiles with nearly isothermal cores (de Blok & Bosma 2002). Again, this could be a consequence of vigorous early winds, which would be consistent with the low observed metallicities of dwarfs.

Finally, a substantial subset of massive ellipticals has soft cores. These galaxies are associated with slow rotation and boxy profiles. Numerical simulations suggest an origin via major mergers (Burkert & Naab 2004). One could plausibly imagine that such mergers resulted in substantial central gas flows and generated luminous starbursts that drove massive winds. Supermassive black hole formation by mergers of black holes of comparable mass would also heat and scour out the dark matter cusp, producing a soft core or even a central minimum in the stellar luminosity density (Lauer et al. 2002).

It is indeed possible that galaxies have undergone a wide variety of formation histories. Some of these, such as massive gas accretion onto disky ellipticals and major mergers to form boxy ellipticals, may have had a strong but possibly indirect influence on the dark matter profiles, including cusp, substructure and concentration.

One could possibly distinguish between these alternative hypotheses by the prediction that a massive early starburst would result in boosting the rate of Type II supernova production relative to the longer time scale for SNIa explosions, and hence lead to an $[\alpha/\text{Fe}]$ enhancement in the stellar core. Boxy ellipticals should therefore display systematically higher $[\alpha/\text{Fe}]$ than disky ellipticals, which often have central cusps, or power-law profiles in surface brightness, and have also presumably undergone more quiescent accretion and extended periods of star formation that led to disk formation.

Another lensing result from the image ratios of quadruple lens images of quasars can only be explained if the massive elliptical lenses have substructure on subgalactic mass scales (Dalal & Kochanek 2002). A similar conclusion is inferred from the bending of radio jets on milli-arcsecond scales (Metcalf 2002). Some massive ellipticals have substructure consistent with the full predicted power of CDM, that is to say of order a few percent of the halo dark matter in inhomogeneities, with much of this in million-solar mass clumps. Some of the massive ellipticals, those with disky isophotes, as well as low-mass ellipticals and bulges, often have central power-law stellar cusps. It would of course be interesting to know if these properties were indeed correlated, and characteristic of the disky ellipticals and spheroids that are thought not to have undergone major mergers.

There is one crucial issue to be addressed if massive winds with outflow rates on the order of the star formation rate are to be considered seriously in the context of galaxy formation. Simulations fail to produce such winds in massive galaxies. One finds that the winds interact with ambient gas, undergo strong radiative cooling, and are quenched. I have suggested that a fundamental problem with all existing wind simulations is the omission of crucial subgrid physics (Silk 2003). In particular, Rayleigh-Taylor instabilities, as the wind interacts with the surrounding matter and the shocked shell of dense, cooled gas decelerates, increase the interstellar medium porosity. This enables the galactic outflows to proceed with enhanced efficiency. Simultaneously, Kelvin-Helmholtz instabilities entrain cold interstellar gas into the outflows. Ablation and evaporation of the cold clouds adds to the mass loading of the wind. The enhanced efficiency means that even massive galaxies can have winds. The star formation rate peaks in the protogalactic phase. This is when the interstellar cold gas density is greatest, and hence cooling is most important. If a wind is indeed driven, the wind outflow is of order the star formation rate.

However, a successful wind requires that there be sufficient energy in the outflow to drive the gas out of the galactic potential well. Normal supernovae of Type II, associated with the starburst that characterizes the elevated star formation rate, may not suffice to generate enough energy if the galaxy escape velocity exceeds $\sim 100 \text{ km s}^{-1}$. However, I have argued that even massive galaxies should have undergone a massive outflow phase. The resolution might lie with hypernovae, with characteristic kinetic energies $\sim 10^{53}$ erg. It has been suggested that hypernovae are associated with the deaths of stars in excess of 30 $M_\odot$, the high energies being generated via release of binding energy from infall onto the forming black hole. There are clues from the observed metallicities in at least one starburst galaxy and in metal-poor halo stars that hypernovae may make an important contribution to the observed chemical abundances. Indeed, hypernovae may even be the dominant contributors to the abundance patterns in extremely metal-poor halo stars (Umeda & Nomoto 2003). If indeed all stars above $\sim 30 M_\odot$ explode as hypernovae in metal-poor environments, the specific energy input into the interstellar medium would be enhanced relative to that of ordinary supernovae by an order of magnitude. This would suffice to drive winds from the most massive protogalaxies.

It has also been suggested that hypernovae are preferentially produced via stellar mergers (Portegies Zwart et al. 2002), which are prime candidates for producing the massive, rapid rotators preferred in hypernova models (Nakamura et al. 2001). Such mergers are likely to be most important in high-density cores associated with starbursts, and might be especially frequent during the protogalactic starburst phase associated with spheroid formation.

5.5 Prospects for Nonbaryonic Dark Matter Detection

CDM, despite the current issues that are being raised concerning substructure and dark matter concentration on small scales, has been remarkably successful in accounting for large-scale structure and in leading to the prediction of the amplitude of the CMB temperature fluctuations. However, detection, direct or indirect, has been elusive, at least for the dominant nonbaryonic dark matter component.

Massive neutrinos are the only form of nonbaryonic dark matter known to exist. However, they are subdominant. The relic neutrino number density is $\frac{3}{11}n_\gamma$, where the relic photon density n_γ is $0.24(kT/\hbar c)^3$ and T is the measured CMB temperature of 2.725 ± 0.002 K. A lower bound to the mass comes from atmospheric $\nu_\mu \rightarrow \nu_e$ oscillations, and is $m_\nu \gtrsim 0.1$ eV. Upper limits come from particle physics and from astronomy. Now, $\Omega_\nu h^2 = \Sigma m_\nu / 91.5$ eV. The neutrinoless double beta decay limit on the electron neutrino mass, combined with recent reactor constraints (using KamLAND data) on oscillations, set an upper bound $\Omega_\nu h^2 \lesssim 0.07$ (Minakata & Sugiyama 2002). A similar, but stronger bound comes from large-scale structure, and in particular the linear power spectrum derived from the 2dF galaxy redshift survey, such that the sum of the neutrino masses must be less than 2.5 eV (Elgarøy et al. 2002; Hannestad 2002). Neutrinos can be comparable to the baryon density in contributing to the matter content of the Universe.

Neutrinos decouple relativistically, since decoupling at $kT_{dec} \approx 1 \text{ MeV} \approx 2m_e c^2$ is controlled by neutron freeze-out. Generic massive weakly interacting particle candidates that once were in thermal equilibrium decouple non-relativistically at $kT_\chi \approx m_\chi c^2/20$, and are suppressed in number density by the corresponding Boltzmann factor relative to the photon number density. There is theoretical prejudice that favors the lightest SUSY particle, which is stable if there is a commonly adopted symmetry among the SUSY family of particle partners to the known bosons and fermions. Constraints from the Large Electron Positron Collider set a lower bound on the particle mass of about 50 GeV. A generic upper bound comes from theory: the SUSY scale, and hence the WIMP mass, cannot be much above 1 TeV if SUSY is to be relevant for accounting for the electroweak scale. Since the cross-section for annihilation decreases in the unitarity limit with increasing mass above 100 GeV, one cannot go to too high a mass, typically less than a few TeV, without overclosing the Universe with WIMPs. Realistic models with the observed CDM density (Melchiorri & Silk 2002) $\Omega_{cdm} = 0.12 \pm 0.04h^{-2}$ and including all coannihilation channels (Edsjö et al. 2003) suggest an upper bound of 1500 GeV.

One may hope to detect CDM in the form of the LSP via its annihilations in the halo. The annihilation products include high-energy positrons, antiprotons, gamma rays, and neutrinos. If the halo is uniform, the predicted fluxes fall a factor of 100 or more below observational limits as set, for example, by the diffuse high galactic latitude EGRET gamma ray flux. However, secondary production of gamma rays via $p-p$ and $p-\alpha$ interactions of cosmic rays results in a spectral energy distribution that is softer than the observed diffuse gamma ray flux, suggesting that a more exotic explanation, such as that from annihilations, may possibly be merited.

High-resolution numerical simulations of realizations of galaxy halos predict strong clumpiness on small scales. This would help boost the expected annihilation signal. A reported high-energy positron flux shows a spectral excess relative to predicted cosmic ray secondary positrons near 50 GeV. With a boost factor of 100 to 1000, this could be due to WIMP annihilations in the halo, the fit to the data certainly being improved by a carefully tuned WIMP mass (Baltz et al. 2002).

One consequence is the prediction of a primary high-energy $\bar{p}$ feature that might be potentially observable with the forthcoming PAMELA and AMS experiments. The possible spectral signatures of the diffuse gamma ray and neutrino fluxes from halo annihilations are more elusive. With regard to the gamma rays, one expects a π^0 decay signature both from cosmic ray interactions with interstellar gas and from the annihilations, although the latter would produce a harder spectrum for massive WIMPs. However, one may hope to identify individual halo clumps as pointlike gamma ray sources at the sensitivity and angular resolution of the *GLAST* satellite, to be launched in 2007. Gamma ray lines are another potential signal, that would be a unique tracer of annihilations.

One concern is that the clumpiness of the dark halo may be greatly overestimated by the numerical simulations. This is because significant astrophysics is omitted from halo modeling. Clumps self-destruct via dynamical interactions with each other and especially with the Galactic tidal field. Nevertheless, even the neutrino flux from annihilations may be detectable with a more extreme strategy that targets the Galactic Center.

Observations strongly suggest that the formation of the spheroid stars is closely coupled to the formation of the central supermassive black hole. One observes that $M_\bullet \approx 10^{-3}M_{sph}$ over a wide dynamic range, with only modest dispersion. In a disk galaxy such as the Milky Way, it is likely that the supermassive black hole at the Galactic Center of $\sim 2.6 \times 10^6 M_\odot$

(Ghez 2004) formed by a combination of gas accretion onto a seed black hole and merging of smaller black holes, rather than by major black hole merging events. Cosmology, and in particular studies of primordial star formation, suggests that the seed black hole masses are likely to be around $100 M_\odot$. Inferences about the primordial IMF lead to the likely presence in the protogalactic halo of large numbers of $100 M_\odot$ black holes (Islam, Taylor, & Silk 2002; Volonteri, Haardt, & Madau 2003). Gas accretion in the gas-rich protogalaxy is then the favored formation model for the central supermassive black hole.

Another option would be for the black hole and accompanying spheroid to form by hierarchical clustering and merging of roughly equal-mass black holes that are embedded in mini-spheroids of stars. This seems implausible for the Milky Way black hole and spheroid because the merging time scales would seem to be very long, and also because of the remarkable observed chemodynamical continuity between thin disk, thick disk, and spheroid stars. This is suggestive of a relatively nonviolent history. The last significant merger would have to have occurred at least 12 Gyr ago and preceded disk formation. The stellar ages and chemical properties of the disk and bulge suggest a coeval formation. In order to account for bar and bulge formation, secular evolution rather than merging is usually preferred in Milky Way-type galaxies, and more generally in late-type disks.

A rather different history may be applicable to massive spheroid formation. Of course, elliptical galaxies may well be older, on average, and form in denser environments. Major mergers that triggered bursts of star formation most likely played an important role in the formation of massive ellipticals, as inferred for example from the high central stellar surface densities and much circumstantial evidence from ultraluminous infrared galaxies. Our halo, however, appears to have retained the kinematic substructure characteristic of minor mergers (Gilmore, Wyse, & Norris 2002).

The Milky Way is an especially interesting case. There is a central bar, and there are also indications from the rotation curve and the gravitational microlensing optical depth that no more than 10% of the mass within two disk scale lengths can be nonbaryonic dark matter. The formation of the supermassive black hole adds a further complication that may modify the dark matter profile close to the center. In fact, we can take advantage of this, if the CDM consists of the lightest stable SUSY particle.

The presence of the supermassive black hole allows a potentially important probe of the dark matter density concentration in the inner galaxy. As the black hole forms within the preexisting dark matter potential well, a dark matter cusp develops within the zone of gravitational influence of the black hole, or about 0.1 pc. The cusp slope is $r^{-9/4}$ for a soft galactic core, and is steeper for an initially power-law profile in the galaxy inner halo. Dark matter annihilations are correspondingly boosted, within a radius between 0.1pc and about 100 Schwarzschild radii, within which the dark matter annihilates in less than a dynamical crossing time. One consequence is that a point source of high-energy neutrinos should be detectable toward the Galactic Center (Gondolo & Silk 1999; Ullio, Zhao, & Kamionkowski 2001). However, this is not the end of the story, for there are also radio and gamma ray signatures.

In fact, the advection-dominated accretion model for Sgr A*, for the flow in the accretion disk around the central black hole in the Milky Way, explains all of the observations except in the low-frequency radio and gamma ray regimes (Quataert & Narayan 1999). An additional nonthermal source of energetic particles is needed. Annihilations boosted by the black hole–enhanced cusp provide an attractive possible explanation.

The annihilations into e^+e^- pairs generate synchrotron emission. Remarkably, the predicted spectral shape, dominated by synchrotron self-absorption, matches that observed. Moreover the unidentified EGRET gamma ray source at the Galactic Center can also be fit spectrally, and the magnetic field strength can then be inferred by comparing the radio and gamma ray fluxes (Bertone, Sigl, & Silk 2002). The inferred field strength depends on the initial profile and the WIMP cross-section at a specified dark matter density, both leading to several orders of magnitude uncertainty in the required magnetic field strength, which in fact spans the equipartition estimate. In practice, the profile uncertainties dominate.

The observed radio and gamma ray luminosities can then be explained as being a consequence of the energy input from annihilations if the initial central core density profile is reasonably soft, as compared to the NFW profile. Detection of a neutrino source by the Northern hemisphere underwater neutrino telescopes ANTARES or NESTOR could eventually provide the smoking gun that confirms annihilations as a significant energy source in the inner 0.1 pc of the Milky Way.

5.6   Conclusions

Dark matter has been remarkably elusive. Despite 70 years over which the problem of the dark matter has been recognized, there has been no confirmation of its nature. However, observations are proceeding at a great pace, both from the observational astronomy side, especially, but not exclusively, via gravitational lensing, and on the experimental side by the enormous progress being made in direct and indirect detection experiments. There is increasing recognition that two of the greatest unresolved issues in physical cosmology, the nature of the dark matter and the formation of the galaxies, are intimately connected. Supermassive black holes and active galactic nuclei are important ingredients that need to be incorporated into our modeling before we can understand how protogalaxies evolve.

Imaging dark matter would be a wonderful achievement to crown our studies of these astrophysical puzzles. If our prejudices about dark matter are correct, annihilations provide a potential signal. Direct detection cannot compete: at the Large Hadron Collider any SUSY evidence will have no direct relevance for dark matter, although, of course, the complementarity is an essential strategy. Even direct detection experiments in deep underground laboratories are searching for a SUSY-inspired scattering signal that may be far below 10^{-10} picobarns, and so will be beyond any possible reach in the foreseeable future, where the goal for ton-scale detectors is of order 10^{-8} picobarns.

The relic density of dark matter particles specifies the annihilation cross-section within model uncertainties, and indirect detection provides our optimal strategy for confirming that the SUSY LSP is the dark matter candidate. It may well be that the clumpiness of the dark halo will facilitate our obtaining these images in the gamma ray domain. The cosmic ray signatures will provide invaluable confirmation, as will detection of high-energy neutrinos from the Sun that are generated by annihilations of WIMPs trapped in the solar core. The supermassive black hole in the center of the Milky Way may present a unique signal amplifier by which we can view the dark matter, in neutrinos as well as in gamma rays. The uncertainties in the halo fine-structure and in how the dark matter aggregated and the central supermassive black hole formed are simply too great to believe any theorist's assurance of the outcome. We must search.

Acknowledgements. I thank G. Bertone, G. Bryan, P. Podsiadlowski, A. Slyz, and J. Taylor for helpful discussions.

References

Adelberger, K. L., Steidel, C. C., Shapley, A. E., & Pettini, M. 2003, ApJ, 584, 45

Baltz, E., Edsjö, J., Freese, K. & Gondolo, P. 2002, Phys. Rev. D, 65, 063511

Bell, E. F., McIntosh, D. H., Katz, N., & Weinberg, M. D. 2003, ApJ, 585, L117

Benson, A. J., Bower, R. G., Frenk, C. S., & White, S. D. M. 2000, MNRAS, 314, 557

Bertone, G., Sigl, G., & Silk, J. 2002, MNRAS, 337, 98

Binney, J., & Evans, N. W. 2001, MNRAS, 327, L27

Binney, J., Gerhard, O. E., & Silk, J. 2001, MNRAS, 321, 471

Bournaud, F., & Combes, F. 2002, A&A, 394, L35

Burkert, A., & Naab, T. 2004, in Carnegie Observatories Astrophysics Series, Vol. 1: Coevolution of Black Holes and Galaxies, ed. L. C. Ho (Cambridge: Cambridge Univ. Press), in press

Dalal, N., & Kochanek, C. S. 2002, ApJ, 572, 25

Davé, R., et al. 2001, ApJ, 552, 528

Debattista, V. P., & Sellwood, J. A. 2000, ApJ, 543, 704

de Blok, W. J. G., & Bosma, A. 2002, A&A, 385, 816

Edsjö, J., Schelke, M., Ullio, P., & Gondolo, P. 2003, J. Cosmol. Astropart. Phys., 04, 001

Elgarø, Ø., et al. 2002, Phys. Rev. Lett., 89, 061301

Fukugita, M., Hogan, C. J., & Peebles, P. J. E. 1998, ApJ, 503, 518

Gerhard, O. E., & Silk, J. 1996, ApJ, 472, 34

Ghez, A. M. 2004, in Carnegie Observatories Astrophysics Series, Vol. 1: Coevolution of Black Holes and Galaxies, ed. L. C. Ho (Cambridge: Cambridge Univ. Press), in press

Gilmore, G., Wyse, R. F. G., & Norris, J. E. 2002, ApJ, 574, L39

Gondolo, P., & Silk, J. 1999, Phys. Rev. Lett., 83, 1719

Hannestad, S. 2002, Phys. Rev. D, 66, 125011

Islam, R. R., Taylor, J. E., & Silk, J. 2003, MNRAS, 340, 647

Kerins, E., Binney, J., & Silk, J. 2002, MNRAS, 332, L29

Klypin, A. A., Zhao, H. S., & Somerville, R. S. 2002, ApJ, 573, 597

Kochanek, C. S. 2003, ApJ, 583, 49

Lauer, T. R., et al. 2002, AJ, 124, 1975

Lawrence, A. 2001, MNRAS, 323, L147

Melchiorri, A., & Silk, J. 2002, Phys. Rev. D, 66, 041301

Metcalf, R. B. 2002, ApJ, 580, 696

Minakata, H., & Sugiyama, H. 2002, Phys. Lett. B, 532, 275

Nakamura, T., Umeda, H., Iwamoto, K., Nomoto, K., Hashimoto, M., Hix, W. R., & Thielemann, F.-K. 2001, ApJ, 555, 880

Navarro, J. F., Frenk, C. S., & White, S. D. M. 1996, ApJ, 462, 563

Penton, S. V., Shull, J. M., & Stocke, J. T. 2000, ApJ, 544, 150

Pettini, M., Rix, S. A., Steidel, C. C., Adelberger, K. L., Hunt, M. P., & Shapley, A. E. 2002, ApJ, 569, 742

Portegies Zwart, S. F., Makino, J., McMillan, S. L. W., & Hut, P. 2002, in Stellar Collisions, Mergers and their Consequences, ed. M. Shara (San Francisco: ASP), 95

Quataert, E., & Narayan, R. 1999, ApJ, 520, 298

Rafikov, R. R., & Draine, B. T. 2001, ApJ, 547, 207

Sanders, R. H., & McGaugh, S. 2002, ARA&A, 40, 263

Sciama, D. W. 2000, MNRAS, 312, 33

Shapley, A., Steidel, C. C., Pettini, M., & Adelberger, K. L. 2003, ApJ, 588, 65

Silk, J. 2003, MNRAS, 343, 249

Simcoe, R., Sargent, W. L. W., & Rauch, M. 2002, ApJ, 578, 737

Sołtan, A. M., Freyberg, M., & Hasinger, G. 2002, A&A, 395, 475

Toft, S., Rasmussen, J., Sommer-Larsen, J., & Pedersen, K. 2002, MNRAS, 335, 799

Ullio, P., Zhao, H. S., & Kamionkowski, M. 2001, Phys. Rev. D, 64, 3504

Umeda, H., & Nomoto, K. 2003, Nature, 422, 871

Volonteri, M., Haardt, F., & Madau, P. 2003, ApJ, 582, 559

Walker, M., & Wardle, M. 1998, ApJ, 498, L125

Wardle, M., & Walker, M. 1999, ApJ, 527, L109
Weinberg, M. D., & Katz, N. 2002, ApJ, 580, 627
Zwicky, F. 1933, Helv. Phys. Acta, 6, 110

6

Status of cosmology on the occasion of the Carnegie Centennial

WENDY L. FREEDMAN[1] and MICHAEL S. TURNER[2]
(1) The Observatories of the Carnegie Institution of Washington
(2) University of Chicago and Fermi National Accelerator Laboratory

Abstract

Cosmology today is free of serious controversies, though not of exciting discoveries and puzzles. Many controversies have been settled: there is agreement about the Hubble constant (72 ± 7 km s^{-1} Mpc^{-1}), a consistent age (13 ± 1.5 Gyr), consensus that the cold dark matter paradigm provides a successful explanation for how structure formed, and evidence for an early period of inflation. The basic features of the Universe are now quantified: it is spatially flat and has the critical density; ordinary matter accounts for $4 \pm 1\%$ of the total mass/energy content with dark matter and dark energy accounting for the other 96%; and the expansion is accelerating not decelerating. This remarkable picture is supported by an interlocking web of increasingly more precise observations. It also raises profound questions: What is the dark matter? What is the mysterious dark energy whose repulsive gravity is causing the Universe to speed up? Did the early Universe undergo a rapid acceleration (inflation)? An ambitious program of new observations, experiments and theoretical investigations is being launched, with the aim of elucidating our understanding of this new cosmology.

6.1   Introduction

Cosmology has attracted more than its share of attention and controversies over the years. They include: Big Bang versus Steady State; expansion age versus age of astrophysical objects; the nature of quasars; the value of the Hubble constant; isothermal versus adiabatic density perturbations; hot versus cold dark matter (CDM); flat versus non-flat; CDM versus topological defects and nonbaryonic dark matter models versus baryons-only models. Today cosmology is attracting attention not for its controversies—currently there are none—but rather for the remarkable progress toward a consistent cosmological model that describes our Universe and the compelling questions now being asked (NRC 2003).

It has always been a challenge to find ideas expansive enough to describe the Universe and to create instruments powerful enough to explore the Universe. This explains at least in part the uneven progress in cosmology as well as the propensity for controversy (without definitive observations it can be difficult to settle issues). It also is at the root of the current boom: we have both powerful ideas, many motivated by advances in elementary particle physics, and powerful probes of the Universe. The latter include a new generation of astronomical instruments, from the *Hubble Space Telescope* (*HST*) and numerous 6 to 10-meter ground-based optical telescopes, to satellite-borne instruments that probe the Universe across the electromagnetic spectrum, as well as laboratory probes, from accelerator experiments to specialized, ultrasensitive detectors searching for dark matter particles.

Fig. 6.1. George Ellery Hale and three of the four telescopes he built, which were the world's largest at the time: the 60-inch and 100-inch telescopes at Mount Wilson and the 200-inch Hale telescope at Mount Palomar.

In 1902, Andrew Carnegie donated a gift of $10 million to the Carnegie Institution of Washington to "encourage investigation, research and discovery in the broadest and most liberal manner, and the application of knowledge to the improvement of mankind." Under the energetic guidance and vision of the solar astronomer and first Carnegie Observatories Director, George Ellery Hale, 60-inch and 100-inch telescopes were built on Mount Wilson, and soon thereafter the construction of the 200-inch Hale telescope on Mount Palomar was initiated (see Fig. 6.1).

The Carnegie Institution and its scientists have made and are continuing to make seminal contributions to our understanding of the origin and evolution of the Universe. In the early part of the century, while at Carnegie, Harlow Shapley studied the distribution of globular clusters in the Milky Way, and displaced the Sun and Solar System from the center of the Universe. Hubble's discovery of Cepheid variables within nebulae led first to the recognition that there were galaxies outside of our Milky Way, and then in 1929, to the discovery of the expansion of the Universe. Walter Baade's bold concept of stellar populations revolutionized the study of stellar and galactic evolution as well as impacting the cosmological distance scale. Allan Sandage has made seminal contributions to measurements of the Hubble constant and the ages of globular clusters. Vera Rubin's optical spectroscopy of galaxies was pivotal in establishing the presence and importance of dark matter in the Universe. With the twin Magellan telescopes now in full operation and even bolder plans for a 20-meter telescope, prospects for future discoveries and breakthroughs remain just as bright.

In this paper we discuss the current status of cosmology. The outline of this review follows from this three-sentence summary of cosmology today:

- We have delineated the basic features of the Universe (*The Cosmological Model*).
- There is still much to understand about the Universe (*The Big Questions*).
- We have excellent prospects for deepening our understanding (*Looking Forward*).

6.2   The Cosmological Model

On the large scales today (> 100 Mpc) and on all scales at early times ($t < 10^6$ yr), our Universe is well described by the isotropic and homogeneous Friedmann-Robertson-Walker (FRW) models of general relativity. The features of the FRW model that well describes our Universe are summarized as follows:

- Spatially flat with the critical density ($\Omega_0 = 1 \pm 0.03$), expanding at a current rate of $H_0 \equiv (\dot{R}/R)_0 = 72 \pm 7$ km s^{-1} Mpc^{-1}, and accelerating with a deceleration parameter of $q_0 \equiv -\ddot{R}/RH_0^2 = -0.67 \pm 0.25$.
- Age $t_0 = 13 \pm 1.5$ Gyr.
- Composition: Baryons $\Omega_b = 0.04 \pm 0.01$, with stars contributing $\Omega_* \approx 0.005$; nonbaryonic dark matter $\Omega_{DM} = 0.29 \pm 0.04$ with massive neutrinos contributing $\Omega_\nu = 0.001 - 0.05$; dark energy (something with very negative pressure, $w \equiv p/\rho < -0.6$) $\Omega_{DE} = 0.67 \pm 0.06$; cosmic microwave background (CMB) photons $\Omega_\gamma \approx 10^{-4}$ with temperature $T_0 = 2.725 \pm 0.001$ K.
- Early, hot, dense primordial soup phase.
- Large-scale structure build up through the gravitational attraction of CDM.
- Evidence for an early phase of accelerated expansion (inflation), which produced seed density perturbations from quantum fluctuations.

This new cosmology builds and extends upon the highly successful hot Big Bang model (standard cosmology), which emerged in the 1970s and describes the evolution of the Universe from the beginning of Big Bang nucleosynthesis (BBN) ($t \approx 10^{-2}$ s) until the present ($t \simeq 14$ Gyr) (see, e.g., Weinberg 1972; Peebles 1993; or Kolb & Turner 1990). A striking success of cosmology today is a set of cosmological parameters with reliable error bars, shown in Table 6.1 from a recent review (Freedman & Turner 2004). The quantity and quality of cosmological data have advanced to the point where essentially all of the parameters are supported by an interlocking web of independent measurements. To illustrate this, we discuss in detail the measurements underlying some of the parameters.

6.2.1   The Hubble Constant

Hubble's discovery of a deceptively simple correlation between galaxy distance and recession velocity 74 years ago did not foreshadow the challenge of obtaining an accurate value for the Hubble constant. The most difficult aspect lies in establishing accurate distances over cosmologically large scales. The importance of pinning down H_0 has only grown with time: not only does it set the scale for all cosmological distances and times, but also its accurate determination is needed to take full advantage of increasingly precise measurements of other cosmological quantities.

Determining an accurate value for H_0 was one of the motivating reasons for building the

Table 6.1. *Our 16 Cosmological Parameters*

Parameter	Value[*]	Description	WMAP[†]
		Ten Global Parameters	
h	0.72 ± 0.07	Present expansion rate[a]	$0.71^{+0.04}_{-0.03}$
q_0	-0.67 ± 0.25	Deceleration parameter[b]	-0.66 ± 0.10^c
t_0	$13 \pm 1.5\,\mathrm{Gyr}$	Age of the Universe[d]	$13.7 \pm 0.2\,\mathrm{Gyr}$
T_0	$2.725 \pm 0.001\,\mathrm{K}$	CMB temperature[e]	
Ω_0	1.03 ± 0.03	Density parameter[f]	1.02 ± 0.02
Ω_b	0.039 ± 0.008	Baryon density[g]	0.044 ± 0.004
Ω_{CDM}	0.29 ± 0.04	Cold dark matter density[g]	0.23 ± 0.04
Ω_ν	$0.001 - 0.05$	Massive neutrino density[h]	
Ω_X	0.67 ± 0.06	Dark energy density[g]	0.73 ± 0.04
w	-1 ± 0.2	Dark energy equation of state[i]	$< -0.8\,(95\%\ \mathrm{cl})$
		Six Fluctuation Parameters	
$\sqrt{S}$	$5.6^{+1.5}_{-1.0} \times 10^{-6}$	Density perturbation amplitude[j]	
$\sqrt{T}$	$< \sqrt{S}$	Gravity wave amplitude[k]	$T < 0.90S\ (95\%\ \mathrm{cl})$
σ_8	0.9 ± 0.1	Mass fluctuations on 8 Mpc[l]	0.84 ± 0.04
n	1.05 ± 0.09	Scalar index[f]	0.93 ± 0.03
n_T	—	Tensor index	
$dn/d\ln k$	-0.02 ± 0.04	Running of scalar index[m]	-0.03 ± 0.02

[*]The 1-σ uncertainties quoted in this table represent our combined analysis of published data.
[†]Bennett et al. 2003.
[a]Freedman et al. 2001; note: $H_0 = 100h\,\mathrm{km\,s^{-1}\,Mpc^{-1}}$.
[b]Supernova results combined with measurements of the total matter density, $\Omega_m = \Omega_\nu + \Omega_b + \Omega_{\mathrm{CDM}}$ and Ω_0, assuming $w = -1$ (Riess et al. 1998; Perlmutter et al. 1999a).
[c]WMAP results combined with Tonry et al. 2003.
[d]Value based upon CMB, globular cluster ages, and current expansion rate (Oswald et al. 1996; Knox, Christensen, & Skordis 2001; Krauss & Chaboyer 2003).
[e]Mather et al. 1999.
[f]Combined analysis of four CMB measurements (Sievers et al. 2003).
[g]Combined analysis of CMB, BBN, H_0, and cluster baryon fraction (Turner 2002).
[h]Lower limit from SuperKamiokande measurements (Fukuda et al. 1998); upper limit from structure formation (Elgarøy et al. 2002).
[i]Supernova measurements, CMB, and large-scale structure (Perlmutter, Turner, & White 1999b).
[j]Contribution of density perturbations to the variance of the CMB quadrupole (with $T = 0$) (Gorski et al. 1996).
[k]Contribution of gravity waves to the variance of the CMB quadrupole (upper limit) (Kinney, Melchiorri, & Riotto 2001).
[l]Variance in values reported is larger than the estimated errors; adopted error reflects this (Lahav et al. 2002).
[m]Deviation of the scalar perturbations from a pure power law (Lewis & Bridle 2002).

HST, and the goal of 10% accuracy for H_0 was designated as one of three "Key Projects." Determining H_0 accurately requires the measurement of distances far enough away that both the small- and large-scale motions of galaxies can be neglected compared to the overall

Hubble expansion. The H_0 Key Project provided a Cepheid calibration of a several distance methods, averaging over the systematics of any one method (Freedman et al. 2001).*

Calibrating five secondary methods with Cepheid distances, Freedman et al. (2001) determined a value of $H_0 = 72 \pm 3\,(\text{stat}) \pm 7\,(\text{sys})$ km s^{-1} Mpc^{-1}. Four of the methods yield a value of H_0 between 70 and 72 km s^{-1} Mpc^{-1}, while the fundamental plane gives $H_0 = 82$ km s^{-1} Mpc^{-1}. Type Ia supernovae are the secondary method, which currently extends out to the greatest distances, about 400 Mpc. The measurements are shown in Figure 6.2.

The largest remaining sources of error result from uncertainties in the distance to the Large Magellanic Cloud, photometric calibration of the *HST* Wide Field and Planetary Camera 2, metallicity calibration of the Cepheid period-luminosity relation, and cosmic scatter in the density (and therefore, velocity) field that could lead to observed variations in H_0 on very large scales. These sources of error are what contribute to the systematic uncertainty quoted above.

An external check of the Cepheid distance scale is provided by a distance independently measured to the NGC 4258, a nearby spiral galaxy with an inner disk containing H_2O masers (Herrnstein et al. 1999). Both radial and transverse motions of the maser system have been measured. Assuming a circular, Keplerian model for the disk, Herrnstein et al. (1999) derive a distance to the galaxy of 7.2 ± 0.3 Mpc, with the error increasing to ± 0.5 Mpc allowing for systematic uncertainties in the model. Maoz et al. (1999) used *HST* to discover a sample of 15 Cepheids in NGC 4258. The maser distance is in very good agreement with the Cepheid distance of $7.8 \pm 0.3\,(\text{stat}) \pm 0.5\,(\text{sys})$ Mpc (Newman et al. 2001), and the metallicity-corrected distance of 8.0 Mpc.

At present, to within the uncertainties, there is broad agreement about the value of the Hubble constant, using completely independent techniques. The most recent applications of the Sunyaev-Zel'dovich (SZ) technique, based on two-dimensional interferometry SZ data for 41 well-observed clusters, as discussed by Reese (2004), yield $H_0 = 61 \pm 3\,(\text{stat}) \pm 18\,(\text{sys})$ km s^{-1} Mpc^{-1}. The systematic uncertainties are still large, but the near-term prospects for this method are improving rapidly as additional clusters are being observed, and higher-resolution X-ray and SZ data are becoming available.

A second method for measuring H_0 at very large distances, also independent of the need for any local calibration, comes from the measurement of time delays in gravitational lenses. As noted by Schechter (2001), the small numbers of well-determined time delays leave open dangers in making decisions about which lenses to include in an analysis or not. Based on three lenses, Schechter finds $H_0 = 62$ km s^{-1} Mpc^{-1}, dominated by a systematic error estimated to be of order 20 km s^{-1} Mpc^{-1}. Kochanek (2002), in an analysis of five lenses (one of which overlaps the sample of Schechter) finds $H_0 = 51 \pm 5\,(\text{stat})$ km s^{-1} Mpc^{-1} for models with flat rotation curves, and $H_0 = 73 \pm 8\,(\text{stat})$ km s^{-1} Mpc^{-1} for models with constant mass-to-light ratios. Kochanek & Schechter (2004) report values of $H_0 = 48 \pm 3\,(\text{stat})$ km s^{-1} Mpc^{-1}, and $H_0 = 71 \pm 3\,(\text{stat})$ km s^{-1} Mpc^{-1} for the same two classes of models, based on four lenses that can be modeled by a single galaxy (and are not members of groups or clusters). The numbers of suitable lenses for this kind of analysis should increase in the near future. A long-term monitoring program is being undertaken by the Magellan consortium, using the 6.5-meter telescopes at Las Campanas.

---

* The Cepheid positions, magnitudes, and periods for each of the 18 galaxies observed as part of the Key Project are available at: http://www.ipac.caltech.edu/H0kp/H0KeyProj.html.

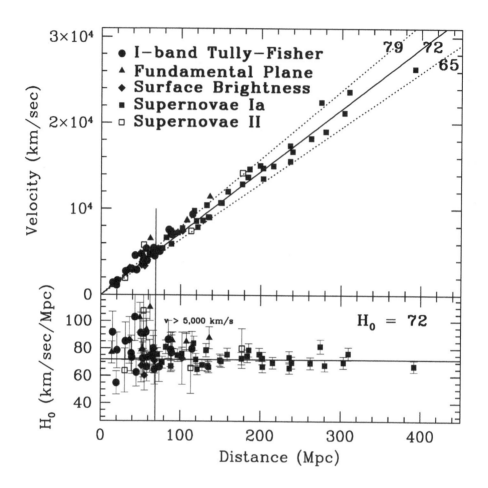

Fig. 6.2. Composite Hubble diagram of velocity versus distance. The Cepheid distances are calibrated relative to a distance to the Large Magellanic Cloud of 50 kpc. The consistency of the five different distance indicators is shown. The lower panel shows the value of the Hubble constant object by object and the convergence to 72 km s^{-1} Mpc^{-1}. The scatter at distances less than 100 Mpc arises due to gravitationally induced "peculiar velocities" that arise from the inhomogeneous distribution of matter.

6.2.2　Age

The time back to the Big Bang depends upon H_0 and the expansion history, which itself depends upon the composition:

$$t_0 = \int_0^\infty \frac{dz}{(1+z)H(z)} = H_0^{-1} \int_0^\infty \frac{dz}{(1+z)[\Omega_m(1+z)^3 + \Omega_{DE}(1+z)^{3(1+w)}]^{1/2}} \quad , \qquad (6.1)$$

where $\Omega_m = \Omega_{CDM} + \Omega_b + \Omega_\nu$ is the total mass density. For a Universe with a Hubble constant of 72 km s^{-1} Mpc^{-1}, $\Omega_m = 1/3$, and $\Omega_{DE} = 2/3$, the time back to the Big Bang is 13 Gyr. Taking account of the uncertainties in H_0 and in the composition, the uncertainty in the age of the Universe is estimated to be about $\pm 1.5\,\mathrm{Gyr}$. The expansion age can also be determined from CMB anisotropy, but without recourse to H_0, and it gives a consistent age, $t_0 = 14 \pm 0.5\,\mathrm{Gyr}$ (Knox et al. 2001).

The expansion age can also be checked for consistency against other cosmic clocks. For example, the best estimates of the age of the oldest stars in the Universe are obtained from globular clusters. Detailed computer models of stellar evolution matched to observations of globular cluster stars yield ages of about 12.5 billion years, with an uncertainty of about 1.5 Gyr (Krauss & Chaboyer 2002). These estimates are also in good agreement with other methods that independently measure the rates of cooling of the oldest white dwarf stars, and techniques that use various radioactive elements as cosmic chronometers (Oswald et al. 1996). Finally, with the assumption that $w = -1$, the Type Ia supernova data can constrain the product of the age and Hubble constant independent of either quantity, $H_0 t_0 = 0.96 \pm 0.04$ (Tonry et al. 2003). This is consistent with the product of the two quantities, $(H_0 = 72 \pm 7\,\mathrm{km\,s^{-1}\,Mpc^{-1}}) \times (t_0 = 13 \pm 1.5\,\mathrm{Gyr}) = 0.96 \pm 0.16$.

In summary, all the ages are consistent with a consensus age of $13 \pm 1.5\,\mathrm{Gyr}$.

6.2.3   Baryon Density

Both the CMB and BBN provide the means for a high-precision measurement of the baryon density. However, they are based upon very different physics: nuclear reactions occurring when the Universe was seconds old (BBN) and gravity-driven acoustic oscillations in the baryon-photon fluid when the Universe was 400,000 yr old (CMB).

The amount of deuterium synthesized in the Big Bang is very sensitive to the baryon density, $(D/H)_p \propto \rho_b^{-1.7}$, and thus a measurement of its primordial abundance can accurately pin down ρ_b. The detection of deuterium Lyman absorption features in six high-redshift ($z \approx 2-4$) quasars (O'Meara et al. 2001; Kirkman et al. 2004) has led to a consistent estimate for the primeval deuterium abundance of $(D/H)_p = 3.0 \pm 0.2 \times 10^{-5}$. Using the most up-to-date calculations of the light-element abundances, this deuterium abundance implies a baryon density, $\rho_b = 3.8 \pm 0.2 \times 10^{-31}\,\mathrm{g\,cm^{-3}}$, or equivalently, $\Omega_b h^2 = 0.02 \pm 0.001$ (Burles, Nollett, & Turner 2001).

The ratio of the heights of the odd to even acoustic peaks in the CMB angular power spectrum depends upon the baryon density. This is a distinctive feature in the CMB power spectrum: the dependence upon peak heights of most other parameters is monotonic. Four experiments—two using microwave interferometers and two using bolometric receivers—have now resolved the first three acoustic peaks (BOOMERanG, MAXIMA, DASI, and CBI). The combined measurements from these experiments are consistent with a baryon density of $\Omega_b h^2 = 0.022 \pm 0.003$ (Sievers et al. 2003).

The BBN and CMB numbers are consistent and point strongly to a low baryon density. The agreement of these two numbers not only makes the case for a low baryon density (inserting the Hubble constant implies $\Omega_b = 0.04 \pm 0.008$), but also speaks to the consistency of the entire cosmological framework, from general relativity and the FRW model to nuclear physics in the early Universe.

Finally, the net absorption of light emitted from very distant quasars by intervening gas, which exists largely in the Lyα forest, provides a third independent accounting of baryons, at a much lower redshift, $z \approx 3-4$ vs. $z \approx 10^{10}$ for BBN and $z \approx 10^3$ for CMB. While additional assumptions must be made (the CDM paradigm to simulate the distribution of clouds, and a lower limit to the flux of ionizing radiation), the result is consistent (McDonald et al. 2001):

$$\Omega_b h^2 > 0.017 \left[\Omega_m h^2 / (0.3)(0.65)^2 \right]^{1/4} \Gamma_{-12}^{1/2} , \tag{6.2}$$

where Γ_{-12} is the photoionization rate in units of $10^{-12} \, s^{-1}$.

What is still lacking is a robust accounting of baryons today; only a small fraction of the baryons are "visible" in the form of stars (about 15%); most are unaccounted for, most likely existing in the form of hot gas somewhere—the intergalactic medium, extended halos, small groups (see, e.g., Fukugita, Hogan, & Peebles 1998).

6.2.4 Matter Density

Determining the mean density of matter in the Universe is a formidable task that has challenged astronomers for almost as long as the Hubble constant has. Because there is inhomogeneity in the distribution of matter on scales approaching 30 Mpc, the matter density must be probed on scales larger than this, for example, using the mass-to-light ratio technique:

$$\langle \rho_m \rangle = \langle M/L \rangle \mathcal{L} \tag{6.3}$$

(see Bahcall, Lubin, & Dorman 1995). However, there remain a number of difficulties inherent in this approach. (1) Mass-to-light ratios for field spirals, which are more representative of the mean galaxy, provide only a lower limit to the mean density because halo mass measurements have yet to converge. (2) While cluster mass-to-light ratios do not suffer from this problem, the typical galaxy is not found in a cluster. And (3) determining the mean luminosity density is difficult and subject to evolutionary effects.

The recent consensus for a value for the matter density and a lower uncertainty, $\Omega_m = 0.3 \pm 0.04$, has resulted from a new approach (Turner 2002; Sievers et al. 2003). The new and previous results agree quite well within their quoted uncertainties. In this new approach, the key inputs are physical parameters that do not depend upon any assumed relationship between mass and light: $\Omega_m h^2$ and $\Omega_b h^2$, deduced from CMB anisotropy measurements; $\Omega_b h^2$, determined from BBN; $\Omega_m h$, determined from the shape of the power spectrum of matter inhomogeneity; and Ω_m / Ω_b, determined from cluster measurements assuming that they provide a fair sample of baryonic and nonbaryonic matter.

It is straightforward to see how these measurements determine Ω_m and Ω_b. The CMB and cluster measurements fix the matter-to-baryon ratio, $\Omega_m / \Omega_b = 7.2 \pm 2$ (CMB, pre-*WMAP*) and 9 ± 1.5 (clusters). The value for $\Omega_b h^2$ (from CMB and BBN) and H_0 fixes $\Omega_b = 0.04 \pm 0.008$; Ω_b, together with the matter-to-baryon ratio, then fixes $\Omega_m = 0.3 \pm 0.04$.

The reliability of this determination of the matter density is tied to the consistency of the input values for Ω_m / Ω_b (CMB and clusters), $\Omega_b h^2$ (BBN and CMB), and H_0 (the various methods discussed above). This simple chain of reasoning and consistent set of measurements of physical parameters make the case that the mean matter density has finally been determined.

6.2.5 CMB Anisotropy

The anisotropy of the CMB arises due to the inhomogeneous distribution of matter and provides a snapshot of the Universe at an earlier, simpler time (redshift 1100 and 400,000 yr after the Big Bang). The temperature is determined to four significant figures, $2.725 \pm 0.001 \, K$, and deviations from a Planckian spectrum are constrained to be less than 0.01% (Mather et al. 1999). Encoded in the small (0.001%) variations in temperature across

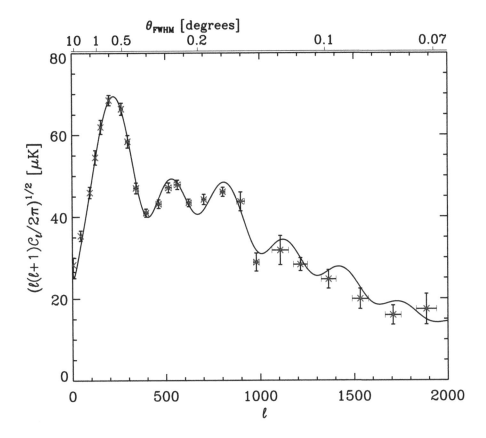

Fig. 6.3. Angular power spectrum of the CMB, incorporating all the pre-*WMAP* data (*COBE*, BOOMERanG, MAXIMA, DASI, CBI, ACBAR, FIRS, VSA, and other experiments). Variance of the multipole amplitude is plotted against multiple number; as indicated by the top scale, multipole ℓ measures the fluctuations on angular scale $\theta \approx 200°/\ell$. Evidence of the baryon-photon oscillations can be seen as the distinct "acoustic peaks." The theoretical curve is the consensus cosmological model. (Figure courtesy of C. Lineweaver.)

the sky (anisotropy) is a wealth of information about the Universe, past, present, and future. Since the discovery of CMB anisotropy in 1992 by the DMR instrument on the *COBE* satellite (Smoot et al. 1992) with its 7° angular resolution and signal-to-noise ratio per pixel of order unity, a host of ground- and balloon-based experiments have been improving both the precision and angular resolution of CMB measurements. The angular (spherical harmonic multipole) power spectrum now extends to $\ell \approx 3000$ (*COBE* probed to $\ell \approx 20$), with each multipole region measured by two or more independent experiments with consistent results; see Figure 6.3. Not only are the same regions of the spectrum probed by different experiments, but also by different techniques (microwave interferometers operating at 30 GHz vs. bolometric receivers operating at frequencies above 100 GHz).

Among the conclusions inferred from the CMB are:

- Spatial flatness of the Universe as indicated by the position of the first acoustic peak ($\ell \simeq 220$), $\Omega_0 = 1.03 \pm 0.03$.
- Low baryon density as indicated by the ratio of the first to second acoustic peaks, $\Omega_b h^2 = 0.022^{+0.004}_{-0.003}$.
- Total matter density, $\Omega_m h^2 = 0.16 \pm 0.04$, as indicated by the heights of the acoustic peaks, which exceeds the baryon density by a factor of 7.
- Independent evidence for dark energy based upon the factor of 3 discrepancy between Ω_m and Ω_0.
- Adiabatic density perturbations, as indicated by the existence of acoustic peaks, with near scale invariance, $n = 1.05 \pm 0.09$.
- Hubble constant ($h = 0.71^{+0.04}_{-0.03}$) and age (13.7 ± 0.2 Gyr) consistent with other independent measurements (these are both *WMAP* results).

6.2.6   Inflationary Beginning

Inflation has been the single most important idea in cosmology for the past 20 years. Twenty years ago there was skepticism about testing inflation, especially in the absence of a standard model. However, over the past five years evidence for inflation has mounted.

While there is no standard model of inflation, inflation makes three basic predictions (Turner 1997):

- Spatially flat Universe (i.e., $\Omega_0 = 1$).
- Nearly scale-invariant spectrum of Gaussian, adiabatic density perturbations.
- Nearly scale-invariant spectrum of gravitational waves.

The evidence for flatness comes from the measurements of CMB anisotropy mentioned above ($\Omega_0 = 1.03 \pm 0.03$) and an independent cosmic inventory of material that adds to the critical density: $(\Omega_m \simeq 1/3) + (\Omega_{DE} \simeq 2/3) = 1$. The discovery of cosmic acceleration provided an important (and successful) test of inflation: before that time, firm estimates for the matter density were falling a factor of 3 short of the critical density, and dark energy suddenly filled the gap.

The second prediction leads to a series of tests: the existence of the harmonically spaced acoustic peaks associated with adiabatic perturbations; spectrum characterized by power-law index $n = 1 \pm \mathcal{O}(0.1)$ and slow variation $dn/d\ln k \approx \mathcal{O}(10^{-3})$; and density perturbations with Gaussian statistics. The angular power spectrum of CMB anisotropy (discussed above) provides evidence for at least three acoustic peaks, has determined $n = 1.05 \pm 0.09$ and $dn/d\ln k = -0.02 \pm 0.04$, and shown that the underlying density perturbations are consistent with being Gaussian. The data are consistent with the four tests of inflation associated with the prediction of density perturbations produced from quantum fluctuations (see Fig. 6.4).

In addition, the density perturbations predicted by inflation, together with the prediction of a flat Universe, gave rise to the CDM scenario for structure formation (see, e.g., Blumenthal et al. 1984). A host of cosmological observations, from measurements of large-scale structure to the sequence of structure formation (galaxies first, clusters next, etc.) to the distribution of the masses of clusters and galaxies, have tested the CDM scenario (see, e.g., Primack 2001). CDM is consistent with a large body of diverse observations, and today most would agree that it plays an important role in structure formation. Shown in Figure 6.5 is a comparison of the predicted CDM power spectrum of density inhomogeneity today with that determined from a variety of observations.

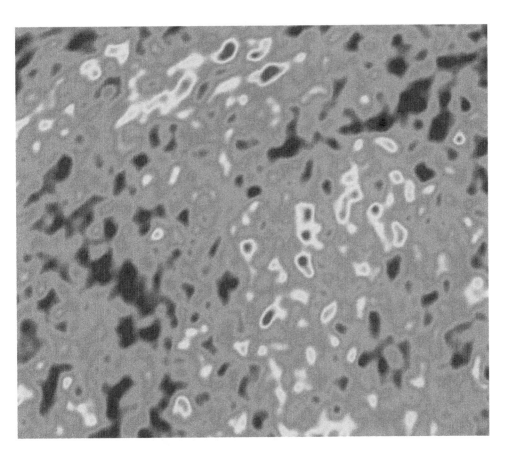

Fig. 6.4. Small portion of the CMB sky as measured by *WMAP*; for reference, the size of the typical hot/cold spots is of order 1°. According to inflation, this image is quantum noise on length scales $\ll 10^{-17}$ cm blown up in size by the expansion of the Universe during inflation and thereafter.

The third basic inflationary prediction—an almost scale-invariant spectrum of gravitational waves—remains untested, and is one of the biggest challenges in cosmology. Like the amplitude of the density perturbations, there is no firm prediction for the amplitude of the gravitational wave spectrum. However, unlike density perturbations, where the existence of structure today leads to a prediction for the required amplitude, there is no phenomenological prediction. Detecting the inflation-produced gravitational waves would not only confirm the third basic prediction, but would also reveal when inflation took place as the amplitude is tied directly to the energy scale of inflation.

6.2.7   *Cosmic Acceleration and Dark Energy*

Two independent lines of evidence lead to the remarkable conclusion that the expansion of the Universe is speeding up, not slowing down. The first, as reviewed by Fil-

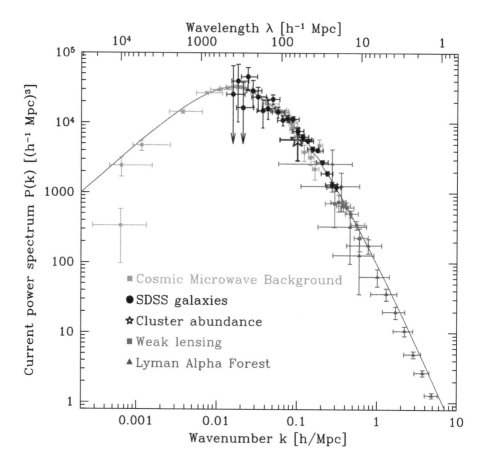

Fig. 6.5. Power spectrum of density inhomogeneity today obtained from a variety of measurements, including large-scale structure, CMB, weak lensing, rich clusters, and the Lyα forest. The curve is the theoretical prediction for the consensus cosmology model. (From Tegmark et al. 2004.)

ippenko (2004), is the direct evidence from Type Ia supernovae. That, too, involves two independent (and very competitive) groups whose results agree: the deceleration parameter is negative (Riess et al. 1998; Perlmutter et al. 1999a; Tonry et al. 2003). Megapixel CCDs and pre-planned scheduling of telescopes for the discovery and follow-up spectroscopy have enabled the discovery of Type Ia supernovae over the redshift interval from zero to beyond one, and data for over 200 supernovae have now been published (see Fig.6.6), and a number of ambitious projects hope to increase this by a factor of a several over the next five years. No evidence for serious systematic effects (for example, the presence of grey dust or evolution of supernovae) has been uncovered, though numerous tests have been carried out. The most recent study indicates that if the equation of state of the dark energy is $w = -1$ (vacuum

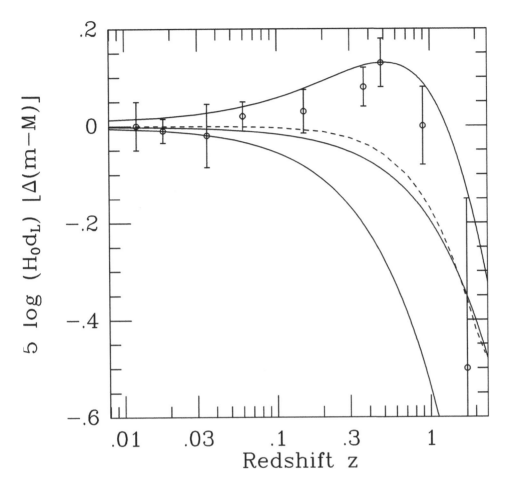

Fig. 6.6. Differential Hubble diagram adapted from Tonry et al. (2003). Data for 200 Type Ia supernovae are binned into 9 points, and shown relative to models with $\Omega_\Lambda = 0.7$, $\Omega_m = 0.3$; $\Omega_\Lambda = 0$, $\Omega_m = 0.3$; and $\Omega_\Lambda = 0$, $\Omega_m = 1$ (top to bottom). The broken curve represents a flat, coasting universe, i.e., $q = 0$ for all z.

energy/cosmological constant), then $\Omega_\Lambda - 1.4\Omega_m = 0.35 \pm 0.14$. Further, if a flat universe is assumed, this yields $\Omega_m = 0.28 \pm 0.06$ and $\Omega_\Lambda = 0.72 \pm 0.06$ (Tonry et al. 2003).

As noted above, the second line of evidence involves the CMB. The position of the first acoustic peak indicates that the total amount of matter and energy is equal to the critical density ($\Omega_0 = 1.03 \pm 0.03$), while the CMB and (other methods) indicate that the amount of that in matter is only about 1/3, leaving 2/3 unaccounted for. In order not to have been "noticed" this additional component must be unclustered, and in order not to interfere with the formation of structure it must evolve much more slowly than matter, which implies a large negative pressure, $p \leq -\rho/2$ ($\rho \propto R^{-3(1+w)}$ where $w = p/\rho$) (Turner 2000). [The argument is a simple one: small density fluctuations only grow during the matter-dominated era; the ratio of dark matter to dark energy in the past grows as $(1+z)^{-3w}$; thus, the more negative the pressure is, the more recent the matter-dominated era ended.]

6.3   The Big Questions

While the new cosmology describes the basic features of the Universe, it does not provide a coherent picture complete with a physical basis. It does frame big questions for cosmologists to answer, questions that involve both astrophysics and particle physics and accelerators and telescopes.

What is the dark matter?

An extremely solid case exists for the bulk of the dark matter not being made of neutrons and protons, but rather being made of a new form of stable (or long-lived) matter. Evidence for neutrino mass has validated this hypothesis, though neutrinos provide only a small fraction of the dark matter ($\Omega_\nu \simeq 0.001 - 0.05$). Finishing the neutrino story will involve laboratory and accelerator experiments. Identifying the nature of the bulk of the dark matter is not only of importance to cosmologists, but also to particle physicists as the two leading candidates, the neutralino and the axion, have deep implications for particle physics (see, e.g., Sadoulet 1999).

The neutralino is the lightest of a new class of particles, the supersymmetric partners of the ordinary particles. The existence of the axion would solve a nagging and profound problem with quantum chromodynamics, the theory of how quarks and gluons interact. Detection schemes for the neutralino illustrate how astrophysics and particles are deeply connected: production at an accelerator (Tevatron at FNAL or LHC at CERN); direct detection of neutralinos that comprise the halo of the Milky Way; detection of the annihilation products of neutralinos in the halo (positrons, gamma-rays, and antiprotons) or in the Sun and Earth (high-energy neutrinos).

What is dark energy and what is the destiny of the Universe?

At present there is no compelling explanation for cosmic acceleration. It could be due to a diffuse dark energy with large negative pressure ($p \leq -\rho/3$) or an indication of the breakdown of general relativity and evidence for new gravitational physics. The connections to fundamental physics are strong: the energy of the quantum vacuum has just the right property—large, negative pressure ($p = -\rho$)—to explain cosmic acceleration. However, at the same, it raises three questions:

- Why is the energy associated with the quantum vacuum at least 55 orders of magnitude smaller than any estimate for it?
- If it is vacuum energy, what is the explanation for the fact that its energy density is small, but not zero?
- If quantum vacuum energy is too small to explain cosmic acceleration, what does?

Then too there is the question of why the acceleration began so recently, and when and if it will end. Since the primary effect of dark energy is upon the expansion rate of the Universe, telescopes that probe the past expansion history by measuring supernova light curves, measuring the growth of structure, and counting galaxies and clusters will play the leading role in shedding light on dark energy.

Where did the baryons come from?

Even the origin of the matter we are made of is still a mystery. To ensure the existence of quark-based matter today, there must be an asymmetry between quarks and antiquarks of a few parts per billion at times earlier than about 10 microseconds (in a baryon symmetric universe, quarks and antiquarks annihilate to a negligible level when the Universe is 10 microseconds old). This asymmetry can arise through nonequilibrium baryon-number and CP-violating interactions in the early Universe (baryogenesis). One possibility is that the needed asymmetry begins as an excess of neutrinos over antineutrinos, which arises early on through neutrino interactions that involve neutrino mass and neutrino CP violation, and later, nonperturbative electroweak processes that violate both baryon and lepton number transmute the neutrino asymmetry into a baryon asymmetry. Further understanding of CP violation and physics beyond the standard model of particle physics is needed, for which accelerators are required.

When did inflation take place? What are future tests of inflation?

The CMB radiation has provided the first solid evidence for inflation and is consistent with its first two basic predictions, spatial flatness and almost scale-invariant, adiabatic density perturbations. Additional measurements of CMB anisotropy will sharpen these tests by a factor of 30 or so, for example by determining Ω_0 and n to 0.1% precision. The biggest test involves detecting the signature of the gravitational waves produced by inflation. Unfortunately, there is no firm prediction for the amplitude (though it has been argued that the contribution of gravity waves to the quadruple anisotropy of the CMB is at least 10^{-3} that of density perturbations; Hoffman & Turner 2001). However, if detected they would not only confirm the third basic prediction of inflation, but also immediately reveal the time when inflation took place. There are two basic ways of getting at the gravity waves: direct detection and through their imprint on CMB anisotropy, especially their polarization signature. Neither the LIGO nor *LISA* detectors will have sufficient sensitivity to reach the level predicted by standard inflationary models (Turner 1997)—but maybe we will be pleasantly surprised. For now, CMB anisotropy offers the most promise, especially the possibility of a satellite to follow *WMAP/Planck* dedicated to CMB polarization. Knox & Song (2002) estimate that the gravity wave polarization signature can be detected if its contribution to the CMB quadrupole is at least 5×10^{-4} that of density perturbations.

How did the Universe begin?

When inflation was proposed over 20 years ago it appeared revolutionary, solving the vexing horizon, flatness, and monopole problems, as well as explaining the origin of the perturbations that seeded all structure in the Universe. With the advent of superstring/M-theory, which holds the promise of unifying the forces and particles as well as a quantum description of gravity, expectations have risen. In particular, inflation does nothing to address the problem of the initial singularity of Big Bang cosmology. Addressing the Big Bang singularity is critical to both having a complete model of cosmology as well as getting at how the Universe began. There have been some small steps toward tackling this tall issue, such as the brane world scenarios now being studied (see, e.g., Lykken 2001; Rubakov 2001; or Steinhardt & Turok 2002).

Does general relativity provide an accurate description of the Universe?

One of the side benefits of a set of interlocking and precision cosmological measurements is that the consistency of the underlying assumptions can be tested. There is no bigger assumption in cosmology than that of the validity of general relativity to describe the Universe across billions of years in time and billions of light years in space. Over the next decade, precision measurements will create many opportunities to test general relativity in completely new ways.

The baryon density is an example of things to come. We now have two precision measurements: the first based on the abundance of deuterium and nuclear physics when the Universe was seconds old ($\Omega_b h^2 = 0.020 \pm 0.001$), and the second based upon the anisotropy of the CMB and gravitational physics when the Universe was 400,000 years old ($\Omega_b h^2 = 0.022^{+0.004}_{-0.003}$). Both measurements assume the validity of general relativity and its prediction for the expansion rate of the Universe. The agreement between these two numbers test general relativity in a completely new regime. We can expect more such tests of gravity theory as well as other physics.

6.4   Astrophysical Cosmology

While the connections between the largest and the smallest scales in Nature largely involve the underlying cosmological model and the earliest moments of creation, cosmology is more than simply determining and understanding the cosmological model. It also encompasses reconstructing the evolution that got us here and addressing the future. Once the cosmological model is fixed, the study of evolution, be it of galaxies or baryons, becomes simpler. Just a simple example; at fixed redshift, with fixed H_0, the age of a flat ($\Omega_\Lambda = 0.7$) Universe and an Einstein-de Sitter universe differ by almost a factor of 2. The emergence of a consistent set of accurate ($\pm 10\%$) cosmological parameters has greatly simplified the astrophysical cosmology problem by fixing the cosmology.

An important aspect of inflation is that it provides not only a means of resolving a number of cosmological issues, but also a natural framework for understanding the growth of structure in the Universe: the CDM paradigm. Numerical simulations, currently with billions of particles, yield power spectra for the large-scale distribution of galaxies that are quantitatively consistent with the SDSS and 2dF galaxy survey data (see, e.g., Dodelson et al. 2001; Tegmark, Hamilton, & Xu 2002). In these models, galaxy formation proceeds hierarchically, with baryons and CDM assembling into galaxies, and over time, the great walls and clusters and filaments we observe today. Deep *HST* observations, revealing the changing morphology of galaxies at different redshifts, and detailed kinematics and chemical abundance analyses of stars within the halo of the Milky Way are consistent with a picture of galaxies undergoing significant evolution, and with frequent mergers and larger galaxies accreting smaller galaxies over time.

These simulations provide an excellent match for a flat universe in which most of the mass is cold and dark, and which today is dominated by a dark energy component; see Figure 6.7. More complicated than the dark matter particles, for which only gravity is important, is the incorporation of the additional astrophysics governing the baryons: gas dynamics, the formation of molecules, their destruction and cooling, radiative transfer, star formation, heating, and feedback. Models are becoming increasingly sophisticated in their treatment of these processes—and this must continue—because current and future observations are, and will continue to be, critical in providing constraints on these models.

At the moment, we know little empirically about the epoch between a redshift of about 1100, the surface of last scattering of the CMB, and a redshift of just greater than 6, where the most distant SDSS quasars have been discovered. Astrophysically speaking, much happened between then and now and a tremendous amount remains to be learned: What were the first

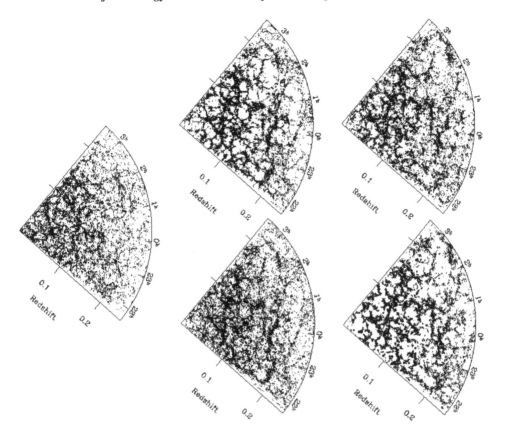

Fig. 6.7. The large-scale distribution of galaxies determined from the 2dF Galaxy Redshift Survey and mock catalogs from numerical simulations of CDM. (Figure courtesy of the 2dF collaboration.)

stars to form and in what systems did they form? How did they affect their environment and the subsequent generation of stars? How did the billion-solar-mass black holes that power the high-redshift quasars form? How did the chemical elements build up? How did clusters form and evolve? Where are the bulk of the baryons? What processes led to the Hubble sequence of galaxies? What was the ionization history of the Universe? The list goes on.

Whether a fuller understanding of the cosmological framework and its connections to fundamental physics comes sooner or later, there is much still to be learned about the evolution of the astrophysical Universe.

6.5   Looking Forward: A Bright Future

The questions before us are grand and cut across both physics and astronomy. But so are the tools we have to answer them. An impressive array of new ground-based telescopes, space-based missions, and accelerator and laboratory-based experiments will soon be tackling today's big questions.

Accelerators and laboratory experiments will help solve the riddle of dark matter and the origin of ordinary matter. Advances in theory promise to put the whole picture together. The precision in cosmological parameters will improve as subsequent years of *WMAP* data are accumulated, as ground-based supernova, gravitational lens, and SZ surveys are expanded, and with the construction of the Large Synoptic Survey Telescope (LSST), the launch of ESA's *Planck* mission, and the launch of the *Supernova Acceleration Probe (SNAP)* satellite.

Our knowledge of astrophysical cosmology and insight into the reionization epoch and galaxy formation will increase dramatically with the building of the Atacama Large Millimeter Array (ALMA) and giant (20 to 30-meter class) optical telescopes, and in space as observations begin to be made by the *Space Infrared Telescope Facility (SIRTF)* and when the *James Webb Space Telescope (JWST)* is launched. *Constellation X* will open a window on the first clusters to form and on black holes. Much will be learned about the epoch of reionization from the LOw Frequency ARray (LOFAR) to measure inhomogeneities in the 21 cm neutral hydrogen density distribution at high redshift ($z > 10$), and from the Square Kilometer Array (SKA) at redshifts $z < 10$.

One hundred years ago the Carnegie Institution of Washington launched a revolution in cosmology by taking astronomy high into the mountains of California. Today, that revolution continues, with ground- and space-based telescopes operating across the electromagnetic spectrum; with neutrino, cosmic ray and gravity wave detectors monitoring the heavens; and with accelerators and laboratory experiments teaching us about cosmological physics. The future of cosmology on the occasion of the Carnegie Centennial appears very bright indeed.

References

Bahcall, N., Lubin, L. M., & Dorman, V. 1995, ApJ, 447, L81

Bennett, C. L., et al. 2003, ApJS, 148, 1

Blumenthal, G. R., Faber, S. M., Primack, J. R., & Rees, M. J. 1984, Nature, 311, 517

Burles, S. M., Nollett, K. M., & Turner, M. S. 2001, ApJ, 552, L1

Dodelson, S., et al. 2001, ApJ, 572, 140

Elgarøy, O., et al. 2002, Phys. Rev. Lett., 89, 061301

Filippenko, A. V. 2004, in Carnegie Observatories Astrophysics Series, Vol. 2: Measuring and Modeling the Universe, ed. W. L. Freedman (Cambridge: Cambridge Univ. Press), in press

Freedman, W. L., et al. 2001, ApJ, 553, 47

Freedman, W. L., & Turner, M. S. 2004, Rev. Mod. Phys., 75, in press

Fukuda, Y., et al. 1998, Phys. Rev. Lett., 81, 1562

Fukugita, M., Hogan, C. J., & Peebles, P. J. E. 1998, ApJ, 503, 518

Gorski, K., et al. 1996, ApJ, 464, L11

Herrnstein, J., Moran, J. M., Greenhill, L. J., Diamond, P., Inoue, M., Nakai, N., Miyoshi, M., Henkel, C., & Riess, A. 1999, Nature, 400, 539

Hoffman, M., & Turner, M. S. 2001, Phys. Rev. D, 64, 02350

Kinney, W., Melchiorri, A., & Riotto, A. 2001, Phys. Rev. D, 63, 023505

Kirkman, D., Tytler, D., Suzuki, N., O'Meara, J. M., & Lubin, D. 2004, ApJS, in press

Knox, L., Christensen, N., & Skordis, C. 2001, ApJ, 563, L95

Knox, L., & Song, Y.-S. 2002, Phys. Rev. Lett., 89, 011303

Kochanek, C. S. 2002, ApJ, 578, 25

Kochanek, C. S., & Schechter, P. L. 2004, in Carnegie Observatories Astrophysics Series, Vol. 2: Measuring and Modeling the Universe, ed. W. L. Freedman (Cambridge: Cambridge Univ. Press), in press

Kolb, E. W., & Turner, M. S. 1990, The Early Universe (Redwood City, CA: Addison-Wesley)

Krauss, L. M., & Chaboyer, B. 2003, Science, 299, 65

Lahav, O., et al. 2002, MNRAS, 333, 961

Lewis, A., & Bridle, S. 2002, Phys. Rev. D, 66, 103511

Lykken, J. 2001, in Boulder 2000, Flavor Physics for the Millennium, ed. J. Rosner (Singapore: World Scientific), 827

Maoz, E., Newman, J. A., Ferrarese, L., Stetson, P. B., Zepf, S. E., Davis, M., Freedman, W. L., & Madore, B. F. 1999, Nature, 401, 351

Mather, J. C., et al. 1999, ApJ, 512, 511

McDonald, P., et al. 2001, ApJ, 562, 52

National Research Council (NRC) 2003, Connecting Quarks to the Cosmos (Washington, DC: National Academies Press)

Newman, J. A., Ferrarese, L., Stetson, P. B., Maoz, E., Zepf, S. E., Davis, M., Freedman, W. L., & Madore, B. F. 2001, ApJ, 553, 562

O'Meara, J. M., Tytler, D., Kirkman, D., Suzuki, N., Prochaska, J. X., Lubin, D., & Wolfe, A. M. 2001, ApJ, 552, 718

Oswald, T. D., Smith, J. A., Wood, M. A., & Hintzen, P. 1996, Nature, 382, 692

Page, L. 2004, in Carnegie Observatories Astrophysics Series, Vol. 2: Measuring and Modeling the Universe, ed. W. L. Freedman (Cambridge: Cambridge Univ. Press), in press

Peebles, P. J. E. 1993, Principles of Physical Cosmology (Princeton: Princeton Univ. Press)

Perlmutter, S., et al. 1999a, ApJ, 517, 565

Perlmutter, S., Turner, M. S., & White, M. 1999b, Phys. Rev. Lett., 83, 670

Primack, J. R. 2002, in Proc. 5th International UCLA Symp. on Sources and Detection of Dark Matter, ed. D. Cline (astro-ph/0205391)

Reese, E. D. 2004, in Carnegie Observatories Astrophysics Series, Vol. 2: Measuring and Modeling the Universe, ed. W. L. Freedman (Cambridge: Cambridge Univ. Press), in press

Riess, A. G., et al. 1998, AJ, 116, 1009

Rubakov, V. A. 2001, Phys. Usp., 44, 871

Sadoulet, B. 1999, Rev. Mod. Phys., 71, S197

Schechter, P. L. 2001, in Gravitational Lensing: Recent Progress and Future Goals, ed. T. G. Brainerd & C. S. Kochanek (San Francisco: ASP), 427

Sievers, J. L., et al. 2003, ApJ, 591, 599

Smoot, G. F., et al. 1992, ApJ, 396, L1

Steinhardt, P. J., & Turok, N. 2002, Science, 296, 1436

Tegmark, M., et al. 2004, ApJ, submitted (astro-ph/0310725)

Tegmark, M., Hamilton, A. J. S., & Xu, Y. 2002, MNRAS, 335, 887

Tonry, J., et al. 2003, ApJ, 594, 1

Turner, M. S. 1997a, in Generation of Cosmological Large-scale Structure, ed. D. N. Schramm & P. Galeotti (Dordrecht: Kluwer), 153

——. 1997b, Phys. Rev. D, 55, R435

——. 2000, in Type Ia Supernovae: Theory and Cosmology, ed. J. C. Niemeyer and J. W. Truran (Cambridge: Cambridge Univ. Press), 101

——. 2002, ApJ, 576, L101

Weinberg, S. 1972, Gravitation and Cosmology (New York: S. Wiley)

6.6　Addendum: *WMAP*

About two months after the Carnegie Symposium, the *Microwave Anisotropy Probe* team presented the results of the first year of data from their all-sky, five-channel CMB anisotropy measurements.* Details about *WMAP* are reported more extensively in by Bennett et al. (2003) and Page (2004); we discuss the *WMAP* results here briefly in the context of this review.

The hallmarks of *WMAP* are all-sky coverage and control of systematics (at present, to better than 0.5%). The *WMAP* angular power spectrum is consistent with the composite compiled from previous experiments (BOOMERanG, DASI, VSA, CBI, ACBAR, and *COBE*), but with systematic uncertainties in the absolute temperature scale and much less sky coverage. The panoply of cosmological parameters determined by *WMAP* are consistent with previous determinations, but with smaller error bars. For example, $\Omega_0 = 1.02 \pm 0.02$,

---

* At the same time, NASA announced that the satellite had been renamed the *Wilkinson Microwave Anisotropy Probe* or *WMAP*, in honor of David Wilkinson, who spent his entire career designing and carrying out experiments to study the CMB, including both *COBE* and *MAP*, as well as training the majority of scientists working on the CMB.

$\Omega_b h^2 = 0.0224 \pm 0.0009$, $n = 0.93 \pm 0.03$, and age $t_0 = 13.7 \pm 0.2\,\text{Gyr}$ (assuming a flat universe and $w = -1$). The consistency speaks volumes about the maturing of cosmology and the robustness of the basic model, and the diminishing error bar and control over systematics bodes well for the future with *Planck* and other experiments on the horizon.

There were two new results from *WMAP*: (1) the ruling out of the first model of inflation, Linde's original model of chaotic inflation, $V(\phi) = \lambda\phi^4$; and (2) the first evidence for the beginning of the reionization of the Universe, at a redshift $z = 20 \pm 10$, in the cross correlation between polarization and temperature anisotropy. The former illustrates the promise of future CMB anisotropy measurements to test inflation and inform us about the underlying physics. The latter result, consistent with what was expected for CDM, combined with the observation of the Gunn-Petersen trough in $z > 6$ quasars implies that the epoch of reionization began (ionized fraction of order 50% or so) around $z \approx 20$ and was complete (ionization fraction of 99.999%) by around $z \approx 6$.

As several cosmologists put it, the most surprising result from *WMAP* was the absence of a surprising result. This is further testimony to the fact that cosmology has reached a level of maturity where the basic features and parameters of the underlying model are supported by several measurements, each with different systematics and a reliable error estimate. However, this era of boring consistency probably will not last for long. Precision measurements raise the stakes as the opportunities for discrepancies increase. In the *WMAP* data there are some intriguing deviations: Both the quadrupole and octupole anisotropies are lower than expected (though the statistical significance is only about 2σ) and at higher values of ℓ there are low points and high points. With precision data, cosmological theories will be increasingly in danger!

7

The extragalactic distance scale

JOSEPH B. JENSEN,[1] JOHN L. TONRY,[2] and JOHN P. BLAKESLEE[3]
(1) Gemini Observatory
(2) University of Hawaii Institute for Astronomy
(3) John Hopkins University

Abstract

Significant progress has been made during the last 10 years toward resolving the debate over the expansion rate of the Universe. The current value of the Hubble parameter, H_0, is now arguably known with an accuracy of 10%, largely due to the tremendous increase in the number of galaxies in which Cepheid variable stars have been discovered. Increasingly accurate secondary distance indicators, many calibrated using Cepheids, now provide largely concordant measurements of H_0 well out into the Hubble flow, and deviations from the smooth Hubble flow allow us to better measure the dynamical structure of the local Universe. The change in the Hubble parameter with redshift provided the first direct evidence for acceleration and "dark energy" in the Universe. Extragalactic distance measurements are central to determining the size, age, composition, and fate of the Universe. We discuss remaining systematic uncertainties, particularly related to the Cepheid calibration, and identify where improvements are likely to be made in the next few years.

7.1   Introduction

The measurement of distances to the "nebulae" early in the twentieth century revolutionized our understanding of the scale of the Universe and provided the first evidence for universal expansion (an overview of the history of cosmology can be found in Longair (this volume). Distance measurements have played a profound role in unraveling the nature of the Universe and the objects in it ever since. Without knowing how far away objects are, we would not be able to learn very much about their sizes, energy sources, or masses. On a universal scale, extragalactic distance measurements lie at the heart of our understanding of the size, age, composition, evolution, and future of the Universe. The value of the Hubble parameter today, H_0, sets the scale of the Universe in space and time, and measuring H_0 depends heavily on accurate extragalactic distance measurements out to hundreds of Mpc. Accurate distance measurements are also needed to map the deviations from the smooth Hubble expansion, or peculiar velocities, which presumably arise gravitationally due to the distribution of mass in the local Universe.

During the last decade several important steps have been taken toward resolving the factor of ~ 2 uncertainty in the scale of the Universe that overshadowed observational cosmology for 30 years. Many of the improvements are the direct results of a better calibration and extension of the Cepheid variable star distance scale. While significant systematic uncertainties remain, Hubble constant measurements made using a wide variety of distance measurement techniques are now converging on values between 60 and 85 $\text{km s}^{-1} \text{Mpc}^{-1}$; very few measurements lie outside this range, with the majority falling between 70 and 75 km s^{-1} Mpc^{-1}. Several recent advances made the improved Cepheid calibration possible. First, the *Hipparcos* satellite made parallax distance measurements accurate to 10% to Cepheid variable stars in the solar neighborhood. *Hipparcos* parallax measurements helped pin down the zeropoint brightness of the Cepheid variable stars, but not enough Cepheids could be observed to properly determine the slope of the period-luminosity (PL) relation. Second, the *Hubble Space Telescope (HST)* provided the spatial resolution and sensitivity to detect Cepheids in galaxies as far away as 25 Mpc, allowing for the first time a calibration of a number of secondary extragalactic distance indicators using Cepheids. Finally, the OGLE microlensing experiment turned up thousands of Cepheids in the Large Magellanic Cloud (LMC) that were used to accurately determine the PL relation. The improved Cepheid calibration of secondary distance indicators has largely yielded concordance on the scale of the Universe, and the uncertainty in H_0 is now arguably close to 10%.

This summary is not intended to be an inclusive review of all the distance measurement techniques and their relative strengths and weaknesses. Instead, we highlight some recent measurements and identify the most significant remaining systematic uncertainties. Cepheids are emphasized, as they are currently used to calibrate most secondary distance indicators. We close by summarizing how planned future projects will improve our knowledge of the expansion rate and eventual fate of the Universe.

7.2   The Cepheid Calibration

The Hubble Constant Key Project (KP) team set out to determine the Hubble constant to 10% or better by reliably measuring Cepheid distances to galaxies reaching distances of $\sim$25 Mpc (Mould et al. 2000; Freedman et al. 2001). The KP sample galaxies included field and cluster spirals, including several in the nearby Virgo, Fornax, and Leo I clusters. The KP team performed V and I band photometry using two independent reduction packages and analysis techniques to understand and control systematic errors as much as possible. Near-IR measurements with NICMOS were used to check the validity of the reddening law adopted by the KP team (Macri et al. 2001). All the KP results have been presented using a distance modulus to the LMC of 18.50 mag. To keep the KP results on a common footing, the KP measurements were all reported using the PL relation determined by Freedman & Madore (1990), which was derived from a relatively limited set of LMC Cepheids. No adjustment to the PL relation was made to the baseline KP measurements for differences in metallicity between Cepheids (Ferrarese et al. 2000b).

The KP team was not alone in taking advantage of *HST*'s spatial resolution and sensitivity to find Cepheids in relatively distant galaxies. Cepheids have also been measured in supernova (SN) host galaxies by A. Sandage, A. Saha, and collaborators. Their observations targeted galaxies in which Type Ia SNe have occurred for the purpose of calibrating SNe as a standard candle. The Sandage and Saha team made use of similar data reduction techniques as the KP team and the same LMC distance. An additional Cepheid measurement in the Leo I galaxy NGC 3368 was made by Tanvir, Ferguson, & Shanks (1999); NGC 3368 later hosted SN 1998bu.

Data for the entire combined sample of 31 Cepheid galaxies from both teams was presented by Ferrarese et al. (2000b) and Freedman et al. (2001) to facilitate comparison and calibration of secondary distance indicators on a common Cepheid foundation (Ferrarese et al. 2000a,b; Gibson et al. 2000; Kelson et al. 2000; Sakai et al. 2000). Mould et al. (2000)

combined these results and found $H_0 = 71 \pm 6$ km s^{-1} Mpc^{-1}. The results of the KP calibration of the SN, Tully-Fisher (TF), fundamental plane (FP), and surface brightness fluctuation (SBF) distance indicators are included in the subsequent sections.

The gravitational lensing data of the OGLE experiment increased the number of good Cepheid measurements in the LMC by more than an order of magnitude ($\sim$650 fundamental-mode Cepheids; Udalski et al. 1999a,b). Freedman et al. (2001) applied the new Cepheid PL relation to the combined Cepheid database. The same LMC distance modulus of $\mu_{LMC} =$ 18.50 mag used in the earlier KP papers was maintained by Freedman et al. They also argued for a modest metallicity correction of -0.2 ± 0.2 mag dex^{-1} in (O/H) (Kennicutt et al. 1998). It is interesting to note that the new greatly improved PL relation has a somewhat different slope in the *I* band, resulting in a distance-dependent offset. Only the brightest, longest-period Cepheids can be detected in the most distant galaxies, so the revised slope will have the largest effect in the most distant galaxies. Adopting the new PL relation reduces the distance moduli of the most distant galaxies by up to $\sim$0.2 mag. The metallicity correction counteracts the shorter distances to some extent, and the change in the resulting Hubble constant when adopting both the metallicity correction and the new PL relation is small (72 ± 8 km s^{-1} Mpc^{-1}). If the new PL relation is adopted without the metallicity correction, the Hubble constant would increase by a few percent (depending on which Cepheid galaxies are used to calibrate a particular secondary distance indicator). An recent independent survey of LMC Cepheids has confirmed the slope of the OGLE PL relation (Sebo et al. 2002). The new results, derived from Cepheids with periods comparable to those of the more-distant KP galaxies, agrees with the Udalski et al. (1999a,b) OGLE PL zeropoint to 0.04 mag (or 2% in H_0), well within the uncertainties of the two measurements.

In the following sections, all Cepheid-based distance measurements will be compared to the Ferrarese et al. (2000b) scale, using the original KP zeropoint and no metallicity correction, or to the Freedman et al. (2001) compilation, which uses the OGLE ("new") PL relation and metallicity correction of -0.2 mag dex^{-1}. In all cases, the LMC distance adopted is 18.50 mag. The metallicity-corrected Freedman et al. (2001) and uncorrected Ferrarese et al. (2000b) KP compilations are not strictly comparable; the metallicity difference between Galactic and LMC Cepheids would result in a $\sim$0.08 mag difference in the distance modulus to the LMC.

7.3   Secondary Distance Indicators and the Hubble Constant

Most secondary indicators derive their zeropoint calibration from Cepheids. We focus here on those techniques that have been calibrated using the common foundation of the KP Cepheid measurements. A few new results that are independent of the Cepheid calibration are presented as well (Table 7.1).

7.3.1   *Type Ia Supernovae*

The brightness of exploding white dwarf SNe can be calibrated using a single parameter (Phillips 1993; Hamuy et al. 1995; Riess, Press, & Kirshner 1996). After correcting their luminosities for decline rates, Type Ia SNe are a very good standard candle with a variance of about 10%. Both the KP and Sandage and Saha teams have calibrated Ia SNe using *HST* Cepheid measurements. While many of the Cepheids observations are identical, the two teams make numerous different choices regarding the detailed analyses. They also make use of different historical SNe to compute the value of H_0. The Sandage and Saha team consistently get larger distances and smaller values of the Hubble constant than the KP team does. The differences are discussed in detail by Parodi et al. (2000) and Gibson et al. (2000). Parodi et al. find $H_0 = 58 \pm 6$ km s^{-1} Mpc^{-1}, while Gibson et al. report $H_0 = 68 \pm 2 \pm 5$ km s^{-1} Mpc^{-1} (the first uncertainty is statistical, the second systematic). Both teams used the same LMC distance and similar reduction software, but included different calibrators and different analysis techniques. Much of the disagreement between the two groups has its origin in the selection and analysis of the individual Cepheid variables and which to exclude. The remaining difference arises from the choice of which historical SN data to trust and which SNe to exclude from the fit to the distant Hubble flow. Hamuy et al. (1996) measured a value of $H_0 = 63 \pm 3 \pm 3$ km s^{-1} Mpc^{-1} using the 30 Ia SNe of the Calan/Tololo survey and four Cepheid calibrators. Ajhar et al. (2001) found that the optical I-band SBF distances to galaxies with Type Ia SNe were entirely consistent when differences between Cepheid calibrators were taken into account. They found that SBFs and SNe give identical values of H_0 [73 km s^{-1} Mpc^{-1} on the original KP system, 75 on the new Freedman et al. (2001) calibration, and 64 on the Sandage and Saha calibration].

SNe at high redshift ($z > 0.5$), and their departure from a linear Hubble velocity, have been used to explore the change in the universal expansion rate with time. The measurements of two collaborations (the High-z Team, led by B. Schmidt, and the Supernova Cosmology Project, led by S. Perlmutter) have provided the best evidence to date that the Universe is expanding at an increasing rate. The implication of the SN data is that $\sim 70\%$ of the energy density in the Universe is in some form of "dark energy" such as vacuum energy, "quintessence," or something even more bizarre (Perlmutter et al. 1997, 1998; Schmidt et al. 1998). The use of SNe to probe the equation of state of the Universe is the topic of other papers in this volume. In general, the distant SNe do not need to be put on an absolute distance scale to study the change in the Hubble parameter with time.

Kim et al. (1997) used the first few SNe discovered by the Supernova Cosmology Project to constrain the Hubble constant. They found that $H_0 < 82$ km s^{-1} Mpc^{-1} in an $\Omega_\Lambda = 0.7, \Omega_m = 0.3$ Universe (as suggested by the distant SN data and cosmic microwave background measurements of the flatness of the Universe).

Tonry et al. (2003, in preparation) have recently compiled a database of all the currently available Type Ia SNe data using a common calibration and consistent analysis techniques. The result is a uniform data set of 209 well-measured SN distances in units of km s^{-1}. An independent distance to any of the galaxies therefore leads immediately to a tie to the Hubble flow out to redshifts greater than one. Six galaxies from the Tonry et al. (2003) database have Cepheid distances determined by the KP team (Freedman et al. 2001). Using the new PL relation and metallicity corrections, the SN data give a Hubble constant of 74 ± 3 km s^{-1} Mpc^{-1}. The exquisite tie to the Hubble flow is shown in Figure 7.1, along with the best fit Hubble constant. The line indicates the evolution of H_0 for an empty ($q_0 = 0$) cosmology, and the deviation from that line at $z \approx 0.5$ to 1 is the best evidence for an accelerating "dark energy" dominated Universe.

%includegraphics[height=85mm,width=115mm]combined.eps

The distant SN data more tightly constrain the product $H_0 t_0$ than H_0 alone. The long, thin error ellipses of the SN data fall along lines of constant $H_0 t_0$ for SNe at $z \approx 0.5$. The Supernova Cosmology Project found $H_0 t_0 = 0.93 \pm 0.06$ (Perlmutter et al. 1997), and the High-z team measured $H_0 t_0 = 0.95 \pm 0.04$ (Tonry et al. 2003). For an age of the Universe of 13 Gyr, this implies a Hubble constant of ~ 70 km s^{-1} Mpc^{-1}. In the future, other constraints

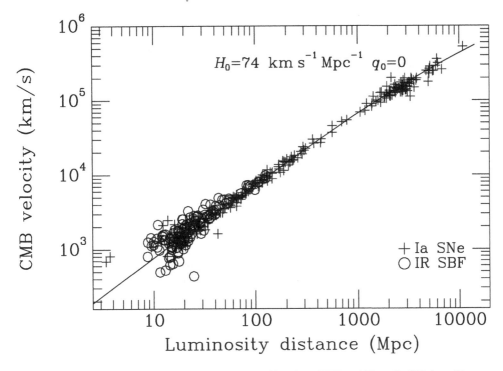

Fig. 7.1. Hubble diagram derived from the combined IR SBF and Type Ia SN data (Jensen et al. 2003; Tonry et al. 2003). The line indicates the expansion velocity for an empty Universe. The fit to the SN data give a Hubble constant of 74 km s^{-1} Mpc^{-1}.

on the age of the Universe, combined with highly accurate values of $H_0 t_0$ from high-redshift SNe at different redshifts, will give us better constraints on the Hubble parameter and how it has changed with time.

7.3.2 Surface Brightness Fluctuations

The amplitude of luminosity fluctuations in dynamically hot systems arises due to statistical fluctuations in the number of stars per resolution element (Tonry & Schneider 1988; Blakeslee, Ajhar, & Tonry 1999). SBFs are distance-dependent: the nearer a galaxy is, the bumpier it appears. The brightness of the fluctuations depends directly on the properties of the brightest stars in a given population, making SBFs a stellar standard candle. Significant SBF surveys have been completed at I, where the effects of age and metallicity are degenerate, and in the near-IR, where fluctuations are brightest and extinction is minimized.

The Hubble constant derived from I-band SBFs, as calibrated by the KP team (Ferrarese et al. 2000a), is $H_0 = 70 \pm 5 \pm 6$ km s^{-1} Mpc^{-1} (random and systematic uncertainties, respectively), using the four distant *HST* measurements of Lauer et al. (1998). The I-band SBF team used the much larger sample of $\sim$300 galaxies and a slightly different calibration to find a somewhat larger Hubble constant of $H_0 = 77 \pm 4 \pm 7$ (Tonry et al. 2000, 2001), using the original KP calibration of Ferrarese et al. (2000b). The I-band SBF survey team fitted a detailed model of the velocity field of the local Universe to get their determination of the Hubble constant. Half the difference between the Ferrarese et al. (2000a) and Tonry et al. (2000) results is due to the velocity field corrections, and the other to differences in the choice of Cepheid calibration galaxies. SBFs can be measured in the bulges of a few spiral galaxies with known Cepheid distances, and the preferred SBF calibration of Tonry et al. uses only galaxies with distances known from both Cepheids and SBFs. As with all the secondary indicators calibrated using Cepheids, moving to the new PL relation would result in an increase in the Hubble constant of 3% including the Freedman et al. (2001) metallicity correction, or 8% using the new PL relation alone.

Infrared measurements using NICMOS on the *HST* have extended SBF measurements beyond 100 Mpc, where deviations from the smooth Hubble flow should be small. Jensen et al. (2001) measured a Hubble constant between 72 and 77 km s^{-1} Mpc^{-1} using the original KP Cepheid calibration. When reanalyzed using an updated calibration and the new OGLE PL relation, they find $H_0 = 77$ km s^{-1} Mpc^{-1} using the Freedman et al. (2001) calibration (including the metallicity correction; Jensen et al. 2003).

7.3.3   *Fundamental Plane*

Elliptical galaxies are very homogeneous in their photometric and dynamical properties. By accurately measuring surface brightness, size, and central velocity dispersion, the position of an elliptical galaxy on the "fundamental plane" (FP) gives an estimate of the distance with an accuracy of $\sim$20%. The FP is an improved version of the Faber-Jackson and $D_N - \sigma$ relations, which are also used to determine distances to elliptical galaxies. Kelson et al. (2000) combined various FP data for the Fornax, Virgo, and Leo I clusters, for which Cepheid distances had been measured. They applied the Cepheid calibration of the FP relation for the three nearby clusters to 11 more distant clusters. The resulting Hubble constant of $82 \pm 5 \pm 10$ km s^{-1} Mpc^{-1} is somewhat higher than the KP estimates using other secondary distance indicators. Adopting the metallicity correction of Kennicutt et al. (1998) would reduce the value of H_0 by 6%. The relative placement of the spiral Cepheid calibrators and elliptical FP galaxies within the three nearby clusters is one of the primary sources of systematic uncertainty ($\sim$5%).

Hudson et al. (2001) have combined a number of FP data sets, making them much more homogeneous by cross checking the photometry and velocity dispersion measurements. When the Hudson et al. data are analyzed using the Ferrarese et al. (2000b) Cepheid calibration, Blakeslee et al. (2002) find that the Hubble constant is consistent with SBFs and with other secondary distance indicators. They find $H_0 = 68 \pm 3$ km s^{-1} Mpc^{-1}, in excellent agreement with the KP calibrations of the other secondary distance indicators. Using the new PL relation (Freedman et al. 2001), the FP Hubble constant is $73 \pm 4 \pm 11$ km s^{-1} Mpc^{-1}, where SBFs have been used to make a direct connection between the FP galaxies and the Cepheid calibrators.

7.3.4   *Tully-Fisher*

Like elliptical galaxies, the photometric and kinematic properties of spiral galaxies are closely related. The rotation velocities and brightnesses of spiral galaxies can be measured, making it possible to estimate the distance to a galaxy. Tully-Fisher (TF) distance measurements are among the most widely used, although the accuracy of an individual measurement is generally taken to be about 20%. Sakai et al. (2000) used the compiled TF data for 21 calibrators with Cepheid distances and applied the results to a large data set of 23 clusters within 10,000 km s^{-1} (Giovanelli et al. 1997). Sakai et al. found that the TF Hubble constant is $H_0 = 71 \pm 4 \pm 7$ km s^{-1} Mpc^{-1}, in very good agreement with the other techniques already mentioned.

7.3.5   Type II Supernovae

The expanding-photospheres method of determining distances to Type II SNe can be calibrated using a zeropoint based on Cepheid distances (although the expanding-photospheres method is a primary distance indicator that does not require a Cepheid calibration, as described in § 7.6). The KP team (Freedman et al. 2001) applied the new Cepheid calibration to the SN measurements of Schmidt et al. (1994) and found $H_0 = 72 \pm 9 \pm 7$ km s^{-1} Mpc^{-1}, in close agreement with the value of $H_0 = 73 \pm 6 \pm 7$ reported by Schmidt et al. A second way of using Type II SNe as a distance estimator has been developed by Hamuy (2001). The expansion velocity for a particular type of "plateau" SN is correlated with its luminosity. The average of four SNe give a Hubble constant of 75 ± 7 km s^{-1} Mpc^{-1} using a Cepheid calibration comparable to the Freedman et al. zeropoint (M. Hamuy, private communication). The best-measured and only modern SN of the four (SN 1999em) gives $H_0 = 66 \pm 12$ km s^{-1} Mpc^{-1}.

7.3.6   Other Distance Indicators Calibrated Using Cepheids

The KP team also presented Cepheid calibrations of several other distance indicators, including the globular cluster luminosity function (GCLF), the planetary nebula luminosity function (PNLF), and the tip of the red giant branch (TRGB). Ferrarese et al. (2000a,b) compiled results for these techniques and compared them to SBFs.

The GCLF technique has been recently shown to be as good a distance indicator as other secondary techniques when appropriate corrections are made for completeness, background sources, and luminosity function width (Kundu & Whitmore 2001; Okon & Harris 2002). To measure reliable distances, globular clusters fainter than the GCLF peak must be detected. Some earlier measurements did not go deep enough to reach the peak luminosity, and the results were less reliable and possibly biased toward smaller distances. By measuring GCLF distances relative to the Virgo cluster, and adopting a distance to Virgo of 16 Mpc as the calibration (which is independent of, but consistent with, the Cepheid calibration), the resulting Hubble constant is near 70 km s^{-1} Mpc^{-1} (Okon & Harris 2002). The Cepheid distance to Virgo from Ferrarese et al. (2000a) is 16.1 Mpc; thus, the agreement between the KP calibration of the GCLF technique and the newer measurements is very good.

Ciardullo et al. (2002) recently made a detailed comparison of the PNLF technique to the Cepheid distance scale. The current PNLF technique makes a correction for metallicity of the host galaxy. Using a distance of 710 kpc to M31 to determine the zeropoint for the PNLF method, they found that the Cepheid and PNLF distances are consistent within the statistical uncertainties of the two methods. The agreement between Cepheids and PNLF in the particular case of NGC 4258 also leads to a ~ 1-σ disagreement with the geometrical maser distance.

In Table 7.1 we summarize a few recent measurements of H_0 and the Cepheid calibration used (when appropriate). Most of the measurements are calibrated directly using the original KP zeropoint or the new OGLE PL relation ("New PL"). Many also include metallicity corrections (indicated by "+Z"). References and additional calibration details are included. Willick & Batra (2001) provide an independent calibration of the Cepheid distance scale using the new OGLE PL relation.

Table 7.1. *A Few Recently Published Hubble Constant Measurements*

Technique	H_0 (km s^{-1} Mpc^{-1})	Cepheid calibration	Reference
Key Project summary	72 ± 8	New PL+Z	Freedman et al. 2001
Cepheids+IRAS flows	85 ± 5	New PL	Willick & Batra 2001
Type Ia Supernovae	59 ± 6	Sandage team	Parodi et al. 2001
	59 ± 6	Sandage team	Saha et al. 2001
	$71 \pm 2 \pm 6$	New PL+Z	Freedman et al. 2001
	74 ± 3	New PL+Z	Tonry et al. 2003, in prep.
	$73 \pm 2 \pm 7$	New PL+Z	Gibson & Stetson 2001
I-band SBFs	$77 \pm 4 \pm 7$	Orig. KP	Tonry et al. 2000
	$70 \pm 5 \pm 6$	New PL+Z	Freedman et al. 2001
	75	New PL+Z	Ajhar et al. 2001
H-band SBFs	$72 \pm 2 \pm 6$	Orig. KP+*I*-SBF	Jensen et al. 2001
	$77 \pm 3 \pm 6$	New PL+Z	Jensen et al. 2003, in prep.
K-band SBFs	71 ± 8	Orig. KP+*I*-SBF	Liu & Graham 2001
Tully-Fisher	$71 \pm 3 \pm 7$	New PL+Z	Freedman et al. 2001
Fundamental Plane	$82 \pm 6 \pm 9$	New PL+Z	Freedman et al. 2001
	$73 \pm 4 \pm 11$	New PL+Z	Blakeslee et al. 2002
Type II Supernovae	$72 \pm 9 \pm 7$	New PL+Z	Freedman et al. 2001
	75 ± 7	New PL+Z	M. Hamuy, private comm.
Globular Custer LF	~ 70	similar to Orig. KP	Okon & Harris 2002
Sunyaev-Zel'dovich	$60 \pm 3 \pm 30\%$	...	Carlstrom et al. 2002
Gravitational lenses	61 to 65	...	Fassnacht et al. 2002
	$59 \pm 12 \pm 3$	...	Treu & Koopmans 2002
Type Ia SNe (theory)	67 ± 9	...	Höflich & Khokhlov, 1996
Type II SNe (theory)	67 ± 9	...	Hamuy 2001

7.4   Systematic Uncertainties in the Cepheid Calibration

The KP and SN calibration teams have provided a large and uniform data set of consistently calibrated Cepheid distances. There are, however, several systematic uncertainties that prevent achieving an accuracy much better than 10% in distance. The primary systematic uncertainties are common to all the Cepheid measurements, and are all similar in magnitude (Freedman et al. 2001). A concerted effort to improve the accuracy in several areas is therefore needed to significantly reduce the uncertainty in H_0 using Cepheid-calibrated secondary distance indicators.

7.4.1   The Distance to the Large Magellanic Cloud

The distance to the LMC is a fundamental rung in the distance ladder. The LMC is large enough and distant enough to contain a wide assortment of stellar types at nearly the same distance, yet close enough for individual stars to be easily resolved. The LMC contains stars and globular clusters spanning a wide range in age. It also hosted the Type II SN 1987A, the best-studied SN ever. The LMC is crucial for the Cepheid calibration because there are not enough Galactic Cepheids with independently determined distances to pin down the zeropoint, the PL relation, and metallicity dependence simultaneously. Only in the LMC do we have a sample of thousands of Cepheids at a common distance.

The fact that the distance to the LMC is not yet well determined is a significant and persistent problem (Walker 1999; Paczyński 2001; Benedict et al. 2002a). There are, unfortunately, still "long" and "short" LMC distance scales. While everyone would probably agree that a distance modulus of $\mu_{LMC} = 18.35$ mag is consistent with the "short scale," and

18.6 mag corresponds to the "long" scale, the distinction between the two is somewhat artificial. It is interesting to note that, while some techniques favor longer or shorter distance scales on average, the measurements cover the range with no appreciable bimodality. Furthermore, one author may state a distance or range of distances as being consistent with the long scale, while another, quoting a distance in the same range, will state that it supports the short scale. We regard the distinction as arbitrary; in reality, there is a continuous range of measurements that span values significantly larger than the statistical uncertainties of the individual measurements.

There have been several compilations of LMC distances recently (e.g., Walker 1999; Mould et al. 2000; Benedict et al. 2002a), and we do not review them here. While it is very helpful to broadly survey all the recently published measurements, it is important to remember that a large number of publications of a particular value does not necessarily indicate correctness. Nor do more recent measurement necessarily deserve more trust than older ones. There is some hope of resolving the debate if we look to some recent measurements that make use of new or improved geometrical techniques (a few are listed in Table 7.2). Hopefully, modest improvements in the near future will resolve the issue of the LMC distance, at least at the 10% level.

The study of the SN 1987A has resulted in several recent geometrical distance determinations (or upper limits), ranging from $\mu_{LMC} < 18.37$ to 18.67 mag (Gould & Uza 1997; Panagia 1999; Carretta et al. 2000; Benedict et al. 2002b). The "light echo" measurements are particularly important because of their insensitivity to adopted extinction values. A recent spectral fitting of the expanding atmosphere of SN 1987A by Mitchell et al. (2002) gives a distance of 18.5 ± 0.2 mag. The SN 1987A measurements are consistent with both $\mu_{LMC} = 18.35$ and 18.50 mag.

One of the most promising techniques today makes use of detached eclipsing binaries (DEBs), for which the orbital parameters can be determined and the geometrical distance derived. The three DEBs that have been observed in the LMC have distances in the range of 18.30 to 18.50 mag. Two measurements are consistent with the short distance scale (18.38 mag by Ribas et al. 2002; 18.30 mag by Guinan et al. 1998) and the other two are larger (18.50 mag by Fitzpatrick et al. 2002; 18.46 mag by Groenewegen & Salaris 2001). We can only conclude that DEBs are consistent with a distance modulus to the LMC of both 18.35 and 18.50 mag. Measurements of many more DEBs in the LMC and other nearby galaxies will be invaluable in helping to resolve the controversy surrounding the distance to the LMC.

Cepheid distance measurements to the LMC generally favor the long scale, although values as low as 18.29 mag (and as high as 18.72 mag) have been reported (Benedict et al. 2002a). A good summary of the Cepheid measurements was presented by Benedict et al. (2002a,b), who find a mean distance modulus of 18.53 mag (average of many measurements and techniques). This is consistent with their own measurement of 18.50 ± 0.13 mag based on new *HST* parallax measurements of δ Cephei. Recently, Keller & Wood (2002) measured $\mu_{LMC} = 18.55 \pm 0.02$ mag, and Di Benedetto (2002) reported a distance modulus of 18.59 ± 0.04 mag. Although Cepheids alone cannot yet rule out the short-scale zeropoint of 18.35 mag, the most recent measurements are weighted toward values nearer to 18.5 mag. Whether or not Cepheid luminosities depend significantly on metallicity is an open question (see the discussion in Freedman et al. 2001). The effect of applying the metallicity correction of -0.2 ± 0.2 mag dex^{-1} (O/H) metallicity would be to decrease the distance to the LMC by 0.08 mag relative to the higher-metallicity Galactic Cepheids. Further work with larger samples of Galactic Cepheids with higher metallicities than are found in the LMC is needed to resolve this issue. One promising line of research suggests that first-overtone Cepheids should be less sensitive to metallicity than the fundamental-mode pulsators. Bono et al. (2002) report overtone Cepheid distances near 18.5 mag, in good agreement with the longer-period fundamental-mode Cepheids.

NGC 4258 is the only other external galaxy besides the LMC with a reliable geometrical distance measurement. The orbital properties of masers around the central black hole in NGC 4258 can be determined to give an absolute geometrical distance of 7.2 ± 0.5 Mpc (Herrnstein et al. 1999). Cepheids have been discovered in NGC 4258, and they provide an independent distance measurement that is approximately 1 Mpc larger (Maoz et al. 1999; Newman et al. 2001); the KP Cepheid calibration (Freedman et al. 2001) is discrepant at the 1-σ level if the distance to the LMC is 18.50 mag. The two measurements would agree if $\mu_{LMC} = 18.31$ mag. If the maser and Cepheid distances are both reliable, they could be used to rule out a distance of 18.50 to the LMC. One alternative possibility is that the Cepheid distance is a bit off; given the relatively small number of Cepheids detected (18), this may not be unreasonable. A second possibility is that the Cepheid metallicity correction should have the opposite sign as that used by the KP, as suggested by the theoretical work of Caputo, Marconi, & Musella (2002). The recent results of Ciardullo et al. (2002) using the PNLF distance to this galaxy suggests that the distance to the LMC should be reduced; both Cepheids and PNLF would be consistent with the maser distance if μ_{LMC} were 18.3 mag.

While the original controversy between long and short distance scales arose primarily due to differences between RR Lyrae variables and Cepheids, the two methods are starting to converge. Statistical parallax measurements favor values between 18.2 and 18.3 mag, while Baade-Wesselink measurements prefer larger distances. The average value of many RR Lyrae measurements is 18.45 mag, and new data based on *HST* parallax distances to Galactic RR Lyrae stars give μ_{LMC} values between 18.38 and 18.53 mag (Benedict et al. 2002a,b). The RR Lyrae distances now agree with Cepheid distances at the 1-σ level.

Red clump stars have been used as a distance indicator in support of the short-distance scale (Benedict et al. 2002a). Red clump measurements span the range in distance modulus, from 18.07 to 18.59 mag (Stanek, Zaritsky, & Harris 1998; Romaniello et al. 2000). Recent near-IR measurements that minimize uncertainties in extinction result in a distance modulus of 18.49 ± 0.03 mag (Alves et al. 2002). Pietrzynski & Gieren (2002) find 18.50 ± 0.05 mag, with a statistical uncertainty of only 0.008 mag. Red clump distance measurements do not yet exclude either the long or the short distance scales.

Many other distance measurement techniques have been applied to the LMC, and we refer the reader to the compilation in Benedict et al. (2002a). Most of the measurements of μ_{LMC} published this year are consistent with a distance modulus between 18.45 and 18.55 mag, with the notable exception of the Cepheid distance to NGC 4258, which implies $\mu_{LMC} = 18.31$ mag (although uncertainties in the metallicity correction to Cepheid distances lessen the significance of the discrepancy, as explained by Caputo et al. 2002). Since the NGC 4258 Cepheid distance is only discrepant at the 1-σ level, we believe that a change from the LMC distance of 18.50 mag is not justified at the present time.

Resolving the debate between the long and short distances to the LMC will not necessarily reduce the uncertainty in the Hubble constant. The differences between techniques have usually exceeded the quoted uncertainties, both systematic and statistical. The systematic uncertainty in μ_{LMC} adopted by the KP team was 0.13 mag. Even if we choose the best re-

Table 7.2. *A Summary of Recently Published LMC Distances*

Technique	μ_{LMC} (mag)	Reference
Cepheids/masers (NGC 4258)	18.31 ± 0.11	Newman et al. 2002
Cepheids (δ Cep)	18.50 ± 0.13	Benedict et al. 2002b
Cepheids (mean of many techniques)	18.53	Benedict et al. 2002a,b
Eclipsing binaries	18.38 ± 0.08	Ribas et al. 2002
	18.50 ± 0.05	Fitzpatrick et al. 2002
	18.46 ± 0.07	Groenewegen & Salaris 2001
	18.30 ± 0.07	Guinan et al. 1998
SN 1987A	18.5 ± 0.2	Mitchell et al. 2002
RR Lyr	18.38 to 18.53	Benedict et al. 2002a
RR Lyr (mean of many techniques) .	18.45 ± 0.08	Benedict et al. 2002a,b
Red clump	18.49 ± 0.03	Alves et al. 2002
Red clump	$18.50 \pm 0.008 \pm 0.05$	Pietrzynski & Gieren 2002

sults from the different techniques that fall closest to the adopted modulus of 18.50 mag, the scatter will likely still be >0.13 mag. Furthermore, it is likely that any individual measurement, taken on its own as the most reliable available, will have a total uncertainty no better than 0.1 mag. Reducing the uncertainty in the LMC distance modulus will require more than the elimination of systematic errors that are not completely accounted for in the current set of uncertainty estimates. It is, however, a crucial step in our progress toward improving the precision of the distance ladder techniques.

7.4.2 *Metallicity Corrections to Cepheid Luminosities*

The possibility that the luminosity of a Cepheid variable star depends on its metallicity is one of the most significant remaining uncertainties. Most of the distant galaxies with known Cepheid distances have metallicities similar to those of the Galactic Cepheids, and significantly higher metallicity than the LMC Cepheids. The magnitude of the metallicity correction is not very important, provided that it is applied consistently. At the present time it is not clear if a metallicity correction is justified or not (Udalski et al. 2001; Caputo et al. 2002; Jensen et al. 2003).

Most of the Cepheid distance measurements published, including the KP papers prior to Freedman et al. (2001), have no metallicity correction applied. For the final KP results, Freedman et al. adopted a metallicity correction of -0.2 mag dex^{-1} in (O/H) (Kennicutt et al. 1998). Since both the old and new Cepheid calibrations use the same LMC distance $\mu_{\mathrm{LMC}} = 18.50$ mag, it should be noted that a direct comparison between the old calibration (Ferrarese et al. 2000b) and the Freedman et al. (2001) calibration, which includes the metallicity correction, requires an offset of ~ 0.08 mag due to the difference between Galactic and LMC Cepheid metallicities. In other words, the difference between the Mould et al. (2000) and Freedman et al. (2001) values of H_0 would be 4% larger than reported if Galactic and LMC Cepheids were metallicity-corrected the same as Cepheids in the more distant galaxies.

The Freedman et al. (2001) distances derived using the new PL calibration can be used without the metallicity correction, with a corresponding increase in H_0. Because of the distance-dependent nature of the change to the new PL relation, the Hubble constant that results from using no metallicity correction depends on which Cepheid calibrators are used to tie to distant secondary techniques. For the case of *I*-band SBFs, the new PL relation without metallicity corrections results in an increase in the Hubble constant of 8% over the original KP calibration, and 5% over the new PL relation with metallicity corrections included.

7.4.3   *Systematic Photometric Uncertainties*

One significant source of systematic error in the data taken with WFPC2 has been the uncertainty in the photometric zeropoint, which the KP team estimate at 0.09 mag. Extinction corrections also contribute to the photometric uncertainties in Cepheid measurements. More details can be found in the KP papers (see Freedman et al. 2001).

Blending of Cepheids with other stars is another potential source of systematic uncertainty. If blending is significant, the Cepheid distances to the most distant galaxies surveyed may be underestimated by 10 to 20% (Mochejska et al. 2000). Gibson, Maloney, & Sakai (2000) found no evidence of a trend in residuals with distance and no difference between the WF and PC camera measurements, which have different spatial sampling and therefore different sensitivities to blending. Ferrarese et al. (2000c) also found no significant offset in the Cepheid photometry due to crowding. They based their uncertainty estimate of 0.02 mag on tests in which they added artificial stars to their images and processed them in the same way as the real stars. The strict criteria for selecting and measuring Cepheids used by the KP team can explain the small effect of blending on the photometry of the artificial stars added.

Although improved spatial resolution helps minimize the effects of blending, it is not complete protection. Physical companions to Cepheids cannot be resolved. Furthermore, the effect of blending on Cepheid distances is not limited to a single companion. The surface brightness fluctuations in the underlying population are a background with structure on the scale of the point-spread function that can make the Cepheid look brighter or fainter. In the most distant galaxies, there will be a slight detection bias in favor of Cepheids superimposed on bright fluctuations. The fluctuations are very red, so not only will the brightness of the Cepheid be overestimated, but the color observed will be too red. The corresponding extinction correction will be larger than it should be, enhancing the overestimate of the Cepheid luminosity and underestimate of the distance.

7.5   Peculiar Velocities

Any measurement of the Hubble constant relies on both distance *and* velocity measurements. Within 50 Mpc, the clumpy distribution of mass leads to peculiar velocities that can be larger than the Hubble expansion velocity. It is therefore critical that recession velocities within 50 Mpc be corrected depending on where the galaxy lies relative to the Virgo cluster, Great Attractor, and so forth. Differences in how these corrections are applied has led to significant differences in measured values of H_0. Half the difference between the KP SBF Hubble constant and that of Tonry et al. (2000) is due to differences in the velocity model. Furthermore, there is some evidence suggesting that the local Hubble expansion rate is slightly larger than the global value, which would be the natural result of the local Universe being slightly less dense than the global average (Zehavi et al. 1998; Jensen et al. 2001). While the evidence is far from conclusive (Giovanelli, Dayle, & Haynes 1999; Lahav 2000), it reinforces the importance of measuring the expansion rate of the Universe as far out as possible. There is obviously great advantage in measuring H_0 at distances large enough to be free of peculiar velocities. Beyond 100 Mpc, even the largest peculiar velocities ($\sim$1500 km s^{-1}) are only a fraction of the Hubble velocity ($\sim$7000 km s^{-1}). The Ia SNe, SBF, Tully-Fisher, and FP Hubble constant measurements made beyond 100 Mpc presented in the previous sections all show the excellent consistency expected from a solid tie to the distant Hubble flow.

7.6   Bypassing the Distance Ladder

Two techniques, gravitational lens time delays and the Sunyaev-Zel'dovich (SZ) effect, promise to provide measurements of H_0 at significant redshifts independent of the calibration of the local distance scale. Both of these techniques are discussed in detail elsewhere in this volume (Kochanek & Schechter 2004; Reese 2004).

Time delay measurements in multiple-image gravitational lens systems can provide a geometrical distance and Hubble constant provided the mass distribution is known in the radial region between the images of the gravitationally lensed quasar (Kochanek 2002, 2003). Recent time-delay measurements give $H_0 \approx 60$ km s^{-1} Mpc^{-1}, and are somewhat lower than H_0 found using Cepheid-calibrated secondary distance indicators. Treu & Koopmans (2002) reported $H_0 = 59^{+12}_{-7} \pm 3$ km s^{-1} Mpc^{-1} for an $\Omega_\Lambda = 0.7, \Omega_m = 0.3$ Universe. Fassnacht et al. (2002) found values between 61 and 65 for the same cosmological model. Cardone et al. (2002) found $H_0 = 58^{+17}_{-15}$ km s^{-1} Mpc^{-1}, in good agreement with the others. Kochanek (2002) showed that values of H_0 between 51 and 73 km s^{-1} Mpc^{-1} are possible; the lower limit corresponds to cold dark matter M/L concentrations, while the upper limit is set by the constant-M/L limit. Now that more accurate time delays have been measured for $\sim$5 systems using radio, optical, and X-ray observations, the distribution of mass in the lensing galaxy is the largest remaining uncertainty in gravitational lens measurements of the Hubble constant.

The SZ effect at submillimeter wavelengths, when combined with X-ray measurements of the hot gas in galaxy clusters, can be exploited to determine the angular diameter distance to the cluster (Carlstrom, Holder, & Reese 2002; Reese et al. 2002). To date, there are 38 SZ distance measurements extending to redshifts of $z = 0.8$. A fit to the 38 measurements yields $H_0 = 60 \pm 3$ km s^{-1} Mpc^{-1}, with an additional systematic uncertainty of order 30% (Carlstrom et al. 2002). The primary systematic uncertainties arise from cluster structure (clumpiness and departures from isothermality) or from point-source contamination. The SZ Hubble constant is also a function of the mass and dark energy density of the Universe; the value presented here assumes $\Omega_m = 0.3$ and $\Omega_\Lambda = 0.7$ (Reese et al. 2002). Future SZ surveys are expected to discover hundreds of new clusters, which should greatly improve our understanding of the density and expansion rate of the Universe to $z \approx 2$ (Carlstrom et al. 2002).

Using models to predict the absolute luminosity of a standard candle is another way to sidestep the issues with empirical distance scale calibrations. Both Type Ia and II SNe, along with SBFs, can be primary distance indicators. We usually choose to calibrate SNe and SBFs empirically using Cepheids because of the acknowledged uncertainties in the many model parameters. The models are good enough, however, to provide some constraints on the distance scale and the quality of the Cepheid calibration.

Recent models of Type Ia SNe are now detailed enough to predict the absolute luminosity of the burst and therefore allow distances to be derived directly. The results reported by Höflich & Khokhlov (1996), for example, show that Ia SN models are consistent with a Hubble constant of 67 ± 9 km s^{-1} Mpc^{-1}. While there are clearly many details in the explosion models that must be carefully checked against observations, it is reassuring that the predictions are in the right ball park. Type II SNe have also been used as primary standard candles using a theoretically calibrated expanding-photosphere technique (Schmidt et al. 1994; Hamuy 2001). Schmidt et al. found $H_0 = 73 \pm 6 \pm 7$ km s^{-1} Mpc^{-1}, independent of the Cepheid calibration. Hamuy's (2001) updated measurement using the same technique yielded 67 ± 7 km s^{-1} Mpc^{-1}.

SBFs are proportional to the second moment of the stellar luminosity function. Standard stellar population models can be integrated to predict SBF magnitudes for populations with particular ages and metallicities, and then compared to observations (Blakeslee, Vazdekis, & Ajhar 2001; Liu, Charlot, & Graham 2000; Liu, Graham, & Charlot 2002). In the *I* band, SBF comparisons with stellar population models would agree with observations better if the original KP Cepheid zeropoint were fainter by 0.2 ± 0.1 mag (Blakeslee 2002), which would make H_0 10% larger. Jensen et al. (2003) found that H-band SBFs were entirely self-consistent with both the Vazdekis (1999, 2001) and Bruzual & Charlot (1993) models when the calibration of Freedman et al. (2001) was used without metallicity corrections. The Hubble constant implied by the IR population models is 8% higher than that determined using the original KP Cepheid calibration of Ferrarese et al. (2000a) and 5% higher than that of Freedman et al. (2001) with the metallicity correction.

7.7   Probability Distributions and Systematic Uncertainties

As distance measurement techniques mature, two things generally happen: first, the number of measurements increases, reducing the statistical uncertainty, and second, the larger data sets make it possible to correct for secondary effects that modify the brightness of the standard candle. The result is a reduction in the statistical uncertainty to the point that systematic effects start to dominate. This is certainly true for Cepheids and the secondary distance indicators calibrated using them. As shown in previous sections, systematic uncertainties in the Cepheid distances dominate statistical uncertainties and the systematic differences between different secondary techniques. Addressing systematic uncertainties is a difficult job that cannot proceed until a sufficiently large number of measurements has been made to understand the intrinsic dispersion in the properties of a standard candle.

Some of the techniques discussed have not yet been applied to enough systems to conclusively say that small number statistics are not an issue, even though the formal statistical uncertainty of an individual measurement might be small. Gravitational lens time delays have only been measured in five systems, for example. SZ measurements are only now reaching large enough samples to start addressing the systematic uncertainties. Fortunately, the sample sizes will increase significantly as surveys to find and measure more lensed quasars and SZ clusters proceed. Supernovae of all types are rare enough that finding enough nearby SNe for calibration purposes has required using limited and often old, unreliable data.

The probability distributions of systematic uncertainties is not always known, and is frequently not Gaussian. For example, the range of H_0 values permitted by the gravitational lens measurements is set by systematic uncertainties in the lens mass distribution. The extremes are rather rigidly limited by constraints on the possible fraction and distribution of dark matter in galaxies. The probability distribution is therefore rather close to a top hat, with "sigma limits" that are not at all Gaussian (e.g., 2-$\sigma < 2 \times \sigma$). It is clearly not appropriate to add errors of this nature in quadrature.

Many researchers have maintained separate accounting of systematic and random uncertainties when possible (cf., Ferrarese et al. 2000a). Even this approach requires the addition of different systematic uncertainties in quadrature, assuming that individual systematic uncertainties are independent and Gaussian in nature. In many cases this is probably justified; in others, simply reporting a range of possible values is more appropriate. The problem with reporting such ranges as a systematic uncertainty is that they are often viewed as being overly pessimistic by the casual reader who regards them as "1-σ" uncertainties. Given the difficulty in comparing very different systematic uncertainties, it is probably premature to judge the SZ and gravitational lens results as being inconsistent with the Cepheid-based distance indicators at the present time when the number of measurements is still rather few and the systematics have not been explored in great detail.

7.8 Future Prospects

The secondary distance indicators, mostly calibrated using Cepheids, are all in good agreement when calibrated uniformly. These imply a Hubble constant between 70 and 75 km s^{-1} Mpc^{-1}. Many other techniques that do not rely on the calibration of the traditional distance ladder generally agree with this result at the 1-σ level; further concordance between independent techniques will require a careful analysis of systematic uncertainties and new survey data that will become available in the next few years. Improvements in the Cepheid calibration will require a better zeropoint and larger samples covering a range of period and metallicity. Near-IR photometry will help reduce uncertainties due to extinction. High-resolution imaging will help reduce blending and allow measurements of Cepheids in more distant galaxies. Improved photometry and excellent resolution will make the ACS and WFPC3 on *HST* powerful Cepheid-measuring instruments.

SIM and *GAIA* are two astrometry satellite missions planned by NASA and ESA that will help reduce systematic uncertainties in the extragalactic distance scale by providing accurate (1%) parallax distances for a significant number of Galactic Cepheids. *SIM* and *GAIA* will allow us to calibrate the Cepheid zeropoint, PL relation, and metallicity corrections without having to rely on the LMC sample or the distance to the LMC. We are optimistic that the debate over the "long" and "short" distance scales for the LMC will soon be behind us. While this will be a significant milestone, it will not help much to reduce the statistical uncertainty in the measured value of the Hubble constant. It will, however, remove one persistent source of systematic error. To achieve a Hubble constant good to 5% using a distance estimator tied to the LMC will require much more accurate geometrical measurements, and a reduction of the other systematic uncertainties as well.

With *SIM* and *GAIA*, the calibration of several variable star distance scales in addition to the Cepheids will be solidified. These include RR Lyrae variables, delta Scuti stars, and shorter-period overtone pulsators. The increased sensitivity and spatial resolution of the next generation of large ground and space telescopes will allow us to detect these fainter variables in the distant galaxies in which only Cepheids are currently detectable. Other variable stars will allow us to resolve questions about bias that arise when only the very brightest members of the Cepheid population are detected and used to determine the distance.

The number of ways to bypass the Cepheid rung of the distance ladder will increase dramatically when *SIM* and *GAIA* allow us to calibrate a number of secondary distance indicators directly from statistical parallax distance measurements to M31, M32, and M33. Techniques like SBF, Tully-Fisher, FP, GCLF, PNLF, and so forth, have already been used to determine accurate relative distances between the Local Group members M31, M32, and M33 and more distant galaxies. By determining their distances directly from statistical parallax measurements, the systematic uncertainties in the Cepheid calibration and LMC distance will be avoided altogether.

Improvements in techniques that bypass the local distance ladder and secondary distance indicators are imminent. Larger samples of well-measured gravitational lens time delays and SZ clusters will help reduce statistical uncertainties and provide insight into the systematics. Better mass models for gravitational lenses will not only lead to better determinations of H_0, but also be valuable in constraining the quantity and distribution of dark matter in galaxies. The increasing number of SZ measurements will lead to a better understanding of galaxy cluster structure and evolution.

The number of direct geometrical distance measurements to nearby galaxies will also increase. The masers detected in NGC 4258 must also exist in other galaxies. A larger sample of detached eclipsing binaries, both in the LMC and in other galaxies, will help overcome the systematic uncertainties and provide more consistent distances. These techniques, like SNe in nearby galaxies, are limited by small-number statistics. An individual measurement may seem reliable, but until more are found, our confidence in them will be limited.

Several new synoptic and survey facilities are currently being planned that will discover many thousands of SNe. The *SNAP* satellite will discover thousands of SNe over its lifetime. It will be able to measure optical and near-IR brightnesses and collect spectra for SN classification. The Large Synoptic Survey Telescope (LSST) and PanSTARRs survey telescopes, currently in the planning stages, will discover hundreds of thousands of Ia SNe every year. The synoptic telescopes will also reveal a multitude of faint variable stars in the Galaxy. With this wealth of data, systematic uncertainties can be addressed and the expansion rate of the Universe determined as a function of redshift to $z > 1$. We will only be limited by our ability to follow up the SN discoveries to determine reliable distances.

Perhaps the best determination of H_0 in the future will come from the combination of multiple joint constraints, just as the conclusions regarding the $\Omega_\Lambda = 0.7, \Omega_m = 0.3$ Universe came from merging the Type Ia SN results with the measurements of $\Omega_{tot} = 1.0$ from the cosmic microwave background experiments. For example, both $H_0 t_0$ and $\Omega_b h^2$ are now known to 5%. Other examples of joint constraints that include H_0 are described elsewhere in this volume.

Acknowledgements. J. Jensen acknowledges the support of the Gemini Observatory, which is operated by the Association of Universities for Research in Astronomy, Inc., on behalf of the international Gemini partnership of Argentina, Australia, Brazil, Canada, Chile, the United Kingdom, and the United States of America.

References

Ajhar, E. A., Tonry, J. L., Blakeslee, J. P., Riess, A. G., & Schmidt, B. P. 2001, ApJ, 559, 584

Alves, D. R., Rejkuba, M., Minniti, D., & Cook, K. H. 2002, ApJ, 573, L51

Benedict, G. F., et al. 2002a, AJ, 123, 473

——. 2002b, AJ, 124, 1695

Blakeslee, J. P. 2002, in A New Era In Cosmology, ed. T. Shanks & N. Metcalfe (Chelsea: Sheridan Books)

Blakeslee, J. P., Ajhar, E. A., & Tonry, J. L. 1999, in Post-Hipparcos Cosmic Candles, ed. A. Heck & F. Caputo (Dordrecht: Kluwer), 181

Blakeslee, J. P., Lucey, J. R., Tonry, J. L., Hudson, M. J., Narayanan, V. K., & Barris, B. J. 2002, MNRAS, 330, 443

Blakeslee, J. P., Vazdekis, A., & Ajhar, E. A. 2001, MNRAS, 320, 193

Bono, G., Groenewegen, M. A. T., Marconi, M., & Caputo, F. 2002, ApJ, 574, L33

Bruzual A., G., & Charlot, S. 1993, ApJ, 405, 538

Caputo, F., Marconi, M., & Musella, I. 2002, ApJ, 566, 833

Cardone, V. F., Capozziello, S., Re, V., & Piedipalumbo, E. 2002, A&A, 382, 792

Carlstrom, J. E., Holder, G. P., & Reese, E. D. 2002, ARA&A, 40, 643

Carretta, E., Gratton, R. G., Clementini, G., & Pecci, F. F. 2000, ApJ, 533, 215

Ciardullo, R., Feldmeier, J. J., Jacoby, G. H., De Naray, R. K., Laychak, M. B., & Durrell, P. R. 2002, ApJ, 577, 31

Di Benedetto, G. P. 2002, AJ, 124, 1213

Fassnacht, C. D., Xanthopoulos, E., Koopmans, L. V. E., & Rusin, D. 2002, ApJ, 581, 823

Ferrarese, L., et al. 2000a, ApJ, 529, 745

——. 2000b, ApJS, 128, 431

Ferrarese, L., Silbermann, N. A., Mould, J. R., Stetson, P. B., Saha, A., Freedman, W. L., & Kennicutt, R. C., Jr. 2000c, PASP, 112, 117

Fitzpatrick, E. L., Ribas, I., Guinan, E. F., DeWarf, L. E., Maloney, F. P., & Massa, D. 2002, ApJ, 564, 260

Freedman, W. L., et al. 2001, ApJ, 553, 47

Freedman, W. L., & Madore, B. F. 1990, ApJ, 365, 186

Gibson, B. K., et al. 2000, ApJ, 529, 723

Gibson, B. K., Maloney, P. R., & Sakai, S. 2000, ApJ, 530, L5

Gibson, B. K., & Stetson, P. B. 2001, ApJ, 547, L103

Giovanelli, R., Dale, D. A., & Haynes, M. P. 1999, ApJ, 525, 25

Giovanelli, R., Haynes, M. P., Herter, T., Bogt, N. P., Da Costa, L. N., Freudling, W., Salzer, J. J., & Wegner, G. 1997, AJ, 113, 22

Gould, A., & Uza, O. 1997, ApJ, 494, 118

Groenewegen, M. A. T., & Salaris, M. 2001, A&A, 366, 752

Guinan, E. F., et al. 1998, ApJ, 509, L21

Hamuy, M. 2001, Ph.D. Thesis, Univ. Arizona

Hamuy, M., Phillips, M. M., Maza, J., Suntzeff, N. B., Schommer, R. A., & Aviles, R. 1995, AJ, 109, 1669

Hamuy, M., Phillips, M. M., Suntzeff, N. B., Schommer, R. A., Maza, J., & Aviles, R. 1996, AJ, 112, 2398

Herrnstein, J. R., et al. 1999, Nature, 400, 539

Höflich, P., & Khokhlov, A. 1996, ApJ, 457, 500

Hudson, M. J., Lucey, J. R., Smith, J. R., Schlegel, D. J., & Davies, R. L. 2001, MNRAS, 327, 265

Jensen, J. B., Tonry, J. L., Barris, B. J., Thompson, R. I., Liu, M. C., Rieke, M. J., Ajhar, E. A., & Blakeslee, J. P. 2003, ApJ, 583, 712

Jensen, J. B., Tonry, J. L., Thompson, R. I., Ajhar, E. A., Lauer, T. R., Rieke, M. J., Postman, M., & Liu, M. C. 2001, ApJ, 550, 503

Keller, S. C., & Wood, P. R. 2002, ApJ, 578,144

Kelson, D. D., et al. 2000, ApJ, 529, 768

Kennicutt, R. C., Jr., et al. 1998, ApJ, 498, 181

Kim, A. G., et al. 1997, ApJ, 476, L63

Kochanek, C. S. 2002, ApJ, 578, 25

——. 2003, ApJ, 583, 49

Kochanek, C. S., & Schechter, P. L. 2004, in Carnegie Observatories Astrophysics Series, Vol. 2: Measuring and Modeling the Universe, ed. W. L. Freedman (Cambridge: Cambridge Univ. Press), in press

Kundu, A., & Whitmore, B. C. 2001, AJ, 121, 2950

Lahav, O. 2000, in Cosmic Flows 1999: Towards an Understanding of Large-Scale Structure, ed. S. Courteau, M. A. Strauss, & J. A. Willick (Chelsea: Sheridan Books), 377

Lauer, T. R., Tonry, J. L., Postman, M., Ajhar, E. A., & Holtzman, J. A. 1998, ApJ, 499, 577

Liu, M. C., Charlot, S., & Graham, J. R. 2000, ApJ, 543, 644

Liu, M. C., & Graham, J. R. 2001, ApJ, 557, L31

Liu, M. C., Graham, J. R., & Charlot, S. 2002, ApJ, 564, 216

Macri, L. M., et al. 2001, ApJ, 549, 721

Maoz, E., Newman, J. A., Ferrarese, L., Stetson, P. B., Zepf, S. E., Davis, M., Freedman, W. L., & Madore, B. F. 1999, Nature, 401, 351

Mitchell, R. C., Baron, E., Branch, D., Hauschildt, P. H., Nugent, P. E., Lundqvist, P., Blinnikov, S., & Pun, C. S. J. 2002, ApJ, 574, 293

Mochejska, B. J., Macri, L. M., Sasselov, D. D., & Stanek, K. Z. 2000, AJ, 120, 810

Mould, J. R., et al. 2000, ApJ, 529, 786

Newman, J. A., Ferrarese, L., Stetson, P. B., Maoz, E., Zepf, S. E., Davis, M., Freedman, W. L., & Madore, B. F., 2001, ApJ, 553, 562

Okon, V. M. M., & Harris, W. E. 2002, ApJ, 567, 294

Paczyński, B. 2001, Acta Astron., 51, 81

Panagia, N. 1999, in IAU Symp. 190, New Views of the Magellanic Clouds, ed. Y.-H. Chu et al. (Dordrecht: Kluwer), 549

Parodi, B. R., Saha, A., Sandage, G. A., & Tammann, G. A. 2000, ApJ, 540, 634

Perlmutter, S. et al. 1997, ApJ, 483, 565

———. 1998, Nature, 391, 51

Phillips, M. M. 1993, ApJ, 413, L105

Pietrzynski, G., & Gieren, W. 2002, AJ, 124, 2633

Reese, E. D. 2004, in Carnegie Observatories Astrophysics Series, Vol. 2: Measuring and Modeling the Universe, ed. W. L. Freedman (Cambridge: Cambridge Univ. Press), in press

Reese, E. D., Carlstrom, J. E., Joy, M., Mohr, J. J., Grego, L., & Holzapfel, W. L. 2002, ApJ, 581, 53

Ribas, I., Fitzpatrick, E. L., Maloney, F. P., & Guinan, E. F. 2002, ApJ, 574, 771

Riess, A. G., Press, W. H., & Kirshner, R. P. 1996, ApJ, 473, 88

Romaniello, M., Salaris, M., Cassisi, S., & Panagia, N. 2000, ApJ, 530, 738

Saha, A., Sandage, A., Tammann, G. A., Dolphin, A. E., Christensen, J., Panagia, N., & Macchetto, F. D. 2001, ApJ, 562, 314

Sakai, S., et al. 2000, ApJ, 529, 698

Sebo, K. M., Rawson, D., Mould, J., Madore, B. F., Putman, M. E., Graham, J. A., Freedman, W. L., Gibson, B. K., & Germany, L. M. 2002, ApJS, 142, 71

Schmidt, B. P., et al. 1998, ApJ, 507, 46

Schmidt, B. P., Kirshner, R. P., Eastman, R. G., Phillips, M. M., Suntzeff, N. B., Hamuy, M., Maza, J., & Aviles, R. 1994, ApJ, 432, 42

Stanek, K. Z., Zaritsky, D., & Harris, J. 1998, ApJ, 500, L141

Tanvir, N. R., Ferguson, H. C., & Shanks, T. 1999, MNRAS, 310, 175

Tonry, J. L., Blakeslee, J. P., Ajhar, E. A., & Dressler, A. 2000, ApJ, 530, 625

Tonry, J. L., Dressler, A., Blakeslee, J. P., Ajhar, E. A., Fletcher, A. B., Luppino, G. A., Metzger, M. R., & Moore, C. B. 2001, ApJ, 546, 681

Tonry, J. L., & Schneider 1988, AJ, 96, 807

Treu, T., & Koopmans, V. E. 2002, MNRAS, 337, L6

Udalski, A., Soszynski, I., Szymanski, M., Kubiak, M., Pietrzynski, G., Wozniak, P., & Zebrun, K. 1999a, Acta Astron., 49, 223

Udalski, A., Szymanski, M., Kubiak, M., Pietrzynski, G., Soszynski, I., Wozniak, P., & Zebrun, K. 1999b, Acta Astron., 49, 201

Udalski, A., Wyrzykowski, L., Pietrzynski, G., Szewczyk, O., Szymanski, M., Kubiak, M., Soszynski, I., & Zebrun, K. 2001, Acta Astron., 51, 221

Vazdekis, A. 1999, ApJ, 513, 224

———. 2001, Ap&SS, 276, 921

Walker, A. 1999, in Post-Hipparcos Cosmic Candles, ed. A. Heck & F. Caputo (Dordrecht: Kluwer), 125

Willick, J. A., & Batra, P. 2001, ApJ, 548, 564

Zehavi, I., Riess, A. G., Kirshner, R. P., & Dekel, A. 1998, ApJ, 503, 483

8

The Hubble constant from gravitational lens time delays

CHRISTOPHER S. KOCHANEK[1] and PAUL L. SCHECHTER[2]
(1) Harvard-Smithsonian Center for Astrophysics
(2) Massachusetts Institute of Technology

Abstract

There are now 10 firm time delay measurements in gravitational lenses. The physics of time delays is well understood, and the only important variable for interpreting the time delays to determine H_0 is the mean surface mass density $\langle \kappa \rangle$ (in units of the critical density for gravitational lensing) of the lens galaxy at the radius of the lensed images. More centrally concentrated mass distributions with lower $\langle \kappa \rangle$ predict higher Hubble constants, with $H_0 \propto 1 - \langle \kappa \rangle$ to lowest order. While we cannot determine $\langle \kappa \rangle$ directly given the available data on the current time delay lenses, we find $H_0 = 48 \pm 3$ km s^{-1} Mpc^{-1} for the isothermal (flat rotation curve) models, which are our best present estimate for the mass distributions of the lens galaxies. Only if we eliminate the dark matter halo of the lenses and use a constant mass-to-light ratio (M/L) model to find $H_0 = 71 \pm 3$ km s^{-1} Mpc^{-1} is the result consistent with local estimates. Measurements of time delays in better-constrained systems or observations to obtain new constraints on the current systems provide a clear path to eliminating the $\langle \kappa \rangle$ degeneracy and making estimates of H_0 with smaller uncertainties than are possible locally. Independent of the value of H_0, the time delay lenses provide a new and unique probe of the dark matter distributions of galaxies and clusters because they measure the total (light + dark) matter surface density.

8.1   Introduction

Fifteen years prior to their discovery in 1979, Refsdal (1964) outlined how gravitationally lensed quasars might be used to determine the Hubble constant. Astronomers have spent the quarter century since their discovery working out the difficult details not considered in Refsdal's seminal papers.

The difficulties encountered fall into two broad categories—measurement and modeling. Time delays can be hard to measure if the fluxes of the images do not vary, or if the images are faint, or if they lie very close to each other. Modeling gravitational potentials with a small number of constraints is likewise difficult, either because the lens geometry is complex or because the data poorly constrain the most important aspects of the gravitational potential. We will argue that these difficulties are surmountable, both in principle and in practice, and that an effort considerably smaller than that of the *HST* Hubble Constant Key Project will yield a considerably smaller uncertainty in the Hubble constant, H_0.

While the number of systems with measured time delays is small, their interpretation implies a value for H_0, which, given our current understanding of the dark matter distributions of galaxies, is formally inconsistent with that obtained using Cepheids. The Key Project value of $H_0 = 72 \pm 8$ km s^{-1} Mpc^{-1} (Freedman et al. 2001) is consistent with the lens data only if the lens galaxies have significantly less dark matter than is expected theoretically or has been measured for other early-type galaxies. While it is premature to argue for replacing the local estimates, we hope to persuade the astronomical community that the time delay result deserves both careful attention and further study.

Interpreting time delays requires a model for the gravitational potential of the lens, and in most cases the uncertainties in the model dominate the uncertainty in H_0. Thus, the main focus of this review will be to explain the dependence of time delays on gravitational potentials. We start in §8.2 by introducing the time delay method and illustrating the physics of time delays with a series of simple models. In §8.3 we review a general mathematical theory of time delays to show that, for most lenses, the only important parameter of the model is the mean surface density of the lens at the radius of the images. In §8.4 we discuss the effects of the environment of the lens on time delays. We review the data on the time delay lenses in §8.5 and their implications for the Hubble constant and dark matter in early-type galaxies in §8.6. The present time delay lenses have a degeneracy between H_0 and the amount of dark matter, so in §8.7 we outline several approaches that can eliminate the degeneracy. Finally, in §8.8 we discuss the future of time delays. Unless otherwise stated, we assume a flat, $\Omega_m = 0.3$, $\Omega_\Lambda = 0.7$ cosmological model.

8.2   Time Delay Basics

The observations of gravitationally lensed quasars are best understood in light of Fermat's principle (e.g., Blandford & Narayan 1986). Intervening mass between a source and an observer introduces an effective index of refraction, thereby increasing the light-travel time. The competition between this Shapiro delay from the gravitational field and the geometric delay due to bending the ray paths leads to the formation of multiple images at the stationary points (minima, maxima, and saddle points) of the travel time (for more complete reviews, see Narayan & Bartelmann 1999 or Schneider, Ehlers, & Falco 1992).

As with glass optics, there is a thin-lens approximation that applies when the optics are small compared to the distances to the source and the observer. In this approximation, we need only the effective potential, $\psi(\vec{x}) = (2/c^2)(D_{ls}/D_s) \int dz \phi$, found by integrating the 3D potential ϕ along the line of sight. The light-travel time is

$$\tau(\vec{x}) = \left[\frac{1+z_l}{c}\right] \left[\frac{D_l D_s}{D_{ls}}\right] \left[\frac{1}{2}\left(\vec{x}-\vec{\beta}\right)^2 - \psi(\vec{x})\right], \tag{8.1}$$

where $\vec{x} = (x,y) = R(\cos\theta, \sin\theta)$ and $\vec{\beta}$ are the angular positions of the image and the source, $\psi(\vec{x})$ is the effective potential, $(\vec{x}-\vec{\beta})^2/2$ is the geometric delay in the small-angle approximation, z_l is the lens redshift, and D_l, D_s, and D_{ls} are angular-diameter distances to the lens, to the source, and from the lens to the source, respectively. The only dimensioned quantity in the travel time is a factor of $H_0^{-1} \simeq 10h^{-1}$ Gyr arising from the H_0^{-1} scaling of the angular-diameter distances.

We observe the images at the extrema of the time delay function, which we find by setting the gradients with respect to the image positions equal to zero, $\vec{\nabla}_x \tau = 0$, and finding all the stationary points $(\vec{x}_A, \vec{x}_B, \cdots)$ associated with a given source position $\vec{\beta}$. The local magnification of an image is determined by the magnification tensor M_{ij}, whose inverse is determined by the second derivatives of the time delay function,

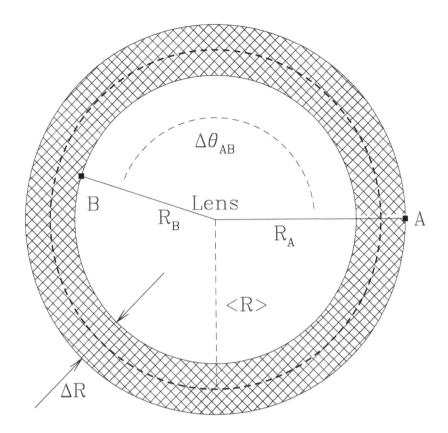

Fig. 8.1. Schematic diagram of a two-image time delay lens. The lens lies at the origin, with two images A and B at radii R_A and R_B from the lens center. The images define an annulus of average radius $\langle R \rangle = (R_A + R_B)/2$ and width $\Delta R = R_A - R_B$, and the images subtend an angle $\Delta\theta_{AB}$ relative to the lens center. For a circular lens $\Delta\theta_{AB} = 180°$ by symmetry.

$$M_{ij}^{-1} = \vec{\nabla}_x \vec{\nabla}_x \tau(\vec{x}) = \begin{pmatrix} 1 - \kappa - \gamma\cos 2\theta_\gamma & \gamma\sin 2\theta_\gamma \\ \gamma\sin 2\theta_\gamma & 1 - \kappa + \gamma\cos 2\theta_\gamma \end{pmatrix}, \tag{8.2}$$

where the convergence $\kappa = \Sigma/\Sigma_c$ is the local surface density in units of the critical surface density $\Sigma_c = c^2 D_s/4\pi G D_l D_{ls}$, and γ and θ_γ define the local shear field and its orientation. The determinant of the magnification tensor is the net magnification of the image, but it is a signed quantity depending on whether the image has positive (maxima, minima) or negative (saddle points) parity.

A simple but surprisingly realistic starting point for modeling lens potentials is the singular isothermal sphere (the SIS model) in which the lens potential is simply

$$\psi(\vec{x}) = bR, \qquad \text{where} \qquad b = 4\pi \frac{D_{ls}}{D_s} \frac{\sigma^2}{c^2} = 1\rlap{.}''45 \left(\frac{\sigma}{225\,\text{km s}^{-1}} \right)^2 \frac{D_{ls}}{D_s} \qquad (8.3)$$

is a deflection scale determined by geometry and σ is the 1D velocity dispersion of the lens galaxy. For $|\vec{\beta}| < b$, the SIS lens produces two colinear images at radii $R_A = |\vec{\beta}| + b$ and $R_B = b - |\vec{\beta}|$ on opposite sides of the lens galaxy (as in Fig. 8.1 but with $\Delta\theta_{AB} = 180°$).[*] The A image is a minimum of the time delay and leads the saddle point, B, with a time delay difference of

$$\Delta t_{SIS} = \tau_B - \tau_A = \frac{1}{2} \left[\frac{1+z_l}{c} \right] \left[\frac{D_l D_s}{D_{ls}} \right] (R_A^2 - R_B^2). \qquad (8.4)$$

Typical time delay differences of months or years are the consequence of multiplying the $\sim 10 h^{-1}$ Gyr total propagation times by the square of a very small angle ($b \approx 3 \times 10^{-6}$ radians so, $R_A^2 \approx 10^{-11}$). The SIS model suggests that lens time delay measurements reduce the determination of the Hubble constant to a problem of differential astrometry. This is almost correct, but we have made two idealizations in using the SIS model.

The first idealization was to ignore deviations of the radial (monopole) density profile from that of an SIS with density $\rho \propto r^{-2}$, surface density $\Sigma \propto R^{-1}$, and a flat rotation curve. The SIS is a special case of a power-law monopole with lens potential

$$\psi(\vec{x}) = \frac{b^2}{(3-\eta)} \left(\frac{R}{b} \right)^{3-\eta}, \qquad (8.5)$$

corresponding to a (3D) density distribution with density $\rho \propto r^{-\eta}$, surface density $\Sigma \propto R^{1-\eta}$, and rotation curve $v_c \propto r^{(2-\eta)/2}$. For $\eta = 2$ we recover the SIS model, and the normalization is chosen so that the scale b is always the Einstein ring radius. Models with smaller (larger) η have less (more) centrally concentrated mass distributions and have rising (falling) rotation curves. The limit $\eta \to 3$ approaches the potential of a point mass. By adjusting the scale b and the source position $|\vec{\beta}|$, we can fit the observed positions of two images at radii R_A and R_B on opposite sides ($\Delta\theta_{AB} = 180°$) of the lens for any value of η.[†] The expression for the time delay difference can be well approximated by (Witt, Mao, & Keeton 2000; Kochanek 2002)

$$\Delta t(\eta) = \tau_B - \tau_A \simeq (\eta - 1)\Delta t_{SIS} \left[1 - \frac{(2-\eta)^2}{12} \left(\frac{\Delta R}{\langle R \rangle} \right)^2 \cdots \right], \qquad (8.6)$$

where $\langle R \rangle = (R_A + R_B)/2 \simeq b$ and $\Delta R = R_A - R_B$ (see Fig. 8.1). While the expansion assumes $\Delta R/\langle R \rangle$ (or $|\vec{\beta}|$) is small, we can usually ignore the higher-order terms. There are two important lessons from this model.

(1) Image astrometry of simple two-image and four-image lenses generally cannot constrain the radial mass distribution of the lens.

(2) More centrally concentrated mass distributions (larger η) predict longer time delays, resulting in a larger Hubble constant for a given time delay measurement.

---

[*] The deflections produced by the SIS lens are constant, $|\vec{x} - \vec{\beta}| = b$, so the total image separation is always $2b$. The outer image is brighter than the inner image, with signed magnifications $M_A^{-1} = 1 - b/R_A > 0$ (a positive parity minimum) and $M_B^{-1} = 1 - b/R_B < 0$ (a negative parity saddle point). The model parameters, $b = (R_A + R_B)/2 = \langle R \rangle$ and $|\vec{\beta}| = (R_A - R_B)/2 = \Delta R/2$, can be determined uniquely from the image positions.

[†] In theory we have one additional constraint because the image flux ratio measures the magnification ratio, $f_A/f_B = |M_A|/|M_B|$, and the magnification ratio depends on η. Unfortunately, the systematic errors created by milli- and microlensing make it difficult to use flux ratios as model constraints (see §8.5).

These problems, which we will address from a different perspective in §8.3, are the cause of the uncertainties in estimates of H_0 from time delays.

The second idealization was to ignore deviations from circular symmetry due to either the ellipticity of the lens galaxy or the local tidal gravity field from nearby objects. A very nice analytic example of a lens with angular structure is a singular isothermal model with *arbitrary* angular structure, where the effective potential is $\psi = bRF(\theta)$, and $F(\theta)$ is an arbitrary function. The model family includes the most common lens model, the singular isothermal ellipsoid (SIE). The time delays for this model family are simply Δt_{SIS}, *independent of the angular structure of the lens* (Witt et al. 2000)! This result, while attractive, does not hold in general, and we will require the results of §8.3 to understand the effects of angular structure in the potential.

8.3 Understanding Time Delays: A General Theory

The need to model the gravitational potential of the lens is the aspect of interpreting time delays that creates the greatest suspicion. The most extreme view is that it renders the project "hopeless" because we will never be able to guarantee that the models encompass the degrees of freedom needed to capture all the systematic uncertainties. In order to address these fears we must show that we understand the specific properties of the gravitational potential determining time delays and then ensure that our parameterized models include these degrees of freedom.

The examples we considered in §8.2 illustrate the basic physics of time delays, but an extensive catalog of (non)parametric models demonstrating the same properties may not be convincing to the skeptic. We will instead show, using standard mathematical expansions of the potential, which properties of the lens galaxy are required to understand time delays with accuracies of a few percent. While we can understand the results of all models for the time delays of gravitational lenses based on this simple theory, full numerical models should probably be used for most detailed, quantitative analyses. Fortunately, there are publically available programs for both the parametric and nonparametric approaches.* Our analysis uses the geometry of the schematic lens shown in Figure 8.1. The two images define an annulus bounded by their radii, R_A and R_B, and with an interior region for $R < R_B$ and an exterior region for $R > R_A$.

The key to understanding time delays comes from Gorenstein, Falco, & Shapiro (1988; see also Saha 2000), who showed that the time delay of a circular lens depends only on the image positions and the *surface density $\kappa(R)$ in the annulus between the images*. The mass of the interior region is implicit in the image positions and accurately determined by the astrometry. From Gauss' law, we know that the radial distribution of the mass in the interior region and the amount or distribution of mass in the exterior region is irrelevant. A useful approximation is to assume that the surface density in the annulus can be *locally* approximated by a power law $\kappa \propto R^{1-\eta}$ and that the mean surface density in the annulus is $\langle \kappa \rangle = \langle \Sigma \rangle / \Sigma_c$. The time delay between the images is (Kochanek 2002)

$$\Delta t = 2\Delta t_{SIS} \left[1 - \langle \kappa \rangle - \frac{1-\eta\langle\kappa\rangle}{12} \left(\frac{\Delta R}{\langle R \rangle} \right)^2 + O\left(\left(\frac{\Delta R}{\langle R \rangle} \right)^4 \right) \right]. \tag{8.7}$$

---

* The *gravlens* and *lensmodel* (Keeton 2004, cfa-www.harvard.edu/~castles) packages include a very broad range of parametric models for the mass distributions of lenses, and the *PixelLens* package (Williams & Saha 2000, ankh-morpork.maths.qmw.ac.uk/~saha/astron/lens/pix/) implements a nonparametric approach.

Thus, the time delay is largely determined by the average density $\langle\kappa\rangle$, with only modest corrections from the local shape of the surface density distribution even when $\Delta R/\langle R\rangle \simeq 1$. For example, the second-order expansion is exact for an SIS lens ($\langle\kappa\rangle = 1/2$, $\eta = 2$) and reproduces the time delay of a point mass lens ($\langle\kappa\rangle = 0$) to better than 1% even when $\Delta R/\langle R\rangle = 1$. This local model also explains the time delay scalings of the global power-law models we discussed in §8.2. A $\rho \propto r^{-\eta}$ global power law has surface density $\langle\kappa\rangle = (3-\eta)/2$ near the Einstein ring, so the leading term of the time delay is $\Delta t = 2\Delta t_{SIS}(1 - \langle\kappa\rangle) = (\eta - 1)\Delta t_{SIS}$, just as in Equation (8.6).

- The time delay is not determined by the global structure of the radial density profile but rather by the surface density near the Einstein ring.

Gorenstein et al. (1988) considered only circular lenses, but a multipole expansion allows us to understand the role of angular structure (Kochanek 2002). An estimate to the same order as in Equation (8.7) requires only the quadrupole moments of the regions interior and exterior to the annulus, provided the strengths of the higher-order multipoles of the potential have the same order of magnitude as for an ellipsoidal density distribution.* This approximation can fail for the lenses in clusters (see §8.4). The complete expansion for Δt when the two quadrupole moments have independent amplitudes and orientations is not very informative. However, the leading term of the expansion when the two quadrupole moments are aligned illustrates the role of angular structure. Given an exterior quadrupole (i.e., an external shear) of amplitude γ_{ext} and an interior quadrupole of amplitude γ_{int} sharing a common axis θ_γ, the quadrupole potential is

$$\psi_2 = \frac{1}{2}\left(\gamma_{ext}R^2 + \gamma_{int}\frac{\langle R\rangle^4}{R^2}\right)\cos 2(\theta - \theta_\gamma) \tag{8.8}$$

if we define the amplitudes at radius $\langle R\rangle$. For images at positions $R_A(\cos\theta_A, \sin\theta_A)$ and $R_B(\cos\theta_B, \sin\theta_B)$ relative to the lens galaxy (see Fig. 8.1), the leading term of the time delay is

$$\Delta t \simeq 2\Delta t_{SIS}(1 - \langle\kappa\rangle)\frac{\sin^2(\Delta\theta_{AB}/2)}{1 - 4f_{int}\cos^2(\Delta\theta_{AB}/2)}, \tag{8.9}$$

where $\Delta\theta_{AB} = \theta_A - \theta_B$ and $f_{int} = \gamma_{int}/(\gamma_{ext} + \gamma_{int})$ is the fraction of the quadrupole due to the interior quadrupole moment γ_{int}. We need not worry about the possibility of a singular denominator—successful global models of the lens do not allow such configurations.

A two-image lens has too few astrometric constraints to fully constrain a model with independent, misaligned internal and external quadrupoles. Fortunately, when the lensed images lie on opposite sides of the lens galaxy ($\Delta\theta_{AB} \simeq \pi + \delta$, $|\delta| \ll 1$), the time delay becomes insensitive to the quadrupole structure. Provided the angular deflections are smaller than the radial deflections ($|\delta|\langle R\rangle \lesssim \Delta R$), the leading term of the time delay reduces to the result for a circular lens, $\Delta t \simeq 2\Delta t_{SIS}(1 - \langle\kappa\rangle)$. There is, however, one limiting case to

---

* If the quadrupole potential, $\psi_2 \propto \cos 2\theta$, has dimensionless amplitude ϵ_2, then it produces ray deflections of order $O(\epsilon_2 b)$ at the Einstein ring of the lens. In a four-image lens the quadrupole deflections are comparable to the thickness of the annulus, so $\epsilon_2 \simeq \Delta R/\langle R\rangle$. In a two-image lens they are smaller than the thickness of the annulus, so $\epsilon_2 \lesssim \Delta R/\langle R\rangle$. For an ellipsoidal density distribution, the $\cos(2m\theta)$ multipole amplitude scales as $\epsilon_{2m} \sim \epsilon_2^m \lesssim (\Delta R/\langle R\rangle)^m$. This allows us to ignore the quadrupole density distribution in the annulus and all higher-order multipoles. It is important to remember that potentials are much rounder than surface densities [with relative amplitudes for a $\cos(m\theta)$ multipole of roughly $m^{-2}:m^{-1}:1$ for potentials:deflections:densities], so the multipoles relevant to time delays converge rapidly even for very flat surface density distributions.

remember. If the images and the lens are colinear, as in a spherical lens, the component of the shear aligned with the separation vector acts like a contribution to the convergence. In most lenses this would be a modest additional uncertainty—in the typical lens these shears must be small, the sign of the effect should be nearly random, and it is only a true degeneracy in the limit that everything is colinear.

A four-image lens has more astrometric constraints and can constrain a model with independent, misaligned internal and external quadrupoles. The quadrupole moments of the observed lenses are dominated by external shear, with $f_{int} \lesssim 1/4$ unless there is more than one lens galaxy inside the Einstein ring. The ability of the astrometry to constrain f_{int} is important because the delays depend strongly on f_{int} when the images do not lie on opposite sides of the galaxy. If external shears dominate, $f_{int} \simeq 0$ and the leading term of the delay becomes $\Delta t \simeq 2\Delta t_{SIS}(1-\langle\kappa\rangle)\sin^2\Delta\theta_{AB}/2$. If the model is isothermal, like the $\psi = rF(\theta)$ models we considered in §8.2, then $f_{int} = 1/4$ and we again find that the delay is independent of the angle, with $\Delta t \simeq 2\Delta t_{SIS}(1-\langle\kappa\rangle)$. The time delay ratios in a four-image lens are largely determined by the angular structure and provide a check of the potential model.

In summary, if we want to understand time delays to an accuracy competitive with studies of the local distance scale (5%–10%), the only important variable is the surface density $\langle\kappa\rangle$ of the lens in the annulus between the images. When models based on the same data for the time delay and the image positions predict different values for H_0, the differences can always be understood as the consequence of different choices for $\langle\kappa\rangle$. In parametric models $\langle\kappa\rangle$ is adjusted by changing the central concentration of the lens (i.e., like η in the global power-law models), and in the nonparametric models of Williams & Saha (2000) it is adjusted directly. The expansion models of Zhao & Qin (2003, 2004) mix aspects of both approaches.

8.4   Lenses Within Clusters

Most galaxies are not isolated, and many early-type lens galaxies are members of groups or clusters, so we need to consider the effects of the local environment on the time delays. Weak perturbations are easily understood since they will simply be additional contributions to the surface density ($\langle\kappa\rangle$) and the external shear/quadrupole (γ_{ext}) we discussed in §8.3. In this section we focus on the consequences of large perturbations.

As a first approximation we can assume that a nearby cluster (or galaxy) can be modeled by an SIS potential, $\Psi_c(\vec{x}) = B|\vec{x}-\vec{x}_c|$, where B is the Einstein radius of the cluster and $\vec{x}_c = R_C(\cos\theta_c, \sin\theta_c)$ is the position of the cluster relative to the primary lens. We can understand its effects by expanding the potential as a series in R/R_c, dropping constant and linear terms that have no observable consequences, to find that

$$\Psi_c \simeq \frac{1}{4}\frac{B}{R_c}R^2 - \frac{1}{4}\frac{B}{R_c}R^2\cos 2(\theta-\theta_c) + O\left(\frac{B}{R_c^2}R^3\right). \tag{8.10}$$

The first term has the form $(1/2)\kappa_c R^2$, which is the potential of a uniform sheet whose surface density $\kappa_c = B/2R_c$ is that of the cluster at the lens center. The second term has the form $(1/2)\gamma_c R^2\cos 2(\theta-\theta_c)$, which is the (external) tidal shear $\gamma_c = B/2R_c$ that would be produced by putting all the cluster mass inside a ring of radius R_c at the cluster center. All realistic lens models need to incorporate a tidal shear term due to objects near the lens or along the line of sight (Keeton, Kochanek, & Seljak 1997), but as we discussed in §8.3 the shear does not lead to significant ambiguities in the time delay estimates. Usually the local shear cannot be associated with a particular object unless it is quite strong ($\gamma_c \approx 0.1$).*

The problems with nearby objects arise when the convergence κ_c becomes large because of a global degeneracy known as the *mass-sheet degeneracy* (Falco, Gorenstein, & Shapiro 1985). If we have a model predicting a time delay Δt_0 and then add a sheet of constant surface density κ_c, then the time delay is changed to $(1 - \kappa_c)\Delta t_0$ without changing the image positions, flux ratios, or time delay ratios. Its effects can be understood from §8.3 as a contribution to the annular surface density with $\langle \kappa \rangle = \kappa_c$ and $\eta = 1$. The parameters of the lens, in particular the mass scale b, are also rescaled by factors of $1 - \kappa_c$, so the degeneracy can be broken if there is an independent mass estimate for either the cluster or the galaxy.* When the convergence is due to an object like a cluster, there is a strong correlation between the convergence κ_c and the shear γ_c that is controlled by the density distribution of the cluster (for our isothermal model $\kappa_c = \gamma_c$). In most circumstances, neglecting the extra surface density coming from nearby objects (galaxies, groups, clusters) leads to an overestimate of the Hubble constant because these objects all have $\kappa_c > 0$.

If the cluster is sufficiently close, then we cannot ignore the higher-order perturbations in the expansion of Equation (8.10). They are quantitatively important when they produce deflections at the Einstein ring radius b of the primary lens, $B(b/R_c)^2$, that are larger than the astrometric uncertainties. Because these uncertainties are small, the higher-order terms quickly become important. If they are important but ignored in the models, the results can be very misleading.

8.5   Observing Time Delays and Time Delay Lenses

The first time delay measurement, for the gravitational lens Q0957+561, was reported in 1984 (Florentin-Nielsen 1984). Unfortunately, a controversy then developed between a short delay ($\simeq 1.1$ years, Schild & Cholfin 1986; Vanderriest et al. 1989) and a long delay ($\simeq 1.5$ years, Press, Rybicki, & Hewitt 1992a,b), which was finally settled in favor of the short delay only after 5 more years of effort (Kundić et al. 1997; also Schild & Thomson 1997 and Haarsma et al. 1999). Factors contributing to the intervening difficulties included the small amplitude of the variations, systematic effects, which, with hindsight, appear to be due to microlensing and scheduling difficulties (both technical and sociological).

While the long-running controversy over Q0957+561 led to poor publicity for the measurement of time delays, it allowed the community to come to an understanding of the systematic problems in measuring time delays, and to develop a broad range of methods for reliably determining time delays from typical data. Only the sociological problem of conducting large monitoring projects remains as an impediment to the measurement of time

---

* There is a small random component of κ contributed by material along the line of sight (Barkana 1996). This introduces small uncertainties in the H_0 estimates for individual lenses (an rms convergence of $0.01 - 0.05$, depending on the source redshift), but is an unimportant source of uncertainty in estimates from ensembles of lenses because it is a random variable that averages to zero.

* For the cluster this can be done using weak lensing (e.g., Fischer et al. 1997 in Q0957+561), cluster galaxy velocity dispersions (e.g., Angonin-Willaime, Soucail, & Vanderriest 1994 for Q0957+561, Hjorth et al. 2002 for RXJ0911+0551) or X-ray temperatures/luminosities (e.g., Morgan et al. 2001 for RXJ0911+0551 or Chartas et al. 2002 for Q0957+561). For the lens galaxy this can be done with stellar dynamics (Romanowsky & Kochanek 1999 for Q0957+561 and PG1115+080, Treu & Koopmans 2002b for PG1115+080). The accuracy of these methods is uncertain at present because each suffers from its own systematic uncertainties. When the lens is in the outskirts of a cluster, as in RXJ0911+0551, it is probably reasonable to assume that $\kappa_c \leq \gamma_c$, as most mass distributions are more centrally concentrated than isothermal.

Table 8.1. *Time Delay Measurements*

System	N_{im}	Δt (days)	Astrometry	Model	Ref.
HE1104–1805	2	161 ± 7	+	"simple"	1
PG1115+080	4	25 ± 2	+	"simple"	2
SBS1520+530	2	130 ± 3	+	"simple"	3
B1600+434	2	51 ± 2	+/–	"simple"	4
HE2149–2745	2	103 ± 12	+	"simple"	5
RXJ0911+0551	4	146 ± 4	+	cluster/satellite	6
Q0957+561	2	417 ± 3	+	cluster	7
B1608+656	4	77 ± 2	+/–	satellite	8
B0218+357	2	10.5 ± 0.2	–	"simple"	9
PKS1830–211	2	26 ± 4	–	"simple"	10
B1422+231	4	(8 ± 3)	+	"simple"	11

N_{im} is the number of images. Δt is the longest of the measured delays and its 1σ error; delays in parenthesis require further confirmation. The "Astrometry" column indicates the quality of the astrometric data for the system: + (good), +/– (some problems), – (serious problems). The "Model" column indicates the type of model needed to interpret the delays. "Simple" lenses can be modeled as a single primary lens galaxy in a perturbing tidal field. More complex models are needed if there is a satellite galaxy inside the Einstein ring ("satellite") of the primary lens galaxy, or if the primary lens belongs to a cluster. References: (1) Ofek & Maoz 2003, also see Gil-Merino, Wistozki, & Wambsganss 2002, Pelt, Refsdal, & Stabell 2002, and Schechter et al. 2002; (2) Barkana 1997, based on Schechter et al. 1997; (3) Burud et al. 2002b; (4) Burud et al. 2000, also Koopmans et al. 2000; (5) Burud et al. 2002a; (6) Hjorth et al. 2002; (7) Kundić et al. 1997, also Schild & Thomson 1997 and Haarsma et al. 1999; (8) Fassnacht et al. 2002; (9) Biggs et al. 1999, also Cohen et al. 2000; (10) Lovell et al. 1998; (11) Patnaik & Narasimha 2001.

delays in large numbers. Even these are slowly being overcome, with the result that the last five years have seen the publication of time delays in 11 systems (see Table 8.1).

The basic procedures for measuring a time delay are simple. A monitoring campaign must produce light curves for the individual lensed images that are well sampled compared to the time delays. During this period, the source quasar in the lens must have measurable brightness fluctuations on time scales shorter than the monitoring period. The resulting light curves are cross correlated by one or more methods to measure the delays and their uncertainties (e.g., Press et al. 1992a,b; Beskin & Oknyanskij 1995; Pelt et al. 1996; references in Table 8.1). Care must be taken because there can be sources of uncorrelated variability between the images due to systematic errors in the photometry and real effects such as microlensing of the individual images (e.g., Koopmans et al. 2000; Burud et al. 2002b; Schechter et al. 2003). Figure 8.2 shows an example, the beautiful light curves from the radio lens B1608+656 by Fassnacht et al. (2002), where the variations of all four lensed images have been traced for over three years. One of the 11 systems, B1422+231, is limited by systematic uncertainties in the delay measurements. The brand new time delay for

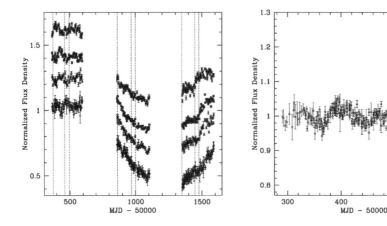

Fig. 8.2. VLA monitoring data for the four-image lens B1608+656. The left panel shows (from top to bottom) the normalized light curves for the B (filled squares), A (open diamonds), C (filled triangles) and D (open circles) images as a function of the mean Julian day. The right panel shows the composite light curve for the first monitoring season after cross correlating the light curves to determine the time delays ($\Delta t_{AB} = 31.5 \pm 1.5$, $\Delta t_{CB} = 36.0 \pm 1.5$ and $\Delta t_{DB} = 77.0 \pm 1.5$ days) and the flux ratios. (From Fassnacht et al. 2002.)

HE1104–1805 (Ofek & Maoz 2003) is probably accurate, but has yet to be interpreted in detail.

We want to have uncertainties in the time delay measurements that are unimportant for the estimates of H_0. For the present, uncertainties of order 3%–5% are adequate (so improved delays are still needed for PG1115+080, HE2149–2745, and PKS1830–211). In a four-image lens we can measure three independent time delays, and the dimensionless ratios of these delays provide additional constraints on the lens models (see §8.3). These ratios are well measured in B1608+656 (Fassnacht et al. 2002), poorly measured in PG1115+080 (Barkana 1997; Schechter et al. 1997; Chartas, Dai, & Garmire 2004) and unmeasured in either RXJ0911+0551 or B1422+231. Using the time delay lenses as very precise probes of H_0, dark matter and cosmology will eventually require still smaller delay uncertainties ($\sim 1\%$). Once a delay is known to 5%, it is relatively easy to reduce the uncertainties further because we can accurately predict when flux variations will appear in the other images and need to be monitored.

The expression for the time delay in an SIS lens (Eqn. 8.4) reveals the other data that are necessary to interpret time delays. First, the source and lens redshifts are needed to compute the distance factors that set the scale of the time delays. Fortunately, we know both redshifts for all 11 systems in Table 8.1. The dependence of the angular-diameter distances on the cosmological model is unimportant until our total uncertainties approach 5% (see §8.8). Second, we require accurate relative positions for the images and the lens galaxy. These uncertainties are always dominated by the position of the lens galaxy relative to the images. For most of the lenses in Table 8.1, observations with radio interferometers (VLA, Merlin, VLBA) and *HST* have measured the relative positions of the images and lenses to accuracies $\lesssim 0\rlap{.}''005$. Sufficiently deep *HST* images can obtain the necessary data for almost any lens, but dust in the lens galaxy (as seen in B1600+434 and B1608+656) can limit the accuracy of the measurement even in a very deep image. For B0218+357 and PKS1830–211, however, the position of the lens galaxy relative to the images is not known to sufficient precision or is disputed (see Léhar et al. 2000; Courbin et al. 2002; Winn et al. 2002).

In practice, we fit models of the gravitational potential constrained by the available data on the image and lens positions, the relative image fluxes, and the relative time delays. When imposing these constraints, it is important to realize that lens galaxies are not perfectly smooth. They contain both low-mass satellites and stars that perturb the gravitational potential. The time delays themselves are completely unaffected by these substructures. However, as we take derivatives of the potential to determine the ray deflections or the magnification, the sensitivity to substructures in the lens galaxy grows. Models of substructure in cold dark matter (CDM) halos predict that the substructure produces random perturbations of approximately $0\rlap{.}''001$ in the image positions (see Metcalf & Madau 2001; Dalal & Kochanek 2002). We should not impose tighter astrometric constraints than this limit. A more serious problem is that substructure, whether satellites ("millilensing") or stars ("microlensing"), significantly affect image fluxes with amplitudes that depend on the image magnification and parity (see, e.g., Wozniak et al. 2000; Burud et al. 2002b; Dalal & Kochanek 2002; Schechter et al. 2003 or Schechter & Wambsganss 2002). Once the flux errors are enlarged to the 30% level of these systematic errors, they provide little leverage for discriminating between models.

We can also divide the systems by the complexity of the required lens model. We define eight of the lenses as "simple," in the sense that the available data suggests that a model consisting of a single primary lens in a perturbing shear (tidal gravity) field should be an adequate representation of the gravitational potential. In some of these cases, an external potential representing a nearby galaxy or parent group will improve the fits, but the differences between the tidal model and the more complicated perturbing potential are small (see §8.4). We include the quotation marks because the classification is based on an impression of the systems from the available data and models. While we cannot guarantee that a system is simple, we can easily recognize two complications that will require more complex models.

The first complication is that some primary lenses have less massive satellite galaxies inside or near their Einstein rings. This includes two of the time delay lenses, RXJ0911+0551 and B1608+656. RXJ0911+0551 could simply be a projection effect, since neither lens galaxy shows irregular isophotes. Here the implication for models may simply be the need to include all the parameters (mass, position, ellipticity $\cdots$) required to describe the second lens galaxy, and with more parameters we would expect greater uncertainties in H_0. In B1608+656, however, the lens galaxies show the heavily disturbed isophotes typical of galaxies undergoing a disruptive interaction. How one should model such a system is unclear. If there was once dark matter associated with each of the galaxies, how is it distributed now? Is it still associated with the individual galaxies? Has it settled into an equilibrium configuration? While B1608+656 can be well fit with standard lens models (Fassnacht et al. 2002), these complications have yet to be explored.

The second complication occurs when the primary lens is a member of a more massive (X-ray) cluster, as in the time delay lenses RXJ0911+0551 (Morgan et al. 2001) and Q0957+561 (Chartas et al. 2002). The cluster model is critical to interpreting these systems (see §8.4). The cluster surface density at the position of the lens ($\kappa_c \gtrsim 0.2$) leads to large corrections to the time delay estimates and the higher-order perturbations are crucial to obtaining a good model. For example, models in which the Q0957+561 cluster was treated simply as an external shear are grossly incorrect (see the review of Q0957+561 models in Keeton et al. 2000). In addition to the uncertainties in the cluster model itself, we must also decide how to include and model the other cluster galaxies near the primary lens. Thus, lenses in clusters have many reasonable degrees of freedom beyond those of the "simple" lenses.

8.6   Results: The Hubble Constant and Dark Matter

With our understanding of the theory and observations of the lenses we will now explore their implications for H_0. We focus on the "simple" lenses PG1115+080, SBS1520+530, B1600+434, and HE2149–2745. We only comment on the interpretation of the HE1104–1805 delay because the measurement is too recent to have been interpreted carefully. We will briefly discuss the more complicated systems RXJ0911+0551, Q0957+561, and B1608+656, and we will not discuss the systems with problematic time delays or astrometry.

The most common, simple, realistic model of a lens consists of a singular isothermal ellipsoid (SIE) in an external (tidal) shear field (Keeton et al. 1997). The model has 7 parameters (the lens position, mass, ellipticity, major axis orientation for the SIE, and the shear amplitude and orientation). It has many degrees of freedom associated with the angular structure of the potential, but the radial structure is fixed with $\langle \kappa \rangle \simeq 1/2$. For comparison, a two-image (four-image) lens supplies 5 (13) constraints on any model of the potential: 2 (6) from the relative positions of the images, 1 (3) from the flux ratios of the images, 0 (2) from the inter-image time delay ratios, and 2 from the lens position. With the addition of extra components (satellites/clusters) for the more complex lenses, this basic model provides a good fit to all the time delay lenses except Q0957+561. Although a naive counting of the degrees of freedom ($N_{dof} = -2$ and 6, respectively) suggests that estimates of H_0 would be underconstrained for two-image lenses and overconstrained for four-image lenses, the uncertainties are actually dominated by those of the time delay measurements and the astrometry in both cases. This is what we expect from §8.3—the model has no degrees of freedom that change $\langle \kappa \rangle$ or η, so there will be little contribution to the uncertainties in H_0 from the model for the potential.

If we use a model that includes parameters to control the radial density profile (i.e., $\langle \kappa \rangle$), for example by adding a halo truncation radius a to the SIS profile [the pseudo-Jaffe model, $\rho \propto r^{-2}(r^2 + a^2)^{-1}$; e.g., Impey et al. 1998; Burud et al. 2002a],* then we find the expected correlation between a and H_0—as we make the halo more concentrated (smaller a), the estimate of H_0 rises from the value for the SIS profile ($\langle \kappa \rangle = 1/2$ as $a \to \infty$) to the value for a point mass ($\langle \kappa \rangle = 0$ as $a \to 0$), with the fastest changes occurring when a is similar to the Einstein radius of the lens. We show an example of such a model for PG1115+080 in Figure 8.3. This case is somewhat more complicated than a pure pseudo-Jaffe model because there is an additional contribution to the surface density from the group to which the lens galaxy belongs. As long as the structure of the radial density profile is fixed (constant a), the uncertainties are again dominated by the uncertainties in the time delay. Unfortunately,

---

* This is simply an example. The same behavior would be seen for any other parametric model in which the radial density profile can be adjusted.

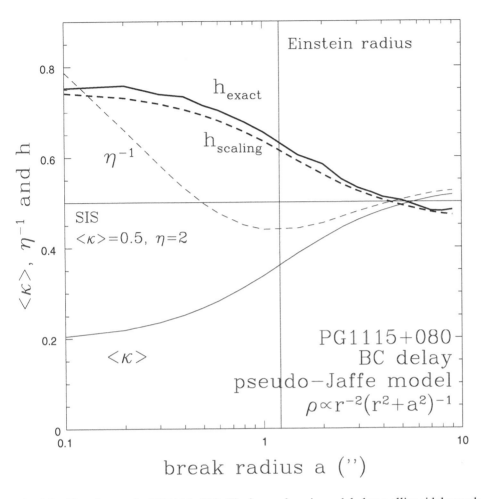

Fig. 8.3. H_0 estimates for PG1115+080. The lens galaxy is modeled as a ellipsoidal pseudo-Jaffe model, $\rho \propto r^{-2}(r^2+a^2)^{-1}$, and the nearby group is modeled as an SIS. As the break radius $a \to \infty$ the pseudo-Jaffe model becomes an SIS model, and as the break radius $a \to 0$ it becomes a point mass. The heavy solid curve (h_{exact}) shows the dependence of H_0 on the break radius for the exact, nonlinear fits of the model to the PG1115+080 data. The heavy dashed curve ($h_{scaling}$) is the value found using our simple theory (§8.3) of time delays. The agreement of the exact and scaling solutions is typical. The light solid line shows the average surface density $\langle \kappa \rangle$ in the annulus between the images, and the light dashed line shows the *inverse* of the logarithmic slope η in the annulus. For an SIS model we would have $\langle \kappa \rangle = 1/2$ and $\eta^{-1} = 1/2$, as shown by the horizontal line. When the break radius is large compared to the Einstein radius (indicated by the vertical line), the surface density is slightly higher and the slope is slightly shallower than for the SIS model because of the added surface density from the group. As we make the lens galaxy more compact by reducing the break radius, the surface density decreases and the slope becomes steeper, leading to a rise in H_0. As the galaxy becomes very compact, the surface density near the Einstein ring is dominated by the group rather than the galaxy, so the surface density approaches a constant and the logarithmic slope approaches the value corresponding to a constant density sheet ($\eta = 1$).

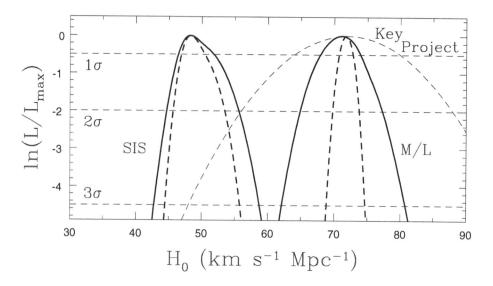

Fig. 8.4. H_0 likelihood distributions. The curves show the joint likelihood functions for H_0 using the four simple lenses PG1115+080, SBS1520+530, B1600+434, and HE2149–2745 and assuming either an SIS model (high $\langle \kappa \rangle$, flat rotation curve) or a constant M/L model (low $\langle \kappa \rangle$, declining rotation curve). The heavy dashed curves show the consequence of including the X-ray time delay for PG1115+080 from Chartas et al. (2004) in the models. The light dashed curve shows a Gaussian model for the Key Project result that $H_0 = 72 \pm 8$ km s^{-1} Mpc^{-1}.

the goodness of fit, $\chi^2(a)$, shows too little dependence on a to determine H_0 uniquely. In general, two-image lenses have too few constraints, and the extra constraints supplied by a four-image lens constrain the angular structure rather than the radial structure of the potential. This basic problem holds for all existing models of the current sample of time delay lenses.

The inability of the present time delay lenses to directly constrain the radial density profile is the major problem for using them to determine H_0. Fortunately, it is a consequence of the available data on the current sample rather than a fundamental limitation, as we discuss in the next section (§8.7). It is, however, a simple trade-off – models with less dark matter (lower $\langle \kappa \rangle$, more centrally concentrated densities) produce higher Hubble constants than those with more dark matter. We do have some theoretical limits on the value of $\langle \kappa \rangle$. In particular, we can be confident that the surface density is bounded by two limiting models. The mass distribution should not be more compact than the luminosity distribution, so a constant mass-to-light ratio (M/L) model should set a lower limit on $\langle \kappa \rangle \gtrsim \langle \kappa \rangle_{M/L} \simeq 0.2$, and an upper limit on estimates of H_0. We are also confident that the typical lens should not have a rising rotation curve at 1–2 optical effective radii from the center of the lens galaxy. Thus, the SIS model is probably the least concentrated reasonable model, setting an upper bound on $\langle \kappa \rangle \lesssim \langle \kappa \rangle_{SIS} = 1/2$, and a lower limit on estimates of H_0. Figure 8.4 shows joint estimates of H_0 from the four simple lenses for these two limiting mass distributions (Kochanek 2004). The results for the individual lenses are mutually consistent and are unchanged by the new

0.149 ± 0.004 day delay for the A_1-A_2 images in PG1115+080 (Chartas et al. 2004). For galaxies with isothermal profiles we find $H_0 = 48 \pm 3$ km s^{-1} Mpc^{-1}, and for galaxies with constant M/L we find $H_0 = 71 \pm 3$ km s^{-1} Mpc^{-1}. While our best prior estimate for the mass distribution is the isothermal profile (see §8.7), the lens galaxies would have to have constant M/L to match Key Project estimate of $H_0 = 72 \pm 8$ km s^{-1} Mpc^{-1} (Freedman et al. 2001).

The difference between these two limits is entirely explained by the differences in $\langle \kappa \rangle$ and η between the SIS and constant M/L models. In fact, it is possible to reduce the H_0 estimates for each simple lens to an approximation formula, $H_0 = A(1 - \langle \kappa \rangle) + B\langle \kappa \rangle(\eta - 1)$. The coefficients, A and $|B| \approx A/10$, are derived from the image positions using the simple theory from §8.3. These approximations reproduce numerical results using ellipsoidal lens models to accuracies of 3 km s^{-1} Mpc^{-1} (Kochanek 2002). For example, in Figure 8.3 we also show the estimate of H_0 computed based on the simple theory of §8.3 and the annular surface density ($\langle \kappa \rangle$) and slope (η) of the numerical models. The agreement with the full numerical solutions is excellent, even though the numerical models include both the ellipsoidal lens galaxy and a group. No matter what the mass distribution is, the five lenses PG1115+080, SBS1520+530, B1600+434, PKS1830-211,* and HE2149-2745 have very similar dark matter halos. For a fixed slope η, the five systems are consistent with a common value for the surface density of

$$\langle \kappa \rangle = 1 - 1.07h + 0.14(\eta - 1)(1 - h) \pm 0.04 \tag{8.11}$$

where $H_0 = 100h$ km s^{-1} Mpc^{-1} and there is an upper limit of $\sigma_\kappa \lesssim 0.07$ on the intrinsic scatter of $\langle \kappa \rangle$. Thus, time delay lenses provide a new window into the structure and homogeneity of dark matter halos, regardless of the actual value of H_0.

There is an enormous range of parametric models that can illustrate how the extent of the halo affects $\langle \kappa \rangle$ and hence H_0—the pseudo-Jaffe model we used above is only one example. It is useful, however, to use a physically motivated model where the lens galaxy is embedded in a standard NFW (Navarro, Frenk, & White 1996) profile halo. The lens galaxy consists of the baryons that have cooled to form stars, so the mass of the visible galaxy can be parameterized using the cold baryon fraction $f_{b,cold}$ of the halo, and for these CDM halo models the value of $\langle \kappa \rangle$ is controlled by the cold baryon fraction (Kochanek 2003). A constant M/L model is the limit $f_{b,cold} \to 1$ (with $\langle \kappa \rangle \simeq 0.2$, $\eta \simeq 3$). Since the baryonic mass fraction of a CDM halo should not exceed the global fraction of $f_b \simeq 0.15 \pm 0.05$ (e.g., Wang, Tegmark, & Zaldarriaga 2002), we cannot use constant M/L models without also abandoning CDM. As we reduce $f_{b,cold}$, we are adding mass to an extended halo around the lens, leading to an increase in $\langle \kappa \rangle$ and a decrease in η. For $f_{b,cold} \simeq 0.02$ the model closely resembles the SIS model ($\langle \kappa \rangle \simeq 1/2$, $\eta \simeq 2$). If we reduce $f_{b,cold}$ further, the mass distribution begins to approach that of the NFW halo without any cold baryons. Figure 8.5 shows how $\langle \kappa \rangle$ and H_0 depend on $f_{b,cold}$ for PG1115+080, SBS1520+530, B1600+434 and HE2149-2745. When $f_{b,cold} \simeq 0.02$, the CDM models have parameters very similar to the SIS model, and we obtain a very similar estimate of $H_0 = 52 \pm 6$ km s^{-1} Mpc^{-1} (95% confidence). If all baryons cool, and $f_{b,cold} = f_b$, then we obtain $H_0 = 65 \pm 6$ km s^{-1} Mpc^{-1} (95% confidence), which is still lower than the Key Project estimates.

---

* PKS1830-211 is included based on the Winn et al. (2002) model of the *HST* imaging data as a single lens galaxy. Courbin et al. (2002) prefer an interpretation with multiple lens galaxies which would invalidate the analysis.

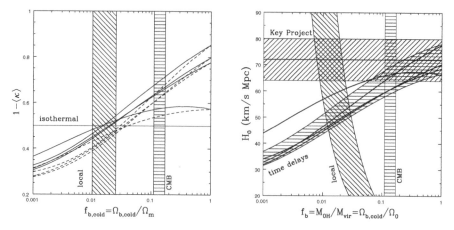

Fig. 8.5. H_0 in CDM halo models. The left panel shows $1 - \langle \kappa \rangle$ for the "simple" lenses (PG1115+080, SBS1520+530, B1600+434, and HE2149–2745) as a function of the cold baryon fraction $f_{b,cold}$. The solid (dashed) curves include (exclude) the adiabatic compression of the dark matter by the baryons. The horizontal line shows the value for an SIS potential. The right panel shows the resulting estimates of H_0, where the shaded envelope bracketing the curves is the 95% confidence region for the combined lens sample. The horizontal band shows the Key Project estimate. For larger $f_{b,cold}$, the density $\langle \kappa \rangle$ decreases and the local slope η steepens, leading to larger values of H_0. The vertical bands in the two panels show the lower bound on f_b from local inventories and the upper bound from the CMB.

We excluded the lenses requiring significantly more complicated models with multiple lens galaxies or very strong perturbations from clusters. If we have yet to reach a consensus on the mass distribution of relatively isolated lenses, it seems premature to extend the discussion to still more complicated systems. We can, however, show that the clusters lenses require significant contributions to $\langle \kappa \rangle$ from the cluster in order to produce the same H_0 as the more isolated systems. As we discussed in §8.5 the three more complex systems are RXJ0911+0551, Q0957+561 and B1608+656.

RXJ0911+0551 is very strongly perturbed by the nearby X-ray cluster (Morgan et al. 2001; Hjorth et al. 2002). Kochanek (2004) found $H_0 = 49 \pm 5$ km s^{-1} Mpc^{-1} if the primary lens and its satellite were isothermal and $H_0 = 67 \pm 5$ km s^{-1} Mpc^{-1} if they had constant mass-to-light ratios. The higher value of $H_0 = 71 \pm 4$ km s^{-1} Mpc^{-1} obtained by Hjorth et al. (2002) can be understood by combining §8.3 and §8.4 with the differences in the models. In particular, Hjorth et al. (2002) truncated the halo of the primary lens near the Einstein radius and used a lower mass cluster, both of which lower $\langle \kappa \rangle$ and raise H_0. The Hjorth et al. (2002) models also included many more cluster galaxies assuming fixed masses and halo sizes.

Q0957+561 is a special case because the primary lens galaxy is the brightest cluster galaxy and it lies nearly at the cluster center (Keeton et al. 2000; Chartas et al. 2002). As a result, the lens modeling problems are particularly severe, and Keeton et al. (2000) found that all previous models (most recently, Barkana et al. 1999; Bernstein & Fischer 1999; and Chae 1999) were incompatible with the observed geometry of the lensed host galaxy. While Keeton et al. (2000) found models consistent with the structure of the lensed host, they covered a range of almost $\pm 25\%$ in their estimates of H_0. A satisfactory treatment of this lens remains elusive.

HE1104–1805 had its delay measured (Ofek & Maoz 2003) just as we completed this review. Assuming the $\Delta t = 161 \pm 7$ day delay is correct, a standard SIE model of this system predicts a very high $H_0 \simeq 90$ km s^{-1} Mpc^{-1}. The geometry of this system and the fact that the inner image is brighter than the outer image both suggest that HE1104–1805 lies in an anomalously high tidal shear field, while the standard model includes a prior to keep the external shear small. A prior is needed because a two-image lens supplies too few constraints to determine both the ellipticity of the main lens and the external shear simultaneously. Since the images and the lens in HE1104–1805 are nearly colinear, the anomalous H_0 estimate for the standard model may be an example of the shear degeneracy we briefly mentioned in §8.3. At present the model surveys needed to understand the new delay have not been made. Observations of the geometry of the host galaxy Einstein ring will resolve any ambiguities due to the shear in the near future (see §8.7).

The lens B1608+656 consists of two interacting galaxies, and, as we discussed in §8.5, this leads to a greatly increased parameter space. Fassnacht et al. (2002) used SIE models for the two galaxies to find $H_0 = 61 - 65$ km s^{-1} Mpc^{-1}, depending on whether the lens galaxy positions are taken from the H-band or I-band lens *HST* images (the statistical errors are negligible). The position differences are probably created by extinction effects from the dust in the lens galaxies. Like isothermal models of the "simple" lenses, the H_0 estimate is below local values, but the disagreement is smaller. These models correctly match the observed time delay ratios.

8.7 Solving the Central Concentration Problem

We can take four approaches to solving the central concentration problem. First, the density profiles of galaxies are not a complete mystery, and we could apply the constraints derived from observations of other (early-type) galaxies to the time delay systems. Second, we could make new observations of the existing time delay lenses in order to obtain additional data that would constrain the density profiles. Third, we could measure the time delays in the systems where the lens galaxies already have well-constrained densities. Fourth, we can use the statistical properties of time delay lenses to break the degeneracies seen in individual lenses.

If we assume that the time delay lenses have the same density structure as other early-type galaxies, then models close to isothermal are favored. For lenses with extended or multi-component sources, the lens models constrain the density distributions and the best fit models are usually very close to isothermal (e.g., Cohn et al. 2001; Winn, Rusin, & Kochanek 2003). Stellar dynamical observations of lenses also favor isothermal models (e.g., Treu & Koopmans 2002a). Stellar dynamical (e.g., Romanowsky & Kochanek 1999; Gerhard et al. 2001) and X-ray (e.g., Loewenstein & Mushotzky 2004) observations of nearby early-type galaxies generally find flat rotation curves on the relevant scales. Finally, weak lensing analyses require significant dark matter on large scales in early-type galaxies (McKay et al. 2002). In general, the data on early-type galaxies seem to prefer isothermal models on the scales relevant to interpreting time delays, while constant M/L models are firmly ruled out. If we must ultimately rely on the assumption that the density profiles of time delay lenses are similar to those of other early-type galaxies, the additional uncertainty added by this assumption will be small and calculable. Moreover, the assumption is no different from the assumptions of homogeneity used in other studies of the distance scale.

We can avoid any such assumptions by determining the density profiles of the time delay lenses directly. One approach is to measure the kinematic properties of the lens galaxy. Since the mass inside the Einstein ring is fixed by the image geometry, the velocity dispersion is controlled by the central concentration of the density. Treu & Koopmans (2002b) apply this method to PG1115+080 and argue that the observed velocity dispersion requires a mass distribution between the isothermal and constant M/L limits with $H_0 = 59^{+12}_{-7}$ km s^{-1} Mpc^{-1}. Note, however, that with this velocity dispersion the lens galaxy does not lie on the fundamental plane, which is very peculiar. A second approach is to use deep infrared imaging to determine the structure of the lensed host galaxy of the quasar (Kochanek, Keeton, & McLeod 2001). The location and width of the Einstein ring depends on both the radial and angular structure of the potential, although the sensitivity to the radial structure of the lens is weak when the annulus bracketing the lensed images is thin ($\Delta R/\langle R \rangle$ small; Saha & Williams 2001). This method will work best for asymmetric two-image lenses ($\Delta R/\langle R \rangle \approx 1$). The necessary data can be obtained with *HST* for most time delay lenses.

We can also focus our monitoring campaigns on lenses already known to have well-constrained density profiles. For the reasons we have already discussed, systems with multi-component sources, well-studied images of the host galaxy or stellar dynamical measurements will have better constrained density profiles than those without any additional constraints. We can also avoid most of the uncertainties in the density profile by measuring the time delays of very low-redshift lenses. When the lens is very close to the observer, the images lie very close to the center of the lens where the stellar mass dominates. A constant M/L model then becomes a very good approximation and we need worry little about the amount or the distribution of the dark matter. The one such candidate at present, Q2237+0305 at $z_l = 0.04$, will have very short delays, but these could be measured by an X-ray monitoring program using the *Chandra* observatory.

Finally, the statistical properties of larger samples of time delay lenses will also help to solve the problem. We already saw in §8.6 that the "simple" time delay lenses must have very similar densities, independent of H_0. This already means that the implications for H_0 no longer depend on individual lenses. In some ways the similarity of the densities is not an advantage—it is actually easier to determine H_0 if the density distributions are inhomogeneous (Kochanek 2004). On the other hand, there are well-defined approaches to using the statistical properties of lens models to estimate parameters that cannot be determined from the models of the individual systems (see Kochanek 2001). The statistics of the problematic flux ratios observed in the lenses (see §8.5) may also provide a means of estimating $\langle \kappa \rangle$. Schechter & Wambsganss (2002) point out that in four-image quasar lenses there is a tendency for the brightest saddle point image to be demagnified compared to reasonable lens models. Microlensing by the stars can naturally explain the observations if the surface density of stars is a small fraction of the total surface density near the images ($\kappa_* \ll \langle \kappa \rangle$), which would rule out constant M/L models where $\kappa_* \simeq \langle \kappa \rangle$.

8.8   Conclusions

The determination of H_0 using gravitational lens time delays has come of age. The last few years have seen a dramatic increase in the number of delay measurements, and there is no barrier other than sociology to rapidly increasing the sample. The interpretation of time delays requires a model for the gravitational potential of the lens. Fortunately, the physics determining time delays is well understood, and the only important variable is the average surface density $\langle \kappa \rangle$ of the lens near the images for which the delay is measured. Unfortunately, there is a tendency in the literature to conceal rather than to illuminate this understanding. Provided a lens does not lie in a cluster where the cluster potential cannot be described by a simple expansion, any lens model that includes the parameters needed to vary the average surface density of the lens near the images and to change the ratio between the quadrupole moment of the lens and the environment includes all the parameters needed to model time delays, to estimate the Hubble constant, and to understand the systematic uncertainties in the results. *All differences between estimates of the Hubble constant for the simple time delay lenses can be understood on this basis.*

Models for the four time delay lenses that can be modeled using a single lens galaxy predict that $H_0 = 48 \pm 3$ km s^{-1} Mpc^{-1} if the lens galaxies have isothermal density profiles with flat rotation curves, and $H_0 = 71 \pm 3$ km s^{-1} Mpc^{-1} if they have constant mass-to-light ratios. The Key Project estimate of $H_0 = 72 \pm 8$ km s^{-1} Mpc^{-1} agrees with the lensing results only if the lenses have little dark matter. We have strong theoretical prejudices and estimates from other observations of early-type galaxies that we should favor the isothermal models over the constant M/L models. We feel that we have reached the point where the results from gravitational lens time delays deserve serious attention and that there is a reasonable likelihood that the local estimates of H_0 are too high. A modest investment of telescope time would allow the measurement of roughly 5–10 time delays per year, and these new delays would rapidly test the current results. Other observations of time delay lenses to measure the velocity dispersions of the lens galaxies or to determine the geometry of the lensed images of the quasar host galaxy can be used to constrain the mass distributions directly. The systematic problems associated with the density profile are soluble not only in theory but also in practice, and the investment of the community's resources would be significantly less that than already invested in the distance scale.

The time delay measurements also provide a new probe of the density structure of galaxies at the boundary between the baryonic and dark matter dominated parts of galaxies (projected distances of 1–2 effective radii). Even if we ignore the actual value of H_0, we can still study the differences in the surface densities. For example, we can show that the present sample of simple lenses must have very similar surface densities. This region is very difficult to study with other probes.

Finally, the time delay measurements can be used to determine cosmological parameters. Time delays basically measure the distance to the lens galaxy, so we can make the same sorts of cosmological measurements as Type Ia supernovae. If the variations in $\langle \kappa \rangle$ between lens galaxies are small, as seem to be indicated by the present data, then the accuracy of the differential measurements will be very good. The present sample has little sensitivity to the cosmological model even with the mass distribution fixed because the time delay uncertainties are still too large and the redshift range is too restricted ($z_l = 0.31$ to 0.72). If we assume that other methods will determine the distance factors more accurately and rapidly, then we can use the time delays to study the evolution of galaxy mass distributions with redshift.

Acknowledgements. CSK thanks D. Rusin and J. Winn for their comments. CSK is supported by the Smithsonian Institution and NASA ATP grant NAG5-9265. PLS is supported by NSF grant AST-0206010.

References

Angonin-Willaime, M.-C., Soucail, G., & Vanderriest, C. 1994, A&A, 291, 411

Barkana, R. 1996, ApJ, 468, 17

———. 1997, ApJ, 489, 21

Barkana, R., Lehár, J., Falco, E. E., Grogin, N. A., Keeton, C. R., & Shapiro, I. I. 1999, ApJ, 520, 479

Bernstein, G., & Fischer, P. 1999, AJ, 118, 14

Beskin, G. M., & Oknyanskij, V. L. 1995, A&A, 304, 341

Biggs, A. D., Browne, I. W. A., Helbig, P., Koopmans, L. V. E., Wilkinson, P. N., & Perley, R. A. 1999, MNRAS, 304, 349

Blandford, R. D., & Narayan, R. 1987, ApJ, 310, 568

Bullock, J. S., Kolatt, T. S., Sigad, Y., Somerville, R. S., Kravtsov, A. V., Klypin, A. A., Primack, J. R., & Dekel, A. 2001, MNRAS, 321, 559

Burud, I., et al. 2000, ApJ, 544, 117

———. 2002a, A&A, 383, 71

———. 2002b, A&A, 391, 481

Chae, K.-H. 1999, ApJ, 524, 582

Chartas, G., Dai, X., & Garmire, G. P. 2004, in Carnegie Observatories Astrophysics Series, Vol. 2: Measuring and Modeling the Universe, ed. W. L. Freedman (Pasadena: Carnegie Observatories, http://www.ociw.edu/ociw/symposia/series/symposium2/proceedings.html)

Chartas, G., Gupta, V., Garmire, G., Jones, C., Falco, E. E., Shapiro, I. I., & Tavecchio, F. 2002, ApJ, 565, 96

Cohen, A. S., Hewitt, J. N., Moore, C. B., & Haarsma, D. B. 2000, ApJ, 545, 578

Cohn, J. D., Kochanek, C. S., McLeod, B. A., & Keeton, C. R. 2001, ApJ, 554, 1216

Courbin, F., Meylan, G., Kneib, J.-P., & Lidman, C. 2002, ApJ, 575, 95

Dalal, N., & Kochanek, C. S. 2002, ApJ, 572, 25

Fassnacht, C. D, Xanthopoulos, E., Koopmans, L. V. E., & Rusin, D. 2002, ApJ, 581, 823

Falco, E. E., Gorenstein, M. V., & Shapiro, I. I. 1985, ApJ, 289, L1

Fischer, P., Bernstein, G., Rhee, G., & Tyson, J. A. 1997, AJ, 113, 521

Florentin-Nielsen, R. 1984, A&A, 138, L19

Freedman, W. L., et al. 2001, ApJ, 553, 47

Gerhard, O., Kronawitter, A., Saglia, R. P., & Bender, R. 2001, AJ, 121, 1936

Gil-Merino, R., Wisotzki, L., & Wambsganss, J. 2002, A&A, 381, 428

Gorenstein, M. V., Falco, E. E., & Shapiro, I. I. 1988, ApJ, 327, 693

Haarsma, D. B., Hewitt, J. N., Lehár, J., & Burke, B. F. 1999, AJ, 510, 64

Hjorth, J., et al., 2002, ApJ, 572, L11

Impey, C. D., Falco, E. E., Kochanek, C. S., Lehár, J., McLeod, B. A., Rix, H.-W., Peng, C. Y., & Keeton, C. R. 1998, ApJ, 509, 551

Keeton, C. R., et al. 2000, ApJ, 542, 74

Keeton, C. R. 2004, ApJ, submitted (astro-ph/0102340)

Keeton, C. R., Kochanek, C. S., & Seljak, U. 1997, ApJ, 482, 604

Kochanek, C. S. 2001, in The Shapes of Galaxies and Their Dark Halos, ed. P. Natarajan (Singapore: World Scientific), 62

———. 2002, ApJ, 578, 25

———. 2003, ApJ, 583, 49

———. 2004, ApJ, submitted (astro-ph/0204043)

Kochanek, C. S., Keeton, C. R., & McLeod, B. A. 2001, ApJ, 547, 50

Koopmans, L. V. E., de Bruyn, A. G., Xanthopoulos, E., & Fassnacht, C. D. 2000, A&A, 356, 391

Kundić, T., et al. 1997, ApJ, 482, 75

Lehár, J., et al. 2000, ApJ, 536, 584

Lovell, J. E. J., Jauncey, D. L., Reynolds, J. E., Wieringa, M. H., King, E. A., Tzioumis, A. K., McCulloch, P. M., & Edwards, P. G. 1998, ApJ, 508, L51

Loewenstein, M., & Mushotzky, R. F. 2004, ApJ, submitted

McKay, T. A., et al. 2002, ApJ, 571, L85

Metcalf, R. B., & Madau, P. 2001, ApJ, 563, 9

Morgan, N. D., Chartas, G., Malm, M., Bautz, M. W., Burud, I., Hjorth, J., Jones, S. E., & Schechter, P. L. 2001, ApJ, 555, 1

Narayan, R., & Bartelmann, M. 1999, in Formation and Structure in the Universe, ed. A. Dekel & J. P. Ostriker (Cambridge: Cambridge Univ. Press), 360

Navarro, J. F., Frenk, C. S., & White, S. D. M. 1996, ApJ, 462, 563

Ofek, E. O., & Maoz, D. 2003, ApJ, 594, 101

Patnaik, A. R., & Narasimha, D. 2001, MNRAS, 326, 1403

Pelt, J., Kayser, R., Refsdal, S., & Schramm, T. 1996, A&A, 305, 97

Pelt, J., Refsdal, S., & Stabell, R. 2002, A&A, 289, L57

Press, W. H., Rybicki, G. B., & Hewitt, J. N. 1992a, ApJ, 385, 404

——. 1992b, ApJ, 385, 416

Refsdal, S. 1964, 128, 307

Romanowsky, A. J., & Kochanek, C. S. 1999, ApJ, 516, 18

Saha, P. 2000, AJ, 120, 1654

Saha, P., & Williams, L. L. R. 2001, AJ, 122, 585

Schechter, P. L., et al. 1997, ApJ, 475, L85

——. 2003, ApJ, 584, 657

Schechter, P. L., & Wambsganss, J. 2002, ApJ, 580, 685

Schild, R., & Cholfin, B. 1986, ApJ, 300, 209

Schild, R., & Thomson, D. J. 1997, AJ, 113, 130

Schneider, P., Ehlers, J., & Falco, E. E. 1992, Gravitational Lenses, (Berlin: Springer-Verlag)

Treu, T., & Koopmans, L. V. E. 2002a, ApJ, 568, L5

——. 2002b, MNRAS, 337, 6

Vanderriest, C., Schneider, J., Herpe, G., Chevreton, M., Moles, M., & Wlerick, G. 1989, A&A, 215, 1

Wang, X., Tegmark, M., & Zaldarriaga, M. 2002, Phys. Rev. D, 65, 123001

Williams, L. L. R., & Saha, P. 2000, AJ, 199, 439

Winn, J. N., Kochanek, C. S., McLeod, B. A., Falco, E. E., Impey, C. D., & Rix, H.-W. 2002, ApJ, 575, 103

Winn, J. N., Rusin, D., & Kochanek, C. S. 2003, ApJ, 587, 80

Witt, H. J., Mao, S., & Keeton, C. R. 2000, ApJ, 544, 98

Wozniak, P. R., Udalski, A., Szymanski, M., Kubiak, M., Pietrzynski, G., Soszynski, I., & Zebrun, K. 2000, ApJ, 540, L65

Zhao, H., & Qin, B. 2003, ApJ, 582, 2

——. 2004, ApJ, submitted (astro-ph/0209304)

9

Measuring the Hubble constant with the Sunyaev-Zel'dovich effect

ERIK D. REESE

The University of California, Berkeley

Abstract

Combined with X-ray imaging and spectral data, observations of the Sunyaev-Zel'dovich effect (SZE) can be used to determine direct distances to galaxy clusters. These distances are independent of the extragalactic distance ladder and do not rely on clusters being standard candles or rulers. Observations of the SZE have progressed from upper limits to high signal-to-noise ratio detections and imaging of the SZE. SZE/X-ray determined distances to galaxy clusters are beginning to trace out the theoretical angular-diameter distance relation. The current ensemble of 41 SZE/X-ray distances to galaxy clusters imply a Hubble constant of $H_0 \approx 61 \pm 3 \pm 18$ km s^{-1} Mpc^{-1}, where the uncertainties are statistical followed by systematic at 68% confidence. With a sample of high-redshift galaxy clusters, SZE/X-ray distances can be used to measure the geometry of the Universe.

9.1   Introduction

Analysis of the Sunyaev-Zel'dovich effect (SZE) and X-ray data provides a method of directly determining distances to galaxy clusters at any redshift. Clusters of galaxies contain hot ($k_B T_e \approx 10$ keV) gas, known as the intracluster medium (ICM), trapped in their potential wells. Cosmic microwave background (CMB) photons passing through a massive cluster interact with the energetic ICM electrons with a probability of $\tau \approx 0.01$. This inverse-Compton scattering preferentially boosts the energy of a scattered CMB photon, causing a small ($\lesssim 1$ mK) distortion in the CMB spectrum, known as the Sunyaev-Zel'dovich effect (Sunyaev & Zel'dovich 1970, 1972). The SZE is proportional to the pressure integrated along the line of sight, $\Delta T \propto \int n_e T_e d\ell$. X-ray emission from the ICM has a different dependence on the density $S_x \propto \int n_e^2 \Lambda_{eH} d\ell$, where Λ_{eH} is the X-ray cooling function. Taking advantage of the different density dependences and with some assumptions about the geometry of the cluster, the distance to the cluster may be determined. SZE and X-ray determined distances are independent of the extragalactic distance ladder and provide distances to high-redshift galaxy clusters. This method does not rely on clusters being standard candles or rulers and relies only on relatively simple properties of highly ionized plasma.

The promise of direct distances has been one of the primary motivations for SZE observations. Efforts over the first two decades after the SZE was first proposed in 1970 (Sunyaev & Zel'dovich 1970, 1972) yielded few reliable detections. Over the last decade, new detectors and observing techniques have allowed high-quality detections and images of the effect for more than 60 clusters with redshifts as high as one. SZE observations are routine enough to build up samples of clusters and place constraints on cosmological parameters.

The SZE offers a unique and powerful observational tool for cosmology. Distances to galaxy clusters yield a measurement of the Hubble constant, H_0. With a sample of high-redshift clusters, SZE and X-ray distances can be used to determine the geometry of the Universe. In addition, the SZE has been used to measure gas fractions in galaxy clusters (e.g., Myers et al. 1997; Grego et al. 2001), which can be used to measure the matter density of the Universe, Ω_m, assuming the composition of clusters represents a fair sample of the universal composition. Upcoming deep, large-scale SZE surveys will measure the evolution of the number density of galaxy clusters, which is critically dependent on the underlying cosmology. In principle, the equation of state of the "dark energy" may be determined from the evolution of the number density of clusters.

In this review, we first outline the properties of the SZE in §9.2 and provide a brief overview of the current state of the observations in §9.3. SZE/X-ray determined distances are discussed in §9.4, briefly discussing the current state of SZE/X-ray distances, sources of systematics, and future potential. SZE surveys are briefly discussed in §9.5 and a summary is given in §9.6. The physics of the SZE is covered in previous reviews (Sunyaev & Zel'dovich 1980; Rephaeli 1995; Birkinshaw 1999), with Birkinshawi (1999) and Carlstrom et al. (2000) providing reviews of the observations. Carlstrom, Holder, & Reese (2002) provide a review with focus on cosmology from the SZE, with special attention to SZE surveys.

9.2 The Sunyaev-Zel'dovich Effect

9.2.1 Thermal Sunyaev-Zel'dovich Effect

The SZE is a small spectral distortion of the CMB spectrum caused by the scattering of the CMB photons off a distribution of high-energy electrons. We only consider the SZE caused by the hot thermal distribution of electrons provided by the ICM of galaxy clusters. CMB photons passing through the center of a massive cluster have a $\tau_e \approx 0.01$ probability of interacting with an energetic ICM electron. The resulting inverse-Compton scattering preferentially boosts the energy of the CMB photon, causing a small ($\lesssim 1$ mK) distortion in the CMB spectrum. To illustrate the small effect, Figure 9.1 shows the SZE spectral distortion for a fictional cluster that is over 1000 times more massive than a typical cluster. The SZE appears as a decrease in the intensity of the CMB at frequencies below $\lesssim 218$ GHz and as an increase at higher frequencies.

The SZE spectral distortion of the CMB, expressed as a temperature change ΔT_{SZE} at dimensionless frequency $x \equiv (h\nu)/(k_B T_{CMB})$, is given by

$$\frac{\Delta T_{SZE}}{T_{CMB}} = f(x)\, y = f(x) \int n_e \frac{k_B T_e}{m_e c^2} \sigma_T \, d\ell, \tag{9.1}$$

where y is the Compton y-parameter, which for an isothermal cluster equals the optical depth times the fractional energy gain per scattering. Here, σ_T is the Thomson cross-section, n_e is the electron number density, T_e is the electron temperature, k_B is the Boltzmann's constant, $m_e c^2$ is the electron rest-mass energy, and the integration is along the line of sight. The frequency dependence of the SZE is

$$f(x) = \left(x \frac{e^x + 1}{e^x - 1} - 4 \right) [1 + \delta_{SZE}(x, T_e)], \tag{9.2}$$

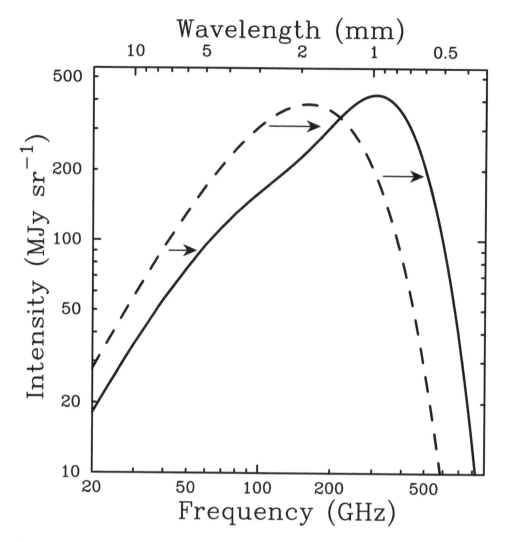

Fig. 9.1. The CMB spectrum, undistorted (dashed line) and distorted by the SZE (solid line). Following Sunyaev & Zel'dovich (1980), to illustrate the effect, the SZE distortion shown is for a fictional cluster 1000 times more massive than a typical massive galaxy cluster. The SZE causes a decrease in the CMB intensity at frequencies $\lesssim 218$ GHz (~ 1.4 mm) and an increase at higher frequencies.

where $\delta_{SZE}(x, T_e)$ is the relativistic correction to the frequency dependence. Note that $f(x) \to -2$ in the nonrelativistic and Rayleigh-Jeans (RJ) limits.

It is worth noting that $\Delta T_{SZE}/T_{CMB}$ is independent of redshift, as shown in Equation 9.1. This unique feature of the SZE makes it a potentially powerful tool for investigating the high-redshift Universe.

Expressed in units of specific intensity, common in millimeter SZE observations, the thermal SZE is

$$\Delta I_{SZE} = g(x)I_0 y, \tag{9.3}$$

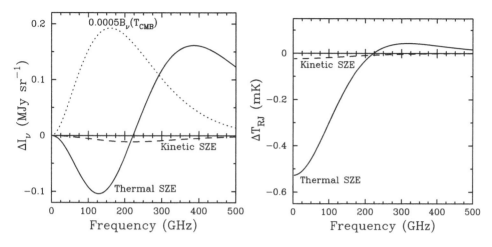

Fig. 9.2. Spectral distortion of the CMB radiation due to the SZE. The left panel shows the intensity, and the right panel shows the Rayleigh-Jeans brightness temperature. The thick solid line is the thermal SZE, and the dashed line is the kinetic SZE. For reference the 2.7 K thermal spectrum for the CMB intensity, scaled by 0.0005, is shown by the dotted line in the left panel. The cluster properties used to calculate the spectra are an electron temperature of 10 keV, a Compton y-parameter of 10^{-4}, and a peculiar velocity of 500 km s^{-1}.

where $I_0 = 2(k_B T_{CMB})^3/(hc)^2$ and the frequency dependence is given by

$$g(x) = \frac{x^4 e^x}{(e^x-1)^2} \left(x\frac{e^x+1}{e^x-1} - 4 \right) [1+\delta_{SZE}(x,T_e)]. \qquad (9.4)$$

ΔT_{SZE} and ΔI_{SZE} are simply related by the derivative of the blackbody with respect to temperature, $|dB_\nu/dT|$.

The spectral distortion of the CMB spectrum by the thermal SZE is shown in Figure 9.2 (solid line) for a realistic massive cluster ($y = 10^{-4}$), in units of intensity (left panel) and RJ brightness temperature (right panel). The RJ brightness is shown because the sensitivity of a radio telescope is calibrated in these units. It is defined simply by $I_\nu = (2k_B\nu^2/c^2)T_{RJ}$, where I_ν is the intensity at frequency ν, k_B is Boltzmann's constant, and c is the speed of light. The CMB blackbody spectrum, $B_\nu(T_{CMB})$, multiplied by 0.0005 (dotted line), is also shown for comparison. Note that the spectral signature of the thermal effect is distinguished readily from a simple temperature fluctuation of the CMB. The kinetic SZE distortion is shown by the dashed curve (§9.2.2). In the nonrelativistic regime, it is indistinguishable from a CMB temperature fluctuation.

The gas temperatures measured in massive galaxy clusters are around $k_B T_e \approx 10$ keV (Mushotzky & Scharf 1997; Allen & Fabian 1998) and are measured to be as high as ~ 17 keV in the galaxy cluster 1E 0657−56 (Tucker et al. 1998). At these temperatures, electron velocities are becoming relativistic, and small corrections are required for accurate interpretation of the SZE. There has been considerable theoretical work to include relativistic corrections to the SZE (Wright 1979; Fabbri 1981; Rephaeli 1995; Rephaeli & Yankovitch 1997; Stebbins 1997; Challinor & Lasenby 1998; Itoh, Kohyama, & Nozawa 1998; Nozawa, Itoh, & Kohyama 1998a; Sazonov & Sunyaev 1998a, 1998b; Challinor & Lasenby 1999; Molnar & Birkinshaw 1999; Dolgov et al. 2001). All of these derivations agree for $k_B T_e \lesssim 15$

keV, appropriate for galaxy clusters. For a massive cluster with $k_B T_e \approx 10$ keV ($k_B T_e / m_e c^2 \approx 0.02$), the relativistic corrections to the SZE are of order a few percent in the RJ portion of the spectrum, but can be substantial near the null of the thermal effect. Convenient analytical approximations to fifth order in $k_B T_e / m_e c^2$ are presented in Itoh et al. (1998).

The measured SZE spectrum of Abell 2163, spanning the decrement and increment with data obtained from different telescopes and techniques, is shown in Figure 9.3 (Holzapfel et al. 1997a; Désert et al. 1998; LaRoque et al. 2004). Also plotted is the best-fit model (solid) consisting of thermal (dashed) and kinetic (dotted) SZE components. The SZE spectrum is a good fit to the data, demonstrating the consistency and robustness of modern SZE measurements.

The most important features of the thermal SZE are: (1) it is a small spectral distortion of the CMB, of order ~ 1 mK, which is proportional to the cluster pressure integrated along the line of sight (Eq. 9.1); (2) it is independent of redshift; and (3) it has a unique spectral signature with a decrease in the CMB intensity at frequencies $\lesssim 218$ GHz and an increase at higher frequencies.

9.2.2   *Kinetic Sunyaev-Zel'dovich Effect*

If the cluster is moving with respect to the CMB rest frame, there will be an additional spectral distortion due to the Doppler effect of the cluster bulk velocity on the scattered CMB photons. If a component of the cluster velocity, v_{pec}, is projected along the line of sight to the cluster, then the Doppler effect will lead to an observed distortion of the CMB spectrum, referred to as the kinetic SZE. In the nonrelativistic limit, the spectral signature of the kinetic SZE is a pure thermal distortion of magnitude

$$\frac{\Delta T_{SZE}}{T_{CMB}} = -\tau_e \left(\frac{v_{pec}}{c} \right), \tag{9.5}$$

where v_{pec} is along the line of sight; that is, the emergent spectrum is still described completely by a Planck spectrum, but at a slightly different temperature, lower (higher) for positive (negative) peculiar velocities (Sunyaev & Zel'dovich 1972; Phillips 1995; Birkinshaw 1999). Figure 9.2 illustrates the kinetic SZE (dashed) for a typical galaxy cluster with a peculiar velocity of 500 km s^{-1}. Figure 9.3 shows the SZE spectrum of the galaxy cluster A2163 along with the best-fit model consisting of thermal (dashed) and kinetic (dotted) SZE components.

Relativistic perturbations to the kinetic SZE are due to the Lorentz boost to the electrons provided by the bulk velocity (Nozawa et al. 1998a; Sazonov & Sunyaev 1998a). The leading term is of order $(k_B T_e / m_e c^2)(v_{pec}/c)$, and for a 10 keV cluster moving at 1000 km s^{-1} the effect is about an 8% correction to the nonrelativistic term. The $(k_B T_e / m_e c^2)^2 (v_{pec}/c)$ term is only about 1% of the nonrelativistic kinetic SZE, and the $(v_{pec}/c)^2$ term is only 0.2%.

9.3   Measurements of the SZE

In the 20 years following the first papers by Sunyaev & Zel'dovich (1970, 1972), there were few firm detections of the SZE despite a considerable amount of effort (see, for a review of early experiments, Birkinshaw 1999). Over the last several years, however, observations of the effect have progressed from low signal-to-noise ratio detections and upper limits to high-confidence detections and detailed images. In this section we briefly review the current state of SZE observations.

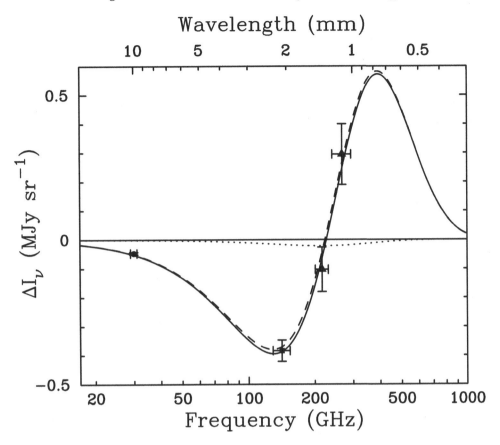

Fig. 9.3. The measured SZE spectrum of Abell 2163. The data point at 30 GHz is from BIMA (LaRoque et al. 2004), at 140 GHz is the weighted average of Diabolo and SuZIE measurements (filled square; Holzapfel et al. 1997a; Désert et al. 1998), and at 218 GHz and 270 GHz from SuZIE (filled triangles; Holzapfel et al. 1997a). Uncertainties are at 68% confidence with the FWHM of the observing bands shown. The best-fit thermal and kinetic SZE spectra are shown by the dashed line and the dotted lines, respectively, with the spectra of the combined effect shown by the solid line. The limits on the Compton y-parameter and the peculiar velocity are $y_0 = 3.71^{+0.36+0.33}_{-0.36-0.16} \times 10^{-4}$ and $v_p = 320^{+880+480}_{-740-440}$ km s^{-1}, respectively, with statistical followed by systematic uncertainties at 68% confidence (Holzapfel et al. 1997a; LaRoque et al. 2004).

The dramatic increase in the quality of the observations is due to improvements both in low-noise detection systems and in observing techniques, usually using specialized instrumentation to control carefully the systematics that often prevent one from obtaining the required sensitivity. The sensitivity of a low-noise radio receiver available 20 years ago should have easily allowed the detection of the SZE toward a massive cluster. Most attempts, however, failed due to uncontrolled systematics. Now that the sensitivities of detector systems have improved by factors of 3 to 10, it is clear that the goal of all modern SZE instruments is the control of systematics. Such systematics include, for example, the spatial and temporal variations in the emission from the atmosphere and the surrounding ground, as well as gain instabilities inherent to the detector system used.

The observations must be conducted on the appropriate angular scales. Galaxy clusters have a characteristic size scale of order a Mpc. For a reasonable cosmology, a Mpc subtends an arcminute or more at any redshift. Low-redshift clusters will subtend a much larger angle; for example, the angular extent of the Coma cluster ($z = 0.024$) is of order a degree (core radius $\sim 10'$; Herbig et al. 1995). The detection of extended low-surface brightness objects requires precise differential measurements made toward widely separated directions on the sky. The large angular scale presents challenges to control offsets due to differential ground pick-up and atmospheric variations.

9.3.1 *Sources of Astronomical Contamination and Confusion*

There are three main sources of astronomical contamination and confusion that must be considered when observing the SZE: (1) CMB primary anisotropies, (2) radio point sources, and (3) dust from the Galaxy and external galaxies. For distant clusters with angular extents of a few arcminutes or less, the CMB anisotropy is expected (Hu & White 1997) and found to be damped considerably on these scales (Church et al. 1997; Subrahmanyan et al. 2000; Dawson et al. 2001; see also Holzapfel et al. 1997b and LaRoque et al. 2004 for CMB limits to SZE contamination). For nearby clusters, or for searches for distant clusters using beams larger than a few arcminutes, the intrinsic CMB anisotropy must be considered. The unique spectral behavior of the thermal SZE can be used to separate it from the intrinsic CMB in these cases. Note, however, that for such cases it will not be possible to separate the kinetic SZE effects from the intrinsic CMB anisotropy without relying on the very small spectral distortions of the kinetic SZE due to relativistic effects.

Historically, the major source of contamination in the measurement of the SZE has been radio point sources. It is obvious that emission from point sources located along the line of the sight to the cluster could fill in the SZE decrement, leading to an underestimate. Radio point sources can also lead to overestimates of the SZE decrement—for example, point sources in the reference fields of single-dish observations. The radio point sources are variable and therefore must be monitored. Radio emission from the cluster member galaxies, from the central dominant (cD) galaxy in particular, is often the largest source of radio point source contamination, at least at high radio frequencies (Cooray et al. 1998; LaRoque et al. 2004).

At frequencies near the null of the thermal SZE and higher, dust emission from extragalactic sources as well as dust emission from our own Galaxy must be considered. At the angular scales and frequencies of interest for most SZE observations, contamination from diffuse Galactic dust emission will not usually be significant and is easily compensated. Consider instead the dusty extragalactic sources such as those that have been found toward massive galaxy clusters with the SCUBA bolometer array (Smail et al. 1997). Spectral indices for these sources are estimated to be $\sim 1.5-2.5$ (Fischer & Lange 1993; Blain 1998). Sources with 350 GHz (850 μm) flux densities greater than 8 mJy are common, and all clusters surveyed had multiple sources with flux densities greater than 5 mJy. This translates into an uncertainty in the peculiar velocity for a galaxy cluster of roughly 1000 km s^{-1} (see Carlstrom et al. 2002 for details).

As with SZE observations at radio frequencies, the analyses of high-frequency observa-

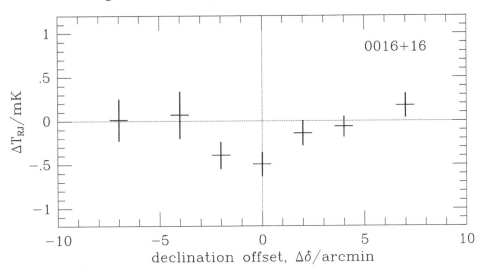

Fig. 9.4. Measurement of the SZE profile across the galaxy cluster Cl 0016 + 16 obtained with the OVRO 40 m telescope (Hughes & Birkinshaw 1998). The observed profile provided confidence in the reliability of SZE detections.

tions must also consider the effects of point sources and require either high dynamic angular range, large spectral coverage, or both to separate the point source emission from the SZE.

9.3.2 Single-dish Observations

The first measurements of the SZE were made with single-dish radio telescopes at centimeter wavelengths. Advances in detector technology made the measurements possible, although early observations appear to have been plagued by systematic errors that led to irreproducible and inconsistent results. Eventually, successful detections using beam-switching techniques were obtained. During this period, the pioneering work of Birkinshaw and collaborators with the OVRO 40 m telescope stands out for its production of results that served to build confidence in the technique (Birkinshaw, Gull, & Northover 1978a, 1978b; Birkinshaw, Hughes, & Arnaud 1991). Figure 9.4 shows the OVRO 40 m scan through the galaxy cluster Cl 0016 + 16. More recently, leading and trailing beam-switching techniques have been used successfully with the OVRO 5 m telescope at 32 GHz to produce reliable detections of the SZE in several intermediate-redshift clusters (Herbig et al. 1995; Myers et al. 1997; Mason, Myers, & Readhead 2001). The SEST 15 m and IRAM 30 m telescopes have been used with bolometric detectors at 140 GHz and chopping mirrors to make significant detections of the SZE in several clusters (Andreani et al. 1996, 1999; Désert et al. 1998; Pointecouteau et al. 1999, 2001). The Nobeyama 45 m telescope has also been been used at 21, 43, and 150 GHz to detect and map the SZE (Komatsu et al. 1999, 2001).

The Sunyaev-Zel'dovich Infrared Experiment (SuZIE) uses its six-element 140 GHz bolometer array to observe in a drift-scanning mode, where the telescope is fixed and the rotation of the Earth moves the beams across the sky. Using this drift-scanning technique, the SuZIE experiment has produced high signal-to-noise ratio strip maps of the SZE emission in several clusters (Holzapfel et al. 1997a; Mauskopf et al. 2000).

Because of the high sensitivity of bolometric detectors at millimeter wavelengths, single-dish experiments are ideally suited for the measurement of the SZE spectrum. By observing at several millimeter frequencies, these instruments should be able to separate the thermal and kinetic SZEs from atmospheric fluctuations and sources of astrophysical confusion.

The measured SZE spectrum of Abell 2163, spanning the decrement and increment with data obtained from different telescopes and techniques, is shown in Figure 9.3 (Holzapfel et al. 1997a; Désert et al. 1998; LaRoque et al. 2004). The SZE spectrum is a good fit to the data, demonstrating the consistency and robustness of modern SZE measurements.

Single-dish observations of the SZE are just beginning to reach their potential, and the future is very promising. The development of large-format, millimeter-wavelength bolometer arrays will increase the mapping speed of current SZE experiments by orders of magnitude. To the extent that atmospheric fluctuations are common across a bolometric array, it will be possible to realize the intrinsic sensitivity of the detectors. Operating from high astronomical sites with stable atmospheres and exceptionally low precipitable water vapor, future large-format bolometer arrays have the potential to produce high signal-to-noise ratio SZE images and search for distant SZE clusters with unprecedented speed.

9.3.3   *Interferometric Observations*

The stability and spatial filtering inherent to interferometry has been exploited to make high-quality images of the SZE. The stability of an interferometer is due to its ability to perform simultaneous differential sky measurements over well-defined spatial frequencies. The spatial filtering of an interferometer also allows the emission from radio point sources to be separated from the SZE emission.

There are several other features that allow an interferometer to achieve extremely low systematics. For example, only signals that correlate between array elements will lead to detected signal. For most interferometers, this means that the bulk of the sky noise for each element will not lead to signal. Amplifier gain instabilities for an interferometer will not lead to large offsets or false detections, although, if severe, they may lead to somewhat noisy signal amplitude. To remove the effects of offsets or drifts in the electronics, as well as the correlation of spurious (noncelestial) sources of noise, the phase of the signal received at each telescope is modulated, and then the proper demodulation is applied to the output of the correlator.

The spatial filtering of an interferometer also allows the emission from radio point sources to be separated from the SZE emission. This is possible because at high angular resolution ($\lesssim 10''$) the SZE contributes very little flux. This allows one to use long baselines, which give high angular resolution, to detect and monitor the flux of radio point sources, while using short baselines to measure the SZE. Nearly simultaneous monitoring of the point sources is important, as they are often time variable. The signal from the point sources is then easily removed, to the limit of the dynamic range of the instrument, from the short-baseline data, which are sensitive also to the SZE.

For the reasons given above, interferometers offer an ideal way to achieve high brightness sensitivity for extended low-surface brightness emission, at least at radio wavelengths. Most interferometers, however, were not designed for imaging low-surface brightness sources. Interferometers have been built traditionally to obtain high angular resolution and thus have employed large individual elements for maximum sensitivity to small-scale emission. As a result, special-purpose interferometric systems have been built for imaging the SZE (Jones et al. 1993; Carlstrom, Joy, & Grego 1996; Padin et al. 2001). All of them have taken advantage of low-noise HEMT amplifiers (Pospieszalski et al. 1995) to achieve high sensitivity.

The first interferometric detection (Jones et al. 1993) of the SZE was obtained with the Ryle Telescope (RT). The RT was built from the 5 Kilometer Array, consisting of eight 13 m telescopes located in Cambridge, England, operating at 15 GHz with East-West configurations. Five of the telescopes can be used in a compact E-W configuration for imaging of the SZE (Jones et al. 1993; Grainge et al. 1993, 1996, 2002a, 2002b; Grainger et al. 2002; Saunders et al. 2003; Jones et al. 2004).

The OVRO and BIMA SZE imaging project uses 30 GHz (1 cm) low-noise receivers mounted on the OVRO* and BIMA† mm-wave arrays in California. They have produced SZE images toward 60 clusters to date (Carlstrom et al. 1996, 2000; Grego et al. 2000, 2001; Patel et al. 2000; Reese et al. 2000, 2002; Joy et al. 2001; LaRoque et al. 2004). A sample of their SZE images is shown in Figure 9.5. All contours are multiples of 2σ of each image, and the full-width at half maximum (FWHM) of the synthesized beam (PSF for this deconvolution) is shown in the lower left-hand corner of each image. Figure 9.5 also clearly demonstrates the independence of the SZE on redshift. All of the clusters shown have similarly high X-ray luminosities, and, as can be seen, the strength of the SZE signals are similar despite the factor of 5 in redshift. The OVRO and BIMA arrays support two-dimensional configurations of the telescopes, including extremely short baselines, allowing good synthesized beams for imaging the SZE of clusters at declinations greater than ~ -15 degrees.

The RT, OVRO, and BIMA SZE observations are insensitive to the angular scales required to image low-redshift ($z \ll 0.1$) clusters. Recently, however, the Cosmic Background Imager (CBI; Padin et al. 2001) has been used to image the SZE in a few nearby clusters (Udomprasert, Mason, & Readhead 2000). The CBI is composed of 13 0.9 m telescopes mounted on a common platform, with baselines spanning 1 m to 6 m. Operating in 10 1 GHz channels spanning $26 - 36$ GHz, it is sensitive to angular scales spanning $3'$ to $20'$. The large field of view of the CBI, 0.75 degrees FWHM, makes it susceptible to correlated contamination from terrestrial sources (i.e., ground emission). To compensate, they have adopted the same observing strategy as for single-dish observations (§9.3.2), by subtracting from the cluster data, data from leading and trailings fields offset by ± 12.5 minutes in Right Ascension from the cluster.

Interferometric observations of the SZE, as for single-dish observations, are just beginning to demonstrate their potential. Upcoming instruments will be over an order of magnitude more sensitive. This next generation of interferometric SZE instruments will conduct deep SZE surveys covering tens, and possibly hundreds, of square degrees. While not as fast as planned large-format bolometric arrays, the interferometers will be able to survey deeper and provide more detailed imaging. In particular, the high resolution and deep imaging provided by future heterogeneous arrays will provide a valuable tool for investigating cluster structure and its evolution. Such studies are necessary before the full potential of large SZE surveys for cosmology can be realized.

---

* An array of six 10.4 m telescopes located in the Owens Valley, CA, operated by Caltech.

† An array of 10 6.1 m mm-wave telescopes located at Hat Creek, CA, operated by the Berkeley-Illinois-Maryland-Association.

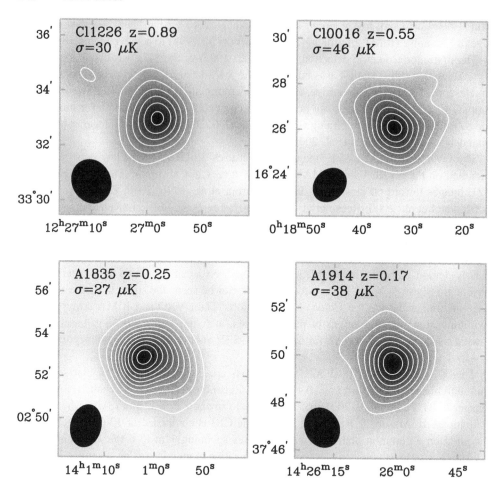

Fig. 9.5. Deconvolved interferometric SZE images for four galaxy clusters over a large red-shift range ($0.17 \leq z \leq 0.89$). The contours are multiples of 2σ, and negative contours are shown as solid lines. The FWHM ellipse of the synthesized beam (PSF) is shown in the lower-left corner of each panel. The rms, σ, appears in the top of each panel. Radio point sources were removed from three of the images shown. The interferometer was able to separate the point source emission from the SZE by using the high-resolution data obtained with long baselines. All of the clusters shown have similarly high X-ray luminosities, and, as can be seen, the strength of the SZE signals are similar despite the factor of 5 in redshift, illustrating the independence of the SZE on redshift.

9.4   The Cosmic Distance Scale from SZE/X-ray Distances

Several years after the SZE was first proposed (Sunyaev & Zel'dovich 1970, 1972), it was recognized that the distance to a cluster could be determined with a measure of its SZE and X-ray emission (Cavaliere, Danese, & de Zotti 1977; Boynton & Murray 1978; Cavaliere & Fusco-Femiano 1978; Gunn, Longair, & Rees 1978; Silk & White 1978; Birkin-shaw 1979). The distance is determined by exploiting the different density dependences of the SZE and X-ray emissions. The SZE is proportional to the first power of the density; $\Delta T_{SZE} \sim \int d\ell n_e T_e$, where n_e is the electron density, T_e is the electron temperature, and $d\ell$

is along the line-of-sight. The distance dependence is made explicit with the substitution $d\ell = D_A d\zeta$, where D_A is the angular-diameter distance of the cluster.

The X-ray emission is proportional to the second power of the density; $S_x \sim \int d\ell n_e^2 \Lambda_{eH}$, where Λ_{eH} is the X-ray cooling function. The angular-diameter distance is solved for by eliminating the electron density,[*] yielding

$$D_A \propto \frac{(\Delta T_0)^2 \Lambda_{eH0}}{S_{x0} T_{e0}^2} \frac{1}{\theta_c},\tag{9.6}$$

where these quantities have been evaluated along the line of sight through the center of the cluster (subscript 0) and θ_c refers to a characteristic scale of the cluster along the line of sight, whose exact meaning depends on the density model adopted. Only the characteristic scale of the cluster in the plane of the sky is measured, so one must relate the characteristic scales along the line of sight and in the plane of the sky. For detailed treatments of this calculation, see Birkinshaw et al. (1991) and Reese et al. (2000, 2002). Combined with the redshift of the cluster and the geometry of the Universe, one may determine the Hubble parameter, with the inverse dependences on the observables as that of D_A. With a sample of galaxy clusters, one fits the cluster distances versus redshift to the theoretical angular-diameter distance relation, with the Hubble constant as the normalization (see, e.g., Fig. 9.6).

9.4.1 Current Status of SZE/X-ray Distances

To date, there are 41 distance determinations to 26 different galaxy clusters from analysis of SZE and X-ray observations. All of these SZE/X-ray distances use *ROSAT* X-ray data and model the cluster gas as a spherical isothermal β model (Cavaliere & Fusco-Femiano 1976, 1978). The *ROSAT* data do not warrant a more sophisticated treatment. In Figure 9.6 we show all SZE-determined distances from high signal-to-noise ratio SZE experiments. The uncertainties shown are statistical at 68% confidence. The theoretical angular-diameter distance relation is shown for three cosmologies assuming $H_0 = 60$ km s^{-1} Mpc^{-1}.

There are currently three homogeneously analyzed samples of clusters with SZE distances: (1) a sample of seven nearby ($z < 0.1$) galaxy clusters observed with the OVRO 5 m telescope (grey solid triangles) that finds $H_0 = 66^{+14+15}_{-11-15}$ (Myers et al. 1997; Mason et al. 2001), (2) a sample of five intermediate-redshift ($0.14 < z < 0.3$) clusters from the RT interferometer that finds $H_0 = 65^{+8+15}_{-7-15}$ (dark grey solid squares) (Jones et al. 2004), and (3) a sample of 18 clusters with $0.14 < z < 0.83$ from interferometric observations by the OVRO and BIMA SZE imaging project, which infers $H_0 = 60^{+4+18}_{-4-13}$ (light grey solid stars) (Reese et al. 2002). The above Hubble constants assume a $\Omega_m = 0.3$, $\Omega_\Lambda = 0.7$ cosmology, and the uncertainties are statistical followed by systematic at 68% confidence. The treatment of uncertainties varies among the three cluster samples.

A fit to the ensemble of 41 SZE-determined distances yields $H_0 \approx 61 \pm 3 \pm 18$ km s^{-1} Mpc^{-1} for an $\Omega_m = 0.3$, $\Omega_\Lambda = 0.7$ cosmology, where the uncertainties are statistical followed by systematic at 68% confidence. Since many of the clusters are at high redshift, the best-fit Hubble constant will depend on the cosmology adopted; the best-fit Hubble constant shifts to 57 km s^{-1} Mpc^{-1} for an open $\Omega_m = 0.3$ universe, and to 54 km s^{-1} Mpc^{-1} for a flat $\Omega_m = 1$ geometry. The systematic uncertainty, discussed below, clearly dominates. The systematic uncertainty is approximate because it is complicated by shared systematics between some

---

[*] Similarly, one could eliminate D_A in favor of the central density, n_{e0}.

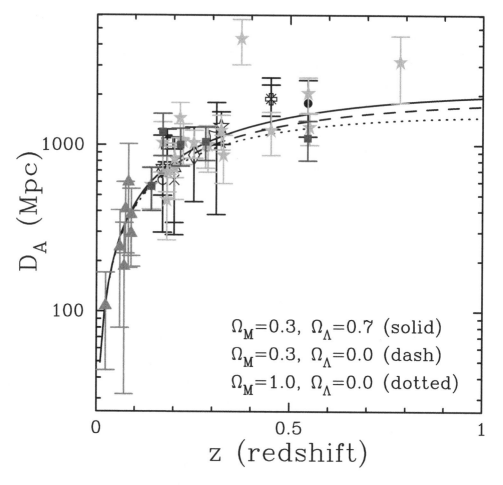

Fig. 9.6. SZE-determined distances versus redshift. The theoretical angular diameter distance relation is plotted for three different cosmologies, assuming $H_0 = 60$ km s^{-1} Mpc^{-1}: Λ—$\Omega_m = 0.3$, $\Omega_\Lambda = 0.7$ (solid line), open—$\Omega_m = 0.3$ (dashed), and flat—$\Omega_m = 1$ (dot-dashed). The clusters are beginning to trace out the angular-diameter distance relation. Three samples are highlighted: seven nearby clusters observed with the OVRO 5 m (grey solid triangles; Myers et al. 1997; Mason et al. 2001); five clusters from Ryle (dark grey solid squares; Grainge et al. 2002b; Saunders et al. 2003; Jones et al. 2004); and 18 clusters from the OVRO/BIMA SZE imaging project (light grey solid stars; Reese et al. 2000, 2002). Additional references: Birkinshaw et al. (1991), Birkinshaw & Hughes (1994), Holzapfel et al. (1997b), Hughes & Birkinshaw (1998), Lamarre et al. (1998), Tsuboi et al. (1998), Andreani et al. (1999), Komatsu et al. (1999), Mauskopf et al. (2000), Patel et al. (2000), Pointecouteau et al. (2001).

distance determinations. For example, including multiple distance determinations to a single cluster overstresses any effects of asphericity and orientation from that galaxy cluster.

Statistical uncertainty includes contributions from the ICM shape parameters, the electron temperature, point sources in the field, and the cooling functions, which depends on T_e, metallicity, and the column density. The largest sources of statistical uncertainty are the

Table 9.1. H_0 *Systematic Uncertainty Budget*

Systematic	Effect (%)
SZE calibration	± 8
X-ray calibration	± 10
N_H	± 5
Asphericity*	± 5
Isothermality	± 10
Clumping	-20
Undetected radio sources	± 12
Kinetic SZE*	± 2
Primary CMB*	$< \pm 1$
Radio halos	-3
Primary beam	± 3
Total	$^{+22}_{-30}$

*Includes $1/\sqrt{18}$ factor for the 18 cluster sample.

ICM shape parameters (from model fitting) and the X-ray determined electron temperature ($D_A \propto T_e^{-2}$). Uncertainty from model fitting is roughly 20% in the distance and is dominated by the uncertainty in the central decrement. The contribution from T_e on the distances varies greatly from $\sim 5\%$ to $\sim 30\%$, with 10%–20% being typical values. As expected, nearby cluster temperatures are more precisely determined than those of distant galaxy clusters.

9.4.2   *Sources of Possible Systematic Uncertainty*

The absolute calibration of both the SZE and X-ray observations directly affects the distance determinations. In addition to the absolute calibration uncertainty from the observations, there are possible sources of systematic uncertainty that depend on the physical state of the ICM and other sources that can contaminate the cluster SZE emission. Table 9.1 summarizes the systematic uncertainties in the Hubble constant determined from 30 GHz interferometric SZE observations of a sample of 18 clusters (Reese et al. 2002), but are typical of most SZE experiments. The entries marked with asterisks are expected to average out for a large sample of clusters and include a $1/\sqrt{18}$ factor reflecting the 18 clusters used in this work. For detailed discussions of systematics see Birkinshaw (1999) and Reese et al. (2000, 2002).

9.4.2.1   *Cluster Structure*

Most clusters do not appear circular in radio, X-ray, or optical wavelengths. Under the assumption of axisymmetric clusters, the combined effect of cluster asphericity and its orientation on the sky conspires to introduce a roughly $\pm 20\%$ random uncertainty in H_0 determined from one galaxy cluster (Hughes & Birkinshaw 1998). When one considers a large, unbiased sample of clusters, with random orientations, the errors due to imposing a spherical model are expected to cancel, resulting in a precise determination of H_0. Numerical simulations using triaxial β models support this assumption (Sulkanen 1999).

Departures from isothermality in the cluster atmosphere may result in a large error in the distance determination from an isothermal analysis. The *ROSAT* band is fairly insensitive to temperature variations, showing a $\sim 10\%$ change in the PSPC count rate for a factor of 2 change in temperature for $T_e > 1.5$ keV gas (Mohr, Mathiesen, & Evrard 1999). A mixture of simulations and studies of nearby clusters suggests a 10% effect on the Hubble parameter due to departures from isothermality (e.g., Inagaki, Suginohara, & Suto 1995; Roettiger, Stone, & Mushotzky 1997).

Clumping of the intracluster gas is a potentially serious source of systematic error in the determination of the Hubble constant. Unresolved clumps in an isothermal intracluster plasma will enhance the X-ray emission by a factor C^2, where

$$C \equiv \frac{\langle n_e^2 \rangle^{1/2}}{\langle n_e \rangle}. \tag{9.7}$$

If significant substructure exists in galaxy clusters, the cluster generates more X-ray emission than expected from a uniform ICM, leading to an underestimate of the angular-diameter distance ($D_A \propto S_{x0}^{-1}$) and therefore an overestimate of the Hubble parameter by a factor C^2. Unlike orientation bias, which averages down for a large sample of clusters, clumping must be measured in each cluster or estimated for an average cluster. There is currently no observational evidence of significant clumping in galaxy clusters. If clumping were significant and had large variations from cluster to cluster, we might expect larger scatter than is seen in the Hubble diagrams from SZE and X-ray distances (Fig. 9.6). In addition, the agreement between SZE (e.g., Grego et al. 2001) and X-ray (e.g., Mohr et al. 1999) determined gas fractions from galaxy clusters also suggests that clumping is not a large effect.

9.4.2.2   Possible SZE Contaminants

Undetected point sources near the cluster mask the central decrement, causing an underestimate in the magnitude of the decrement and therefore an underestimate of the angular diameter distance. Point sources in reference fields and for interferometers, the complicated synthesized beam shapes, may cause overestimates of the angular diameter distance. Massize clusters typically have central dominant (cD) galaxies, which are often radio bright. Therefore, it is likely that there is a radio point source near the center of each cluster. Typical radio sources have falling spectra, roughly $\alpha \approx 1$, where $S_\nu \propto \nu^{-\alpha}$. At 30 GHz, possible undetected point sources just below the detection threshold of the observations introduce a $\sim 10\%$ uncertainty.

Cluster peculiar velocities with respect to the CMB introduce an additional CMB spectral distortion known as the kinetic SZE (see §9.2.2). For a single isothermal cluster, the ratio of the kinetic SZE to the thermal SZE is

$$\left| \frac{\Delta T_{kinetic}}{\Delta T_{thermal}} \right| = 0.2 \frac{1}{|f(x)|} \left(\frac{v_{pec}}{1000 \text{ km s}^{-1}} \right) \left(\frac{10 \text{ keV}}{T_e} \right), \tag{9.8}$$

where v_{pec} is the peculiar velocity along the line of sight. At low frequencies (Rayleigh-Jeans regime), $f(x) \approx -2$ and the kinetic SZE is $\sim 10\%$ of the thermal SZE. Recent observational evidence suggests a typical one-dimensional rms peculiar velocity of ~ 300 km s^{-1} (Watkins 1997), and recent simulations found similar results (Colberg et al. 2000). In general, the kinetic effect is $\lesssim 10\%$ that of the thermal SZE, except near the thermal null at ~ 218 GHz where the kinetic SZE dominates (see Fig. 9.2). Cluster peculiar velocities are randomly distributed, so when averaged over an ensemble of clusters, the effect from peculiar velocities should cancel.

CMB primary anisotropies have the same spectral signature as the kinetic SZE. The effects of primary anisotropies on cluster distances depend strongly on the beam size of the SZE observations and the typical angular scale of the clusters being observed (nearby versus distant clusters); the CMB effects on the inferred Hubble constant should average out over an ensemble of clusters. Recent BIMA observations provide limits on primary anisotropies on scales of a few arcminutes (Holzapfel et al. 2000; Dawson et al. 2001). On these scales, CMB primary anisotropies are an unimportant ($\lesssim 1\%$) source of uncertainty. For nearby clusters, or for searches for distant clusters using beams larger than a few arcminutes, the intrinsic CMB anisotropy must be considered. The unique spectral signature of the thermal SZE can be used to separate it from primary CMB anisotropy. However, it will not be possible to separate primary CMB anisotropies from the kinetic SZE without relying on the very small spectral distortions of the kinetic SZE due to relativistic effects.

The SZE may be masked by large-scale diffuse nonthermal radio emission in clusters of galaxies, known as radio halos. If present, radio halos are located at the cluster centers, have sizes typical of galaxy clusters, and have a steep radio spectrum ($\alpha \approx 1-3$; Hanish 1982; Moffet & Birkinshaw 1989; Giovannini et al. 1999; Kempner & Sarazin 2001). Because halos are rare, little is known about their nature and origin, but they are thought to be produced by synchrotron emission from an accelerated or reaccelerated population of relativistic electrons (e.g., Jaffe 1977; Dennison 1980; Roland 1981; Schlickeiser, Sievers, & Thiemann 1987). Conservative and simplistic modeling of the possible effects of these halos implies a $\sim 3\%$ overestimate on the inferred Hubble parameter from radio halos (Reese et al. 2002). Reese et al. (2002) show Very Large Array NVSS contours overlaid on 30 GHz interferometric SZE images, suggesting that radio halos have little impact on SZE observations and therefore on SZE/X-ray distances.

Imprecisely measured beam shapes affect the inferred central decrements and therefore affect the Hubble constant. For interferometric observations, the primary beam is determined from holography measurements. Conservative and simple modeling of the effects of the primary beam suggests that the effect on the Hubble constant is a few percent ($\lesssim 3\%$) at most. In theory, this is a controllable systematic with detailed measurements of beam shape and is currently swamped by larger potential sources of systematic uncertainty.

9.4.3   Future of SZE/X-ray Distances

The prospects for improving both the statistical and systematic uncertainties in the SZE distances in the near future are promising. Note, from Equation 9.6, that the error budget in the distance determination is sensitive to the absolute calibration of the X-ray and SZE observations. Currently, the best absolute calibration of SZE observations is $\sim 2.5\%$ at 68% confidence, based on observations of the brightness of the planets Mars and Jupiter. Efforts are now underway to reduce this uncertainty to the 1% level (2% in H_0). Uncertainty in the X-ray intensity scale also adds another shared systematic. The accuracy of the *ROSAT* X-ray intensity scale is debated, but a reasonable estimate is believed to be $\sim 10\%$. It is hoped that the calibration of the *Chandra* and *XMM-Newton* X-ray telescopes will greatly reduce this uncertainty.

Possible sources of systematic uncertainty are summarized in Table 9.1. These values come from 30 GHz interferometric SZE observations and *ROSAT* data for a sample of 18 galaxy clusters (Reese et al. 2002), but are typical of most SZE distance determinations. The largest systematic uncertainties are due to departures from isothermality, the possibility of clumping, and possible point source contamination of the SZE observations (for detailed discussion of systematics, see, e.g., Birkinshaw 1999; Reese et al. 2000, 2002). *Chandra* and *XMM-Newton* are already providing temperature profiles of galaxy clusters (e.g., Markevitch et al. 2000; Nevalainen, Markevitch, & Forman 2000; Tamura et al. 2001). The unprecedented angular resolution of *Chandra* will provide insight into possible small-scale structures in clusters. In addition, multiwavelength studies by existing radio observatories, for example the Very Large Array, can shed light on the residual point source contamination of the radio wavelength SZE measurements. Therefore, though currently at the 30% level, many of the systematics can and will be addressed through both existing X-ray and radio observatories and larger samples of galaxy clusters provided from SZE surveys.

The beauty of the SZE and X-ray technique for measuring distances is that it is completely independent of other techniques, and that it can be used to measure distances at high redshifts directly. Since the method relies on the well-understood physics of highly ionized plasmas, it should be largely independent of cluster evolution. Inspection of Figure 9.6 already provides confidence that a large survey of SZE distances consisting of perhaps a few hundred clusters with redshifts extending to one and beyond would allow the technique to be used to trace the expansion history of the Universe, providing a valuable independent check of the recent determinations of the geometry of the Universe from Type Ia supernovae (Riess et al. 1998; Perlmutter et al. 1999) and CMB primary anisotropy experiments (Stompor et al. 2001; Netterfield et al. 2002; Pryke et al. 2002; Spergel et al. 2003).

9.5   SZE Surveys

A promising way of finding high-redshift clusters is to perform deep, large-scale SZE surveys, taking advantage of the redshift independence of the SZE. Such surveys will provide large catalogs of galaxy clusters, many of which will be at high redshift. The redshift evolution of the number density of galaxy clusters is critically dependent on the underlying cosmology, and in principle can be used to determine the equation of state of the "dark energy" (e.g., Haiman, Mohr, & Holder 2001; Holder et al. 2000; Holder, Haiman, & Mohr 2001). Figure 9.7 illustrates the dependence of the evolution of cluster number density on cosmology (Holder et al. 2000). All three cosmologies are normalized to the local cluster abundance. Notice that for low-Ω_m cosmologies, there are a significant number of galaxy clusters with $z > 1$.

There are a number of dedicated SZE experiments under construction that will perform deep, large-scale SZE surveys. In the next few years, there are a number of interferometric approaches that should find hundreds of galaxy clusters. Bolometer array based SZE experiments should find roughly thousands of clusters just one year after the interferometric experiments. The following generation of bolometer array based SZE experiments should measure tens of thousands of galaxy clusters on a roughly five-year time scale. The possible systematics that could affect the yields of SZE surveys are presently too large to realize the full potential of a deep SZE survey covering thousands of square degrees. These systematics include, for example, the uncertainties on the survey mass detection limit owing to unknown cluster structure and cluster gas evolution. These systematics can begin to be addressed through detailed follow-up observations of a moderate-area SZE survey. High-resolution SZE, X-ray, and weak lensing observations will provide insights into the evolution and struc-

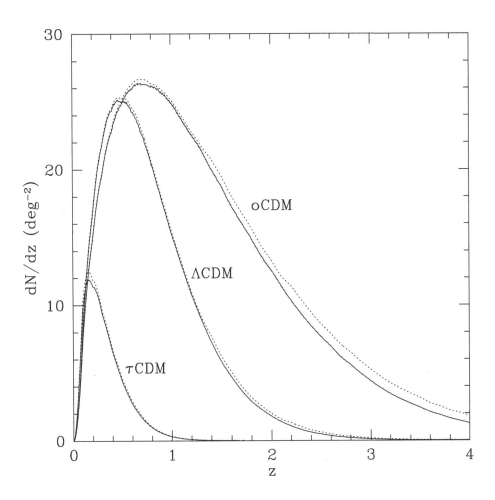

Fig. 9.7. Expected number counts of galaxy clusters from an upcoming dedicated interferometric SZE survey array. The curves are very different for the three different cosmologies, having been normalized to the local cluster abundance. The two sets of curves are slightly different treatments of the mass limits. Notice that for low-Ω_m cosmologies, there are a significant number of galaxy clusters with $z > 1$. The cosmologies in the figure are (Ω_m, Ω_Λ, σ_8): ΛCDM (0.3, 0.7, 1.0); OCDM (0.3, 0.0, 0.6); and τCDM (1.0, 0.0, 1.0). (See Holder et al. 2000, from which this figure is adapted, for details.)

ture of the cluster gas. Numerical simulations directly compared and normalized to the SZE yields should provide the necessary improvement in our understanding of the mass function.

9.6   Summary

Soon after it was proposed, it was realized that SZE observations could yield distances to galaxy clusters when combined with X-ray data. SZE/X-ray distances are independent of the extragalactic distance ladder and do not rely on clusters being standard candles or rulers. The promise of direct distances helped motivate searches for the SZE in galaxy clusters. It is only in recent years that observations of the small ($\lesssim$ 1 mK) SZE signal have yielded reliable detections and imaging, mostly due to advances in observational techniques.

SZE/X-ray distances have progressed from distance determinations one cluster at a time to samples of tens of galaxy clusters. To date, there are 41 SZE/X-ray determined distances to 26 different galaxy clusters. The combined 41 cluster distances imply a Hubble parameter of $\sim 61 \pm 3 \pm 18$ km s^{-1} Mpc^{-1}, where the approximate uncertainties are statistical followed by systematic at 68% confidence. Shared systematics between the determined distances make it difficult to determine them precisely. For example, multiple distance determinations to a single galaxy cluster overstress any asphericity and orientation effects. There are three homogeneously analyzed samples of clusters consisting of five, seven, and 18 galaxy clusters. Even with these small samples, systematics clearly dominate. Systematic uncertainties are approachable with current X-ray and radio observatories.

SZE/X-ray determined direct distances to galaxy clusters are beginning to trace out the theoretical angular-diameter distance relation. It is clear that with a large sample of galaxy clusters at high redshift, SZE/X-ray distances will be able to determine the geometry of the Universe. In fact, if for some (highly unlikely) reason there was a systematic offset with the SZE/X-ray distances, the shape of the curve could still be determined and cosmological parameters inferred.

The redshift independence of the SZE will be exploited with surveys of large patches of sky with next generation, dedicated SZE experiments. Such surveys will provide large catalogs of clusters, many of which will be at high redshift ($z > 1$), providing large enough samples to use SZE/X-ray distances to determine both the Hubble constant and the geometry of the Universe. In addition, the cluster yields from deep, large-scale SZE surveys are critically dependent on the cosmology, potentially allowing a determination of the equation of state of the "dark energy." High-resolution SZE, X-ray, and optical (including weak lensing) observations will provide insights into the evolution and structure of the cluster gas necessary to exploit fully the evolution in cluster yields from SZE surveys.

Acknowledgements. The author is grateful for financial support from NASA Chandra Postdoctoral Fellowship PF 1-20020.

References

Allen, S. W., & Fabian, A. C. 1998, MNRAS, 297, L57

Andreani, P., et al. 1996, ApJ, 459, L49

——. 1999, ApJ, 513, 23

Birkinshaw, M. 1979, MNRAS, 187, 847

——. 1999, Phys. Rep., 310, 97

Birkinshaw, M., Gull, S. F., & Northover, K. J. E. 1978a, Nature, 275, 40

——. 1978b, MNRAS, 185, 245

Birkinshaw, M., & Hughes, J. P. 1994, ApJ, 420, 33

Birkinshaw, M., Hughes, J. P., & Arnaud, K. A. 1991, ApJ, 379, 466

Blain, A. W. 1998, MNRAS, 297, 502

Boynton, P. E., & Murray, S. S. 1978, HEAO B Guest Observer proposal

Carlstrom, J. E., Holder, G. P., & Reese, E. D. 2002, ARA&A, 40, 643

Carlstrom, J. E., Joy, M., & Grego, L. 1996, ApJ, 456, L75

Carlstrom, J. E., Joy, M. K., Grego, L., Holder, G. P., Holzapfel, W. L., Mohr, J. J., Patel, S., & Reese, E. D. 2000, Physica Scripta Volume T, 85, 148

Cavaliere, A., Danese, L., & de Zotti, G. 1977, ApJ, 217, 6

Cavaliere, A., & Fusco-Femiano, R. 1976, A&A, 49, 137

——. 1978, A&A, 70, 677

Challinor, A., & Lasenby, A. 1998, ApJ, 499, 1

——. 1999, ApJ, 510, 930

Church, S. E., Ganga, K. M., Ade, P. A. R., Holzapfel, W. L., Mauskopf, P. D., Wilbanks, T. M., & Lange, A. E. 1997, ApJ, 484, 523

Colberg, J. M., White, S. D. M., MacFarland, T. J., Jenkins, A., Pearce, F. R., Frenk, C. S., Thomas, P. A., & Couchman, H. M. P. 2000, MNRAS, 313, 229

Cooray, A. R., Grego, L., Holzapfel, W. L., Joy, M., & Carlstrom, J. E. 1998, AJ, 115, 1388

Dawson, K. S., Holzapfel, W. L., Carlstrom, J. E., Joy, M., LaRoque, S. J., & Reese, E. D. 2001, ApJ, 553, L1

Dennison, B. 1980, ApJ, 239, L93

Désert, F. X., et al. 1998, NewA, 3, 655

Dolgov, A. D., Hansen, S. H., Pastor, S., & Semikoz, D. V. 2001, ApJ, 554, 74

Fabbri, R. 1981, Ap&SS, 77, 529

Fischer, M. L., & Lange, A. E. 1993, ApJ, 419, 433

Giovannini, G., Tordi, M., & Feretti, L. 1999, NewA., 4, 141

Grainge, K., Grainger, W. F., Jones, M. E., Kneissl, R., Pooley, G. G., & Saunders, R. 2002a, MNRAS, 329, 890

Grainge, K., Jones, M., Pooley, G., Saunders, R., Baker, J., Haynes, T., & Edge, A. 1996, MNRAS, 278, L17

Grainge, K., Jones, M., Pooley, G., Saunders, R., & Edge, A. 1993, MNRAS, 265, L57

Grainge, K., Jones, M. E., Pooley, G., Saunders, R., Edge, A., Grainger, W. F., & Kneissl, R. 2002b, MNRAS, 333, 318

Grainger, W. F., Das, R., Grainge, K., Jones, M. E., Kneissl, R., Pooley, G. G., & Saunders, R. D. E. 2002, MNRAS, 337, 1207

Grego, L., Carlstrom, J. E., Joy, M. K., Reese, E. D., Holder, G. P., Patel, S., Cooray, A. R., & Holzapfel, W. L. 2000, ApJ, 539, 39

Grego, L., Carlstrom, J. E., Reese, E. D., Holder, G. P., Holzapfel, W. L., Joy, M. K., Mohr, J. J., & Patel, S. 2001, ApJ, 552, 2

Gunn, J. E., Longair, M. S., & Rees, M. J., ed. 1978, Observational Cosmology (Sauverny: Observatoire de Geneve)

Haiman, Z., Mohr, J. J., & Holder, G. P. 2001, ApJ, 553, 545

Hanisch, R. J. 1982, A&A, 116, 137

Herbig, T., Lawrence, C. R., Readhead, A. C. S., & Gulkis, S. 1995, ApJ, 449, L5

Holder, G., Haiman, Z. ., & Mohr, J. J. 2001, ApJ, 560, L111

Holder, G. P., Mohr, J. J., Carlstrom, J. E., Evrard, A. E., & Leitch, E. M. 2000, ApJ, 544, 629

Holzapfel, W. L., et al. 1997b, ApJ, 480, 449

Holzapfel, W. L., Ade, P. A. R., Church, S. E., Mauskopf, P. D., Rephaeli, Y., Wilbanks, T. M., & Lange, A. E. 1997a, ApJ, 481, 35

Holzapfel, W. L., Carlstrom, J. E., Grego, L., Holder, G., Joy, M., & Reese, E. D. 2000, ApJ, 539, 57

Hu, W., & White, M. 1997, ApJ, 479, 568

Hughes, J. P., & Birkinshaw, M. 1998, ApJ, 501, 1

Inagaki, Y., Suginohara, T., & Suto, Y. 1995, PASJ, 47, 411

Itoh, N., Kohyama, Y., & Nozawa, S. 1998, ApJ, 502, 7

Jaffe, W. J. 1977, ApJ, 212, 1

Jones, M., et al. 1993, Nature, 365, 320

Jones, M. E., et al. 2004, MNRAS, submitted (astro-ph/0103046)

Joy, M., et al. 2001, ApJ, 551, L1

Kempner, J. C., & Sarazin, C. L. 2001, ApJ, 548, 639

Komatsu, E., Kitayama, T., Suto, Y., Hattori, M., Kawabe, R., Matsuo, H., Schindler, S., & Yoshikawa, K. 1999, ApJ, 516, L1

Komatsu, E., Matsuo, H., Kitayama, T., Kawabe, R., Kuno, N., Schindler, S., & Yoshikawa, K. 2001, PASJ, 53, 57

Lamarre, J. M., et al. 1998, ApJ, 507, L5

LaRoque, S. J., Reese, E. D., Carlstrom, J. E., Holder, G., Holzapfel, W. L., Joy, M., & Grego, L. 2004, ApJ, submitted (astro-ph/0204134)

Markevitch, M., et al. 2000, ApJ, 541, 542

Mason, B. S., Myers, S. T., & Readhead, A. C. S. 2001, ApJ, 555, L11

Mauskopf, P. D., et al. 2000, ApJ, 538, 505

Moffet, A. T., & Birkinshaw, M. 1989, AJ, 98, 1148

Mohr, J. J., Mathiesen, B., & Evrard, A. E. 1999, ApJ, 517, 627

Molnar, S. M., & Birkinshaw, M. 1999, ApJ, 523, 78

Mushotzky, R. F., & Scharf, C. A. 1997, ApJ, 482, L13

Myers, S. T., Baker, J. E., Readhead, A. C. S., Leitch, E. M., & Herbig, T. 1997, ApJ, 485, 1

Netterfield, C. B., et al. 2002, ApJ, 571, 604

Nevalainen, J., Markevitch, M., & Forman, W. 2000, ApJ, 536, 73

Nozawa, S., Itoh, N., & Kohyama, Y. 1998a, ApJ, 508, 17

——. 1998b, ApJ, 507, 530

Padin, S., et al. 2001, ApJ, 549, L1

Patel, S. K., et al. 2000, ApJ, 541, 37

Perlmutter, S., et al. 1999, ApJ, 517, 565

Phillips, P. R. 1995, ApJ, 455, 419

Pointecouteau, E., Giard, M., Benoit, A., Désert, F. X., Aghanim, N., Coron, N., Lamarre, J. M., & Delabrouille, J. 1999, ApJ, 519, L115

Pointecouteau, E., Giard, M., Benoit, A., Désert, F. X., Bernard, J. P., Coron, N., & Lamarre, J. M. 2001, ApJ, 552, 42

Pospieszalski, M. W., Lakatosh, W. J., Nguyen, L. D., Lui, M., Liu, T., Le, M., Thompson, M. A., & Delaney, M. J. 1995, IEEE MTT-S Int. Microwave Symp., 1121

Pryke, C., Halverson, N. W., Leitch, E. M., Kovac, J., Carlstrom, J. E., Holzapfel, W. L., & Dragovan, M. 2002, ApJ, 568, 46

Reese, E. D., et al. 2000, ApJ, 533, 38

Reese, E. D., Carlstrom, J. E., Joy, M., Mohr, J. J., Grego, L., & Holzapfel, W. L. 2002, ApJ, 581, 53

Rephaeli, Y. 1995, ApJ, 445, 33

Rephaeli, Y., & Yankovitch, D. 1997, ApJ, 481, L55

Riess, A. G., et al. 1998, AJ, 116, 1009

Roettiger, K., Stone, J. M., & Mushotzky, R. F. 1997, ApJ, 482, 588

Roland, J. 1981, A&A, 93, 407

Saunders, R., et al. 2003, MNRAS, 341, 937

Sazonov, S. Y., & Sunyaev, R. A. 1998a, ApJ, 508, 1

——. 1998b, Astron. Lett., 24, 553

Schlickeiser, R., Sievers, A., & Thiemann, H. 1987, A&A, 182, 21

Silk, J., & White, S. D. M. 1978, ApJ, 226, L103

Smail, I., Ivison, R. J., & Blain, A. W. 1997, ApJ, 490, L5

Spergel, D., et al. 2003, ApJS, 148, 175

Stebbins, A. 1997, preprint: astro-ph/9709065

Stompor, R., et al. 2001, ApJ, 561, L7

Subrahmanyan, R., Kesteven, M. J., Ekers, R. D., Sinclair, M., & Silk, J. 2000, MNRAS, 315, 808

Sulkanen, M. E. 1999, ApJ, 522, 59

Sunyaev, R. A. & Zel'dovich, Y. B. 1970, Comments Astrophys. Space Phys., 2, 66

——. 1972, Comments Astrophys. Space Phys., 4, 173

——. 1980, ARA&A, 18, 537

Tamura, T., et al. 2001, A&A, 365, L87

Tsuboi, M., Miyazaki, A., Kasuga, T., Matsuo, H., & Kuno, N. 1998, PASJ, 50, 169

Tucker, W., et al. 1998, ApJ, 496, L5

Udomprasert, P. S., Mason, B. S., & Readhead, A. C. S. 2000, in Constructing the Universe with Clusters of Galaxies, ed. F. Durret & G. Gerbal (Paris: IAP), E48

Watkins, R. 1997, MNRAS, 292, L59

Wright, E. L. 1979, ApJ, 232, 348

10

How much is there of what?

Measuring the mass density of the Universe

VIRGINIA TRIMBLE
University of California, Irvine and University of Maryland

Abstract
The density of the Universe is, by general agreement, one, in suitable units and with somewhat uncertain partition among a number of constituents, of which perhaps 20%-30% is positive-pressure matter. The present contribution addresses methods of measurement, mainstream results, the more popular constituents, unresolved issues, and some less popular candidates and views, not all quite in this order.

10.1   Introduction

Modern cosmology takes as given general relativity as the correct description of spacetime, gravity, and their relationship (at least back to the time of big bang nucleosynthesis and before) and the interpretation of the velocity-distance relation (Hubble's law) as the expansion of a relativistic spacetime. I will tell you if we get to a paragraph where suspension of belief in this foundation is required.

Hubble (1934) himself reported a density for the Universe of $(1.3 - 1.6) \times 10^{-30}$ g cm^{-3}, based, he said, on an average nebular mass of $(6-8) \times 10^8 \, M_\odot$ and a nebular density of $(20-30) Mpc^{-3}$. Hubble, incidentally, always said and wrote "nebulae." Galaxies was Shapley's word.

Hey, not half bad, you may say, until you remember, first, that his value of H_0 was 536 km s^{-1} Mpc^{-1} (accurate to within $\pm 10\%$, just like all modern values), so that the critical density was 5.4×10^{-28} g cm^{-3}, and, second that Zwicky (1933) had already published a velocity dispersion for the Coma cluster implying a mass-to-light ratio for its members nearly a factor 100 larger than the 3 $M_\odot/L_\odot$ assumed by Hubble to get from galaxy brightnesses to galaxy masses. If you would like to say that he was measuring the density in luminous, stellar material and got 0.3% of the closure density, you will again be impressed (this time correctly) by the nearness to modern numbers.

But darker clouds were already gathering, with Sinclair Smith (1936) reporting an average mass per nebula of $2 \times 10^{11} \, M_\odot$ for members of the Virgo cluster whose radial velocities he had measured, Eric Holmberg (1937) finding an average of $10^{11} \, M_\odot$ from velocity differences for a handful of binary galaxies (in work inspired by Knut Lundmark, another undersung hero of this wild territory), and Horace Babcock's (1939) rotation curve for M31 with a last, fuzzy data point at 380 km s^{-1}, implying a mass of $10^{11} \, M_\odot$. All of these were for a Hubble parameter of 500 km s^{-1} Mpc^{-1}, so you can multiply them by some favorite number between 5 and 10. Notice that all the methods—cluster velocity dispersions, pairwise velocity differences, and rotation curves—are still in use.

Another modern method belongs to this pre-historic period, gravitational lensing of the sort we now call strong, of one galaxy by another, advanced by Fritz Zwicky (1937) first as a way of confirming general relativity, second as a way of studying distant galaxies that would otherwise be too faint to see, and third as an independent way of measuring the masses of the galaxies that function as lenses.

Then, of course, there was a war and a period during which a large fraction of creative young astronomers concentrated on stellar physics (because of the 1939 breakthrough on hydrogen fusion as the dominant energy source and the post-war advent of digital computers), but, by 1961, at the time of a pair of conferences associated with the Berkeley IAU General Assembly (Neyman, Page, & Scott 1961; McVittie 1962), it was generally accepted that the large velocity dispersions of clusters of galaxies, especially the ones that look relaxed and smooth, constitute a problem that astronomers should be thinking about.

Over the next few years (a) Abell (1965) estimated the density of the Universe to be 10^{-30} g cm^{-3} for $H_0 = 75$ km s^{-1} Mpc^{-1} (that is, 10% of closure density) if his superclusters had masses of $10^{16} M_\odot$ and filled 10% of space, (b) what we now call weak lensing slipped quietly into the literature (Gunn 1967, though it was initially entitled "a fundamental limit on the accuracy of angular measurements in observational cosmology" and portions of the idea credited to unpublished thoughts of Richard Feynman), and (c) the third primary method of measuring masses for large structures was conceived. A 1965 NRL rocket flight had seen X-rays coming from the general direction of the Coma cluster. Felten et al. (1966) tried to account for these as both nonthermal and thermal emission and ended up favoring the latter. They asked the question, could all the "missing mass" of the cluster (4×10^{48} g) be in hot gas, and, if so, what temperature must the gas have in order to emit all the X-rays? Their answers were yes, and about 10^8 K, close to what you would suppose from the cluster velocity dispersion of 1050 km s^{-1}. This consideration lay nearly dormant until the *Einstein Observatory* began to return a few cluster images with some angular resolution in both surface brightness and temperature (Fabricant, Lecar, & Gorenstein 1980). Thus a review soon after (Forman & Jones 1982) could say with some confidence that a handful of nearby clusters indeed have mass-to-light ratios of 100 or thereabouts and that X-ray gas makes up only about 10% of the total.

Meanwhile, by 1974 (Einasto et al. 1974; Ostriker, Peebles, & Yahil 1974) dark matter had overtaken its competitors as the most likely explanation for large apparent M/L's on all scales. For additional historical material and references see Trimble (1987, 1995, 2001)

10.2   Methods

If you want to measure density, you need two things, a volume and the mass within it. As a rule, volume is easier, provided that someone tells you the correct value of H_0 (Dr. Freedman at this meeting). Not all volumes are spherical, which can be source of systematic error, due, for instance, to end-on prolate clusters being over-represented among X-ray or lensing samples. How large a volume do you need? Well, the whole observable Universe is generally not a good choice (as one sometimes has to explain to students and science journalists), but neither is one too small to be representative. Luckily, the Universe seems not to be fractal or hierarchically structured beyond a few hundred Mpc, and voids and superclusters like the ones near us simply repeat. This means that de Vaucouleurs (1970) was wrong when he said that it was incredibly improbable that he should be writing just when, after centuries of growth in our ideas of cosmic horizons, the largest structures had finally been found. Incredibly improbable, perhaps, but apparently true. Nothing more will be said about volumes here, and we will carry a spare $h = H_0/100$ km s^{-1} Mpc^{-1} around with us for use when needed.

Now, what about measuring masses? This is a good deal harder. You still need somebody to tell you H_0, and there may not be any electromagnetic radiation coming from the volume you want to know about (Kraan-Korteweg et al. 1999). All of the tactics mentioned in § 10.1 are still in use, and some additional ones might be contemplated. They are mentioned here with a cross reference if a speaker focused on a particular one and with just one or two recent literature citations that enable you to get on-system if necessary otherwise. These are: (1) Rotation curves (Sofue & Rubin 2001), (2) binary galaxies (Junqueira & Chan 2002), (3) velocity dispersions in clusters (Rines et al. 2002), (4) X-ray emission from clusters (N. Bahcall; Schuecker et al. 2002), (5) structure on scales larger than clusters, often in combination with some simulation that begins with a primordial fluctuation spectrum (Seljak; Shanks; Annis; Percival et al. 2001; Susperregi 2002), (6) an X-ray cluster variant in which one assumes the cosmic baryon density to have the BBN value and the ratio of total to baryon mass to be that given by clusters (Allen, Schmidt, & Fabian 2002), (7) gravitational lensing (Blandford 2001; Kochanek, Ellis), both strong (Claeskens & Surdej 2002) and weak, by clusters (King et al. 2002), single galaxies (Brainerd, Blandford, & Smail 1996), and by structures larger than clusters (Wittman et al. 2000).

One can imagine (at least) four additional approaches to mass measurements for galaxies and clusters, which have not so far added much. First is escape velocity. If you see a star or galaxy with much more than 1.4 times the local circular velocity, and it seems to have been around a long time, you might reasonably deduce that there is a good deal of mass still further outside than your location, and the largest observed speed tells you how much. This really applies only to the Milky Way and confirms numbers from other data. Second is gravitational redshift. Unfortunately, for typical galaxies and clusters, the wavelength displacement is only about one part in 10^6.

Third is the Sachs-Wolfe (1967) effect. The idea is that photons drop down into gravitational potential wells and have to climb back out to get to us, taking sufficient time that the well has changed with the expansion of the Universe (getting shallower if it is still a linear perturbation) and/or growth of the perturbation (getting deeper). The two possible changes fight each other, but at least interesting limits may be possible (Boughn & Crittenden 2003).

Fourth is back-calculation from velocity data—the large-scale deviations around smooth Hubble flow, known for more than a quarter century (Rubin et al. 1976a,b) and attributed to structures like the Great Attractor (Burstein et al. 1986), then unattributed (Mathewson, Ford, & Buchhorn 1992), and then perhaps attributed again (Vauglin et al. 2002).

Of course you could just assume a fixed mass-to-light ratio, after the fashion of Opik (1922) and Hubble (1934), though if your purpose in attempting a mass measurement is to be able to say something about M/L ratios, this is probably not a good choice. The closest thing found in modern dark matter studies is the "maximal disk" assumption (Bissantz, & Gerhard 2002, who are for; Debattista, Corsini, & Aguerri 2002, who are against).

And, finally, there are global approaches, for which the effective volume is much of the observable Universe, such as luminosity distance vs. redshift for Type Ia supernovae (Perlmutter, Phillips) and analysis of fluctuations in the cosmic microwave background (Spergel; Lange; Readhead; de Bernardis et al. 2002) . Comparison of a simulation of formation of large-scale structure with the real world is also a sort of global method, and the one that arguably led to widespread recognition that the Universe was probably not closed by positive-pressure matter of any sort. At the time this was widely called "the failure of standard cold dark matter," and Hamilton et al. (1991) were among a number of groups that pointed it out more or less simultaneously.

10.3   Results

These are conveniently divisible into (a) numbers close to what most of us now expect for M/L and Ω_m, (b) reconciliation of discordant numbers for what should be the same quantity (the mass of a particular cluster from velocities, X-rays, and lensing, or cluster vs. more global values for total matter density), and (c) minority views in favor of either very small matter densities or values close to unity and larger divergences.

The broadest-brush version of the expected, best-buy, result has not changed enormously since the papers of Ostriker et al. (1974) and Einasto et al. (1974). That is, the mass-to-light ratio in solar units included within a given scale length increases more or less monotonically from 0.5 to a few for star clusters and galactic disks to something like 10 for the inner parts of galaxies, probed with rotation curves, globular clusters, and halo gas, to 20–30 for the extreme reaches of galaxies and binary pairs, to 100 or so for groups, and tapers off at 200–300 for rich clusters and superclusters, showing no signs of rising higher when you look globally. What is clearer now than then is that there is a real range of M/L values at each length scale, and not just error bars. Some dwarf spheroidal companion galaxies (perhaps even some globular clusters) have their own dark matter halos, and some do not (Braine et al. 2001; Kleyna et al. 2001; Bromm and Clark 2002; Fellhauer & Kroupa 2002; Prada & Burkert 2002). And while there are apparently no truly empty halos (Haynes et al. 2002; Miralles et al. 2002), some are less full of luminous baryons than others and so can have very large M/L's (Sahu et al. 1998; Tully et al. 2002). Historically, these have been difficult or impossible to find, but weak lensing surveys of "empty fields" are beginning to fill in the inventory.

10.3.1   *Conventional Recent Results*

On the whole, galaxies have continued to yield galaxy-type numbers: 7.4×10^{10} $M_\odot$ for the Milky Way inside our orbit (Olling & Merrifield 2001, an M/L of 3–4), $M/L_B =$ 22 for M81 out to 21 kpc (Schroder et al. 2002) and $(1-2) \times 10^{12}\ M_\odot$ for the whole Local Group and M81 group (Gallart et al. 2001; Karachentsev et al. 2002a,b). Binary galaxies (e.g., $M/L = 17.6$ for a bunch of pairs of large ellipticals; Junqueira & Chan 2002) also fall where they have been accustomed to fall, between inner galaxy and cluster numbers.

Both masses and M/L ratios get larger as you look at larger assemblages (Adams & Mazure 2002, an application of early-release SDSS data). Their mean value of cluster M/L_B is $360h$. Girardi et al. (2002) also record the increase of M/L with scale going from 0.5 to $300 \times 10^{12}\ M_\odot$ for the entities considered.

Other concordant numbers (given the various bands used for L and the sometimes unstated values of H_0) include a mean of $239h_{50}$ (Hoekstra et al. 2002) and one of 470 ± 180 (Dye et al. 2002). A King et al. (2002) value of $M/L_K = 40$ sounds as if it might have crept down from the previous paragraph, as indeed it has, since it pertains only to the inner core of the cluster (coming from weak lensing also in the K band). Superclusters have, of course, even larger masses in excess (as Abell concluded) of $2 \times 10^{16}\ h^{-1}\ M_\odot$ (Caretta et al. 2002).

For the global numbers, one is perhaps entitled to be surprised at their stability. A majority of panelists at a discussion within the 1997 IAU General Assembly in Kyoto voted for a set very close to what the majority of speakers at this conference are voting for (65, 0.3, 0.7, 0.1, etc., and it is left as an exercise for the reader to attach these to the symbols normally used to indicate the quantities). This point of view was thoroughly dominating the literature by 1998, when I recorded more than 20 papers reporting $\Omega_m = 0.3\pm0.1$, of which Eke et al. (1998) was particularly well written. Indeed a sort of Monrovian era of good feeling might be said to have set in when Lineweaver (1999) and Bahcall et al. (1999) concluded that you could bring together numbers from bound clusters of galaxies, larger-scale structures, the apparent brightnesses of supernovae (Perlmutter; Phillips; Hamuy), and "early" (meaning anything before January 2003) measurements of the acoustic peaks in the spectrum of fluctuations of the CMB and get mutually consistent results with the 0.3/0.7 split between stuff with positive pressure and stuff with negative pressure.

On the whole, this concordance has persisted. Numbers for Ω_m published in the last year that overlap 0.3 at the $1-2\sigma$ level include, with the probes used, 0.32–0.38 (Susperregi 2002, analysis of four redshift surveys), 0.35$\pm$0.1 (Borgani et al. 2001, *ROSAT* clusters), 0.25$\pm$0.1 (Roukema, Mamon, & Bajtlik 2002, clustering of QSOs in the 2dF survey), 0.3$\pm$0.04 (Allen et al. 2002, relaxed X-ray clusters in the *Chandra* database but assuming that they have their fair share of baryons, which is perhaps unfair), 0.20$\pm$0.03 (Percival et al. 2001, 2dF again, assuming scale invariant primordial fluctuations), 0.2$\pm$0.1 (Hoekstra et al. 2001, weak lensing by groups), 0.17 (Bahcall & Comerford 2001, clusters), and, the most honest errors bars we've seen, $\Omega_m\, h^2 = 0.02$–0.3 (Schuecker et al. 2002, X-ray clusters), plus, of course, nearly all of the analyses of CMB data presented at this meeting and by Miller, Nichol, & Batuski (2001), Pryke et al. (2002, DASI), Elgarøy, Gramann, & Lahav (2002), Abroe et al. (2002, Maxima), de Bernardis et al. (2002, Boomerang and DASI), and the one to read, if you're reading only one, 0.28 ± 0.11 (Bridle et al. 2001), again bringing together data from supernovae, microwave background, and large-scale structure.

10.3.2 *Discrepancies and Reconciliations*

Neta Bahcall at this meeting and elsewhere has noted that the numbers coming from global considerations tend to be somewhat larger than the best values of Ω_m from recognizable clusters of galaxies and that this may constitute a problem. The alternative is that, just as there are some very diffuse baryons that we are only now learning to spot, there may be some very diffuse CDM that we have not quite learned how to spot.

A second issue is particular clusters for which different methods of analysis yield different total masses. A typical example is RX J1347–1145, for which Cohen & Kneib (2002) conclude that the virial mass is the smallest, the X-ray mass is the largest, and the lensing mass comes in the middle. They suggest that the cause might be a merger-in-progress, which has shocked the gas (making it hotter than it "ought" to be) while the two separate velocity distributions have not yet had a chance to discover that they now belong to a single, more massive cluster. Other discrepancies are typically in the same order, typically not very large, and typically can be explained or explained away (Rines et al. 2002 on A2199; Czoske et al. 2002 and Machacek et al. 2002 on A2218; Allen, Schmidt, & Fabian 2001, Hoekstra et al. 2002, and Athreya et al. 2002 on MS1008–1224).

I do not really think that these systematic disagreements constitute "a problem," since they are generally in the direction you would expect from on-going mergers and how the various methods probe gravitating matter. They may, however, mean that one cannot use agreement of lensing numbers with the others as a confirmation of general relativity in comparison to some other theory of gravity with the same weak-field limit. Remember that a Newtonian argument also yields gravitational bending of light, by half the GR angle. Theories that do not have Newtonian gravity as the weak-field, small-velocity limit, may, however, still be ruled out (Edery 1999).

Great was the wailing and gnashing of teeth when, into a world already disenchanted with standard (closure density) CDM came the first few Type Ia supernovae, with luminosity distances that seemed to require Ω_m much larger than Ω_Λ (Perlmutter et al. 1997). Have faith! said we of unbounded (this is a pun) optimism. It will all go away. And sure enough it did (Garnavich et al. 1998; Riess et al. 1998, 2001; Perlmutter et al. 1998, 1999).

Next, there remain some measurements of Ω_m by various methods, published within the last few years, that do not fall within the $1-2\sigma$ band of the official value 0.3. A few are significantly smaller, e.g. Reiprich & Böhringer (2002, with a best value of 0.12 and a firm upper limit of 0.31 from X-ray clusters and the assumption of a Press-Schechter luminosity function), 0.13 (Arnaud, Aghanim, & Neuman 2002, more X-ray clusters, but reaching to $z = 0.8$), and an upper limit of 0.1 from three clusters with X-ray and weak-lensing data (Gray et al. 2002).

Most discordant values, however, are well above the consensus band, perhaps only because, if a quantity has to be positive, there is a lot more phase space above 0.3 than below. Thus Ω_m close to one (and Ω_Λ close to zero) has been found a better, or at least as good, a fit as 0.3/0.7 to the following data sets:

(a) The K-magnitude redshift relation for strong radio galaxies, on the assumption that ones at $z = 0$ are more massive than the ones at large redshift (Inskip et al. 2002).

(b) The clustering perpendicular to and along the line of sight of QSOs in the 2dF survey (Hoyle et al. 2002, an application of the Alcock-Paczyński test).

(c) The contribution of gravitational lensing to quasar counts (Croom & Shanks 1999).

(d) Non-linear clustering (Szapudi et al. 1999).

(e) And more (cited in Trimble & Aschwanden 2000 and earlier papers in that series), including Pen (1997), who used the classic angular diameter-redshift test like the one that led Loh & Spillar (1986) astray a decade before. Just this year, finally, one of these $\Theta - z$ measurements has been found inconsistent with a large matter density (Lima & Alcaniz 2002).

Finally, there are alternative theories of gravity, motivated at least partly by the desire to avoid having any dark (or anyhow non-baryonic) material in galaxies. If you would like to chase these down, possible starting points are Mannheim (2001) on conformal or Weyl gravity (he is for it) and Milgrom (2002) on Modified Newtonian Dynamics (MOND; he invented it and is still in favor of it). The underlying principles of MOND (a fixed minimum gravitational acceleration) are simple enough that we bears of very little brain can see how it might be tested, by comparing mass measurements made by different techniques (and therefore how it might fail). Those of conformal gravity are not.

Authors who insist on doing cosmologically sensitive calculations, like the distribution of v/c ratios for QSO jets (Wang et al. 2001) under the assumption that $H_0 = 100$ km s^{-1} Mpc^{-1} and $q_0 = 1/2$ probably also belong in here. So, perhaps, does Harrison (1993), who pointed out that, at the time, it was not so easy as you might suppose to exclude a universe with $H_0 = 10$ km s^{-1} Mpc^{-1}, $\Omega_m = 10$, plus very large deviations from Hubble flow. This is, I think, no longer true.

And if observed redshifts are not the (correctly done) sum of cosmic expansion plus peculiar motion through space (Arp 1971; Bell 2002) then all bets are off.

10.4 What is it All?

Neutrinos, now that their rest masses have been (more or less) measured, are no longer a "dark matter candidate." Rather, they are simply part of the inventory, to be included in calculations where appropriate. Other speakers (e.g., Silk; Sadoulet) will have told you about the respectable candidates for the rest of the non-baryonic, non-leptonic $\Omega_m = 0.25$ or thereabouts, but it probably falls to me to tell you about the less conventional ones that have made it into the literature in the past few years. There are enormous numbers of these. Some are variants on the main stream, for instance (a) self-interacting dark matter, which had a brief vogue (Spergel & Steinhardt 2000) before shattering on the reefs of conflicting observations (Yoshida et al. 2000), (b) decaying dark matter, one very specific version of which (Sciama 1990) made a prediction that was not confirmed by a search for the expected product photons (Davidsen et al. 1991), but other versions of which remain viable and perhaps even useful for the "cusp problem" (Cen 2001), and (c) annihilating dark matter (Craig & Davis 2001).

Other candidates are at least tied to portions of physics that one feels one ought to understand, for instance cosmic strings (Spergel & Pen 1997), decaying domain walls (Dvali, Liu, & Vachaspati 1998), and topological defects in general (Digal et al. 2000), and a whole zoo of scalar fields and particles (Cormier & Holman 2000; Dabrowski & Schunck 2000; Goodman 2000; Hu & Peebles 2000).

Still others have been put forward by people of whom one thinks so highly that surely there must be something in the ideas, even if one isn't quite sure what it is. Examples include small, cold gas clouds (Gerhard & Silk 1996), dust with charge = mass in the $c = G = 1$ units of GR (Bonnor 1996), and white dwarfs that were never luminous stars (Lynden-Bell & Tout 2001).

And this still leaves 25 or so apparently distinct proposals for the dominant dark matter that I have spotted over the years and cited in the Ap9x and Ap200x series (with another 10 or so waiting for Ap02). A quick and dirty way to decide whether you should take them seriously is by noting whether they were published in *Astrophysics and Space Science* (perhaps not), in *Physical Review Letters* (probably yes, at least if you are a particle physicist), or one of the more-tightly refereed journals of astronomy and astrophysics, meaning either that at least it didn't contradict existing observations at the time, or that the referee got worn down. Here is a subset of my favorites, deliberately not ordered by that variable.

- Solid hydrogen (White 1996)
- Rydberg matter (which, by the way means atoms in highly excited orbits like $n = 109$, not something at 13.6 eV (Badiei & Holmlid 2002)
- Q balls (Kusenko & Steinhardt 2001)
- Clusters of MACHOs with cluster mass $4 \times 10^4 \, M_\odot$ (Kerins 1997)
- DAEMONS (Drobyshevski 2000)
- Decaying B-balls (Enqvist & MacDonald 1998)
- Planck-mass remnants from the evaporation of primordial black holes (Alexeyev et al. 2002)
- A vector-based theory of gravity (Jeffries 1996)
- Spin = 0, charge = 0 bosons of mass 10^{-34} eV with a time-dependent scalar field (Israelit 1996)

- Short-lived particles that are created continuously from vacuum quantum fluctuations during the gravitational formation of galaxies (Majernik 1996)
- Point-like masses (in the MACHO range) with attached gas halos, so that they produce non-gray lensing events and have escaped detection (Bozza et al. 2001)

A comparable number of papers put observational or laboratory limits on various candidates from axions and nuclearites to black holes of both very large and very small masses.

10.5   Conclusions

It is difficult to summarize the cosmic density situation as it is today in any way very different from what Ivan King said in 1977 or I did in 1987: there is an enormous body of evidence indicative of gravitating mass which does not emit or absorb its fair share of light (which evidence we can abbreviate with the phrase "dark matter"), but we haven't a clue what most of it is. I should perhaps just add that there are a good many more things than there were in 1977 that we now know that it is not, though the inventory of remaining candidates is strange, wondrous, and growing.

Acknowledgements. I am grateful to Wendy Freedman for the invitation to participate in the 100th anniversary of the Carnegie Institution, to Susan Lehr for help in getting the manuscript into statutory form, and to the Peter Gruber Foundation for arranging their board meeting in such a way as to cover part of my travel expenses!

References

Abell, G. O. 1965, ARA&A, 3, 1
Abroe, M. E., et al. 2002, MNRAS, 334, 11
Adams, C., & Mazure, A. 2002, A&A, 381, 420
Alexeyev, S. O., Barrow, A., Bowdole, G., Sazhin, M. V., & Khovanskaya, O. S. 2002, Astron. Lett., 28, 428
Allen, S. W., Schmidt, R. W., & Fabian, A. C. 2001, MNRAS, 328, L37
——. 2002, MNRAS, 334, L11
Arnaud, M., Aghanim, N., & Neumann, D. M. 2002, A&A, 389, 1
Arp, H. C. 1971, Astrophys. Lett., 7, 221
Athreya, R. M., Mellier, Y., van Waerbeke, L., Pelló, R., Fort, B., & Dantel-Fort, M. 2002, A&A, 384, 743
Babcock, H. W. 1939, Lick Obs. Bull., 19, 41
Badiei, S., & Holmlid, L. 2002, MNRAS, 333, 360
Bahcall, N. A., & Comerford, J. M. 2001, ApJ, 565, L5
Bahcall, N. A., Ostriker, J. P., Perlmutter, S., & Steinhardt, P. J. 1999, Science, 284, 1481
Bell, M. B. 2002, ApJ, 566, 705
Bissantz, N., & Gerhard, O. 2002, MNRAS, 330, 591
Blandford, R. D. 2001, PASP, 113, 1309
Bonnor, W. B. 1996, MNRAS, 282, 1467
Borgani, S., et al. 2001, ApJ, 561, 13
Boughn, S., & Crittenden, R. G. 2003, in The Emergence of Cosmic Structure, ed. S. S. Holt & C. Reynolds (New York: AIP), 67
Bozza, V., Jetzer, Ph., Mancini, L., & Scarpetta, G. 2001, A&A, 382, 6
Braine, J., Duc, P.-A., Lisenfeld, U., Charmandaris, V., Vallejo, O., Leon, S., & Brinks, E. 2001, A&A, 378, 51
Brainerd, T. G., Blandford, R. D., & Smail, I. 1996, ApJ, 466, 623
Bridle, S. L., Zehavi, I., Dekel, A., Lahav, O., Hobson, M. P., & Lasenby, A. N. 2001, MNRAS, 321, 333
Bromm, V., & Clarke, C. J. 2002, ApJ, 566, L1
Burstein, D. S., Davies, R. L., Dressler, A., Faber, S. M., & Lynden-Bell, D. 1986, in Galaxy Distances and Deviations from Universal Expansion, ed. B. F. Madore & R. B. Tully (Dordrecht: Reidel), 123
Caretta, C. A., Maia, M. A. G., Kawasaki, W., & Willmer, C. N. A. 2002, AJ, 123, 1200
Cen, R. 2001, ApJ, 549, L195
Claeskens, J.-J., & Surdej, J. 2002, A&AR, 10, 263

Cohen, J. G., & Kneib, J.-P. 2002, ApJ, 573, 524

Cormier, D. & Holman, R. 2002, Phys. Rev. Lett., 84, 5936

Craig, M. W., & Davis, M. 2001, NewA, 6, 425

Croom, S. M., & Shanks, T. 1999, MNRAS, 307, L17

Czoske, O., Moore, B., Kneib, J.-P., & Soucail, G. 2002, A&A, 386, 31

Dabrowski, M. P., & Schunck, F. E. 2000, ApJ, 535, 300

Davidsen, A. F., Kriss, G. A., Ferguson, H. C., Blair, W. P., Bowers, C. W., & Kimble, R. A. 1991, Nature, 351, 128

Debattista, V. P., Corsini, E. M., & Aguerri, J. A. L. 2002, MNRAS, 332, 65

de Bernardis, P., et al. 2002, ApJ, 564, 559

de Vaucouleurs, G. 1970, Science, 167, 1203

Digal, S., Ray, R., Sengupta, S., & Srivastava, A. M. 2000, Phys. Rev. Lett., 84, 826

Drobyshevski, E. M. 2000, MNRAS, 311, L1

Dvali, G., Liu, H., & Vachaspati, T. 1998, Phys. Rev. Lett., 80, 2281

Dye, S., et al. 2002, A&A, 386, 12

Edery, A. 1999, Phys. Rev. Lett., 83, 3990

Einasto, J., Saar, E., Kaasik, A., & Chernin, A. D. 1974, Nature, 250, 209

Eke, V. R., Cole, S., Frenk, C. S., & Patrick H. J. 1998, MNRAS, 298, 1145

Elgarøy, Ø., Gramann, M., & Lahav, O. 2002, MNRAS, 333, 93

Enqvist, K., & MacDonald, J. 1998, Phys. Rev. Lett., 81, 3071

Fabricant, D., Lecar, M., & Gorenstein, P. 1980, ApJ, 241, 552

Fellhauer, M., & Kroupa, P. 2002, MNRAS, 330, 642

Felten, J. E., Gould, R. J., Stein, W. A., & Woolf, N. J. 1966, ApJ, 146, 955

Forman, W., & Jones, C. 1982, ARA&A, 20, 547

Gallart, C., Martínez-Delgado, D., Gómez-Flechoso, M. A., & Mateo, M. 2001, AJ, 121, 2572

Garnavich, P. M., et al. 1998, ApJ, 493, L53

Gerhard, O. E., & Silk, J. 1996, ApJ, 472, 34

Girardi, M., Manzato, P., Mezzetti, M., Giuricin, G., & Limboz, F. 2002, ApJ, 569, 720

Goodman, J. 2000, NewA, 5, 103

Gray, M. E., Taylor, A. N., Meisenheimer, K., Dye, S., Wolf, C., & Thommes, E. 2002, ApJ, 568, 141

Gunn, J. E. 1967, ApJ, 147, 61

Hamilton, A. J. S., Matthews, A., Kumar, P., & Lu, E. 1991, ApJ, 374, L1

Harrison, E. 1993, ApJ, 405, L1

Haynes, T., Cotter, G., Baker, J. C., Eales, S., Jones, M. E., Rawlings, S., & Saunders, R. 2002, MNRAS, 334, 262

Hoekstra, H., et al. 2001, ApJ, 548, L5

Hoekstra, H., Franx, M., Kuijken, K., & van Dokkum, P. G. 2002, MNRAS, 333, 911

Holmberg, E. 1937, Annals of Obs. Lund, 6, 1

Hoyle, F., Outram, P. J., Shanks, T., Boyle, B. J., Croom, S. M., & Smith, R. J. 2002, MNRAS, 332, 311

Hu, W., & Peebles, P. J. E. 2000, ApJ, 528, L61

Hubble, E. P. 1934, ApJ, 79, 8

Inskip, K. J., Best, P. N., Longair, M. S., & MacKay, D. J. C. 2002, MNRAS, 329, 277

Israelit, M. 1996, Ap&SS, 240, 331

Jeffries, C. 1996, Ap&SS, 238, 201

Junqueira, S., & Chan, R. 2002, Ap&SS, 279, 271

Karachentsev, I. D., et al. 2002a, A&A, 383, 812

——. 2002b, A&A, 389, 812

Kerins, E. J. 1997, A&A, 322, 709

King, I. R. 1977, in The Evolution of Galaxies and Stellar Populations, ed. B. M. Tinsley & R. B. Larson (New Haven: Yale Univ. Obs.), 1

King, L. J., Clowe, D. I., Lidman, C., Schneider, P., Erben, T., Kneib, J.-P., & Meylan, G. 2002, A&A, 385, L5

Kleyna, J. T., Wilkinson, M. I., Evans, N. W., & Gilmore, G. 2001, ApJ, 563, L115

Kraan-Korteweg, R. D., van Driel, W., Briggs, F., Binggeli, B., & Mostefaoui, T. I. 1999, A&AS, 135, 255

Kusenko, A., & Steinhardt, P. J. 2001, Phys. Rev. Lett., 87, 1301

Lima, J. A. S., & Alcaniz, J. S. 2002, ApJ, 566, 15

Lineweaver, C. H. 1999, Science, 284, 1503

Loh, E. D., & Spillar, E. J. 1986, ApJ, 307, L1

Lynden-Bell, D., & Tout, C. A. 2001, ApJ, 558, 1

Machacek, M. E., Bautz, M. W., Canizares, C., & Garmire, G. P. 2002, ApJ, 567, 188

Majernik, V. 1996, Ap&SS, 240, 133

Mannheim, P. D. 2001, ApJ, 561, 1

Mathewson, D. S., Ford, V. L., & Buchhorn, M. 1992, ApJ, 389, L5

McVittie, G. C., ed. 1962, IAU Symp. 15, Problems in Extragalactic Research, (New York: Macmillan)

Milgrom, M. 2002, ApJ, 571, L81

Miller, C. J., Nichol, R. C., & Batuski, D. J. 2001, Science, 292, 2362

Miralles, J.-M., et al. 2002, A&A, 388, 68

Neyman, J., Page, T., & Scott, E. 1961, AJ, 66, 633

Olling, R. P., & Merrifield, M. R. 2001, MNRAS, 326, 164

Opik, E. 1922, ApJ, 55, 406

Ostriker, J. P., Peebles, P. J. E., & Yahil, A. 1974, ApJ, 193, L1

Pen, U.-L. 1997, NewA, 2, 309

Percival, W. J., et al. 2001, MNRAS, 327, 1297

Perlmutter, S., et al. 1997, ApJ, 483, 565

——. 1998, Nature, 391, 51

——. 1999, ApJ, 517, 565

Prada, F., & Burkert, A. 2002, ApJ, 564, L69

Pryke, C., Halverson, N. W., Leitch, E. M., Kovac, J., Carlstrom, J. E., Holzapfel, W. L., & Dragovan, M. 2002, ApJ, 568, 46

Reiprich, T. H., & Böhringer, H. 2002, ApJ, 567, 716

Riess, A. G., et al. 1998, AJ, 116, 1009

——. 2001, ApJ, 580, 49

Rines, K., Geller, M. J., Diaferio, A., Mahdavi, A., Mohr, J. J., & Wegner, G. 2002, AJ, 124, 1266

Roukema, B. F., Mamon, G. A., & Bajtlik, S. 2002, A&A, 382, 397

Rubin, V. C., Ford, W. K., Jr., Thonnard, N., Roberts, M. S., & Graham, J. A. 1976, AJ, 81, 687

Rubin, V. C., Thonnard, N., Ford, W. K., Jr., & Roberts, M. S. 1976, AJ, 81, 719

Sachs, R. K., & Wolfe, A. M. 1967, ApJ, 147, 73

Sahu, K. C., et al. 1998, ApJ, 492, L125

Schroder, L. L., Brodie, J. P.,, Kissler-Patig, M., Huchra, J. P., & Phillips, A. C. 2002, AJ, 123, 2473

Schuecker, P., Guzzo, L., Collins, C. A., & Böhringer, H. 2002, MNRAS, 335, 807

Sciama, D. W. 1990, Comm. Ap., 15, 71

Smith, S. 1936, ApJ, 83, 23

Sofue, Y., & Rubin, V. 2001, ARA&A, 39, 137

Spergel, D. N., & Pen, U.-L. 1997, ApJ, 491, L67

Spergel, D. N., & Steinhardt, P. J. 2000, Phys. Rev. Lett., 84, 3760

Susperregi, M. 2002, ApJ, 563, 473

Szapudi, I., Quinn, T., Stadel, J., & Lake, G. 1999, ApJ, 517, 54

Trimble, V. 1987, ARA&A, 25, 425

——. 1995, in Dark Matter, ed. S. S. Holt & C. L. Bennett (New York: AIP), 57

——. 2001, in Gravitational Lensing, ed. T. G. Brainerd & C. S. Kochanek (San Francisco: ASP), 1

Trimble, V., & Aschwanden, M. A. 2000, PASP, 112, 434

Tully, R. B., Somerville, R. S., Trentham, N., & Verheijen, M. A. W. 2002, ApJ, 569, 573

Vauglin, I., et al. 2002, A&A, 387, 1

Wang, W. H., Hong, X. Y., Jiang, D. R., Venturi, T., Chen, Y. J., & An, T. 2001, A&A, 380, 123

White, R. S. 1996, Ap&SS, 240, 75

Wittman, D. M., Tyson, J. A., Kirkman, D., Dell'Antonio, I., & Bernstein, G. 2000, Nature, 405, 143

Yoshida, N., Springel, V., White, S. D. M., & Tormen, G. 2000, ApJ, 544, L87

Zwicky, F. 1933, Helv. Phys. Acta, 6, 110

——. 1937, Phys. Rev., 51, 290 & 679

11

Big Bang Nucleosynthesis: probing the first 20 minutes

GARY STEIGMAN
Departments of Physics and of Astronomy, The Ohio State University

Abstract

Within the first 20 minutes of the evolution of the hot, dense, early Universe, astrophysically interesting abundances of deuterium, helium-3, helium-4, and lithium-7 were synthesized by the cosmic nuclear reactor. The primordial abundances of these light nuclides produced during Big Bang Nucleosynthesis (BBN) are sensitive to the universal density of baryons and to the early-Universe expansion rate which at early epochs is governed by the energy density in relativistic particles ("radiation") such as photons and neutrinos. Some 380 kyr later, when the cosmic background radiation (CBR) radiation was freed from the embrace of the ionized plasma of protons and electrons, the spectrum of temperature fluctuations imprinted on the CBR also depended on the baryon and radiation densities. The comparison between the constraints imposed by BBN and those from the CBR reveals a remarkably consistent picture of the Universe at two widely separated epochs in its evolution. Combining these two probes leads to new and tighter constraints on the baryon density at present, on possible new physics beyond the standard model of particle physics, as well as identifying some challenges to astronomy and astrophysics. In this review the current status of BBN will be presented along with the associated estimates of the baryon density and of the energy density in radiation.

11.1   Introduction

The present Universe is observed to be expanding and filled with radiation (the 2.7 K cosmic background radiation; CBR) as well as with "ordinary matter" (baryons), "dark matter," and "dark energy." As a consequence, the early Universe must have been hot and dense. Sufficiently early in its evolution, the universal energy density would have been dominated by relativistic particles ("radiation dominated"). During its early evolution the Universe passed through a brief epoch when it functioned as a cosmic nuclear reactor, synthesizing the lightest nuclides: D, ^{3}He, ^{4}He, and ^{7}Li. These relics from the distant past provide a unique window on the early evolution of the Universe, as well as being valuable probes of the standard models of cosmology and particle physics. Comparing the predicted primordial abundances with those inferred from observational data tests the standard models and may uncover clues to their modifications and/or to extensions beyond them. It is clear that Big Bang Nucleosynthesis (BBN), one of the pillars of modern cosmology, has a crucial role to play as the study of the evolution of the Universe enters a new, data-rich era.

As with all science, cosmology depends on the interplay between theoretical ideas and observational data. As new and better data become available, models may need to be refined, revised, or even replaced. A consequence of this is that any *review* such as this one is merely a signpost along the road to a better understanding of our Universe. While details of the current "standard" model, along with some of its more popular variants to be discussed here, may need to be revised or rejected in the future, the underlying physics to be described here can provide a useful framework and context for understanding those changes. Any quantitative conclusions to be reached today will surely need to be modified in the light of new data. This review is, then, a status report on the standard model, highlighting its successes as well as exposing the current challenges it faces. While we may rejoice in the consistency of the standard model, there is still much work, theoretical and observational, to be done.

11.2   An Overview of BBN

To set a context for the confrontation of theoretical predictions with observational data it is useful to review the physics and cosmology of the early evolution of the Universe, touching on the specifics relevant for the synthesis of the light nuclides during the first ~ 20 minutes. In this section is presented an overview of this evolution along with the predicted primordial abundances, first in the standard model and then for two examples of nonstandard models which involve variations on the early-Universe expansion rate (Steigman, Schramm, & Gunn 1977) or asymmetries between the number of neutrinos and antineutrinos (e.g., Kang & Steigman 1992, and references therein).

11.2.1   Early Evolution

Discussion of BBN can begin when the Universe is a few tenths of a second old and the temperature is a few MeV. At such an early epoch the energy density is dominated by the relativistic (R) particles present, and the Universe is said to be "radiation-dominated." For sufficiently early times, when the temperature is a few times higher than the electron rest-mass energy, these are photons, $e^\pm$ pairs, and, for the standard model of particle physics, three flavors of left-handed (i.e., one helicity state) neutrinos (and their right-handed antineutrinos).

$$\rho_R = \rho_\gamma + \rho_e + 3\rho_\nu = \frac{43}{8}\rho_\gamma, \tag{11.1}$$

where ρ_γ is the energy density in CBR photons (which, today, have redshifted to become the CBR photons at a temperature of 2.7 K).

In standard BBN (SBBN) it is assumed that the neutrinos are fully decoupled prior to $e^\pm$ annihilation and do not share in the energy transferred from the annihilating $e^\pm$ pairs to the CBR photons. In this approximation, in the post-$e^\pm$ annihilation Universe, the photons are hotter than the neutrinos by a factor $T_\gamma/T_\nu = (11/4)^{1/3}$, and the relativistic energy density is

$$\rho_R = \rho_\gamma + 3\rho_\nu = 1.68\rho_\gamma. \tag{11.2}$$

During these radiation-dominated epochs the age (t) and the energy density are related by $\frac{32\pi G}{3}\rho_R t^2 = 1$, so that once the particle content (ρ_R) is specified, the age of the Universe is known (as a function of the CBR temperature T_γ). In the standard model,

$$\text{Pre}-e^\pm \text{ annihilation}: \ t\,T_\gamma^2 = 0.738 \text{ MeV}^2 \text{ s}, \tag{11.3}$$

Post–$e^{\pm}$ annihilation : $t\, T_{\gamma}^2 = 1.32\text{ MeV}^2$ s. (11.4)

Also present at these early times are neutrons and protons, albeit in trace amounts compared to the relativistic particles. The relative abundance of neutrons and protons is determined by the charged-current weak interactions.

$$p+e^- \longleftrightarrow n+\nu_e, \quad n+e^+ \longleftrightarrow p+\bar{\nu}_e, \quad n \longleftrightarrow p+e^- +\bar{\nu}_e. \qquad (11.5)$$

As time goes by and the Universe expands and cools, the lighter protons are favored over the heavier neutrons and the neutron-to-proton ratio decreases, initially following the equilibrium form $(n/p)_{eq} \propto \exp(-\Delta m/T)$, where $\Delta m = 1.29$ MeV is the neutron-proton mass difference. As the temperature drops the two-body collisions in Equation 11.5 become too slow to maintain equilibrium and the neutron-to-proton ratio, while continuing to decrease, begins to deviate from (*exceeds*) this equilibrium value. For later reference, we note that if there is an *asymmetry* between the numbers of ν_e and $\bar{\nu}_e$ ("neutrino degeneracy"), described by a chemical potential μ_e (such that for $\mu_e > 0$ there are more ν_e than $\bar{\nu}_e$), then the equilibrium neutron-to-proton ratio is modified to $(n/p) \propto \exp(-\Delta m/T - \mu_e/T)$. In place of the neutrino chemical potential, it is convenient to introduce the dimensionless degeneracy parameter $\xi_e \equiv \mu_e/T$, which is invariant as the Universe expands.

Prior to $e^{\pm}$ annihilation, at $T \approx 0.8$ MeV when the Universe is ~ 1 second old, the two-body reactions regulating n/p become slow compared to the universal expansion rate and this ratio "freezes in," although, in reality, it continues to decrease, albeit more slowly than would be the case for equilibrium. Later, when the Universe is several hundred seconds old, a time comparable to the neutron lifetime ($\tau_n = 885.7 \pm 0.8$ s), the n/p ratio resumes falling exponentially: $n/p \propto \exp(-t/\tau_n)$. Since there are several billion CBR photons for every nucleon (baryon), the abundances of any complex nuclei are entirely negligible at these early times.

Notice that since the n/p ratio depends on the competition between the weak interaction rates and the early-Universe expansion rate (as well as on a possible neutrino asymmetry), any deviations from the standard model (e.g., $\rho_R \to \rho_R + \rho_X$ or $\xi_e \neq 0$) will change the relative numbers of neutrons and protons available for building more complex nuclides.

11.2.2 Building the Elements

At the same time that neutrons and protons are interconverting, they are also colliding among themselves to create deuterons: $n+p \longleftrightarrow D+\gamma$. However, at early times, when the density and average energy of the CBR photons are very high, the newly formed deuterons find themselves bathed in a background of high-energy gamma rays capable of photodissociating them. Since there are more than a billion photons for every nucleon in the Universe, before the deuteron can capture a neutron or a proton to begin building the heavier nuclides, the deuteron is photodissociated. This bottleneck to BBN persists until the temperature drops sufficiently so that there are too few photons energetic enough to photodissociate the deuterons before they can capture nucleons to launch BBN. This occurs after $e^{\pm}$ annihilation, when the Universe is a few minutes old and the temperature has dropped below 80 keV (0.08 MeV).

Once BBN begins in earnest, neutrons and protons quickly combine to form D, ^{3}H, ^{3}He, and ^{4}He. Here, at ^{4}He, there is a different kind of bottleneck. There are no stable mass-5 nuclides. To jump this gap requires ^{4}He reactions with D or ^{3}H or ^{3}He, all of which are positively charged. The Coulomb repulsion among these colliding nuclei suppresses the reaction rates, ensuring that virtually all of the neutrons available for BBN are incorporated in ^{4}He (the most tightly bound of the light nuclides), and also that the abundances of the heavier nuclides are severely depressed below that of ^{4}He (and even of D and ^{3}He). Recall that ^{3}H is unstable, decaying to ^{3}He. The few reactions that manage to bridge the mass-5 gap lead mainly to mass-7 (^{7}Li or ^{7}Be, which, later, when the Universe has cooled further, will capture an electron and decay to ^{7}Li); the abundance of ^{6}Li is 1 to 2 orders of magnitude below that of the more tightly bound ^{7}Li. Finally, there is another gap at mass-8. This absence of any stable mass-8 nuclei ensures there will be no astrophysically interesting production of any heavier nuclides.

The primordial nuclear reactor is short-lived, quickly encountering an energy crisis. Because of the falling temperature and the Coulomb barriers, nuclear reactions cease rather abruptly as the temperature drops below ~ 30 keV, when the Universe is ~ 20 minutes old. This results in "nuclear freeze-out," since no already existing nuclides are destroyed (except for those that are unstable and decay) and no new nuclides are created. In ~ 1000 seconds BBN has run its course.

11.2.3   *The SBBN-predicted Abundances*

The primordial abundances of D, ^{3}He, and ^{7}Li(^{7}Be) are rate limited, depending sensitively on the competition between the nuclear reaction rates (proportional to the nucleon density) and the universal expansion rate. As a result, these nuclides are all potential baryometers. As the Universe expands, the nucleon density decreases so it is useful to compare it to that of the CBR photons: $\eta \equiv n_{\rm N}/n_\gamma$. Since this ratio turns out to be very small, it is convenient to introduce

$$\eta_{10} \equiv 10^{10}(n_{\rm N}/n_\gamma) = 274\Omega_{\rm b}h^2\,, \tag{11.6}$$

where $\Omega_{\rm b}$ is the ratio of the present values of the baryon and critical densities and h is the present value of the Hubble parameter in units of 100 km s^{-1} Mpc^{-1} . As the Universe evolves (post-$e^\pm$ annihilation) this ratio is accurately preserved so that η at the time of BBN should be equal to its value today. Testing this relation over ten orders of magnitude in redshift, over a timespan of some 10 billion years, can provide a confirmation of, or pose a challenge to the standard model.

In contrast to the other light nuclides, the primordial abundance of ^{4}He (mass fraction Y) is relatively insensitive to the baryon density, but since virtually all neutrons available at BBN are incorporated in ^{4}He, Y does depend on the competition between the weak interaction rates (largely fixed by the accurately measured neutron lifetime) and the universal expansion rate. The higher the nucleon density, the earlier can the D bottleneck be breached. Since at early times there are more neutrons (as a fraction of the nucleons), more ^{4}He will be synthesized. This latter effect is responsible for a very slow (logarithmic) increase in Y with η. Given the standard model relation between time and temperature and the measured nuclear and weak cross sections and decay rates, the evolution of the light-nuclide abundances may be calculated and the relic, primordial abundances predicted as a function of the one free parameter, the nucleon density or η. These predictions for SBBN are shown in Figure 11.1.

Not shown on Figure 11.1 are the relic abundances of ^{6}Li, ^{9}Be, ^{10}B, and ^{11}B; for the same range in η, all of them lie offscale, in the range $10^{-20} - 10^{-13}$. The results shown here are

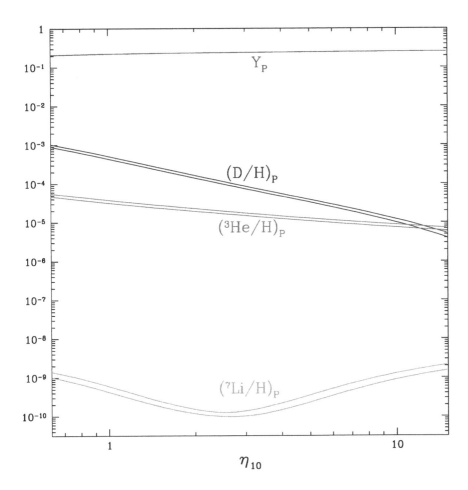

Fig. 11.1. The SBBN-predicted primordial abundances of D, ^{3}He, and ^{7}Li (by number with respect to hydrogen), and the ^{4}He mass fraction Y as a function of the nucleon abundance η_{10}. The widths of the bands reflect the theoretical uncertainties.

from the BBN code developed and refined over the years by my colleagues at The Ohio State University (OSU). They are in excellent agreement with the published results of the Chicago group (Burles, Nollett, & Turner 2001). Notice that the abundances appear in Figure 11.1 as bands. These reflect the theoretical uncertainties in the predicted abundances. For the OSU code the errors in D/H and ^{3}He/H are at the $\sim 8\%$ level, while they are much larger, $\sim 12\%$, for ^{7}Li. Burles et al. (2001), in a reanalysis of the relevant published cross sections, have reduced the theoretical errors by roughly a factor of 3 for D and ^{3}He and a factor of 2 for ^{7}Li. The reader may not notice the band shown for ^{4}He, since the uncertainty in Y, dominated by the very small uncertainty in the neutron lifetime, is at only the $\sim 0.2\%$ level ($\sigma_Y \approx 0.0005$).

Based on the discussion above it is easy to understand the trends shown in Figure 11.1. D and ^{3}He are burned to ^{4}He. The higher the nucleon density, the faster this occurs, leaving behind fewer nuclei of D or ^{3}He. The very slight increase of Y with η is largely due to BBN starting earlier at higher nucleon density (more complete burning of D, ^{3}H, and ^{3}He to ^{4}He) and higher neutron-to-proton ratio (more neutrons, more ^{4}He). The behavior of ^{7}Li is more interesting. At relatively low values of $\eta_{10} \lesssim 3$, mass-7 is largely synthesized as ^{7}Li [by ^{3}H(α,γ)^{7}Li reactions], which is easily destroyed in collisions with protons. So, as η increases at low values, destruction is faster and ^{7}Li/H *decreases*. In contrast, at relatively high values of $\eta_{10} \gtrsim 3$, mass-7 is largely synthesized as ^{7}Be [via ^{3}He(α,γ)^{7}Be reactions], which is more tightly bound than ^{7}Li and, therefore, harder to destroy. As η increases at high values, the abundance of ^{7}Be *increases*. Later in the evolution of the Universe, when it is cooler and neutral atoms begin to form, ^{7}Be will capture an electron and β-decay to ^{7}Li.

11.2.4 Nonstandard BBN

The predictions of the primordial abundance of ^{4}He depend sensitively on the early expansion rate (the Hubble parameter H) and on the amount—if any—of a $\nu_e - \bar{\nu}_e$ asymmetry (the ν_e chemical potential μ_e or the neutrino degeneracy parameter ξ_e). In contrast to ^{4}He, the BBN-predicted abundances of D, ^{3}He and ^{7}Li are determined by the competition between the various two-body production/destruction rates and the universal expansion rate. As a result, the D, ^{3}He, and ^{7}Li abundances are sensitive to the post-$e^{\pm}$ annihilation expansion rate, while that of ^{4}He depends on *both* the pre- and post-$e^{\pm}$ annihilation expansion rates; the former determines the "freeze-in" and the latter modulates the importance of β-decay (see, e.g., Kneller & Steigman 2003). Also, the primordial abundances of D, ^{3}He, and ^{7}Li, while not entirely insensitive to neutrino degeneracy, are much less affected by a nonzero ξ_e (e.g., Kang & Steigman 1992). Each of these nonstandard cases will be considered below. Note that the abundances of at least two different relic nuclei are needed to break the degeneracy between the baryon density and a possible nonstandard expansion rate resulting from new physics or cosmology, and/or a neutrino asymmetry.

11.2.4.1 Additional Relativistic Energy Density

The most straightforward variation of SBBN is to consider the effect of a nonstandard expansion rate $H' \neq H$. To quantify the deviation from the standard model it is convenient to introduce the "*expansion rate factor*" (or speedup/slowdown factor) S, where

$$S \equiv H'/H = t/t'. \tag{11.7}$$

Such a nonstandard expansion rate might result from the presence of "extra" energy contributed by new, light (relativistic at BBN) particles "X". These might, but need not, be additional flavors of active or sterile neutrinos. For X particles that are decoupled, in the sense that they do not share in the energy released by $e^{\pm}$ annihilation, it is convenient to account for the extra contribution to the standard-model energy density by normalizing it to that of an "equivalent" neutrino flavor (Steigman et al. 1977),

$$\rho_X \equiv \Delta N_\nu \rho_\nu = \frac{7}{8}\Delta N_\nu \rho_\gamma. \tag{11.8}$$

For SBBN, $\Delta N_\nu = 0$ ($N_\nu \equiv 3 + \Delta N_\nu$) and for each such additional "neutrino-like" particle (i.e., any two-component fermion), if $T_X = T_\nu$, then $\Delta N_\nu = 1$; if X should be a scalar, $\Delta N_\nu = 4/7$. However, it may well be that the X have decoupled even earlier in the evolution of the Universe and have failed to profit from the heating when various other particle-antiparticle pairs annihilated (or unstable particles decayed). In this case, the contribution to ΔN_ν from each such particle will be < 1 $(< 4/7)$. Henceforth we drop the X subscript. Note that, in principle, we are considering any term in the energy density that scales like "radiation" (i.e., decreases with the expansion of the Universe as the fourth power of the scale factor). In this sense, the modification to the usual Friedman equation due to higher dimensional effects, as in the Randall-Sundrum model (Randall & Sundrum 1999a,b; see also Cline, Grojean, & Servant 1999; Binetruy et al. 2000; Bratt et al. 2002), may be included as well. The interest in this latter case is that it permits the possibility of an apparent *negative* contribution to the radiation density ($\Delta N_\nu < 0$; $S < 1$). For such a modification to the energy density, the pre-$e^\pm$ annihilation energy density in Equation 11.1 is changed to

$$(\rho_R)_{pre} = \frac{43}{8} \left(1 + \frac{7\Delta N_\nu}{43}\right) \rho_\gamma. \tag{11.9}$$

Since any *extra* energy density ($\Delta N_\nu > 0$) speeds up the expansion of the Universe ($S > 1$), the right-hand side of the time-temperature relation in Equation 11.3 is smaller by the square root of the factor in parentheses in Equation 11.9.

$$S_{pre} \equiv (t/t')_{pre} = (1 + \frac{7\Delta N_\nu}{43})^{1/2} = (1 + 0.163\Delta N_\nu)^{1/2}. \tag{11.10}$$

In the post-$e^\pm$ annihilation Universe the extra energy density is diluted by the heating of the photons, so that

$$(\rho_R)_{post} = 1.68(1 + 0.135\Delta N_\nu)\rho_\gamma \tag{11.11}$$

and

$$S_{post} \equiv (t/t')_{post} = (1 + 0.135\Delta N_\nu)^{1/2}. \tag{11.12}$$

While the abundances of D, ^{3}He, and ^{7}Li are most sensitive to the baryon density (η), the ^{4}He mass fraction (Y) provides the best probe of the expansion rate. This is illustrated in Figure 11.2 where, in the $\Delta N_\nu - \eta_{10}$ plane, are shown isoabundance contours for D/H and Y_P (the isoabundance curves for ^{3}He/H and for ^{7}Li/H, omitted for clarity, are similar in behavior to that of D/H). The trends illustrated in Figure 11.2 are easy to understand in the context of the discussion above. The higher the baryon density (η_{10}), the faster primordial D is destroyed, so the relic abundance of D is *anticorrelated* with η_{10}. But, the faster the Universe expands ($\Delta N_\nu > 0$), the less time is available for D destruction, so D/H is positively, albeit weakly, correlated with ΔN_ν. In contrast to D (and to ^{3}He and ^{7}Li), since the incorporation of all available neutrons into ^{4}He is not limited by the nuclear reaction rates, the ^{4}He mass fraction is relatively insensitive to the baryon density, but it is very sensitive to both the pre- and post-$e^\pm$ annihilation expansion rates (which control the neutron-to-proton ratio). The faster the Universe expands, the more neutrons are available for ^{4}He. The very slow increase of Y_P with η_{10} is a reflection of the fact that for a higher baryon density, BBN begins earlier, when there are more neutrons. As a result of these complementary correlations, the pair of primordial abundances $y_D \equiv 10^5(D/H)_P$ and Y_P, the ^{4}He mass fraction, provide observational constraints on both the baryon density (η) and on the universal expansion rate factor S (or on ΔN_ν) when the Universe was some 20 minutes old. Comparing these to similar constraints from when the Universe was some 380 Kyr old, provided by the *WMAP* observations of the CBR polarization and the spectrum of temperature fluctuations, provides

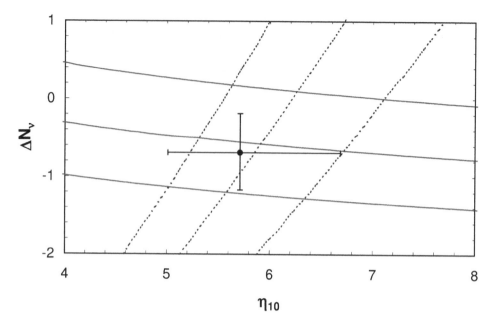

Fig. 11.2. Isoabundance curves for D and ^{4}He in the $\Delta N_\nu - \eta_{10}$ plane. The solid curves are for ^{4}He (from top to bottom: $Y = 0.25, 0.24, 0.23$). The dotted curves are for D (from left to right: $y_D \equiv 10^5(D/H) = 3.0, 2.5, 2.0$). The data point with error bars corresponds to $y_D = 2.6\pm0.4$ and $Y_P = 0.238\pm0.005$; see the text for discussion of these abundances.

a test of the consistency of the standard models of cosmology and of particle physics and further constrains the allowed range of the present-Universe baryon density (e.g., Barger et al. 2003a,b; Crotty, Lesgourgues, & Pastor 2003; Hannestad 2003; Pierpaoli 2003).

11.2.4.2 Neutrino Degeneracy
The baryon-to-photon ratio provides a dimensionless measure of the universal baryon asymmetry, which is very small ($\eta \lesssim 10^{-9}$). By charge neutrality the asymmetry in the charged leptons must also be of this order. However, there are no observational constraints, save those to be discussed here (see Kang & Steigman 1992; Kneller et al. 2002, and further references therein), on the magnitude of any asymmetry among the neutral leptons (neutrinos). A relatively small asymmetry between electron type neutrinos and antineutrinos ($\xi_e \gtrsim 10^{-2}$) can have a significant impact on the early-Universe ratio of neutrons to protons, thereby affecting the yields of the light nuclides formed during BBN. The strongest effect is on the BBN ^{4}He abundance, which is neutron limited. For $\xi_e > 0$, there is an excess of neutrinos (ν_e) over antineutrinos ($\bar{\nu}_e$), and the two-body reactions regulating the neutron-to-proton ratio (Eq. 11.5) drive down the neutron abundance; the reverse is true for $\xi_e < 0$. The effect of a nonzero ν_e asymmetry on the relic abundances of the other light nuclides is much weaker. This is illustrated in Figure 11.3, which shows the D and ^{4}He isoabundance curves in the $\xi_e - \eta_{10}$ plane. The nearly horizontal ^{4}He curves reflect the weak dependence of Y_P on the baryon density, along with its significant dependence on the neutrino asymmetry. In contrast, the nearly vertical D curves reveal the strong dependence of y_D on the baryon density

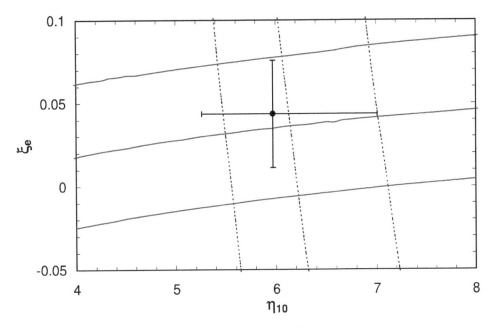

Fig. 11.3. Isoabundance curves for D and ^{4}He in the $\xi_e - \eta_{10}$ plane. The solid curves are for ^{4}He (from top to bottom: Y_P = 0.23, 0.24, 0.25). The dotted curves are for D (from left to right: $y_D \equiv 10^5$(D/H) = 3.0, 2.5, 2.0.) The data point with error bars corresponds to y_D = 2.6±0.4 and Y_P = 0.238±0.005; see the text for discussion of these abundances.

and its weak dependence on any neutrino asymmetry (^{3}He/H and ^{7}Li/H behave similarly: strongly dependent on η, weakly dependent on ξ_e). This complementarity between y_D and Y_P permits the pair $\{\eta, \xi_e\}$ to be determined once the primordial abundances of D and ^{4}He are inferred from the appropriate observational data.

11.3 Primordial Abundances

It is clear from Figures 11.1 – 11.3 that tests of the consistency of SBBN, along with constraints on any new physics, will be data-driven. While D (and/or ^{3}He and/or ^{7}Li) largely constrain the baryon density and ^{4}He plays a similar role for ΔN_ν and/or for ξ_e, there is an interplay among η_{10}, ΔN_ν, and ξ_e, which is quite sensitive to the adopted abundances. For example, a *lower* primordial D/H *increases* the BBN-inferred value of η_{10}, leading to a *higher* predicted primordial ^{4}He mass fraction. If the primordial ^{4}He mass fraction derived from the data is "low," then a low upper bound on ΔN_ν (or a nonzero lower bound on ξ_e) will be inferred. It is therefore crucial to avoid biasing any conclusions by *underestimating* the present uncertainties in the primordial abundances derived from the observational data.

The four light nuclides of interest, D, ^{3}He, ^{4}He, and ^{7}Li follow very different evolutionary paths in the post-BBN Universe. In addition, the observations leading to their abundance determinations are also very different. Neutral D is observed in absorption in the UV; singly ionized ^{3}He is observed in emission in Galactic H $_{\text{II}}$ regions; both singly and doubly ionized ^{4}He are observed in emission via recombinations in extragalactic H $_{\text{II}}$ regions; ^{7}Li is observed in absorption in the atmospheres of very metal-poor halo stars. The different histories and observational strategies provide some insurance that systematic errors affecting the inferred primordial abundances of any one of the light nuclides are unlikely to distort the inferred abundances of the others.

11.3.1   Deuterium

The post-BBN evolution of D is straightforward. As gas is incorporated into stars the very loosely bound deuteron is burned to ^{3}He (and beyond). Any D that passes through a star is destroyed. Furthermore, there are no astrophysical sites where D can be produced in an abundance anywhere near that observed (Epstein, Lattimer, & Schramm 1976). As a result, as the Universe evolves and gas is cycled through generations of stars, deuterium is only destroyed. Therefore, observations of the deuterium abundance anywhere, anytime, provide *lower* bounds on its primordial abundance. Furthermore, if D can be observed in "young" systems, in the sense of very little stellar processing, the observed abundance should be very close to the primordial value. Thus, while there are extensive data on deuterium in the solar system and the local interstellar medium of the Galaxy, it is the handful of observations of deuterium absorption in high-redshift, low-metallicity QSO absorption-line systems (QSOALS), which are potentially the most valuable. At sufficiently high redshifts and low metallicities, the primordial abundance of deuterium should reveal itself as a "deuterium plateau."

Inferring the primordial D abundance from the QSOALS has not been without its difficulties, with some abundance claims having been withdrawn or revised. Presently there are $\sim$ half a dozen QSOALS with reasonably firm deuterium detections (Burles & Tytler 1998a,b; D'Odorico, Dessauges-Zavadsky, & Molaro 2001; O'Meara et al. 2001; Pettini & Bowen 2002; Kirkman et al. 2004). However, there is significant dispersion among the derived abundances, and the data fail to reveal the anticipated deuterium plateau (Fig. 11.4 – 11.6; see also Steigman 2004). Furthermore, subsequent observations of the D'Odorico et al. (2001) QSOALS by Levshakov et al. (2002) revealed a more complex velocity structure and led to a revised—and uncertain—deuterium abundance. This sensitivity to poorly constrained velocity structure in the absorbers is also exposed in the analyses of published QSOALS data by Levshakov and collaborators (Levshakov, Kegel, & Takahara 1998a,b, 1999), which lead to consistent, but somewhat higher, deuterium abundances than those inferred from "standard" data reduction analyses.

Indeed, the absorption spectra of D I and H I are identical, except for a wavelength/velocity offset resulting from the heavier reduced mass of the deuterium atom. An H I "interloper," a low-column density cloud shifted by ~ 81 km s^{-1} with respect to the main absorbing cloud, would masquerade as D I . If this is not accounted for, a D/H ratio which is too high would be inferred. Since there are more low-column density absorbers than those with high H I column densities, absorption-line systems with somewhat lower H I column density (e.g., Lyman-limit systems) are more susceptible to this contamination than are the higher H I column density absorbers (e.g., damped Lyα absorbers). However, for the damped Lyα absorbers, an accurate determination of the H I column density requires an accurate placement of the continuum, which could be compromised by interlopers. This might lead to an overestimate of the H I column density and a concomitant underestimate of D/H (J. Linsky, private communication). As will be seen, there is the possibility that each of these effects may have contaminated the current data. Indeed, complex velocity structure in the D'Odorico et al. (2001) absorber (see Levshakov et al. 2002) renders it of less value in

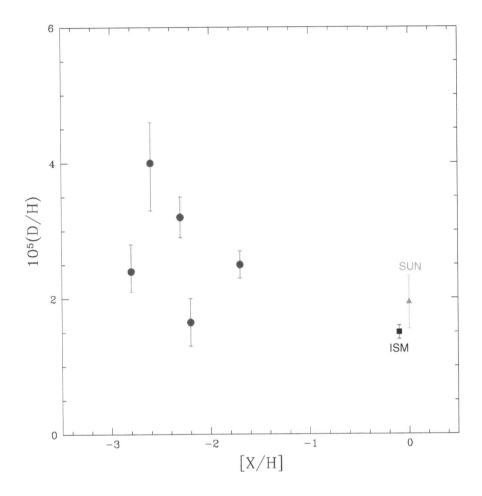

Fig. 11.4. The deuterium abundance, D/H, versus metallicity, "X"(usually, X = Si), from observations (as of early 2003) of QSOALS (filled circles). Also shown for comparison are the D abundances for the local ISM (filled square) and the solar system ("Sun"; filled triangle).

constraining primordial deuterium, and it will not be included in the estimates presented here.

In Figure 11.4 are shown the extant data (circa June 2003) for D/H as a function of metallicity from the work of Burles & Tytler (1998a,b), O'Meara et al. (2001), Pettini & Bowen (2002), and Kirkman et al. (2004). Also shown for comparison are the local interstellar medium (ISM) D/H (Linsky & Wood 2000) and that for the presolar nebula as inferred from solar system data (Geiss & Gloeckler 1998).

On the basis of our discussion of the post-BBN evolution of D/H, a "deuterium plateau" at low metallicity was expected. If, indeed, one is present, it is hidden by the dispersion in the current data. Given the possibility that interlopers may affect both the D_I and the

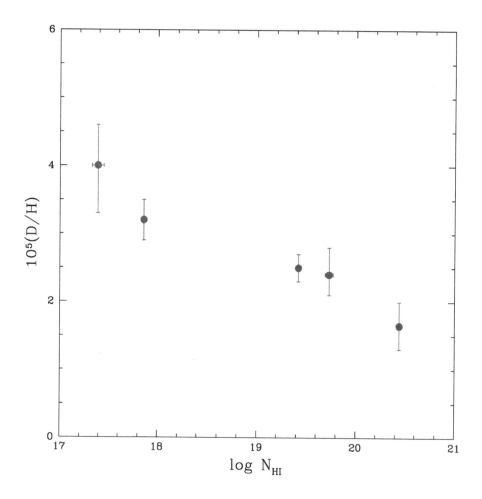

Fig. 11.5. The deuterium abundance, D/H, versus the H I column density in the absorbers, N(H I), for the same QSOALS as in Figure 11.4.

H I column density determinations, it is interesting to plot D/H as a function of N(H I). This is shown in Figure 11.5, where there is some (limited) evidence that D/H is higher in the Lyman-limit systems than in the damped Lyα absorbers.

To decide how to utilize this confusing data it may be of value to consider the observations chronologically. Of the set chosen here, Burles & Tytler (1998a,b) studied the first two lines of sight. For PKS 1937−1009 they derived $y_D \equiv 10^5(\text{D/H}) = 3.25 \pm 0.3$ (Burles & Tytler 1998a), while for Q1009+299 they found $y_D = 3.98^{+0.59}_{-0.67}$ (Burles & Tytler 1998b). These two determinations are in excellent agreement with each other ($\chi^2 = 1.0$), leading to a mean abundance $\langle y_D \rangle = 3.37 \pm 0.27$. Next, O'Meara et al. (2001) added the line of sight to HS 0105+1619, finding a considerably lower abundance $y_D = 2.54 \pm 0.23$. Indeed, while the weighted mean for these three lines of sight is $\langle y_D \rangle = 2.88$, the χ^2 has ballooned to 6.4 (for two degrees of freedom). Absent any evidence that one or more of

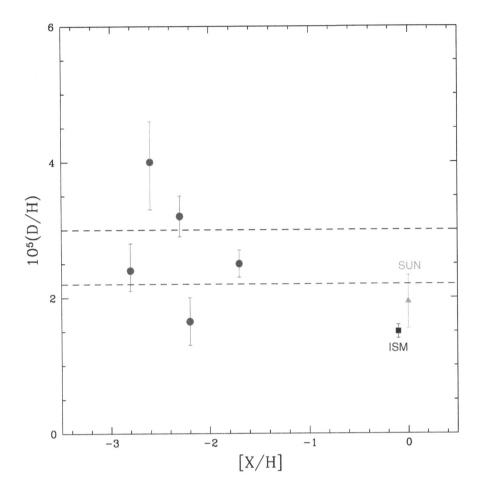

Fig. 11.6. As in Figure 11.4. The dashed lines represent the $\pm 1\sigma$ band calculated from the mean and its dispersion ($(D/H)_P = 2.6 \pm 0.4 \times 10^{-5}$; see the text).

these abundances is in error, O'Meara et al. adopt the mean, and, for the error in the mean, they take the dispersion about the mean (0.72) divided by the square root of the number of data points: $\langle y_D \rangle = 2.88 \pm 0.42$. One year later Pettini & Bowen (2002) published their *HST* data on the line of sight toward Q2206−199, finding a surprisingly low value of $y_D = 1.65 \pm 0.35$. Including this determination reduces the mean to $\langle y_D \rangle = 2.63$, but the dispersion in y_D grows to 1.00 and $\chi^2 = 16.3$ for three degrees of freedom. Clearly, either one or more of these determinations is in error, or the variation among the high-redshift, low-metallicity deuterium abundances is larger than anticipated from our understanding of its evolution (Jedamzik & Fuller 1997). Using the mean and its dispersion (to fix the error), as of the time of the Carnegie Symposium, the best estimate for the primordial D abundance was $\langle y_D \rangle = 2.63 \pm 0.50$. Shortly thereafter, in early 2003, the data of Kirkman et al. (2004) appeared for the line of sight toward Q1243+3047. For this line of sight they find $y_D =$

$2.42^{+0.35}_{-0.25}$. This abundance lies between the lowest and the higher previous values, reducing the overall dispersion to 0.88, while hardly changing the mean from $y_D = 2.63$ to 2.60. While the total χ^2 is still enormous, increasing slightly to 16.6, the reduced χ^2 decreases from 5.4 to 4.2. This is still far too large, suggesting that one or more of these determinations may be contaminated, or that there may actually be real variations in D/H at high redshifts and low metallicities. Notice (see Fig. 11.5) that the largest D/H estimates are from the two absorbers with the lowest H I column densities (Lyman-limit systems), where interlopers *might* contribute to the inferred D I column densities, while the lowest abundances are from the higher H I column density (damped Lyα) absorbers, where interlopers *might* affect the wings of the H I lines used to fix the H I column densities. Absent any further data supporting, or refuting, these possibilities, there is no *a priori* reason to reject any of these determinations.

To utilize the current data, the weighted mean D abundances for these five lines of sight and the dispersion are used to infer the abundance of primordial deuterium (and its uncertainty) adopted in this review: $y_D = 2.6 \pm 0.4$. Note that, given the large dispersion, two-decimal place accuracy seems to be wishful thinking at present. For this reason, in quoting the primordial D abundance inferred from the observational data I have purposely chosen to quote values to only one decimal place. This choice is consistent too with the $\sim 3\% - 8\%$ theoretical uncertainty (at fixed η) in the BBN-predicted abundance. In Figure 11.6 are shown the data, along with the corresponding 1σ band. It is worth remarking that using the same data Kirkman et al. (2004) derive a slightly higher mean D abundance: $y_D = 2.74$. The reason for the difference is that they first find the mean of $\log(y_D)$ and then use it to compute the mean D abundance ($y_D \equiv 10^{\langle \log(y_D) \rangle}$).

11.3.2 Helium-3

The post-BBN evolution of ^{3}He is considerably more complex and model dependent than that of D. Interstellar ^{3}He incorporated into stars is burned to ^{4}He (and beyond) in the hotter interiors, but preserved in the cooler, outer layers. Furthermore, while hydrogen burning in cooler, low-mass stars is a net producer of ^{3}He (Iben 1967; Rood 1972; Dearborn, Schramm, & Steigman 1986; Vassiliadis & Wood 1993; Dearborn, Steigman, & Tosi 1996) it is unclear how much of this newly synthesized ^{3}He is returned to the interstellar medium and how much of it is consumed in post-main sequence evolution (e.g., Sackmann & Boothroyd 1999a,b). Indeed, it is clear that when the data (Geiss & Gloeckler 1998; Rood et al. 1998; Bania, Rood, & Balser 2002) are compared to a large variety of chemical evolution models (Rood, Steigman, & Tinsley 1976; Dearborn et al. 1996; Galli et al. 1997; Palla et al. 2000; Chiappini, Renda, & Matteucci 2002), agreement is only possible for a very delicate balance between net production and net destruction of ^{3}He. For a recent review of the current status of ^{3}He evolution, see Romano et al. (2004). Given this state of affairs it is not possible to utilize ^{3}He as a baryometer, but it may perhaps be used to provide a consistency check. To this end, the abundance inferred by Bania et al. (2002) from an H II region in the outer Galaxy, where post-BBN evolution might have been minimal, is adopted here: $y_3 \equiv 10^5(^3\text{He/H}) = 1.1 \pm 0.2$.

11.3.3 Helium-4

Helium-4 is the second most abundant nuclide in the Universe after hydrogen. In post-BBN evolution gas cycling though stars has its hydrogen burned to helium, increasing the ^{4}He abundance above its primordial value. As with deuterium, a ^{4}He "plateau" is ex-

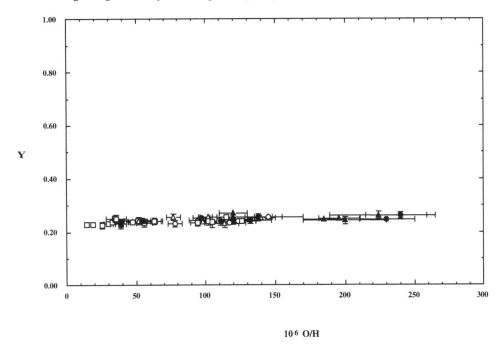

Fig. 11.7. The ⁴He mass fraction, Y, inferred from observations of low-metallicity, extra-galactic H ɪɪ regions versus the oxygen abundance derived from the same data. (Figure courtesy of K. A. Olive.)

pected at sufficiently low metallicity. Although ⁴He is observed in the Sun and in Galactic H ɪɪ regions, the crucial data for inferring its primordial abundance is from observations of the helium and hydrogen emission (recombination) lines from low-metallicity, extragalactic H ɪɪ regions. The present inventory of such regions studied for their helium content is approaching of order 100. Thus, it is not surprising that even with modest observational errors for any individual H ɪɪ region, the statistical uncertainty in the inferred primordial abundance may be quite small. In this situation, care must be taken with hitherto ignored or unaccounted for corrections and systematic errors or biases.

In Figure 11.7 is shown a compilation of the data used by Olive & Steigman (1995) and Olive, Skillman, & Steigman (1997), along with the independent data set obtained by Izotov, Thuan, & Lipovetsky (1997) and Izotov & Thuan (1998). To track the evolution of the ⁴He mass fraction, Y is plotted versus the H ɪɪ region oxygen abundance. These H ɪɪ regions are all metal poor, ranging from $\sim 1/2$ down to $\sim 1/40$ of solar (for a solar oxygen abundance of O/H $\approx 5 \times 10^{-4}$; Allende-Prieto, Lambert, & Asplund 2001). A key feature of Figure 11.7 is that for sufficiently low metallicity the Y versus O/H relation approaches a ⁴He plateau! Since Y increases with metallicity, the relic abundance can either be bounded from above by the lowest metallicity regions, or the Y versus O/H relation may be extrapolated to zero metallicity. The extrapolation is quite small, so that whether the former or the latter approach is adopted the difference in the inferred primordial abundance is small: $|\Delta Y| \lesssim 0.001$.

While the data shown in Figure 11.7 reveal a well-defined primordial abundance for ⁴He, the scale hides the very small statistical errors as well as the tension between the two

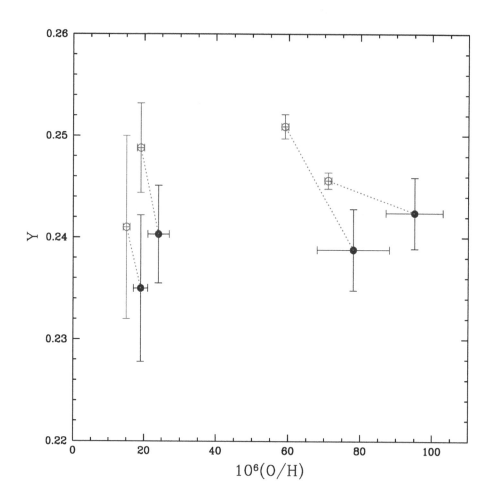

Fig. 11.8. The Peimbert et al. (2002) reanalysis of the ^{4}He abundance data for four of
the Izotov & Thuan (1998) H II regions. The open circles are the Izotov & Thuan (1998)
abundances, while the filled circles are from Peimbert et al. (2002).

groups' helium abundances. Olive & Steigman (1995) and Olive et al. (1997) find Y_P =
0.234 ± 0.003, but Izotov et al. (1997) and Izotov & Thuan (1998) derive $Y_P = 0.244 \pm 0.002$.
Although it is difficult to account for all of the difference, much of it is traceable to the different ways the two groups correct for the contribution to the emission lines from collisional
excitation of neutral helium and also to Izotov and collaborators rejecting some helium emission lines *a posteriori* when they yield "too low" an abundance. Furthermore, for either data
set, there are additional corrections for temperature, for temperature and density fluctuations,
and for ionization, which when applied can change the inferred primordial ^{4}He abundance
by more than the quoted statistical errors (see, e.g., Steigman, Viegas, & Gruenwald 1997;
Viegas, Gruenwald, & Steigman 2000; Gruenwald, Steigman, & Viegas 2002; Peimbert,
Peimbert & Luridiana 2002; Sauer & Jedamzik 2002).

For example, Peimbert et al. (2002) recently reanalyzed the data from four of the Izotov & Thuan (1998) H II regions, employing their own H II region temperatures and accounting for temperature fluctuations. Peimbert et al. (2002) derive systematically lower helium abundances, as shown in Figure 11.8. From this very limited sample Peimbert et al. suggest that the Izotov & Thuan (1998) estimate for the primordial ^{4}He mass fraction might have to be reduced by as much as ~ 0.007. Peimbert et al. go further, combining their redetermined helium abundances for these four H II regions with an accurate determination of Y in a more metal-rich H II region (Peimbert, Peimbert, & Ruiz 2000). Although these five data points are consistent with zero slope in the Y – O/H relation, leading to a primordial abundance $Y_P = 0.240\pm0.001$, this extremely small data set is also consistent with $\Delta Y \approx 40$(O/H), leading to a smaller primordial estimate of $Y_P \approx 0.237$.

It seems clear that until new data address the unresolved systematic errors afflicting the derivation of the primordial helium abundance, the true errors must be much larger than the statistical uncertainties. In an attempt to account for this, here I follow Olive, Steigman, & Walker (2000) and adopt a compromise mean value along with a larger uncertainty: $Y_P = 0.238 \pm 0.005$.

11.3.4 *Lithium-7*

Lithium-7 is fragile, burning in stars at a relatively low temperature. As a result, the majority of any interstellar ^{7}Li cycled through stars is destroyed. For the same reason, it is difficult for stars to create new ^{7}Li and/or to return any newly synthesized ^{7}Li to the ISM before it is destroyed by nuclear burning. In addition to synthesis in stars, the intermediate-mass nuclides ^{6}Li, ^{7}Li, ^{9}Be, ^{10}B, and ^{11}B can be synthesized via cosmic ray nucleosynthesis, either by alpha-alpha fusion reactions, or by spallation reactions (nuclear breakup) in collisions between protons and alpha particles and CNO nuclei. In the early Galaxy, when the metallicity is low, the post-BBN production of lithium is expected to be subdominant to that from BBN abundance. As the data in Figure 11.9 reveal, only relatively late in the evolution of the Galaxy does the lithium abundance increase. The data also confirm the anticipated "Spite plateau" (Spite & Spite 1982), the absence of a significant slope in the Li/H versus [Fe/H] relation at low metallicity due to the dominance of BBN-produced ^{7}Li. The plateau is a clear signal of the primordial lithium abundance. Notice, also, the enormous *spread* among the lithium abundances at higher metallicity. This range in Li/H likely results from the destruction/dilution of lithium on the surfaces of the observed stars while they are on the main sequence and/or lithium destruction during their pre-main sequence evolution, implying that it is the *upper envelope* of the Li/H versus [Fe/H] relation that preserves the history of Galactic lithium evolution. Note, also, that at low metallicity the dispersion is much narrower, suggesting that corrections for depletion/dilution are (may be) much smaller for the Population II stars.

As with the other relic nuclides, the dominant uncertainties in estimating the primordial abundance of ^{7}Li are not statistical, but systematic. The lithium observed in the atmospheres of cool, metal-poor, Population II halo stars is most relevant for determining the BBN ^{7}Li abundance. Uncertainties in the lithium equivalent width measurements, in the temperature scales for the cool Population II stars, and in their model atmospheres dominate the overall error budget. For example, Ryan et al. (2000), using the Ryan, Norris, & Beers (1999) data, infer [Li]$_P \equiv 12 + \log$(Li/H) $= 2.1$, while Bonifacio & Molaro (1997) and Bonifacio, Molaro, & Pasquini (1997) derive [Li]$_P = 2.2$, and Thorburn (1994) finds [Li]$_P = 2.3$. From

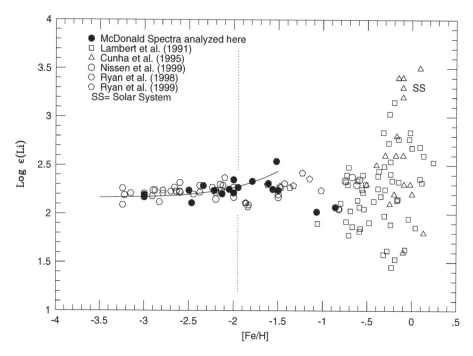

Fig. 11.9. A compilation of the lithium abundance data as a function of metallicity from stellar observations (courtesy of V. V. Smith). $\epsilon(Li) \equiv 10^{12}(Li/H)$, and [Fe/H] is the usual logarithmic metallicity relative to solar. Note the "Spite plateau" in Li/H for [Fe/H] $\lesssim -2$.

recent observations of stars in a metal-poor globular cluster, Bonifacio et al. (2002) derive $[Li]_P = 2.34 \pm 0.056$. As may be seen from Figure 11.9, the indication from the preliminary data assembled by V. V. Smith (private communication) favors a Spite plateau at $[Li]_P \approx 2.2$.

In addition to these intrinsic uncertainties, there are others associated with stellar structure and evolution. The metal-poor halo stars that define the primordial lithium plateau are very old. As a result, they have had time to disturb the prestellar lithium that could survive in their cooler, outer layers. Mixing of these outer layers with the hotter interior where lithium has been (can be) destroyed will dilute or deplete the surface lithium abundance. Pinsonneault et al. (1999, 2002) have shown that rotational mixing may decrease the surface abundance of lithium in these Population II stars by $0.1 - 0.3$ dex while still maintaining the rather narrow *dispersion* among the plateau abundances (see also Chaboyer et al. 1992; Theado & Vauclair 2001; Salaris & Weiss 2002). Pinsonneault et al. (2002) adopted for a baseline (Spite plateau) estimate $[Li] = 2.2 \pm 0.1$, while for an overall depletion factor 0.2 ± 0.1 dex was chosen. Adding these contributions to the log of the primordial lithium abundance *linearly*, an estimate $[Li]_P = 2.4 \pm 0.2$ was derived. In the comparison between theory and observation below, I will adopt the Ryan et al. (2000) estimate $[Li]_P = 2.1 \pm 0.1$, but I will also consider the implications of the Pinsonneault et al. (2002) value.

11.4   Confrontation of Theory with Data

Having reviewed the basic physics and cosmological evolution underlying BBN and summarized the observational data leading to a set of adopted primordial abundances, the predictions may now be confronted with the data. There are several possible approaches that might be adopted. The following option is chosen here. First, concentrating on the predictions of SBBN, deuterium will be used as the baryometer of choice to fix the baryon-to-photon ratio η. This value and its uncertainty are then used to "predict" the ^{3}He, ^{4}He, and ^{7}Li abundances, which are compared to those adopted above. This comparison can provide a test of the consistency of SBBN as well as identify those points of "tension" between theory and observation. This confrontation is carried further to consider the two extensions beyond the standard model [$S \neq 1$ ($\Delta N_\nu \neq 0$); $\xi_e \neq 0$].

11.4.1   Testing the Standard Model

For SBBN, the baryon density corresponding to the D abundance adopted here ($y_D = 2.6 \pm 0.4$) is $\eta_{10} = 6.1^{+0.7}_{-0.5}$, corresponding to $\Omega_b = 0.022^{+0.003}_{-0.002}$. This is in outstanding agreement with the estimate of Spergel et al. (2003), based largely on the new CBR (*WMAP*) data (Bennett et al. 2003): $\Omega_b = 0.0224 \pm 0.0009$. For the baryon density determined by D, the SBBN-predicted abundance of ^{3}He is $y_3 = 1.0 \pm 0.1$, which is to be compared to the outer-Galaxy abundance of $y_3 = 1.1 \pm 0.1$, which is suggested by Bania et al. (2002) to be nearly primordial. Again, the agreement is excellent.

The tension between the data and SBBN arises with ^{4}He. Given the very slow variation of Y_P with η, along with the very high accuracy of the SBBN-predicted abundance, the primordial abundance is tightly constrained: $Y_{SBBN} = 0.248 \pm 0.001$. This should be compared with our adopted estimate of $Y = 0.238 \pm 0.005$ (Olive et al. 2000). Agreement is only at the $\sim 5\%$ level. This tension is shown in Figure 11.10. This apparent challenge to SBBN is also an opportunity. As already noted, while the ^{4}He abundance is insensitive to the baryon density, it is very sensitive to new physics (i.e., nonstandard universal expansion rate and/or neutrino degeneracy).

There is tension, too, when comparing the SBBN-predicted abundance of ^{7}Li with the (very uncertain) primordial abundance inferred from the data. For SBBN the expected abundance is [Li]$_P = 2.65^{+0.09}_{-0.11}$. This is to be compared with the various estimates above that suggested [Li]$_P \approx 2.2 \pm 0.1$. In Figure 11.11 is shown the analog of Figure 11.10 for lithium and deuterium. Depending on the assessment of the uncertainty in the primordial abundance inferred from the observational data, the conflict with SBBN may or may not be serious. In contrast to ^{4}He, ^{7}Li is more similar to D (and to ^{3}He) in that its BBN-predicted abundance is relatively insensitive to new physics. As a result, this tension, if it persists, could be a signal of interesting new astrophysics (e.g., have the halo stars depleted or diluted their surface lithium?).

11.4.2   Nonstandard Expansion Rate: $S \neq 1$ ($\Delta N_\nu \neq 0$)

The excellent agreement between the SBBN-predicted baryon density inferred from the primordial-D abundance and that derived from the CBR and large scale structure (Spergel et al. 2003), and also the agreement between predicted and observed D and ^{3}He suggest that the tension with ^{4}He, if not observational or astrophysical in origin, may be a sign of new physics. As noted earlier, Y_P is sensitive to the early-Universe expansion rate (while D, ^{3}He, and ^{7}Li are less so). A faster expansion ($S > 1$, $\Delta N_\nu > 0$) leads to a higher predicted primor-

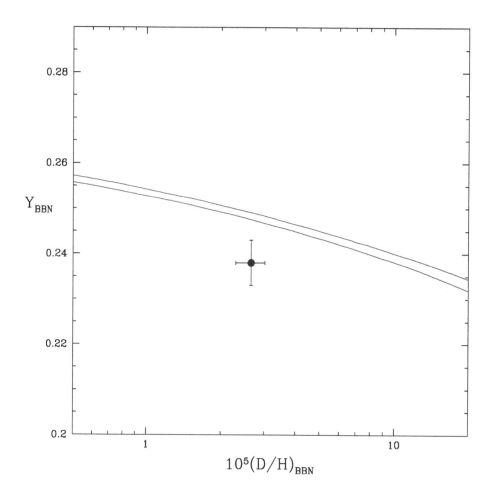

Fig. 11.10. The SBBN-predicted relation between the primordial abundances of D and ^{4}He (mass fraction) is shown by the band, whose thickness represents the uncertainties in the predicted abundances. Also shown by the point and error bars are the adopted primordial abundances of D and ^{4}He (see the text).

dial abundance of ^{4}He, and *vice versa* for $S < 1$ ($\Delta N_\nu < 0$). In Figure 11.12 is shown the same Y_P versus y_D band as for SBBN in Figure 11.10, along with the corresponding bands for the nonstandard cases of a faster expansion ($\Delta N_\nu = 4$) and a slower expansion ($\Delta N_\nu = 2$). It can be seen that the data "prefer" a slower than standard early-Universe expansion rate. If both η and ΔN_ν are allowed to be free, it is possible (not surprisingly) to accommodate the adopted primordial abundances of D and ^{4}He (see Fig. 11.2). Given the similar effects of $\Delta N_\nu \neq 0$ on the BBN-predicted D, ^{3}He, and ^{7}Li abundances, while it is possible to maintain the good agreement (from SBBN) for ^{3}He, the tension between ^{7}Li and D cannot be relieved. In Figure 11.13 are shown the 1-, 2-, and 3-σ BBN contours in the $\eta - \Delta N_\nu$ plane

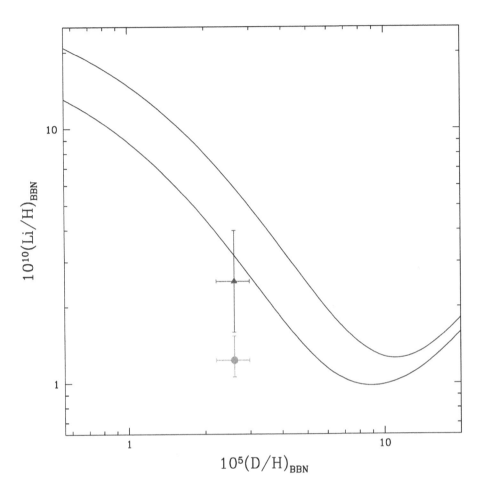

Fig. 11.11. The SBBN-predicted relation between the primordial abundances of D and ^{7}Li is shown by the band, whose thickness reflects the uncertainties in the predicted abundances. The data points are for the primordial abundance of D adopted here along with the Ryan et al. (2000) Li abundance (filled circle) and the Pinsonneault et al. (2002) Li abundance (filled triangle).

derived from the adopted values of y_D and Y_P. Although the best-fit point is at $\Delta N_\nu = -0.7$ (and $\eta_{10} = 5.7$), it is clear that SBBN ($N_\nu = 3$) is acceptable.

The CBR temperature anisotropy spectrum and polarization are also sensitive to the early-Universe expansion rate (see, e.g., Barger et al. 2003a, and references therein). There is excellent overlap between the $\eta - \Delta N_\nu$ confidence contours from BBN as shown in Figure 11.13 and from the CBR (Barger et al. 2003a). In Figure 11.14 are shown the confidence contours in the $\eta - \Delta N_\nu$ plane for a joint BBN – CBR fit (Barger et al. 2003a). Again, while the best fit value for ΔN_ν is negative (driven largely by the adopted value for Y_P), $\Delta N_\nu = 0$ is quite acceptable.

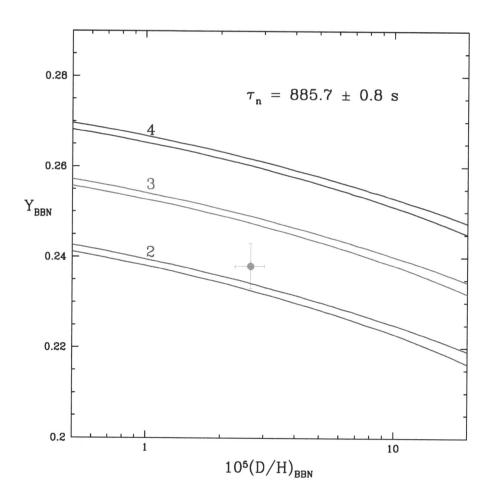

Fig. 11.12. As in Figure 11.10 for $N_\nu = 2, 3, 4$, which correspond to $S = 0.915, 1, 1.078$.

11.4.3   *Neutrino Asymmetry ($\xi_e \neq 0$)*

The tension between D and ^{4}He can also be relieved by nonstandard neutrino physics (see Fig. 11.3). Although the asymmetry (difference between the numbers of particles and antiparticles) in charged leptons, tied to that in the baryons by charge neutrality of the Universe, must be very small, the neutrino asymmetry is unconstrained observationally. Of relevance to BBN is the asymmetry between the electron neutrinos and the electron antineutrinos (ξ_e), which regulates the pre-BBN neutron-to-proton ratio through the reactions in Equation 11.5. In Figure 11.15 are shown the 1- and 2-σ contours in the $\eta - \xi_e$ plane for BBN (for $N_\nu = 3$) and the adopted abundances of D and ^{4}He. As seen before for $\Delta N_\nu \neq 0$, while a fit to the data can be achieved for $\xi_e \neq 0$, the data are not inconsistent with $\xi_e = 0$. Furthermore, as is shown in Figure 11.15, BBN constrains the allowed range for neutrino asymmetry to be very small. For further implications for neutrino physics and for

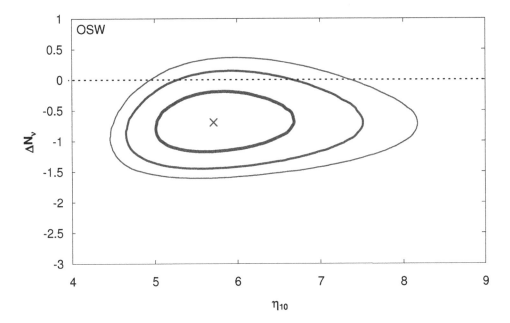

Fig. 11.13. The 1-, 2-, and 3-σ contours in the $\eta - \Delta N_\nu$ plane for BBN and the adopted D and ^{4}He abundances.

a discussion of the case where *both* ΔN_ν and ξ_e are free to differ from zero, see Barger et al. (2003b).

11.5   Summary and Conclusions

Given the standard models of cosmology and particle physics, SBBN predicts the primordial abundances of D, ^{3}He, ^{4}He, and ^{7}Li, which may be compared with the observational data. Of the light nuclides, deuterium is the baryometer of choice, while ^{4}He is an excellent chronometer. The universal density of baryons inferred from SBBN and the adopted primordial D abundance is in excellent (exact!) agreement with that derived from non-BBN, mainly CBR data (Spergel et al. 2003): η_{10}(SBBN) $= 6.10^{+0.67}_{-0.52}$; η_{10}(CBR) $= 6.14 \pm 0.25$. For this baryon density, the predicted primordial abundance of ^{3}He is also in excellent agreement with the (very uncertain) value inferred from observations of an outer-Galaxy H II region (Bania et al. 2002). In contrast, the SBBN-predicted mass fraction of ^{4}He for the concordant baryon density is $Y_P = 0.248 \pm 0.001$, while that inferred from observations of recombination lines in metal-poor, extragalactic H II regions is lower (Olive et al. 2000): $Y_P^{obs} = 0.238 \pm 0.005$. Since the uncertainties in the observationally inferred primordial value are likely dominated by systematics, this $\sim 2\sigma$ difference may not be cause for (much) concern. Finally, there appears to be a more serious issue concerning the predicted and observed lithium abundances. While the predicted abundance is [Li]$_P \approx 2.6 \pm 0.1$, current observations of metal-poor halo stars suggest a considerably smaller value $\approx 2.2 \pm 0.1$.

It has been seen that the tension between D and ^{4}He (or between the baryon density and ^{4}He) can be relieved by either of two variations of the standard model (slower than standard early expansion rate; nonzero chemical potential for the electron neutrino). However, in

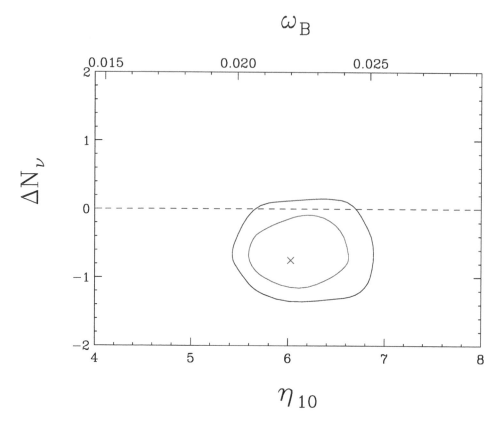

Fig. 11.14. The 1- and 2-σ contours in the $\eta - \Delta N_\nu$ plane for the joint BBN – CBR (*WMAP*) fit (Barger et al. 2003a).

neither of these cases does the BBN-predicted ^{7}Li abundance move any closer to that inferred from the observations.

In the current, data-rich era of cosmological research, BBN continues to play an important role. The spectacular agreement in the baryon density inferred from processes occurring at widely separated epochs confirms the general features of the standard models of cosmology and particle physics. The tensions with ^{4}He and ^{7}Li provide challenges, and opportunities, to cosmology, to astrophysics, and to particle physics. To outline these challenges and opportunities, let us consider each of the light nuclides in turn.

For deuterium the agreement between SBBN and non-BBN determinations is perfect. This may be surprising given the unexpectedly large dispersion among the handful of extant D abundance determinations at high redshifts and low metallicities. Here, the challenge is to observers and theorists. Clearly more data are called for. Perhaps new data will reduce the dispersion. In that case it can be anticipated that the SBBN-predicted baryon density will approach the accuracy of that currently available from non-BBN data. On the other hand, newer data may support the dispersion, suggesting unexpectedly large variations in the D abundance at evolutionary times earlier than expected (Jedamzik & Fuller 1997). Perhaps there is more to be learned about early chemical evolution.

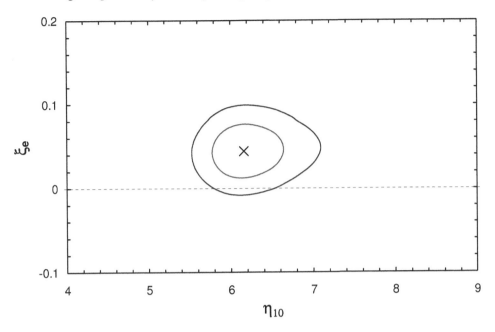

Fig. 11.15. The 1-, 2-, and 3-σ contours in the $\eta - \xi_e$ plane for BBN ($N_\nu = 3$) and the adopted D and ^{4}He abundances (Barger et al. 2003b).

From studies of ^{3}He in Galactic H II regions (Balser et al. 1997; Bania et al. 2002) it appears that in the course of Galactic chemical evolution there has been a very delicate balance between post-BBN production and destruction. If either had dominated, a gradient of the ^{3}He abundance with galactocentric distance should have been seen in the data (see Romano et al. 2004, and references therein). So far, none is. Clearly, more data and a better understanding of the lower mass stars, which should dominate the production and destruction of ^{3}He, would be of value.

The very precise value of the baryon density inferred either from D and SBBN or from non-BBN data, coupled with the very weak dependence of the SBBN abundance of ^{4}He on the baryon density, leads to a very precise prediction of its primordial mass fraction. Although there exists a very large data set of ^{4}He abundance determinations, the observational situation is confused at present. It seems clear that while new data would be valuable, quality is much more important than quantity. Data that can help resolve various corrections for temperature, for temperature and density fluctuations, for ionization corrections, would be of greater value than merely collecting more data that are incapable of addressing these issues. Because of the very large data set(s), the *statistical* uncertainty in the derived primordial mass fraction is very small, $\sigma_{Y_P} \approx 0.002 - 0.003$, while uncertain systematic corrections are much larger $\gtrsim 0.005$. At this point it is systematics, not statistics that dominate the uncertainty in the primordial helium abundance. In this context it is worth considering non-emission line observations that might provide an independent abundance determination. Just such an alternative, the so-called R-parameter method using globular cluster stars was proposed long ago by Iben (1968) and by Iben & Faulkner (1968). It too has many systematic uncertainties associated with its application, but they are different from those for the emission-line studies. Very recently, Cassisi, Salaris, & Irwin (2003), using new stellar models and nuclear reactions rates, along with better data, find $Y_P = 0.243 \pm 0.006$. This is in much better agreement with the expected value (within $\lesssim 1\sigma$) and should stimulate further investigations.

The apparent conflict between the predicted and observed abundances of ^{7}Li, if not simply traceable to the statistical and systematic uncertainties, suggests a gap in our understanding of the structure and evolution of the very old, metal-poor, halo stars. It would appear from the comparison between the predicted and observed abundances that lithium may have been depleted or diluted from the surfaces of these stars by $\sim 0.2 - 0.4$ dex. Although a variety of mechanisms for depletion/dilution exist, the challenge is to account for such a large reduction without at the same time producing a large dispersion around the Spite plateau.

The wealth of observational data accumulated over the last decade or more have propelled the study of cosmology from youth to maturity. BBN has played, and continues to play, a central role in this process. There have been many successes, but much remains to be done. Whether the resolution of the current challenges are observational or theoretical, the future is bright.

Acknowledgements. I am grateful to all the colleagues with whom I have worked, in the past as well as at present, for all I have learned from them; I thank them all. Many of the quantitative results (and figures) presented here are from recent collaborations with V. Barger, J. P. Kneller, J. Linsky, D. Marfatia, K. A. Olive, R. J. Scherrer, S. M. Viegas, and T. P. Walker. I thank V. V. Smith for permission to use Figure 11.9. My research is supported at OSU by the DOE through grant DE-FG02-91ER40690.

References

Allende-Prieto, C., Lambert, D. L., & Asplund, M. 2001, ApJ, 556, L63

Balser, D., Bania, T., Rood, R. T., & Wilson, T. 1997, ApJ, 483, 320

Bania, T., Rood, R. T., & Balser, D. 2002, Nature, 415, 54

Barger, V., Kneller, J. P., Lee, H.-S., Marfatia, D., & Steigman, G. 2003a, Phys. Lett. B, 566, 8

Barger, V., Kneller, J. P., Marfatia, D., Langacker, P., & Steigman, G. 2003b, Phys. Lett. B, 569, 123

Bennett, C. L., et al. 2003, ApJS, 148, 1

Binetruy, P., Deffayet, C., Ellwanger, U., & Langlois, D. 2000, Phys. Lett. B, 477, 285

Bonifacio, P., et al. 2002, A&A, 390, 91

Bonifacio, P., & Molaro, P. 1997, MNRAS, 285, 847

Bonifacio, P., Molaro, P., & Pasquini, L. 1997, MNRAS, 292, L1

Bratt, J. D., Gault, A. C., Scherrer, R. J., & Walker, T. P. 2002, Phys. Lett. B, 546, 19

Burles, S., Nollett, K. M., & Turner, M. S. 2001, Phys. Rev. D, 63, 063512

Burles, S., & Tytler, D. 1998a, ApJ, 499, 699

——. 1998b, ApJ, 507, 732

Cassisi, S., Salaris, M., & Irwin, A. W. 2003, ApJ, 588, 862

Chaboyer, B. C., Deliyannis, C. P., Demarque, P., Pinsonneault, M. H., & Sarajedini, A. 1992, ApJ, 388, 372

Chiappini, C., Renda, A., & Matteucci, F. 2002, A&A, 395, 789

Cline, J. M., Grojean, C., & Servant, G. 1999, Phys. Rev. Lett., 83, 4245

Crotty, P., Lesgourgues, J., & Pastor, S. 2003, Phys. Rev. D, 67, 123005

Dearborn, D. S. P., Schramm, D. N., & Steigman, G. 1986, ApJ, 203, 35

Dearborn, D. S. P., Steigman, G., & Tosi, M. 1996, ApJ, 465, 887 (erratum: ApJ, 473, 570)

D'Odorico, S., Dessauges-Zavadsky, M., & Molaro, P. 2001, A&A, 368, L21

Epstein. R., Lattimer, J., & Schramm, D. N. 1976, Nature, 263, 198

Galli, D., Stanghellini, L., Tosi, M., & Palla, F. 1997, ApJ, 477, 218

Geiss, J., & Gloeckler, G. 1998, Space Sci. Rev., 84, 239

Gruenwald, R., Steigman, G., & Viegas, S. M. 2002, ApJ, 567, 931

Hannestad, S. 2003, JCAP, 5, 4

Iben, I., Jr. 1967, ApJ, 147, 624

——. 1968, Nature, 220, 143

Iben, I., Jr., & Faulkner, J. 1968, ApJ, 153, 101

Izotov, Y. I., & Thuan, T. X. 1998, ApJ, 500, 188

Izotov, Y. I., Thuan, T. X., & Lipovetsky, V. A. 1997, ApJS, 108, 1

Jedamzik, K., & Fuller, G. 1997, ApJ, 483, 560

Kang, H.-S., & Steigman, G. 1992, Nucl. Phys. B, 372, 494

Kirkman, D., Tytler, D., Suzuki, N., O'Meara, J. M., & Lubin, D. 2004, ApJS, submitted (astro-ph/0302006)

Kneller, J. P., Scherrer, R. J., Steigman, G., & Walker, T. P. 2001, Phys. Rev. D, 64, 123506

Kneller, J. P., & Steigman, G. 2003, Phys. Rev. D, 67, 063501

Levshakov, S. A., Dessauges-Zavadsky, M., D'Odorico, S., & Molaro, P. 2002, ApJ, 565, 696 [see also the preprint(s) astro-ph/0105529 (v1 & v2)]

Levshakov, S. A., Kegel W. H., & Takahara, F. 1998a, ApJ, 499, L1

——. 1998b, A&A, 336, L29

——. 1999, MNRAS, 302, 707

Linsky, J. L., & Wood, B. E. 2000, in IAU Symp. 198, The Light Elements and Their Evolution, ed. L. da Silva, M. Spite, & J. R. Medeiros (San Francisco: ASP), 141

Olive, K. A., & Steigman, G. 1995, ApJS, 97, 49

Olive, K. A., Skillman, E., & Steigman, G. 1997, ApJ, 483, 788

Olive, K. A., Steigman, G., & Walker, T. P. 2000, Phys. Rep., 333, 389

O'Meara, J. M., Tytler, D., Kirkman, D., Suzuki, N., Prochaska, J. X., Lubin, D., & Wolfe, A. M. 2001, ApJ, 552, 718

Palla, F., Bachiller, R., Stanghellini, L., Tosi, M, & Galli, D. 2000, A&A, 355, 69

Peimbert, A., Peimbert, M., & Luridiana, V., 2002, ApJ, 565, 668

Peimbert, M., Peimbert, A., & Ruiz, M. T. 2000, ApJ, 541, 688

Pettini, M., & Bowen, D. V. 2001, ApJ, 560, 41

Pierpaoli, E. 2003, MNRAS, 342, L63

Pinsonneault, M. H., Steigman, G., Walker, T. P., & Narayanan, V. K. 2002, ApJ, 574, 398 (PSWN)

Pinsonneault, M. H., Walker, T. P., Steigman, G., & Narayanan, V. K. 1999, ApJ, 527, 180

Randall, L. & Sundrum, R. 1999a, Phys. Rev. Lett., 83, 3370

——. 1999b, Phys. Rev. Lett., 83, 4690

Romano, D., Tosi, M., Matteucci, F., & Chiappini, C. 2004, MNRAS, in press

Rood, R. T. 1972, ApJ, 177, 681

Rood, R. T., Bania, T. M., Balser, D. S., & Wilson, T. L. 1998, Space Sci. Rev., 84, 185

Rood, R. T., Steigman, G., & Tinsley, B. M. 1976, ApJ, 207, L57

Ryan, S. G., Beers, T. C., Olive, K. A., Fields, B. D., & Norris, J. E. 2000, ApJ, 530, L57

Ryan, S. G., Norris, J. E., & Beers, T. C. 1999, ApJ, 523, 654

Sackmann, I.-J., & Boothroyd, A. I. 1999a, ApJ, 510, 217

——. 1999b, ApJ, 510, 232

Salaris, M., & Weiss, A. 2002, A&A, 388, 492

Sauer, D., & Jedamzik, K. 2002, A&A, 381, 361

Spergel, D. N., et al. 2003, ApJS, 148, 175

Spite, M., & Spite, F. 1982, Nature, 297, 483

Steigman, G. 2004, in The Dark Universe: Matter, Energy, and Gravity, ed. M. Livio (Baltimore: STScI), in press (astro-ph/0107222)

Steigman, G., Schramm, D. N., & Gunn, J. E. 1977, Phys. Lett. B, 66, 202

Steigman, G., Viegas, S. M., & Gruenwald, R. 1997, ApJ, 490, 187

Theado, S., & Vauclair, S. 2001, A&A, 375, 70

Thorburn, J. A. 1994, 421, 318

Vassiliadis, E., & Wood, P. R. 1993, ApJ, 413, 641

Viegas, S. M., Gruenwald, R., & Steigman, G. 2000, ApJ, 531, 813

12

Cosmological results from the 2dF Galaxy Redshift Survey

MATTHEW COLLESS
Research School of Astronomy and Astrophysics, The Australian National University

Abstract

The 2dF Galaxy Redshift Survey (2dFGRS) has produced a three-dimensional map of the distribution of 221,000 galaxies covering 5% of the sky and reaching out to a redshift $z \approx 0.3$. This is first map of the large-scale structure in the local Universe to probe a statistically representative volume, and provides direct evidence that the large-scale structure of the Universe grew through gravitational instability. Measurements of the correlation function and power spectrum of the galaxy distribution have provided precise measurements of the mean mass density of the Universe and the relative contributions of cold dark matter, baryons, and neutrinos. The survey has produced the first measurements of the galaxy bias parameter and its variation with galaxy luminosity and type. Joint analysis of the 2dFGRS and cosmic microwave background power spectra gives independent new estimates for the Hubble constant and the vacuum energy density, and constrains the equation of state of the vacuum.

12.1 Introduction

The 2dF Galaxy Redshift Survey (2dFGRS) was made possible by the 2-degree Field (2dF) fiber spectrograph, which was specifically conceived as a tool for performing a massive redshift survey to precisely measure fundamental cosmological parameters. The state-of-the-art redshift surveys of the early 1990's, such as the Las Campanas Redshift Survey (Shectman et al. 1996) and the *IRAS* Point Source Catalog redshift survey (Saunders et al. 2000), either did not cover sufficiently large volumes to be statistically representative of the large-scale structure, or covered large volumes too sparsely to provide precise measurements. An order-of-magnitude increase in the survey volume and sample size was needed to enter the regime of "precision cosmology," and this became the foundational goal of the 2dFGRS.

The 2dF spectrograph can observe 400 objects simultaneously over a $2°$-diameter field of view (Taylor & Gray 1990; Lewis et al. 2002a), and was first placed on the Anglo-Australian Telescope (AAT) in November 1995. The first spectra were taken in mid-1996, and scheduled observations with 2dF at full functionality began in September 1997. The first major redshift survey observing run occurred in October 1997, with the survey passing 50,000 redshifts in mid-1999 (Colless 1999) and 100,000 redshifts in mid-2000. The first 100,000 redshifts and spectra were released publicly in June 2001 (Colless et al. 2001), and the 200,000-redshift mark was achieved toward the end of 2001. The survey observations were completed in April 2002, after 5 years and 272 nights on the AAT. The final survey is an

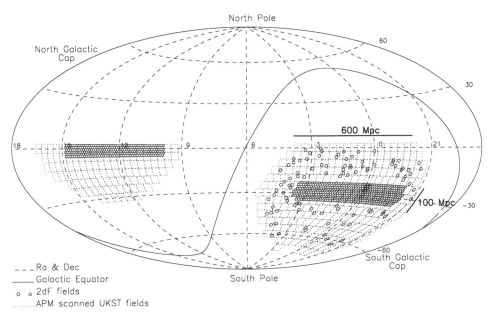

North Pole

North Galactic
Cap

60

30

600 Mpc

100 Mpc

−30

−60
South Galactic
Cap

South Pole

_ _ _ Ra & Dec
_____ Galactic Equator
o o 2dF fields
.... APM scanned UKST fields

Fig. 12.1. A map of the sky showing the locations of the two 2dFGRS survey strips (NGP strip at left, SGP strip at right) and the random fields. Each 2dF field in the survey is shown as a small circle; the sky survey plates from which the source catalog was constructed are shown as dotted squares. The scale of the strips at the mean redshift of the survey is indicated.

order of magnitude larger than any previous redshift survey, and comparable to the ongoing redshift survey of the Sloan Digital Sky Survey (Bernardi, this volume).

The source catalog for the 2dFGRS was a revised and extended version of the APM galaxy catalog (Maddox et al. 1990), which was created by scanning the photographic plates of the UK Schmidt Telescope Southern Sky Survey. The survey targets were chosen to be galaxies with extinction-corrected magnitudes brighter than $b_J = 19.45$ mag. The galaxies were distinguished from stars by the APM image classification algorithm described by Maddox et al., conservatively tuned to include all galaxies at the expense of also including a 5% contamination by stars.

The main survey regions were two declination strips, one in the southern Galactic hemisphere spanning $80° \times 15°$ around the South Galactic Pole (the SGP strip), and the other in the northern Galactic hemisphere spanning $75° \times 10°$ along the celestial equator (the NGP strip); in addition, there were 99 individual 2dF "random" fields spread over the southern Galactic cap (see Fig. 12.1). The large volume that is sparsely probed by the random fields allows the survey to measure structure on scales greater than would be permitted by the relatively narrow widths of the main survey strips. In total, the survey covers approximately 1800 deg^2, and has a median redshift depth of $z = 0.11$. An adaptive tiling algorithm was used to optimally place the 900 2dF fields over the survey regions, giving a highly complete and uniform sample of galaxies on the sky.

Redshifts were measured from 2dF spectra that covered the range from 3600 Å to 8000 Å at a resolution of 9.0 Å. Redshift measurements were obtained both from cross-correlation

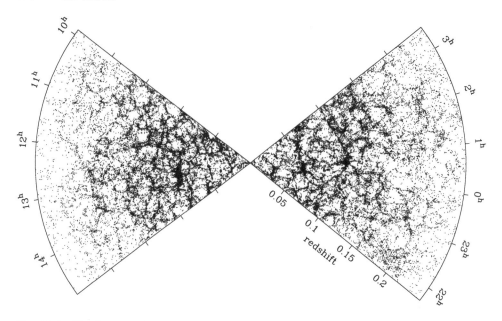

Fig. 12.2. The large-scale structures in the galaxy distribution are shown in this 3°-thick slice through the 2dFGRS map. The slice cuts through the NGP strip (at left) and the SGP strip (at right), and contains 63,000 galaxies.

with template spectra and from fitting emission lines. All redshifts were visually checked and assigned a quality parameter Q in the range 1–5; accepted redshifts ($Q \geq 3$) were found to be 98% reliable and to have a typical uncertainty of 85 km s^{-1}. The overall redshift completeness for accepted redshifts was 92%, although this varied with magnitude. The variation in the redshift completeness with position and magnitude is fully accounted for by the survey completeness mask (Colless et al. 2001; Norberg et al. 2002b).

Figure 12.2 shows a thin slice through the three-dimensional map of over 221,000 galaxies produced by the 2dFGRS. This 3°-thick slice passes through both the NGP strip (at left) and the SGP strip (at right). The decrease in the number of galaxies toward higher redshifts is an effect of the survey selection by magnitude—only intrinsically more luminous galaxies are brighter than the survey magnitude limit at higher redshifts. The clusters, filaments, sheets and voids making up the large-scale structures in the galaxy distribution are clearly resolved. The fact that there are many such structures visible in the figure is a qualitative demonstration that the survey volume comprises a representative sample of the Universe; the small amplitude of the density fluctuations on large scales is quantified by the power spectrum, as discussed in the next section.

12.2   The Large-scale Structure of the Galaxy Distribution

In cosmological models where the initial density fluctuations form a Gaussian random field, such as most inflationary models, the large-scale structure of the galaxy distribution in the linear regime is completely characterized in statistical terms by just two quantities: the mean density and the rms fluctuations in the density as a function of scale. The latter are quantified either through the two-point correlation function or the power spectrum, which

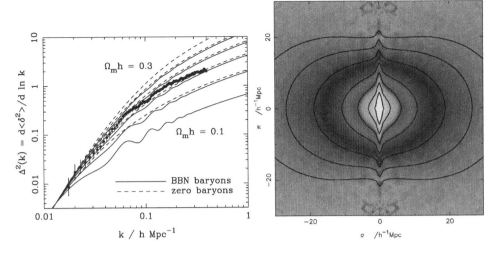

Fig. 12.3. Large-scale structure statistics from the 2dFGRS. The *left* panel shows the dimensionless power spectrum $\Delta^2(k)$ (Percival et al. 2001; Peacock et al. 2004). Overlaid are the predicted linear-theory CDM power spectra with shape parameters $\Omega h = 0.1, 0.15, 0.2,$ 0.25, and 0.3, with the baryon fraction predicted by Big Bang nucleosynthesis (solid curves) and with zero baryons (dashed curves). The *right* panel shows the two-dimensional galaxy correlation function, $\xi(\sigma, \pi)$, where σ is the separation across the line of sight and π is the separation along the line of sight (adapted from Hawkins et al. 2004). The grayscale image is the observed $\xi(\sigma, \pi)$, and the contours show the best-fitting model.

are Fourier transforms of each other. However, a redshift survey does not determine the real-space positions of the galaxies, but rather the redshift-space positions, where the line-of-sight component is not the distance to the galaxy but the galaxy's velocity. This velocity is the combination of the Hubble velocity (which *is* directly related to the distance) and the galaxy's peculiar velocity (the motion produced by the gravitational attraction of the local mass distribution).

The statistical properties of the large-scale structure of the galaxy distribution observed in redshift space are summarized in Figure 12.3, which shows both the correlation function and the power spectrum obtained from the 2dFGRS. The structure on very large scales (several tens to hundreds of Mpc) is best represented by the power spectrum; on smaller scales, where peculiar velocities become more significant and the shape of the power spectrum (as well as the amplitude) differs between redshift space and real space, the redshift-space structure is most clearly shown in the two-dimensional correlation function (see §12.4 below).

The power spectrum, shown in the left panel of Figure 12.3, is well determined from the 2dFGRS on scales less than about $400\,h^{-1}\,\mathrm{Mpc}$ (wavenumbers $k > 0.015$), and its shape is little affected by nonlinear evolution of the galaxy distribution on scales greater than about $40\,h^{-1}\,\mathrm{Mpc}$ ($k < 0.15$). Over this decade in scale, the power spectrum is well fitted by a cold dark matter (CDM) model having a shape parameter $\Gamma = \Omega_m h = 0.20 \pm 0.03$ (Percival et al. 2001). For a Hubble constant around $70\,\mathrm{km\,s^{-1}\,Mpc^{-1}}$ (i.e., $h \approx 0.7$), this implies a mean mass density $\Omega_m \approx 0.3$. The power spectrum also shows some evidence for acoustic oscillations produced by baryon-photon coupling in the early Universe (see §12.5).

The right panel of Figure 12.3 shows the redshift-space two-point correlations as a function of the separations along and across the line of sight, and reveals two main deviations from circular symmetry due to peculiar velocity effects. On intermediate scales, for transverse separations of a few tens of Mpc, the contours of the correlation function are flattened along the line of sight due to the coherent infall of galaxies as structures form in the linear regime. The detection of this effect in the 2dFGRS is a clear confirmation that large-scale structure grows by the gravitational amplification of density fluctuations (Peacock et al. 2001), and allows a direct measurement of the mean mass density of the Universe (see §12.5). The other effect is the stretching of the contours along the line of sight at small transverse separations. This is the finger-of-God effect due to the large peculiar velocities of collapsed structures in the nonlinear regime.

12.3   The Bias of the Galaxy Distribution

A fundamental issue in employing redshift surveys of galaxies as probes of cosmology is the relationship between the observed galaxy distribution and the underlying mass distribution, which is what cosmological models most directly predict. Some bias of the galaxies with respect to the mass is expected on theoretical grounds, but the nature and extent of the effect was not previously well determined. The large size of the 2dFGRS has allowed a thorough investigation of this question.

The simplest model for galaxy biasing postulates a linear relation between fluctuations in the galaxy distribution and fluctuations in the mass distribution: $\delta n/n = b\delta\rho/\rho$. In this case the galaxy power spectrum is related to the mass power spectrum by $P_g(k) = b^2 P_m(k)$. Such a relationship is expected to hold in the linear regime (up to stochastic variations). The first-order relationship between galaxies and mass can therefore be determined by comparing the measured galaxy power spectrum to the matter power spectrum based on a model fit to the cosmic microwave background (CMB) power spectrum, linearly evolved to $z = 0$ and extrapolated to the smaller scales covered by the 2dFGRS power spectrum. Applying this approach, Lahav et al. (2002) find that the linear bias parameter for an L^* galaxy at zero redshift is $b(L^*, z = 0) = (0.96 \pm 0.08)\exp[-\tau + 0.5(n-1)]$, where τ is the optical depth due to reionization and n is the spectral index of the primordial mass power spectrum.

An alternative way of determining the bias employs the higher-order correlations between galaxies in the intermediate, quasi-linear regime. The higher-order correlations are generated by nonlinear gravitational collapse, and so depend on the clustering of the dominant dark matter rather than the galaxies. Thus the stronger the higher-order clustering, the higher the dark matter normalization, and the lower the bias. An analysis of the bispectrum (the Fourier transform of the three-point correlation function) by Verde et al. (2002) yields $b(L^*, z = 0) = 0.92 \pm 0.11$, a result based solely on the 2dFGRS. Moreover, including a second-order quadratic bias term does not improve the fit of the bias model to the observed bispectrum.

For the blue-selected 2dFGRS sample, it therefore seems that L^* galaxies are nearly unbiased tracers of the low-redshift mass distribution. However, this broad conclusion masks some very interesting variations of the bias parameter with galaxy luminosity and type (Fig. 12.4). Norberg et al. (2001, 2002a) show conclusively that the bias parameter varies with luminosity, ranging from $b = 1.5$ for bright galaxies to $b = 0.8$ for faint galaxies. The relation between bias and luminosity is well represented by the simple linear relation $b/b^* = 0.85 + 0.15L/L^*$. They also find that, at all luminosities, early-type galaxies have a higher bias than late-type galaxies. A detailed comparison of the clustering of passive and

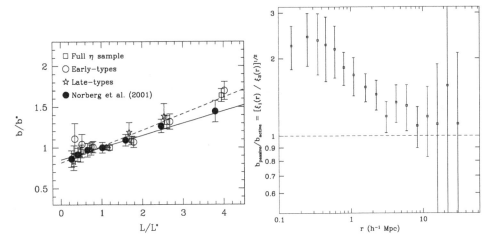

Fig. 12.4. Variations in the bias parameter with luminosity and spectral type. The *left* panel shows the variation with luminosity of the galaxy bias on a scale of $\sim 5\,h^{-1}$ Mpc, relative to an L^* galaxy (Norberg et al. 2002a). The bias variations of the full 2dFGRS sample are compared to subsamples with early and late spectral types, and to earlier results by Norberg et al. (2001). The *right* panel shows the relative bias of passive and actively star-forming galaxies as a function of scale, over the range 0.2–20 h^{-1} Mpc (Madgwick et al. 2004).

actively star-forming galaxies by Madgwick et al. (2004) shows that at small separations, the passive galaxies cluster much more strongly, and the relative bias ($b_{\text{passive}}/b_{\text{active}}$) is a decreasing function of scale. On the largest scales, however, the relative bias tends to a constant value of around 1.3.

12.4   Redshift-space Distortions

The redshift-space distortion of the clustering pattern can be modeled as the combination of coherent infall on intermediate scales and random motions on small scales. The compression of structures along the line of sight due to coherent infall is quantified by the distortion parameter $\beta \simeq \Omega^{0.6}/b$ (Kaiser 1987; Hamilton 1992). The random motions are adequately modeled by an exponential distribution, $f(v) = 1/(a\sqrt{2})\exp(-\sqrt{2}|v|/a)$, where a is the pairwise peculiar velocity dispersion (also called σ_{12}).

The initial analysis of a subset of the 2dFGRS by Peacock et al. (2001) obtained best-fit values of $\beta(L_s, z_s) = 0.43 \pm 0.07$ and $a = 385$ km s^{-1} at an effective weighted survey luminosity $L_s = 1.9L^*$ and survey redshift $z_s = 0.17$. A more sophisticated reanalysis of the full 2dFGRS by Hawkins et al. (2004) obtains $\beta(L_s, z_s) = 0.49 \pm 0.09$ and $a = 506 \pm 52$ km s^{-1}, with $L_s = 1.4L^*$ and $z_s = 0.15$ (right panel of Fig. 12.3). These results, using different fitting methods, are consistent, although the earlier result underestimates the uncertainties by 20%. Applying corrections based on the variation in the bias parameter with luminosity and a constant galaxy clustering model (Lahav et al. 2002) to the Hawkins et al. value for the distortion parameter yields $\beta(L^*, z = 0) = 0.47 \pm 0.08$.

Madgwick et al. (2004) extend this analysis to a comparison of the active and passive galaxies, where the two-dimensional correlation function, $\xi(\sigma, \pi)$, reveals differences in both the bias parameter on large scales and the pairwise velocity dispersion on small scales

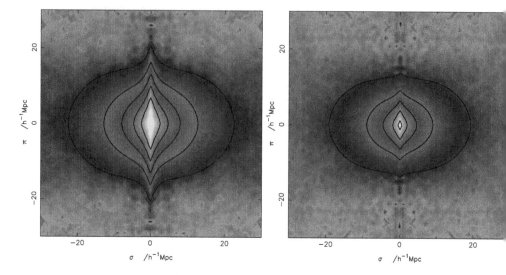

Fig. 12.5. The two-dimensional galaxy correlation function, $\xi(\sigma, \pi)$, for passive (left) and actively star-forming (right) galaxies (Madgwick et al. 2004). The grayscale image is the observed $\xi(\sigma, \pi)$, and the contours show the best-fitting model.

(Fig. 12.5). The distortion parameter is $\beta_{passive} \simeq \Omega_m^{0.6}/b_{passive} = 0.46 \pm 0.13$ for passive galaxies and $\beta_{active} \simeq \Omega_m^{0.6}/b_{active} = 0.54 \pm 0.15$ for active galaxies; over the range 8–$20\,h^{-1}$ Mpc the effective pairwise velocity dispersions are 618 ± 50 km s^{-1} and 418 ± 50 km s^{-1} for passive and active galaxies, respectively.

12.5   The Mass Density of the Universe

The 2dFGRS provides a variety of ways to measure the mean mass density of the Universe, along with the relative amounts of dark matter, baryons, and neutrinos.

Fitting the shape of the galaxy power spectrum in the linear regime with a model including both CDM and baryons (Percival et al. 2001), and assuming that the Hubble constant is $h = 0.7$ with a 10% uncertainty, yields a total mass density for the Universe of $\Omega_m = 0.29 \pm 0.07$ and a baryon fraction of $15\% \pm 7\%$ (i.e., $\Omega_b = 0.044 \pm 0.021$). This analysis used 150,000 galaxies; a preliminary reanalysis of the complete final sample of 221,000 galaxies with the additional constraint that $n = 1$ yields $\Omega_m = 0.26 \pm 0.05$ and $\Omega_b = 0.044 \pm 0.016$ (Peacock et al. 2004; left panel of Fig. 12.6). Including neutrinos as a further constituent of the mass allows an upper limit to be placed on their contribution to the total density, based on the allowable degree of suppression of small-scale structure due to the free streaming of neutrinos out of the initial density perturbations (right panel of Fig. 12.6). Elgarøy et al. (2002) obtain an upper limit on the neutrino mass fraction of 13% at the 95% confidence level (i.e., $\Omega_\nu < 0.034$). This translates to an upper limit on the total neutrino mass (summed over all species) of $m_\nu < 1.8\,\text{eV}$.

An alternative approach to deriving the total mass density is to use the measurements in the quasi-linear regime of the redshift-space distortion parameter $\beta \simeq \Omega_m^{0.6}/b$, in combination with estimates of the bias parameter b (Peacock et al. 2001; Hawkins et al. 2004). Using the

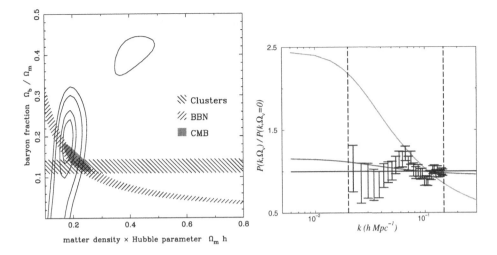

Fig. 12.6. Determinations of the mean mass density, Ω_m, and the baryon and neutrino mass fractions. The *left* panel shows the likelihood surfaces obtained by fitting the full 2dFGRS power spectrum for the shape parameter, $\Omega_m h$, and the baryon fraction, Ω_b/Ω_m (Peacock et al. 2004; cf. Percival et al. 2001). The fit is over the well-determined linear regime ($0.02 < k < 0.15\,h\,\mathrm{Mpc}^{-1}$) and assumes a prior on the Hubble constant of $h = 0.7 \pm 0.07$. The *right* panel shows the ratio of the power spectra for $\Omega_\nu = 0.05$ (top curve) and $\Omega_\nu = 0.01$ (middle curve) to the power spectrum for $\Omega_\nu = 0$ (horizontal line), with amplitudes fitted to the 2dFGRS redshift-space power spectrum (vertical bars). The power spectrum models assume $\Omega_m = 0.3$, $h = 0.7$, and $\Omega_b h^2 = 0.02$; the dashed vertical lines indicate the range in k used in the fits.

Lahav et al. (2002) estimate for b gives $\Omega_m = 0.31 \pm 0.11$, while the Verde et al. (2002) value for b gives $\Omega_m = 0.23 \pm 0.09$.

12.6 Joint LSS-CMB Estimates of Cosmological Parameters

Stronger constraints on these and other fundamental cosmological parameters can be obtained by combining the power spectrum of the present-day galaxy distribution from the 2dFGRS with the power spectrum of the mass distribution at very early times derived from observations of the anisotropies in the CMB. A general analysis of the combined CMB and 2dFGRS data sets (Efstathiou et al. 2002) shows that, at the 95% confidence level, the Universe has a near-flat geometry ($\Omega_k \approx 0 \pm 0.05$), with a low total matter density ($\Omega_m \approx 0.25 \pm 0.08$) and a large positive cosmological constant ($\Omega_\Lambda \approx 0.75 \pm 0.10$, consistent with the independent estimates from observations of high-redshift supernovae).

If the models are limited to those with flat geometries (Percival et al. 2002), then tighter constraints emerge (see Table 12.1). In this case the best estimate of the matter density is $\Omega_m = 0.31 \pm 0.06$, and the physical densities of CDM and baryons are $\omega_c = \Omega_c h^2 = 0.12 \pm 0.01$ and $\omega_b = \Omega_b h^2 = 0.022 \pm 0.002$; the latter agrees very well with the constraints from Big Bang nucleosynthesis. This analysis also provides an estimate of the Hubble constant ($H_0 = 67 \pm 5\,\mathrm{km\,s}^{-1}\,\mathrm{Mpc}^{-1}$) that is independent of, but in excellent accord with, the results from the *Hubble Space Telescope* Key Project. Comparing the uncertainties on the various parameters

Table 12.1. *Cosmological parameters from joint fits to the CMB and 2dFGRS power spectra, assuming a flat geometry (Percival et al. 2002).*

Parameter	Results: scalar only		Results: with tensor component	
	CMB	CMB + 2dFGRS	CMB	CMB + 2dFGRS
$\Omega_b h^2$	0.0205 ± 0.0022	0.0210 ± 0.0021	0.0229 ± 0.0031	0.0226 ± 0.0025
$\Omega_c h^2$	0.118 ± 0.022	0.1151 ± 0.0091	0.100 ± 0.023	0.1096 ± 0.0092
h	0.64 ± 0.10	0.665 ± 0.047	0.75 ± 0.13	0.700 ± 0.053
n_s	0.950 ± 0.044	0.963 ± 0.042	1.040 ± 0.084	1.033 ± 0.066
n_t	–	–	0.09 ± 0.16	0.09 ± 0.16
r	–	–	0.32 ± 0.23	0.32 ± 0.22
Ω_m	0.38 ± 0.18	0.313 ± 0.055	0.25 ± 0.15	0.275 ± 0.050
$\Omega_m h$	0.226 ± 0.069	0.206 ± 0.023	0.174 ± 0.063	0.190 ± 0.022
$\Omega_m h^2$	0.139 ± 0.022	0.1361 ± 0.0096	0.123 ± 0.022	0.1322 ± 0.0093
Ω_b / Ω_m	0.152 ± 0.031	0.155 ± 0.016	0.193 ± 0.048	0.172 ± 0.021

Note: the best-fit parameters and rms errors are obtained by marginalizing over the likelihood distribution of the remaining parameters. Results are given for scalar-only and scalar+tensor models, and for the CMB power spectrum only and the CMB and 2dFGRS power spectra jointly.

in the CMB-only and CMB+2dFGRS columns of Table 12.1 shows the very significant improvements that are obtained by combining the CMB and 2dFGRS data sets.

Joint fits to the 2dFGRS and CMB power spectra also constrain the equation of state parameter $w = p_{vac}/\rho_{vac}c^2$ for the dark energy. Percival et al. (2002) find that in a flat universe the joint power spectra, together with the Hubble Key Project estimate for H_0, imply an upper limit of $w < -0.52$ at the 95% confidence level.

12.7   The Galaxy Population

Alongside these cosmological studies, the 2dFGRS has also produced a wide range of results on the properties of the galaxy population, and provided strong new constraints for models of galaxy formation and evolution. Highlights in this area to date include:

(1) Precise determinations of the optical and near-IR galaxy luminosity functions (Cole et al. 2001; Norberg et al. 2002b).
(2) A detailed characterization of the variations in the luminosity function with spectral type (Folkes et al. 1999; Madgwick et al. 2002).
(3) A determination of the bivariate distribution of galaxies over luminosity and surface brightness (Cross et al. 2001).
(4) A constraint on the space density of rich clusters of galaxies from the velocity dispersion distribution for identified clusters (De Propris et al. 2002).
(5) Separate radio luminosity functions for AGNs and star-forming galaxies (Sadler et al. 2002; Magliocchetti et al. 2002).
(6) Constraints on the star formation history of galaxies from the mean spectrum of galaxies in the local Universe (Baldry et al. 2002).
(7) A measurement of the environmental dependence of star formation rates of galaxies in clusters (Lewis et al. 2002b).
(8) A comparison of the field and cluster luminosity functions for galaxies with difference spectral types (De Propris et al. 2003).

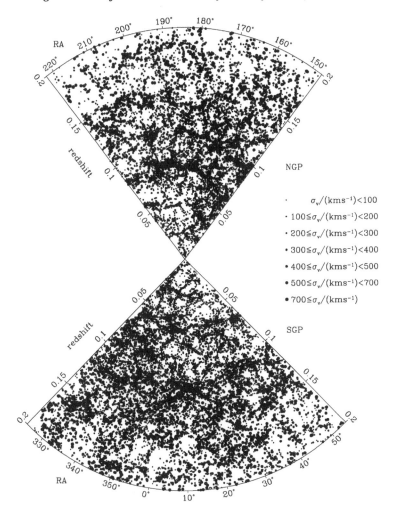

Fig. 12.7. A redshift slice showing the distribution of groups and clusters identified within the 2dFGRS using a three-dimensional friends-of-friends algorithm in position and redshift space (Eke et al. 2004). The number of members found in each cluster is shown by different gray shading; the estimated velocity dispersion is indicated by the size of the dot.

The next step will be to further investigate the correlations between these properties and the local environment of each galaxy, quantified through the local galaxy density or the new group and cluster catalog that has been constructed from the positions and velocity information in the 2dFGRS (Eke et al. 2004; see Fig. 12.7).

12.8   Conclusions

The measurement of cosmological parameters from the 2dFGRS has made a significant contribution to shaping the current consensus model for the fundamental properties of the Universe that has emerged from a range of independent observations, including the measurements of the CMB anisotropies, the distances to high-redshift supernovae, and Big Bang nucleosynthesis. The results obtained to date only represent a fraction of the information that can be extracted from the 2dFGRS on the properties of galaxies and their relation to the large-scale structure of the galaxy distribution. Much more is still to emerge from analysis of the survey, and from combining the 2dFGRS with other large surveys and with detailed follow-up observations.

Further information on the 2dF Galaxy Redshift Survey can be found on the WWW at http://www.mso.anu.edu.au/2dFGRS.

Acknowledgements. These results are presented on behalf of the 2dFGRS team: Ivan K. Baldry, Carlton M. Baugh, Joss Bland-Hawthorn, Sarah Bridle, Terry Bridges, Russell Cannon, Shaun Cole, Matthew Colless, Chris Collins, Warrick Couch, Nicholas Cross, Gavin Dalton, Roberto De Propris, Simon P. Driver, George Efstathiou, Richard S. Ellis, Carlos S. Frenk, Karl Glazebrook, Edward Hawkins, Carole Jackson, Bryn Jones, Ofer Lahav, Ian Lewis, Stuart Lumsden, Steve Maddox, Darren Madgwick, Peder Norberg, John A. Peacock, Will Percival, Bruce A. Peterson, Will Sutherland, and Keith Taylor. The 2dFGRS was made possible through the dedicated efforts of the staff of the Anglo-Australian Observatory, both in creating the 2dF instrument and in supporting it on the telescope.

References

Baldry, I., et al. 2002, ApJ, 569, 582

Cole, S., et al. 2001, MNRAS, 326, 255

Colless, M. M. 1999, Phil. Trans. R. Soc. Lond. A, 357, 105

Colless, M. M., et al. 2001, MNRAS, 328, 1039

Cross, N., et al. 2001, MNRAS, 324, 825

De Propris, R., et al. 2002, MNRAS, 329, 87

——. 2003, MNRAS, 342, 725

Efstathiou, G., et al. 2002, MNRAS, 330, L29

Eke, V., et al. 2004, in preparation

Elgarøy, Ø., et al. 2002, Phys. Rev. Lett., 89, 061301

Folkes, S., et al. 1999, MNRAS, 308, 459

Hamilton, A. J. S. 1992, ApJ, 385, L5

Hawkins, E., et al. 2004, MNRAS, in press (astro-ph/0212375)

Kaiser, N. 1987, MNRAS, 227, 1

Lahav, O., et al. 2002, MNRAS, 333, 961

Lewis, I., et al. 2002a, MNRAS, 333, 279

——. 2002b, MNRAS, 334, 673

Maddox, S. J., Efstathiou, G., Sutherland, W. J., & Loveday, J. 1990, MNRAS, 242, 43P

Madgwick, D. S., et al. 2002, MNRAS, 333, 133

——. 2004, in preparation

Magliocchetti, M., et al. 2002, MNRAS, 333, 100

Norberg, P., et al. 2001, MNRAS, 328, 64

——. 2002a, MNRAS, 332, 827

——. 2002b, MNRAS, 336, 907

Peacock, J. A., et al. 2001, Nature, 410, 169

——. 2004, in preparation

Percival, W. J., et al. 2001, MNRAS, 327, 1297

——. 2002, MNRAS, 337, 1068

Sadler, E. M., et al. 2002, MNRAS, 329, 227

Saunders, W., et al. 2000, MNRAS, 317, 55

Shectman, S. A., Landy, S. D., Oemler, A., Tucker, D. L., Lin, H., Kirshner, R. P., & Schechter, P. L. 1996, ApJ, 470, 172

Taylor, K., & Gray P. 1990, Proc. SPIE, 1236, 290

Verde, L., et al. 2002, MNRAS, 335, 432

13

Large-scale structure in the Sloan Digital Sky Survey

MARIANGELA BERNARDI
Carnegie Mellon University

Abstract

The primary observational goals of the Sloan Digital Sky Survey are (1) to obtain CCD imaging of 10,000 deg^2 of the north Galactic cap in five passbands, with a limiting magnitude in the r band of 22.5, (2) to obtain spectroscopic redshifts of 10^6 galaxies and 10^5 quasars, and (3) to obtain similar data for three ~ 200 deg^2 stripes in the south Galactic cap, with repeated imaging to allow co-addition and variability studies in at least one of these stripes. The resulting photometric and spectroscopic galaxy data sets allow one to map the large-scale structure traced by optical galaxies over a wide range of scales to unprecedented precision. Results relevant to the large-scale structure of our Universe include: a flat model with a cosmological constant $\Omega_\Lambda = 0.7$ provides a good description of the data; the galaxy-galaxy correlation function shows departures from a power law that are statistically significant; and galaxy clustering is a strong function of galaxy type.

13.1   Introduction to the SDSS

The Sloan Digital Sky Survey (SDSS; York et al. 2000) is the result of an international collaborative effort that includes scientists from the U.S., Japan, and Germany (see http://www.sdss.org for details). In brief, the survey uses a dedicated 2.5 meter telescope located at the Apache Point Observatory in New Mexico. Images are obtained by drift scanning with a mosaic camera of 30 2048x2048 CCDs positioned in six columns and five rows (Gunn et al. 1998), which gives a field of view of 3×3 deg^2, with a spatial scale of 0.''4 pix^{-1} in five bandpasses (u, g, r, i, z) with central wavelengths 3560, 4680, 6180, 7500, and 8870 Å (Fukugita et al. 1996). The effective exposure time is 54.1 seconds through each CCD. The SDSS image processing software provides several global photometric parameters for each object, which are obtained independently in each of the five bands. The data are flux calibrated by comparison with a set of overlapping standard-star fields calibrated with a 0.5 m "Photometric Telescope."

The SDSS takes spectra only for a target subsample of calibrated imaging data (Strauss et al. 2002). Spectra are obtained using a multi-object spectrograph, which observes 640 objects at once. The wavelength range of each spectrum is $3800-9200$ Å. The instrumental dispersion is $\log_{10} \lambda = 10^{-4}$ dex pix^{-1}, which corresponds to 69 km s^{-1} per pixel. Each spectroscopic plug plate, 1.5 degrees in radius, has 640 fibers, each $3''$ in diameter. Two fibers cannot be closer than $55''$ due to the physical size of the fiber plug. Typically ~ 500 fibers per plate are used for galaxies, ~ 90 for QSOs, and the remaining for sky spectra and spectrophotometric standard stars.

At the time of writing, the SDSS had imaged roughly $\sim 4,500$ deg^2; approximately 265,000 galaxies and $\sim 35,000$ QSOs had both photometric and spectroscopic information. The first 460 deg^2 and 50,000 spectra have been made public in an Early Data Release (see Stoughton et al. 2002, which includes many technical details of the survey), and roughly four times this will be made available in early 2003.

Data from the multi-waveband SDSS has already made significant contributions to our knowledge of the structure of our Milky Way galaxy and its satellites, correlations between galaxy observables, such as luminosity, size, velocity dispersion, color, chemical composition, star formation rate, etc., and how these depend on galaxy environment, active galactic nuclei, high-redshift quasars, the Lyα forest, and the epoch of reionization. But in this article I will focus exclusively on published results from the SDSS on the large-scale structure of the Universe.

13.2   Galaxy Clustering

In the most successful theoretical models, galaxies grew by gravitational instability from initial seed fluctuations that left their imprint on the cosmic microwave background. The statistics of these initial fluctuations are expected to be Gaussian, so that complete information about these fluctuations is encoded in the shape of the power spectrum $P(k)$ of the initial density fluctuation field. Nonlinear gravitational instability is expected to modify the shape of $P(k)$, and to make the fluctuation field at the present time rather non-Gaussian. These changes are expected to be less severe on large scales, although, because gravity must compete with the expansion of the Universe, what is meant by "large" depends on the amplitude of the initial fluctuations and on the background cosmology. Thus, the large-scale distribution of galaxies at the present time encodes a wealth of cosmological information. One of the principal scientific goals of the SDSS collaboration—a goal that is shared by the 2dFGRS collaboration (see the review by Colless 2004)—is to extract this information. On smaller scales, the clustering is sensitive to the nonlinear gastrophysics of galaxy formation. A generic prediction of most galaxy formation models is that clustering should be a strong function of galaxy type: more luminous galaxies are expected to be more strongly clustered. The SDSS database is ideally suited to quantifying how clustering depends on galaxy properties.

With this in mind, I first discuss measures of clustering in the SDSS angular photometric catalogs. Although these lack the three-dimensional information present in redshift surveys, so most of the clustering signal is washed out by projection effects, angular catalogs are competitive because they have so many more galaxies than spectroscopic surveys. Section 13.3 presents the angular two-point functions, $\omega(\theta)$ and C_ℓ, measured in various apparent magnitude-limited catalogs drawn from the SDSS database. These can be thought of as measurements of the three-dimensional power spectrum $P(k)$ through different windows. It then shows the result of inverting these measurements to derive constraints on the shape and amplitude of $P(k)$. Constraints on $P(k)$ that were obtained more directly from the angular data, without first estimating $w(\theta)$ or C_ℓ, are also described.

Section 13.4 presents results from the three-dimensional catalogs. These are considerably sparser, since spectra are only taken for objects with r-band magnitudes less than about 17.5, whereas the photometry is complete to $r < 22.5$ mag. Measurements of clustering in these are complicated by the fact that we only measure the redshift of a galaxy, not the comoving distance to it—the measured redshift depends both on the distance to the object and the component of its motion along the line-of-sight. Therefore, measures of clustering in redshift space are distorted compared to clustering in real space. If motions are driven by gravity alone, then the difference depends on cosmology in a predictable way—at least on very large scales. Although the data available at present do not probe these large scales, when the survey is complete, the SDSS data set will provide an exquisite test of whether or not gravitational instability is the sole source of large-scale motions. On the smaller scales (< 15 Mpc) probed by the present data, galaxy clustering is a strong function of galaxy type—this is highlighted in Section 13.4. Moreover, the SDSS measurements clearly show that the two-point correlation function of galaxies, long described as a simple power law, does in fact show a statistically significant feature on scales of a few Mpc.

One of the great virtues of the accurate multi-band photometry of the SDSS is that it allows one to make reasonably precise estimates of galaxy redshifts for most objects even when spectra are not available. Measurements of clustering in these photometric redshift catalogs provide the benefit of large galaxy numbers associated with the photometric catalogs, while the photometric redshift estimate can be used to reduce the amount by which the clustering signal is washed-out by projection. Moreover, since the photometric catalog is considerably deeper than the spectroscopic one, it allows one to probe the evolution of clustering out to considerably higher redshifts. These measurements offer a promising way of estimating the evolution of clustering out to redshifts of order unity.

For want of space, I only present results from the lowest order measures of clustering: two-point statistics. Higher-order clustering measures, such as the moments of counts-in-cells (Szapudi et al. 2002) and the void distribution, the bispectrum, the n-point correlation functions, and topological measures such as the genus (Hoyle, Vogeley, & Gott 2002) and other Minkowski functionals have also been, or currently are being, studied. The high quality of the SDSS data also allows various measurements of the weak gravitational lensing effect: McKay et al. (2004) describe galaxy-galaxy shear measurements, and projects to study galaxy-galaxy and galaxy-quasar magnification bias are underway. Also, Nichol et al. (2000) and Bahcall et al. (2003) describe what has been learned from galaxy clusters in the SDSS, and what the future holds for such studies.

13.3 Angular Clustering

In theory, the two-point correlation function $w(\theta)$ and the angular-power spectrum C_ℓ are Fourier (actually Legendre) transforms of one another. Therefore, in theory, they contain the same information. In practice, incomplete sky coverage and other complications mean that the measured values of these two quantities are not equivalent, so the SDSS collaboration has measured both.

13.3.1 *The Angular Correlation Function* $w(\theta)$

In studies of large-scale structure, galaxies are treated as points, and the statistics of point processes are used to quantify galaxy clustering. One of the simplest of these statistics is the two-point correlation function, which measures the excess number of (galaxy) pairs, relative to an unclustered (Poisson) distribution, as a function of pair separation. Operationally, the two-point correlation function is estimated by generating an unclustered random catalog with the same geometry as the survey, and then measuring

$$w(\theta) \equiv \frac{DD - 2DR - RR}{RR}, \tag{13.1}$$

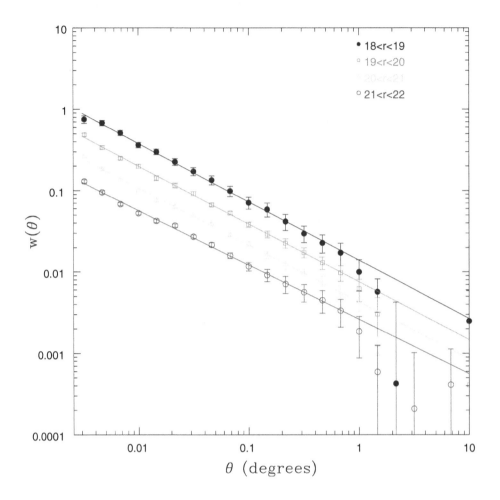

Fig. 13.1. The angular correlation function $\omega(\theta)$ in different magnitude-limited catalogs drawn from the SDSS database. (From Connolly et al. 2002.)

where *DD*, *DR* and *RR* are data-data, data-random, and random-random pair counts in bins of $\theta + \delta\theta$ in the data and random catalogs.

Previous measurements of $\omega(\theta)$, primarily from wide-field photographic plate surveys of the sky, have shown that $\omega(\theta)$ is rather well fit (at least at small separations) by a power law: $\omega(\theta) = (\theta/\theta_0)^{1-\gamma}$, with $\gamma \approx 1.7$. The clustering amplitude, characterized by θ_0, is expected to depend on the depth of a magnitude-limited survey such as the SDSS. This is because the galaxy distribution is expected to be clustered isotropically in three dimensions. A photometric catalog projects out the radial component of the pair separation; the same angular separation can result from galaxies that have vastly different radial separations. Since the clustering amplitude is smaller on large separations, a deeper catalog contains more pairs that are close in the direction perpendicular to the line-of-sight but are well separated along the line-of-sight, thus diluting the overall clustering signal.

Figure 13.1 shows $w(\theta)$ for SDSS galaxies in a number of different apparent magnitude bins. The solid lines show power-law fits to the data, over the range $1' < \theta < 30'$ (the fits use the full covariance matrix from Scranton et al. 2002). Notice that the angular clustering signal on large scales is small: at one degree, $w(\theta) \approx 0.013$ for galaxies with $18 < r^* < 19$ mag. Therefore, sky position-dependent errors in photometric calibration could dominate the signal. Scranton et al. (2002) describe the results of a battery of tests designed to quantify, and where possible correct for, the effects of photometric errors, stellar contamination, seeing, extinction, sky brightness, bright foreground objects, and optical distortions in the camera itself. These tests highlight one of the great features of the SDSS data set—its uniformity.

Notice that the fainter catalogs, which contain galaxies out to greater distances, have a smaller angular clustering amplitude. The precise scaling with apparent magnitude depends on cosmology: a flat universe with $\Lambda = 0.7$ provides a much cleaner scaling than does one in which $\Omega_m = 1$ (but we have not shown this here). As a rough guide to the scales involved, note that the median redshift of galaxies with $18 < r^* < 19$ mag is $z_m = 0.18$ (this median redshift is 0.24, 0.33, and 0.43 for the successively fainter galaxy catalogs). In a flat universe with $\Lambda = 0.7$, one arcminute at $z = 0.18$ corresponds to a distance of 154 h^{-1}kpc, so that 1 h^{-1}Mpc subtends about 0.11 degrees. Clearly, this estimate of $w(\theta)$ probes clustering on rather small scales. The next section describes an estimate of the clustering strength on larger scales.

13.3.2 The Angular Power Spectrum C_ℓ

There are three good reasons for computing the angular power spectrum C_ℓ in addition to the angular correlation function $w(\theta)$. First, on large scales, where the Gaussian approximation is most likely to apply, the C_ℓ estimators retain all of the information contained in the angular clustering signal. Therefore, they represent a lossless compression of the full data set. Second, although both $w(\theta)$ and C_ℓ are obtained by averaging the three-dimensional power spectrum $P(k)$ over window functions, say $W_\theta(k)$ and $W_\ell(k)$, the second of these, W_ℓ, is considerably narrower. This is advantageous if, as I will do shortly, one wishes to invert the measured two-dimensional statistic so as to constrain the form of $P(k)$. Narrow window functions are particularly important since small-scale clustering is expected to be highly non-Gaussian; if the window function is broad, then one must worry about aliasing from small-scale power. And third, it is possible to produce measurements of C_ℓ in which errors are uncorrelated.

Briefly, the measurement is made by dividing the sky patch into N square pixels, each 12.5 on a side and computing the density fluctuation $\delta_i = n_i/\bar{n}_i - 1$ in each pixel. These values of δ_i can be grouped into a vector d, the covariance matrix of which is

$$C \equiv \left\langle dd' \right\rangle = S + N, \qquad \text{where} \qquad S = \sum_i p_i P_i. \qquad (13.2)$$

Here N, assumed to be a diagonal matrix, denotes the contribution to C that comes from the fact that the galaxy distribution is discrete (this is sometimes called the shot-noise contribution), p_i denotes the parameters that specify the amplitude of the power spectrum, and the P_i are matrices that are specified by the survey geometry in terms of Legendre polynomials.

The next step is to determine the values of p_i from the observed data vector d. This involves repeatedly multiplying and inverting $N \times N$ matrices, which is computationally expensive. Therefore, the Karhunen-Loève method is used to compress the information

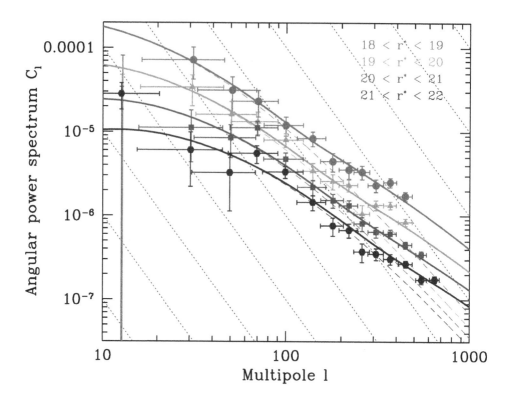

Fig. 13.2. The angular power spectrum C_ℓ in different magnitude-limited catalogs drawn from the SDSS database. As discussed in the text, a crude estimate of the underlying three-dimensional power spectrum is obtained by shifting the same curve vertically and horizontally by an amount that depends on the survey depth: apparently fainter and more distant galaxies should be shifted farther up (because there is more averaging along the line-of-sight that has suppressed fluctuations) and to the left (because as the survey depth increases a given angular scale ℓ corresponds to larger spatial scales). Dotted lines show the direction of this shift when the survey depth is changed. (From Tegmark et al. 2002.)

content of the map before estimating the power spectrum parameters. The actual estimates are made using a quadratic estimator, which effectively Fourier transforms the sky map, squares the Fourier modes in the ith power spectrum band, and averages the results together. The details of this procedure are described in Tegmark et al. (2002).

The results are shown in Figure 13.2. A multipole ℓ corresponds roughly to an angular scale $\theta \approx 180°/\ell$, so that $\ell = 600$, for galaxies at $z = 0.18$, corresponds roughly to a spatial scale of order $3\,h^{-1}\mathrm{Mpc}$.

13.3.3   Inversion to the Three-dimensional $P(k)$

The previous sections presented estimates of the angular correlation function and power spectrum from the SDSS database. These measurements can be used to derive constraints on the three-dimensional power spectrum. This is possible because the angular power spectrum is related to the three-dimensional power spectrum by

$$C_\ell = \int_0^\infty \frac{dk}{k} k^3 P(k) W_\ell(k), \quad \text{where} \quad W_\ell(k) = \frac{2}{\pi} \left[\int_0^\infty dr\, f(r)\, j_\ell(kr) \right]^2. \quad (13.3)$$

Here f is the probability distribution for the comoving distance r to a random galaxy in the survey (which, in a photometric survey, is *not* measured), j_ℓ is a spherical Bessel function, and we have ignored the fact that the power spectrum evolves with redshift. (Strictly speaking, this expression also makes the standard assumption that clustering does not depend on luminosity; the next section shows that the data do not support this assumption, but the quantitative effect on the following analysis is small.) To see what the definition above implies, note that for large values of ℓ, corresponding to small angular scales, $j_\ell(kr)$ is sharply peaked around $kr = \ell$. Assuming the unknown $f(r)$ varies smoothly, we can set it equal to $f(\ell/k)$ and take it out of the integral above, leaving an integral over j_ℓ only, which can be evaluated analytically. Thus, in this approximation, $\ell^3 W_\ell(k) \to [(\ell/k) f(\ell/k)]^2$, and

$$C_\ell \to \int_0^\infty \frac{dk}{k} \frac{k^3 P(k)}{\ell^3} \left[\frac{\ell}{k} f\left(\frac{\ell}{k}\right) \right]^2 \approx \frac{k_\ell^3 P(k_\ell)}{\ell^3} \int_0^\infty \frac{dk}{k} \left[\frac{\ell}{k} f\left(\frac{\ell}{k}\right) \right]^2, \quad (13.4)$$

where the second approximation comes from assuming that the term in square brackets is sharply peaked about its mean value k_ℓ. This term depends on the distribution of comoving distances. To see how, define $r_* \equiv \int dr\, r f(r)$. Then the assumption that f is peaked means we should set $r_* \equiv \beta\ell/k_\ell$, where β is a constant of order unity. Thus, $C_\ell \approx (k_\ell/\ell)^3 P(k_\ell) \approx (\beta/r_*)^3 P(\beta\ell/r_*)$. In other words, C_ℓ is a smoothed version of $P(k)$, which is shifted vertically (by a factor r_*^3) and horizontally (by r_*) on a log-log plot, by an amount that depends on the depth of the sample.

On small scales, the angular correlation function is also related to the power spectrum by a window function:

$$w(\theta) = \int \frac{dk}{k} k^2 P(k) W_\theta(k), \quad \text{where} \quad W_\theta(k) = \frac{1}{2\pi} \int dr\, J_0(kr\theta) f^2(r) \quad (13.5)$$

and J_0 is a Bessell function. In contrast to the window associated with C_ℓ, this W_θ oscillates around zero, so that it is harder to associate a single wavenumber k with the angular $w(\theta)$. Nevertheless, one can still develop techniques for inverting the measured $w(\theta)$ and C_ℓ to obtain the form of $P(k)$. A number of these methods are summarized in Dodelson et al. (2002).

Clearly, the results of the inversion are sensitive to the assumed form for $f(r)$, which, in turn, depends on the magnitude limit of the sample and cosmology. (This is because, at fixed redshift, a flat model with $\Lambda > 0$ has more volume than when $\Lambda = 0$. Therefore, if the sample depth is characterized by a median redshift, which is observable at least in principle, then the typical physical separation between galaxies at that redshift is larger in a flat model with $\Lambda > 0$.) Figure 13.3 shows the result of inverting $w(\theta)$ and C_ℓ in the four apparent magnitude-limited samples presented earlier; the inversion assumed a flat model with $\Omega_\Lambda = 0.7$. Over the range of scales where they overlap, the estimates agree with one another.

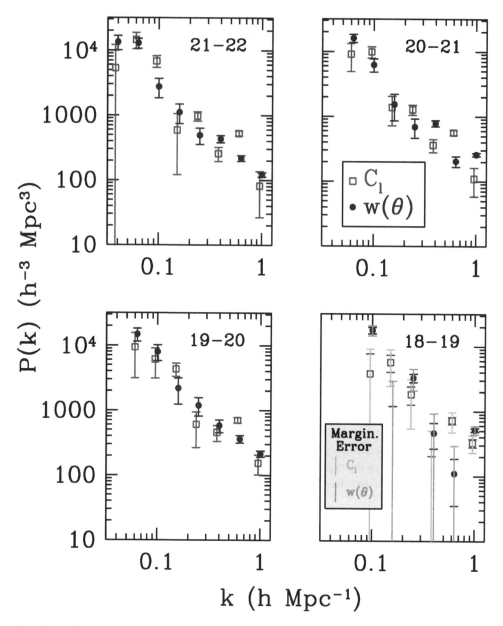

Fig. 13.3. Comparison of $P(k)$ obtained by inverting $\omega(\theta)$ and C_ℓ in four apparent magnitude bins. Error bars are unmarginalized; marginalizing over the non-zero covariances between k bins makes the error bars much larger. (Adapted from Dodelson et al. 2002.)

The implications for $P(k)$ are usually expressed as constraints on the parameters $\sigma_{8g} = b\sigma_8$ and Γ, which describe the amplitude (i.e., an up-down shift of all the points in Fig. 13.3) and the shape [how far to the right does $P(k)$ peak]. The subscript g indicates that the measured $P(k)$ is of the galaxy distribution rather than the dark matter, and the factor b comes from the standard assumption (consistent with numerical simulations of clustering on large scales) that the power in the two distributions differs only by a multiplicative linear bias factor. As the figure shows, the strongest constraints on the shape parameter come from the faintest galaxies (i.e., the magnitude bin $21 < r < 22$ mag): $\Gamma = 0.14^{+0.11}_{-0.06}$ (95% C.L.). The shape of $P(k)$ also depends on the baryon fraction Ω_b/Ω_m: increasing this ratio suppresses power on scales smaller than the peak, and analysis of the full data set will set interesting limits on this parameter also.

13.3.4 Direct Estimates of $P(k)$

The previous subsection described estimates of the three-dimensional $P(k)$ that were derived by first measuring projected quantities $w(\theta)$ and C_ℓ. Since these are essentially smoothed versions of $P(k)$, an alternative procedure is to circumvent the initial measurement of projected quantities, and to work instead with quantities that optimize the signal-to-noise of the data set. This is the KL approach taken by Szalay et al. (2003), who first expand the projected galaxy distribution on the sky over a set of Karhunen-Loève eigenfunctions, and then use a maximum likelihood analysis to derive constraints on the shape and amplitude of $P(k)$. For a flat universe with a cosmological constant, they find $\Gamma = 0.188 \pm 0.04$ and $\sigma_{8g} = 0.915 \pm 0.06$ (statistical errors only). Since $\Gamma \approx \Omega_m h$, if we use the *HST* measurement of the Hubble constant to set $h = 0.7$, then the SDSS results imply $\Omega_m \approx 0.27$.

13.4 Clustering in z Space

The spectroscopic sample provides galaxy redshifts, and hence a reasonably accurate distance measurement, so that, in contrast to the angular photometric catalogs, a much stronger clustering signal can be measured. Moreover, because the redshift is available, it is possible to derive an accurate estimate of the intrinsic luminosity of each galaxy. This allows one to estimate how clustering depends on intrinsic, rather than apparent, properties of galaxies, such as luminosity and rest-frame color. This is important because, in magnitude-limited surveys like the SDSS, the most luminous galaxies are visible at the greatest distances, whereas the least luminous galaxies are only visible nearby. Therefore, the power on the largest scales is dominated by the clustering of the most luminous galaxies, whereas the power on smaller scales comes from a mix of galaxy types. If clustering depends on luminosity, then one must account for the changing mix of galaxy types at each scale when estimating the shape of the power spectrum.

As mentioned previously, peculiar velocities distort clustering statistics in redshift space. One way of accounting for these distortions is to measure the correlation function as a function of the separation parallel and perpendicular to the line-of-sight: $\xi_2(r_p, \pi)$. Only separations parallel to the line-of-sight π are affected by peculiar motions, so that

$$w_p(r_p) \equiv 2 \int_0^\infty d\pi \, \xi_2(r_p, \pi) = 2 \int_0^\infty d\pi \, \xi\left(\sqrt{r_p^2 + \pi^2}\right) \tag{13.6}$$

is independent of redshift-space distortions. Since

$$P(k_p, k_\pi) \equiv \int dr_p \int d\pi \, \xi_2(r_p, \pi) \exp(-ik_p r_p - ik_\pi \pi), \tag{13.7}$$

the quantity $P(k_p, 0)$, being the Fourier transform of $w_p(r_p)$, is also distortion free.

Measurements of the distortion-free correlation function and power spectrum both show that more luminous galaxies are more strongly clustered than less luminous galaxies (Fig. 13.4).

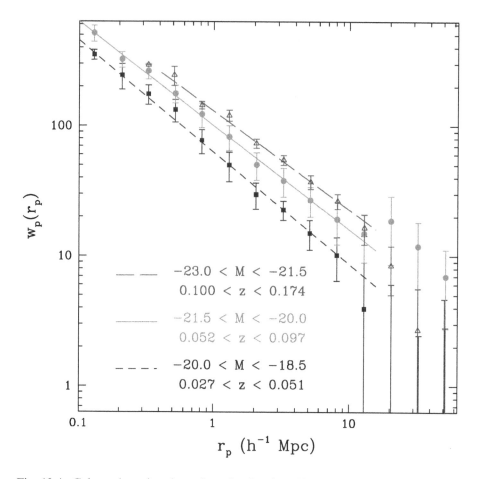

Fig. 13.4. Galaxy clustering depends on luminosity. Changing the luminosity changes the amplitude, but not the slope, of the correlation function. (From Zehavi et al. 2002.)

Whereas the amplitude of $\xi(r)$ appears to depend strongly on luminosity, the shape is approximately independent of L. On the other hand, the shape of the correlation function depends strongly on color: redder galaxies have steeper correlation functions (Fig. 13.5).

As discussed by Budavári et al. (2004), these trends are qualitatively consistent with the following simple model. Suppose there are two types of galaxies, each with its own clustering pattern (say, a red population with a steeper correlation function than the blue population). Then the correlation function of the entire sample will be a weighted sum of the two populations, the weighting being determined by the relative numbers of the two types. Next, suppose that the amplitude of the correlation function in subsamples of each population depends on luminosity, and that the scaling with luminosity is similar for the two populations (on scales larger than 1 Mpc, this is a good description of the SDSS data; on smaller scales, clustering strength increases with luminosity for blue galaxies, whereas red galaxies show the opposite trend). Finally, suppose that the luminosity functions of these two populations have similar shapes, at least at the luminous end. These requirements guarantee that the correlation functions of subsamples defined by luminosity will always have the same

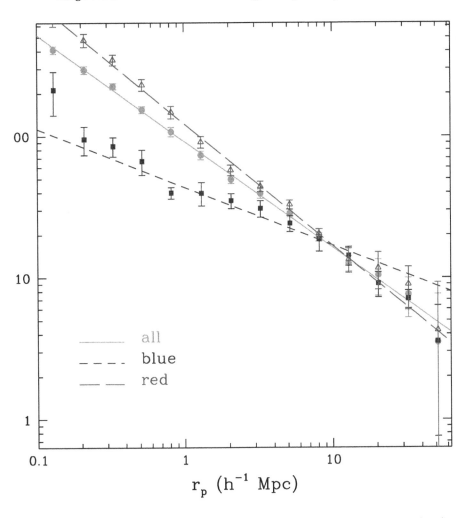

Fig. 13.5. Galaxy clustering depends on color. Changing the color changes the slope and amplitude of $\xi(r)$. (From Zehavi et al. 2002.)

shape, whereas subsamples defined differently will have different shapes. Explaining why this should be so is an interesting challenge for galaxy formation models.

A first study of clustering using photometric redshifts provides qualitatively similar results (Budavári et al. 2004). This is extremely encouraging because the use of photometric redshifts allows one to go considerably fainter than the spectroscopic data set allows. In particular, photometric redshifts offer a cost-effective way of probing clustering out to redshifts of order unity. That this is possible at all is a tribute to the accuracy of the SDSS photometry.

Since the full $\xi(r_p, \pi)$ is sensitive to peculiar velocities, whereas $w_p(r_p)$ is not, a comparison of the two provides a measurement of galaxy peculiar velocities. On the small scales, to which the present data is most sensitive, the dependence of clustering on luminosity and type constrains the velocity dispersions of the halos that different galaxy types populate. The SDSS data show that early-type galaxies populate halos with larger velocity dispersions

(Zehavi et al. 2002), in qualitative agreement with the fact that such galaxies are much more common in massive clusters than in the field.

Since the first measurements of Totsuji & Kihara (1969), the galaxy correlation function has been characterized as a power law. A second look at Figure 13.4 suggests that, while a power law is indeed a good description, it may not be perfect. For example, there will soon be enough data to see if the inflection at $\sim 2h^{-1}$Mpc is more than just a statistical fluctuation. In most *ab initio* models of $\xi(r)$, power laws are purely fortuitous—they are not generic. Explaining the positions of the bumps and wiggles and their dependence on galaxy type, and hence extracting information from these features in $\xi(r)$ will become a rich area of research.

Acknowledgements. Funding for the creation and distribution of the SDSS Archive has been provided by the Alfred P. Sloan Foundation, the Participating Institutions, the National Aeronautics and Space Administration, the National Science Foundation, the U.S. Department of Energy, the Japanese Monbukagakusho, and the Max Planck Society. The SDSS Web site is http://www.sdss.org/.

References

Bahcall, N. A., et al. 2003, ApJ, 585, 182
Budavári, T., et al. 2004, in preparation
Colless, M. 2004, in Carnegie Observatories Astrophysics Series, Vol. 2: Measuring and Modeling the Universe, ed. W. L. Freedman (Cambridge: Cambridge Univ. Press), in press
Connolly, A. J., et al. 2002, ApJ, 579, 42
Dodelson, S., et al. 2002, ApJ, 572, 140
Fukugita, M., Ichikawa, T., Gunn, J. E., Doi, M., Shimasaku, K., & Schneider, D. P. 1996, AJ, 111, 1748
Gunn, J. E., et al. 1998, AJ, 116, 3040
Hoyle, F., Vogeley, M. S., & Gott, J. R., III 2002, ApJ, 580, 663
McKay, T. A., et al. 2004, ApJ, submitted (astro-ph/0108013)
Nichol, R., et al. 2000, HEAD, 32, 14.02
Scranton, R., et al. 2002, ApJ, 579, 48
Stoughton, C., et al. 2002, AJ, 123, 485 (erratum: 123, 3487)
Strauss, M. A., et al. 2002, AJ, 124, 1810
Szalay, A. S., et al. 2003, ApJ, 591, 1
Szapudi, I., et al. 2002, ApJ, 570, 75
Tegmark, M., et al. 2002, ApJ, 571, 191
Totsuji, H., & Kihara, T. 1969, PASJ, 21, 221
York, D. G., et al. 2000, AJ, 120, 1579
Zehavi, I., et al. 2002, ApJ, 571, 172

14

LIGO at the threshold of science operations

ALBERT LAZZARINI
LIGO Laboratory, California Institute of Technology

Abstract

LIGO held its first science run during the two and one-half week period from 23 August to 9 September 2002. The strain sensitivity of the interferometers during this observational period was $h(f) \approx 10^{-20}\,\mathrm{Hz}^{-\frac{1}{2}}$ at $f \approx 300\,\mathrm{Hz}$. The useful instrumental bandwidth spanned the decade $100\,\mathrm{Hz} \lesssim \mathrm{f} \lesssim 1000\,\mathrm{Hz}$. During the 17-day period, the three km-scale LIGO interferometers (2 and 4 km in WA and 4 km in LA) and the 600 m GEO interferometer (outside Hanover, Germany) operated as a network of detectors for part of the time.

The LIGO Scientific Collaboration is presently analyzing the first data run (S1) for a number of classes of sources of gravitational radiation. Working groups within the collaboration are performing searches for signatures coming from four classes of gravitational wave (GW) sources. These include: (1) continuous wave sources (GW counterparts to electromagnetic pulsars), (2) burst or transient sources, such as GW emission from supernovae, (3) inspiral and coalescence of compact binary systems, and (4) a stochastic GW background. The emphasis in this first science run will be to develop the analysis techniques and software pipelines that will be used to analyze data continuously during periods of extended observation in future science runs. At the same time, the S1 data quality are such that it should be possible to provide improved *direct observational* limits on gravitational radiation with these fundamentally new instruments.

14.1   Introduction

In the last few years a number of new gravitational wave (GW) detectors, using long-baseline laser interferometry, have begun to enter into operation. These include the Laser Interferometer Gravitational Wave Observatory (LIGO; Abramovici et al. 1992; Barish & Weiss 1999) detectors located in Hanford, WA and Livingston, LA, built by a Caltech-MIT collaboration, the GEO-600 detector near Hanover, Germany, built by an British-German collaboration (Danzmann et al. 1995), the VIRGO detector near Pisa, Italy, built by an Italian-French collaboration (Caron et al. 1995), and the Japanese TAMA-300 detector in Tokyo (Tsubono 1995). While none of these instruments is yet performing at its design sensitivity, many have begun making dedicated data collecting runs and performing GW search analyses on the data.

In particular, from 23 August to 9 September 2002, the LIGO Hanford and LIGO Livingston Observatories (LHO* and LLO†) took coincident science data (referred to as S1).

---

* http://www.ligo-wa.caltech.edu
† http://www.ligo-la.caltech.edu

Table 14.1. *Operational Duty Cycles of Different Coincidence Modes during S1*

	Locked Time (Hr)	Duty Cycle (%)
Single		
H1 (4 km)	235	57.6
H2 (2 km)	298	73.1
L1(4 km)	170	41.7
Double Coincidence		
H1+L1	116	28.4
H2+L1	131	32.1
H1+H2	188	46.1
Triple Coincidence		
H1+H2 +L1	95.7	23.4

The LHO site contains two, identically oriented interferometers, one having 4 km long measurement arms (referred to as H1), and the other having 2 km long arms (H2); the LLO site contains a single, 4 km long interferometer (L1). GEO also took data in coincidence with the LIGO detectors during that time, although with significantly poorer sensitivity.

The LIGO Scientific Collaboration* is presently analyzing the first data run (S1) for a number of classes of sources of gravitational radiation. Working groups within the collaboration are conducting searches for signatures coming from four classes of GW sources. These include: (1) continuous wave sources, such as periodic sources [GW counterparts to electromagnetic (EM) pulsars], (2) burst or transient sources, such as GW emissions from supernovae or other transient events, (3) inspiral and coalescence of compact binary systems, (4) a stochastic GW background. The emphasis in this first science run will be develop the analysis techniques and software pipelines that will be used to analyze data continuously during periods of extended observation in future science runs. At the same time, the S1 data quality are such that it should be possible to provide improved *direct observational* limits on gravitational radiation with these fundamentally new instruments.

Figure 14.1 presents a composite graph showing the amplitude spectral densities for the three LIGO interferometers taken during the S1 run. The LA 4 km machine was the most sensitive, achieving strain sensitivities of $h(300\,\text{Hz}) \approx 3 \times 10^{-21}\,\text{Hz}^{-\frac{1}{2}}$. Table 14.1 presents the operational duty cycles for the different coincidence modes during S1. During S1, LIGO interferometers had better broad-band sensitivities than any prior GW detector. In addition, the number of machines operating simultaneously was unprecedented.

14.2   S1 Data Analysis Activities

At the time of this writing, the LIGO Scientific Collaboration is still processing the data, and it is not yet possible to publicize the results. The discussion below will concentrate on techniques being applied in the data analysis.

* http://www.ligo.org

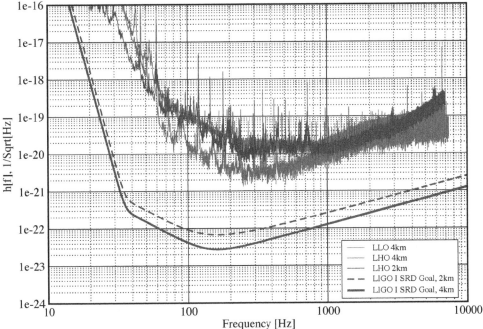

Fig. 14.1. Spectra of instrumental sensitivities for the three LIGO interferometers during the S1 science run. The solid curve corresponds to the design goal for the 4 km long interferometers in WA and LA. The dashed curve gives the goal for the shorter, 2 km long instrument in WA.

In all cases, data analysis pipelines have been set up to provide production capability to process data end-to-end. These pipelines also provide support for Monte Carlo simulations that are required to validate and to calibrate the search efficiencies. Figure 14.2 presents a block diagram schematic of the analysis flow for the burst event search. It is representative of a prototypical analysis pipeline.

Data streams from an interferometer are processed by a data analysis flow that generates candidate events. These events are subsequently analyzed for coincidence among as many interferometers as were operating at the epoch when the candidate event was produced. The GW (strain) channel is processed by *event trigger generator* filter(s) that perform optimal filtering of the strain channel using either template waveforms based on astrophysical source models or parameterizable waveforms (e.g., wavelets) that are correlated continuously against the interferometer output. When the event trigger exceeds a threshold determined by Monte Carlo calibration of the pipeline, a putative event is identified for further post-processing downstream.

A variety of auxiliary channels are used to monitor instrumental and environmental characteristics at the same time the strain channel is analyzed. Transient or off-nominal behavior

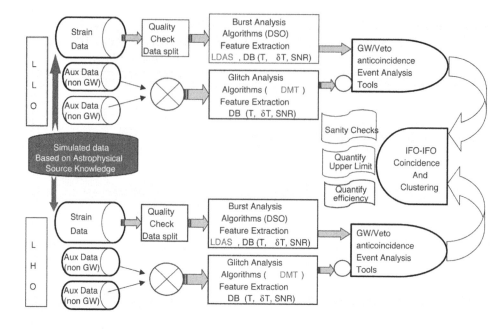

Fig. 14.2. The burst analysis pipeline as a prototypical analysis technique. Notation: DSO — a search algorithm library; LDAS — LIGO Data Analysis System environment; DB — relational database used to archive event metadata; DMT — data monitoring tool suite used to look at non-strain channel data in near real time; IFO — interferometer; LHO — LIGO Hanford Observatory; LLO — LIGO Livingston Observatory.

of these other channels is used to *veto* candidate events during epochs of detection when the instrument was not operating in its quiescent state. Candidates that survive these vetoes are then processed for coincidence among the several interferometers. Further, consistency checks are enforced to verify, for example, that the coincident events are consistent in all interferometers: comparable signal amplitude, duration, and frequency content are required of a coincidence for it to be considered a possible astrophysical event.

Interpretation of the results requires that models of source distributions, source strengths, and waveform characteristics be injected into the data stream using Monte Carlo techniques. In this way the detection efficiency can be determined for different classes of signals.

14.2.1   Classes of GW Sources

Different types of GW sources produce signals with specific signatures encoded in the waveforms. These signatures are exploited using digital signal processing techniques to identify candidate events. Figure 14.3 presents a simulated time-frequency map showing these signatures. Searches for the different source types are discussed in the following subsections.

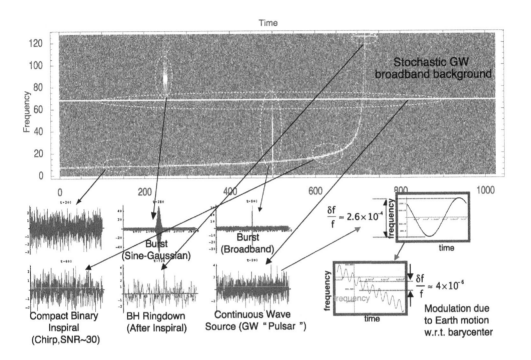

Fig. 14.3. Simulated time-frequency map showing the various classes of GW sources and their characteristic evolution in time. Inspirals produce chirps with a characteristic upswing in frequency. Periodic sources are expected to produce monochromatic signals with an impressed FM signature from the barycentric motion of the Earth. The daily rotation produces a $\sim 10^{-6}$ effect and the orbital motion produces a $\sim 10^{-4}$ effect. Bursts are expected to produce short-duration broad-band signals. The stochastic background is stationary, isotropic, and broad band. Optimal filters exploit the time-dependent evolution of signal phase (or frequency) to discriminate against background noise.

14.2.1.1 Searches for Periodic Sources of GW

Periodic sources are narrow-band coherent signals that extend over the entire period of observation. For sufficiently long observations, the deterministic and well-characterized frequency modulations (FM) imposed by the barycentric motion of the Earth around the Sun can be exploited to verify the extraterrestrial nature of such a source. On the scale of the S1 run, only a portion of the yearly FM cycle is detectable, and the FM signatures correspond to monotonic drifts in signal frequency upon which is superimposed the daily modulations caused by Earth rotation.

The first analyses of LIGO data will search for evidence of GW emission from specific sources with known EM counterparts. Figure 14.4 presents a "landscape" of sensitivities achieved during S1 and possible sources of GW associated with known EM counterparts. *Di-*

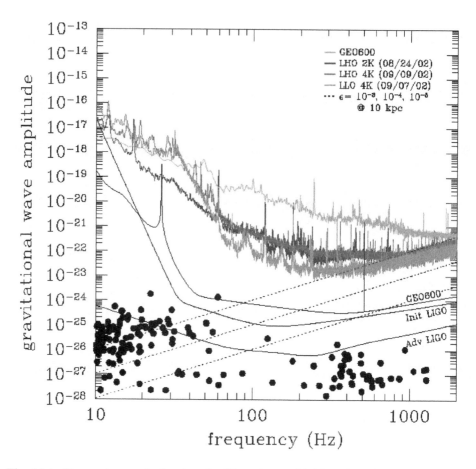

Fig. 14.4. Composite graph showing the S1 strain sensitivities of the four interferometers participating in the science run. The labeled smooth curves correspond to design sensitivities. Data and curves are scaled to show the critical amplitude, $h_c(f) \equiv 4.2\sqrt{S_h(f)/T_{obs}}$ detectable with 99% confidence during the duration of S1, $T_{obs} \approx 17\,\mathrm{d}$. The individual dots correspond to known EM millisecond pulsars. The abscissa is *twice* the EM frequency. The ordinate corresponds the the signal that would be generated *if* observed spin-downs were attributable entirely to GW emission. This provides an astrophysical upper limit derived from energy conservation arguments. The dashed lines (from top to bottom) give the signal strength as a function of frequency for rotating compact stars with equatorial ellipticities of $\epsilon = 10^{-3}, 10^{-4}$, and 10^{-5} at a distance of 10 kpc, corresponding approximately to sources at the center of the Galaxy. (Figure courtesy of R. Depuis, Univ. of Glasgow/GEO.)

rected or *targeted* searches looking at such sources will be employed to validate the analysis techniques and to set direct observational upper limits at progressively more interesting levels. Broad-band interferometers will allow simultaneous observation and parallel analysis of many sources—something not previously possible with resonant cryogenic bars. Previously published observational limits for periodic sources are $h_{max}(f = 1283.86\,\mathrm{Hz}) \leq 3 \times 10^{-20}$ for PSR1939+2134 (Hough et al. 1983) using an interferometer, and $h_{max}(f = 921.35\,\mathrm{Hz}) \leq$

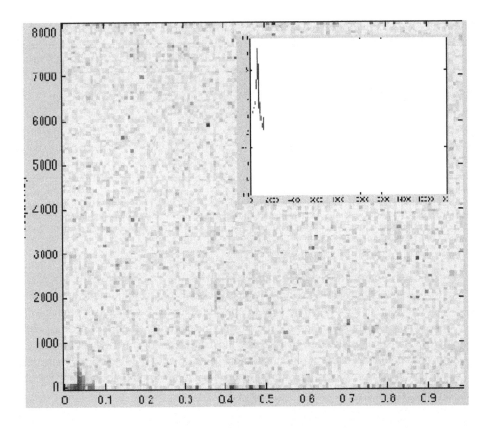

Fig. 14.5. Output of tfclusters for a time-frequency image of a Zwerger-Muller supernova waveform (at $\{t, f\} = \{0.05\text{s}, 0 \to 500\,\text{Hz}\}$).

3×10^{-24} for an untargeted search toward the center of the Galaxy at a frequency constrained by the resonant bar detector (Astone et al. 2002).

In the future the plan is to expand the searches to include other periodic sources, such as low-mass X-ray binaries (e.g., Sco-X1). In addition, a blind search technique is being developed to search for potential GW sources having no known EM counterparts. Such a search targets large portions of the sky and does a parametric search over intrinsic source characteristics (e.g., spin down rates) and extrinsic parameters, such as sky location and their effect on the signal FM characteristics (Brady et al. 1998; Brady & Creighton 2000). Examples of such sources include rapidly rotating, newly formed neutron stars (Andersson 1998; Friedman & Morsink 1998; Lindblom, Owen, & Morsink 1998; Andersson, Kokkotas, & Schutz 1999; Bildsten & Ushomirsky 2000; Jones 2001; Lindblom & Owen 2001; Arras et al. 2003).

Table 14.2. *Rate Estimates for Inspiral of NS/NS, NS/BH, and BH/BH Binaries*[†]

	NS/NS	NS/BH	BH/BH in Field	BH/BH in Clusters
$\mathcal{R}_{gal}$ (yr^{-1})	10^{-6}–5×10^{-4}	$\lesssim 10^{-7}$–10^{-4}	$\lesssim 10^{-7}$–10^{-5}	$\sim 10^{-6}$–10^{-5}
D_I (Mpc)	20	43	100	100
$\mathcal{R}_I$ (yr^{-1})	3×10^{-4} – 0.3	$\lesssim 4 \times 10^{-4}$ – 0.6	$\lesssim 4 \times 10^{-3}$ – 0.6	~ 0.04 – 0.6
D_{WB} (Mpc)	300	650	$z = 0.4$	$z = 0.4$
$\mathcal{R}_{WB}$ (yr^{-1})	1 – 800	$\lesssim 1$ – 1500	$\lesssim 30$ – 4000	~ 300 – 4000

[†]Table adapted from Cutler & Thorne (2002); refer to this reference for the primary sources of these estimates. The table shows estimated rates $\mathcal{R}_{gal}$ in the Galaxy (with masses $\sim 1.4 M_\odot$ for NS and $\sim 10 M_\odot$ for BH), the distances $\mathcal{D}_I$ and $\mathcal{D}_{WB}$ to which initial interferometers and advanced wide-band (WB) interferometers can detect them, and corresponding estimates of detection rates $\mathcal{R}_I$ and $\mathcal{R}_{WB}$. The third row, $\mathcal{R}_I$, corresponds to the present generation of LIGO interferometers.

14.2.1.2   Searches for Burst Sources of GW

Burst sources have no deterministic phase or frequency evolution and thus template-based modeling of source properties is not applicable. Instead, techniques predicated on *novelty detection* are being developed. By novelty detection is meant a search algorithm that employs statistical methods to characterize the data stream over periods of time that are much longer than the expected duration on burst events, then to use these prior characteristics of the data to set thresholds looking for excess signal power on time scales of fractions of a second.

For the S1 run, two techniques have been developed. One approach searches in the time domain and detects the amplitude fluctuations of the signal on short time scales. The other approach utilizes time-frequency maps of the strain channel to look for clustering of contiguous pixels that exceed a predetermined threshold. The former approach implements an algorithm, termed *slope filter*, which looks for large changes of slope over short periods of time. The algorithm is a subset of a more general algorithm developed by the Orsay group (Pradier et al. 2001). The latter approach is being used in two similar implementations. The first of these, termed *excess power* (Anderson et al. 2000, 2001), tessellates the $t - f$ $t - f$ plane into "postage-stamp" patches of constant $\Delta f \Delta t$ and looks for excess power fluctuations. The second algorithm, termed *tfclusters* (Sylvestre 2002), has a clustering algorithm that is able to identify and detect groups of contiguous pixels that have arbitrary shape. Both approaches have been shown capable of detecting excess power embedded in a noisy signal. Figure 14.5 depicts an image of the $t - f$ output of tfclusters for an injected Zerger-Müller (1997) supernova waveform. These same techniques are suitable for detecting black hole ringdowns.

A subgroup of the burst search team is also developing search methodologies for so-called *externally* triggered searches, whereby astrophysical triggers (e.g., gamma-ray burst events) can be used to localize in time searches for coincidences among multiple interferometers. By such techniques, it is possible to provide upper limits, in lieu of detection, of the amount of GW energy associated with externally triggered events seen by other detectors (Finn, Mohanty, & Romano 1999).

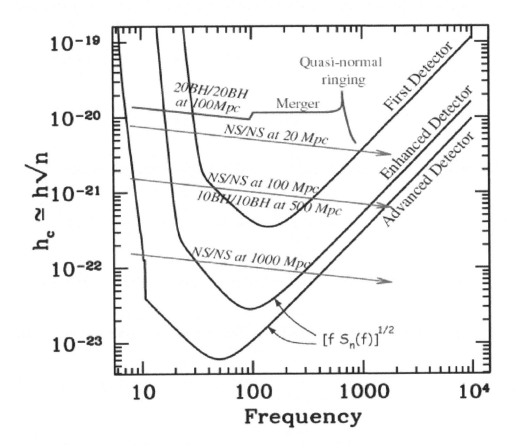

Fig. 14.6. Plot showing sensitivity curves for first and advanced generation instruments. The trajectories correspond to the power spectra for signals arising from the indicated sources.

14.2.1.3 Searches for Compact Binary Inspirals

Waveforms associated with the inspiral of $M_1 + M_2 \lesssim 6 M_\odot$ compact binary systems are the most well-studied sources that can be detected in the LIGO band. What is much less well known is the *rate* with which such events occur in nature. Table 14.2 presents best-estimate rates for binary systems composed of neutron star (NS) pairs, black hole (BH) pairs, and mixed systems (NS/BH) (Cutler & Thorne 2001).

The event rate for NS+NS coalescences in the Galaxy is constrained by the number of observed binary pulsars for which the energy loss due to GW emission will drive the systems to merger in a time that is shorter than $1/H_0$. The resulting constraints on the merger rate in the Galaxy, $10^{-6} \mathrm{yr}^{-1} \lesssim \mathcal{R}_{\mathrm{gal}} \lesssim 5 \times 10^{-4} \mathrm{yr}^{-1}$, extrapolates to a NS/NS event rate for the current generation of instruments that is much less than one per year. Refer to the third row of Table 14.2 for predicted event rates for initial LIGO interferometers.

Information on binary systems is inferred from observations of radio pulsars. Unfortunately, there are no known NS/BH binaries in which the NS is also a pulsar. Therefore, it is necessary to rely on much less certain estimates based on simulations of the evolution of a population of progenitor binary systems to determine the number of systems that lead to compact NS/BH binaries.

It is important to note that the uncertainties in these predicted rates span *3* orders of magnitude. However, it appears to be the case that a one-year observation with current LIGO interferometers is likely to not observe these types of events. The best direct observational limit to date, placed with data from the LIGO 40 m prototype interferometer (Allen et al. 1999), is $\mathcal{R} \leq 4400 \, yr^{-1}$ (90% confidence level) for the Galaxy. S1 will provide a significantly improved observational limit.

The detection technique relies on optimal Wiener filtering in the frequency domain. The interferometer strain data are correlated with theoretically derived signal waveforms (templates), weighted by the reciprocal of the instrument noise spectral density to produce a time series of signal-to-noise ratio. The templates, $\tilde{T}(\vec{\alpha} : f)$, are characterized by a parameter vector $\vec{\alpha}$:

$$SNR(\vec{\alpha} : t) = \int_0^\infty df \frac{\tilde{h}(f)\tilde{T}^\star(\vec{\alpha} : f)}{S_n(f)} e^{-2\pi i f t}. \tag{14.1}$$

The SNR time series is processed by a thresholding algorithm to identify putative events to be further analyzed.

Figure 14.6 shows the dependence of coalescence signals on frequency. The ordinate gives the so-called "characteristic" strain of an inspiral signal. This quantity is the square root of the signal power spectral density multiplied by the number of cycles in the waveform at that frequency: it is a measure of the total power radiated at any given frequency during a coalescence event. This shows that the initial generation of LIGO interferometers can detect coalescence events for NS+NS systems to $D \approx 100$ Mpc. Also shown on the graph is the signal strength calculated for more massive $(10M_\odot + 10M_\odot)$ BH systems at distances $D \approx 500$ Mpc. For more massive systems, the merger and ringdown phases of the coalescence are detectable in the LIGO band. However, the details of the merger phase are presently unknown and require much more effort in numerical relativity simulations before templates can be provided for the data analysis. For the S1 run, systems more massive than NS+NS will not be considered. However, there is a considerable effort in the collaboration to implement template-based searches by the time of the S2 run to look for more massive systems. These searches will be predicated on analytical models that extend the waveform space beyond NS+NS systems (Buonanno, Chen, & Vallisneri 2003a,b).

14.2.1.4 *Searches for a Stochastic Gravitational Wave Background*

Gravitational waves of cosmological origin produce a stochastic background analogous to the relic microwave background radiation, but arising at a much earlier epoch. The search for such a signal is performed by cross-correlating the strain signals from pairs of interferometers and introducing an optimal Wiener filter to maximize detection signal-to-noise ratio (Michelson 1987; Christensen 1992; Flanagan 1993; Allen & Romano 1999).

Strain sensitivity plots

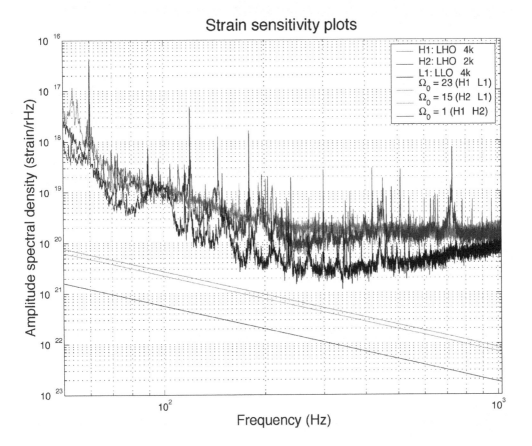

Fig. 14.7. Superposition of the S1 sensitivities of the three LIGO interferometers with predicted sensitivities of different interferometer pairs to an $\Omega_{GW}(f) = $ constant stochastic background.

The background is stochastic in the sense that its properties are only statistically characterized. Its spectral properties are described by the dimensionless quantity

$$\Omega_{\text{gw}}(f) \equiv \frac{f}{\rho_{\text{critical}}} \frac{d\rho_{\text{gw}}}{df} , \tag{14.2}$$

where ρ_{gw} is the energy density in gravitational waves and

$$\rho_{\text{critical}} = \frac{3c^2 H_0^2}{8\pi G} \tag{14.3}$$

is the energy density required (today) to close the Universe. H_0 is the Hubble expansion rate in the present epoch (Bennett et al. 2003):

$$H_0 \equiv h_{100} 100 \text{ km s}^{-1} \text{Mpc}^{-1} \approx 3.24 \times 10^{-18} h_{100} \text{ s}^{-1} \tag{14.4}$$

$$= 71^{+4}_{-3} \text{ km s}^{-1} \text{Mpc}^{-1} \tag{14.5}$$

Overlap Reduction Function
(LIGO-LA and other detectors)

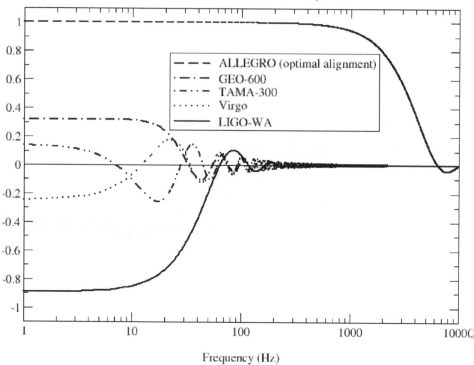

Frequency (Hz)

Fig. 14.8. Overlap reduction function between LIGO Livingston and the other major interferometers plus the LSU cryogenic resonant bar detector ALLEGRO (in an optimal alignment of 72° East of North).

Note that $\Omega_{\mathrm{gw}}(f)\,h_{100}^2$ is independent of the actual Hubble expansion rate, and has been used extensively in the literature when H_0 was not known very accurately. $\Omega_{\mathrm{gw}}(f)$ is related to the one-sided power spectral density of GW strain $S_{\mathrm{gw}}(f)$ via *

$$S_{\mathrm{gw}}(f) = \frac{3H_0^2}{10\pi^2} f^{-3}\Omega_{\mathrm{gw}}(f).$$ (14.6)

Thus, for a stochastic GW background with $\Omega_{\mathrm{gw}}(f)$ = constant, the power in gravitational waves falls off as $1/f^3$.

The spectrum $\Omega_{\mathrm{gw}}(f)$ completely specifies the statistical properties of a stochastic background of gravitational radiation, provided we make several additional assumptions. Namely, we assume that the stochastic background is (1) isotropic, (2) unpolarized, (3) stationary, and (4) Gaussian. Anisotropic or non-Gaussian backgrounds [e.g., due to an incoherent superposition of GW from a large number of unresolved white dwarf binary star systems in our

---

* $S_{\mathrm{gw}}(f)$ is defined by $\lim_{T\to\infty}\frac{1}{T}\int_0^T |h(t)|^2\,dt = \int_0^\infty S_{\mathrm{gw}}(f)\,df$, where $h(t)$ is the GW strain in a single detector due to the stochastic background signal.

Table 14.3. *Summary of Upper Limits on* $\Omega_0 h_{100}^2$ [†]

Observational Technique	Observed Limit	Frequency Domain (Hz)	Ref. [‡]
Cosmic Microwave Background	$\Omega_{gw}(f) h_{100}^2 \leq 7 \times 10^{-11} \left(\frac{H_0}{f}\right)^2$	$H_0 < f < 30\, H_0$	1
Radio Pulsar Timing	$\Omega_{gw}(f) h_{100}^2 \leq 4.8 \times 10^{-9} \left(\frac{f}{f_0}\right)^2$	$f_0 = 10^{-8}$	2
Big Bang Nucleosynthesis	$\int_{f>10^{-8}\,\text{Hz}} d\ln f\, \Omega_{gw}(f) h_{100}^2 \leq 10^{-5}$	$f > 10^{-8}$	3
Room Temp. Resonant Bar (correlation)	$\Omega_{gw}(f) h_{100}^2 \leq 3000$	$f = 900$	4
Interferometers	$\Omega_{gw}(f) h_{100}^2 \leq 3 \times 10^5$	$100 < f < 1000$	5
Cryogenic Resonant Bar (single)	$\Omega_{gw}(f) h_{100}^2 \leq 300$ $\Omega_{gw}(f) h_{100}^2 \leq 5000$	$f = 907$ $f = 1875$	6
Cryogenic Resonant Bar (correlation)	$\Omega_{gw}(f) h_{100}^2 \leq 60$	$f = 907$	7

[†] The upper portion of the Table lists limits derived from astrophysical observations. The lower portion lists limits obtained from prior direct measurement. [‡] References: (1) Wright et al. 1992, Bennett et al. 1996; (2) Kaspi, Taylor, & Ryba 1994; (3) Kolb & Turner 1990; (4) Hough et al. 1975; (5) Compton, Nicholson, & Schutz 1994; (6) Astone et al. 1996, 1999a; (7) Astone et al. 1999b, 2000.

own Galaxy, or a "popcorn" stochastic signal produced by GW from supernova explosions (Blair & Ju 1996; Blair et al. 1997)] will require different data analysis techniques than what is being used for the S1 analysis; see, for example, Allen & Ottewill (1997) and Drasco & Flanagan (2003) for detailed discussions of these different techniques.

Cross correlating interferometer signals with an optimal Wiener filter allows one to detect the presence of a correlated signal at levels several orders of magnitude weaker than the noise spectral density. Figure 14.7 presents theoretical predictions for the magnitudes of stochastic signals arising from a Ω_{GW} = constant background that was at the limits of detectability during S1.

Under the assumptions discussed above, the optimally filtered estimate of the stochastic background is

$$\Omega_{GW}^{estimate} \approx \int_0^\infty df \frac{\gamma(|f|)\tilde{s}_1^*(f)\tilde{s}_2(f)}{|f|^3 \tilde{S}_{n1}(|f|)\tilde{S}_{n2}(|f|)}, \tag{14.7}$$

where $\tilde{s}_i(f)$ are the strain signals from the interferometers and $\tilde{S}_{ni}(f)$ are their noise spectral densities. The function $\gamma(|f|)$ is a geometrical-form factor describing the frequency-dependent response of an antenna pair due to their space-time separation (Flanagan 1993). The response is for an isotropic, unpolarized irradiation by a stochastic GW background; refer to Figure 14.8, which shows $\gamma(|f|)$ for a number of interferometer pairs and also for the LA 4 km interferometer + ALLEGRO cryogenic bar. It is expected that the S1 run will produce a limit for Ω_{GW} that is better than previous direct determinations of this quantity

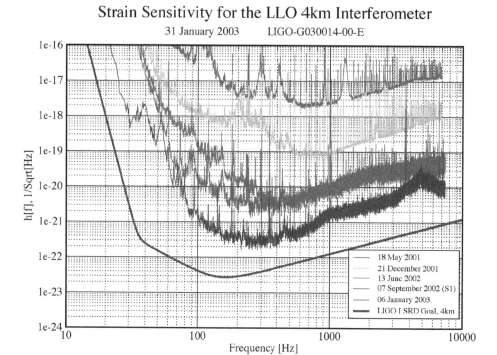

Fig. 14.9. Cascade plot of strain sensitivity improvements for the LA 4 km interferometer, showing progress in performance over the past ∼20 months. The data set] second from the bottom shows the S1 sensitivity. The bottom data set was measured about one month before the start of S2. The solid curve gives the ultimate initial LIGO design performance.

in the decade frequency band $50\,\mathrm{Hz} \lesssim f \lesssim 500\,\mathrm{Hz}$. Table 14.3 presents a comprehensive summary of what is known or has been inferred about the stochastic GW background.

14.3 Summary

LIGO has begun scientific operation and the results for the first science run, S1, are in the process of being analyzed by the collaboration. The primary focus of the S1 run has been to develop the analysis pipelines that will be used to process the strain data for a variety of GW sources. At the same time, it is expected that the results of this analysis will achieve *direct* observational limits on the flux of gravitational radiation from these source classes, which improve on presently published limits.

At the time of this writing, a longer science run, S2, is in progress and will end in mid-April 2003. The instruments are performing ∼ 10 times *better* than they were for S1. Over the past 18 months there has been a breathtaking pace of improvement in the sensitivity of the machines (refer to Fig. 14.9). The LIGO instruments should approach their design sensitivities within a year.

14.4   Note added in proof

The preliminary results from the S1 data analysis were presented at the AAAS Symposium, *Looking Beyond Earth*, at Denver, CO in February 2003 and an update will be presented at the Spring Meeting of the APS at Philadelphia, PA in April 2003. They may be briefly summarized as follows:

- The upper limit for burst events occurring during the S1 observation was set at $\mathcal{R}_{bursts} \lesssim 2$ events day^{-1} (90% confidence level) for signals with peak strain amplitude $h \approx 4 \times 10^{-17}$ having millisecond duration. The strain sensitivity for this rate limit is ~ 5 times less stringent than has been reported by the cryogenic resonant bar consortium (IGEC).
- The search for compact binary inspirals covered the entire Milky Way galaxy, including its two satellites, the Large and Small Magellanic Clouds. This represents the first time a direct GW search extended beyond our Galaxy. The maximum range detectable for a signal-to-noise ratio = 8 event corresponded to ~ 210 kpc for the LA 4 km interferometer. The search covered the mass range for individual binary components $1\,M_\odot \leq M_{1,2} \leq 3\,M_\odot$. The upper limit in event rate was determined to be $\mathcal{R}_{inspirals} \leq 164$ events yr^{-1} per Milky Way equivalent galaxy (90% confidence level).
- The search for periodic sources targeted the millisecond EM pulsar J1939+2134 ($D = 3.6$ kpc). At 95% confidence using a Bayesian analysis, the maximum detectable signal from this EM source that would have been detected is $h_{CW} \leq 1.4 \times 10^{-22}$.
- The search for a stochastic GW background achieved an upper limit sensitivity for the LA 4 km – WA 4 km interferometer pair, corresponding to $\Omega_{GW}(f) \leq 73$ (90% confidence level) over the frequency interval 40Hz $\leq f \leq$ 314Hz.

Acknowledgements. The author is indebted to his colleagues within the LIGO Laboratory and the LIGO Scientific Collaboration, without whose collective participation the S1 run and the subsequent results could not have been achieved. LIGO Laboratory operates under cooperative agreement PHY-0107417. This document has been assigned the LIGO Laboratory document number LIGO-P020035-00.

References

Abramovici, A., et al. 1992, Science, 256, 325
Allen, B., et al. 1999, Phys. Rev. Lett., 83, 1498
Allen, B., & Ottewill, A. C. 1997, Phys. Rev. D, 56, 545
Allen, B., & Romano, J. D. 1999, Phys. Rev. D, 59, 102001
Anderson, W. G., Brady, P. R., Creighton, J. D., & Flanagan, É. 2000, Int. Journ. Mod. Phys. D, 9 303
——. 2001, Phys. Rev. D, 63, 042003
Andersson, N. 1998, ApJ, 502, 708
Andersson, N., Kokkotas, K., & Schutz, B. F. 1999, ApJ, 510, 846
Arras, P., Flanagan, É. É., Morsink, S. M., Schenk, A. K., Teukolsky, S. A., & Wasserman, I. 2003, ApJ, 591, 1129
Astone, P., et al. 1996, Phys. Lett. B, 385, 421
——. 2002, Phys. Rev. D, 65, 022001
——. 1999a, A&A, 343, 19
——. 1999b, A&A, 351, 811
Astone, P., Ferrari, V., Maggiore, M., & Romano, J. D. 2000, Int. Journ. Mod. Phys. D, 9, 361
Barish, B. C., & Weiss, R. 1999, Phys. Today, 52, 44 (http://www.ligo.caltech.edu)
Bennett, C. L., et al. 1996, ApJ, 464, L1
——. 2003, ApJS, 148, 1
Bildsten, L., & Ushomirsky, G. 2000, ApJ, 529, L33
Blair, D., Burman, R., Ju, L., Woodings, S., Mulder, M., & Zadnik, M. G. 1997, preprint
Blair, D., & Ju, L. 1996, MNRAS, 283, 648

Brady., P. R., & Creighton, T. 2000, Phys. Rev. D, 61, 082001

Brady., P. R., Creighton, T., Cutler, C., & Schutz, B. F. 1998, Phys. Rev. D, 57, 2101

Buonanno, A., Chen, Y., & Vallisneri, M. 2003a, Phys. Rev. D, 67, 104025

———. 2003b, Phys. Rev. D, 67, 024016

Caron, B.. et al. 1995, in First Edoardo Amaldi Conference on Gravitational Wave Experiments, ed. E. Coccia, G. Pizzella, & F. Ronga (Singapore: World Scientific), 86

Christensen, N. 1992, Phys. Rev. D, 46, 5250

Compton, K., Nicholson, D., & Schutz, B. F. 1994, Proceedings of the Seventh Marcel Grossman Meeting on General Relativity, ed. R. T. Jantzen, G. M. Keiser, & R. Ruffini (River Edge, New Jersey: World Scientific), 1078

Cutler, C., & Thorne, K. 2001, in Proceedings of GR16, Durban, South Africa

Danzmann, K., et al. 1995, in First Edoardo Amaldi Conference on Gravitational Wave Experiments, ed. E. Coccia, G. Pizzella, & F. Ronga (Singapore: World Scientific), 100

Drasco, S., & Flanagan, É. É. 2002, in Proceedings of the Ninth Marcel Grossmann Meeting on General Relativity, ed. V. G. Gurzadyan, R. T. Jantzen, & R. Ruffini (River Edge, New Jersey: World Scientific), in press

Ferrari, V. 1997, in Proceedings of the 12th Italian Conference on General Relativity and Gravitational Physics, ed. M. Bassan et al. (Singapore: World Scientific), 149

Finn, L. S., Mohanty, S. D., & Romano, J. D. 1999, Phys. Rev. D, 60, 121101

Flanagan, É. É. 1993, Phys. Rev. D, 48, 2389

Friedman, J. L., & Morsink, S. M. 1998, ApJ, 502, 714

Hough, J., Pugh, J. R., Bland, R., & Drever, R. W. P. 1975, Nature, 254, 498

Hough, J., Ward, H., Munley, A. J., Newton, G. P., Meers, B. J., Hoggan, S., Kerr, G. A., & Drever, R. W. P. 1983, Nature, 303, 216

Jones, P. B. 2001, Phys. Rev. D, 64, 084003

Kaspi, V. M., Taylor, J. H., & Ryba, M. F. 1994, ApJ, 428, 713

Kolb, E. W., & Turner, M. S. 1990, in The Early Universe (Reading, MA: Addison Wesley)

Lindblom, L., & Owen, B. J. 2002, Phys. Rev. D, 65, 063006

Lindblom, L., Owen, B. J., & Morsink, S. M. 1998, Phys. Rev. Lett., 80, 4843

Michelson, P. F. 1987, MNRAS, 227, 933

Pradier, T., Arnaud, N., Bizouard, M.-A., Cavalier, F., Davier, M., & Hello, P. 2001, Phys. Rev. D, 63, 042002

Sylvestre, J. 2002, Phys. Rev. D, 66, 102004

Tsubono, K. 1995, in First Edoardo Amaldi Conference on Gravitational Wave Experiments, ed. E. Coccia, G. Pizzella, & F. Ronga (Singapore: World Scientific), 112

Wright, E. L., et al. 1992, ApJ, 396, L13

Zwerger, T., & Müller, E. 1997, A&A, 320, 209

15

Why is the Universe accelerating?

SEAN M. CARROLL

*Enrico Fermi Institute, Department of Physics, and Center for Cosmological Physics,
University of Chicago*
Kavli Institute for Theoretical Physics, University of California, Santa Barbara

Abstract

The universe appears to be accelerating, but the reason why is a complete mystery. The simplest explanation, a small vacuum energy (cosmological constant), raises three difficult issues: why the vacuum energy is so small, why it is not quite zero, and why it is comparable to the matter density today. I discuss these mysteries, some of their possible resolutions, and some issues confronting future observations.

15.1 Introduction

Recent astronomical observations have provided strong evidence that we live in an accelerating universe. By itself, acceleration is easy to understand in the context of general relativity and quantum field theory; however, the very small but nonzero energy scale seemingly implied by the observations is completely perplexing. In trying to understand the universe in which we apparently live, we are faced with a problem, a puzzle, and a scandal:

- The *cosmological constant problem:* why is the energy of the vacuum so much smaller than we estimate it should be?
- The *dark energy* puzzle:* what is the nature of the smoothly distributed, persistent energy density that appears to dominate the universe?
- The *coincidence scandal:* why is the dark energy density approximately equal to the matter density today?

Any one of these issues would represent a serious challenge to physicists and astronomers; taken together, they serve to remind us how far away we are from understanding one of the most basic features of the universe.

The goal of this article is to present a pedagogical (and necessarily superficial) introduction to the physics issues underlying these questions, rather than a comprehensive review; for more details and different points of view see Sahni & Starobinski (2000), Carroll (2001), or Peebles & Ratra (2003). After a short discussion of the issues just mentioned, we will turn to mechanisms that might address any or all of them; we will pay special attention to the dark energy puzzle, only because there is more to say about that issue than the others.

---

* "Dark energy" is not, strictly speaking, the most descriptive name for this substance; lots of things are dark, and everything has energy. The feature that distinguishes dark energy from ordinary matter is not the energy but the pressure, so "dark pressure" would be a better term. However, it is not the existence of the pressure, but the fact that it is negative—tension rather than ordinary pressure—that drives the acceleration of the universe, so "dark tension" would be better yet. And we would have detected it long ago if it had collected into potential wells rather than being smoothly distributed, so "smooth tension" would be the best term of all, not to mention sexier. I thank Evalyn Gates, John Beacom, and Timothy Ferris for conversations on this important point.

We will close with an idiosyncratic discussion of issues confronting observers studying dark energy.

15.2   The Mysteries

15.2.1   *Classical Vacuum Energy*

Let us turn first to the issue of why the vacuum energy is smaller than we might expect. When Einstein proposed general relativity, his field equation was

$$R_{\mu\nu} - \frac{1}{2}Rg_{\mu\nu} = 8\pi G T_{\mu\nu} \,, \tag{15.1}$$

where the left-hand side characterizes the geometry of spacetime and the right-hand side the energy sources; $g_{\mu\nu}$ is the spacetime metric, $R_{\mu\nu}$ is the Ricci tensor, R is the curvature scalar, and $T_{\mu\nu}$ is the energy-momentum tensor. (I use conventions in which $c = \hbar = 1$.) If the energy sources are a combination of matter and radiation, there are no solutions to (15.1) describing a static, homogeneous universe. Since astronomers at the time believed the universe was static, Einstein suggested modifying the left-hand side of his equation to obtain

$$R_{\mu\nu} - \frac{1}{2}Rg_{\mu\nu} + \Lambda g_{\mu\nu} = 8\pi G T_{\mu\nu} \,, \tag{15.2}$$

where Λ is a new free parameter, the cosmological constant. This new equation admits a static, homogeneous solution for which Λ, the matter density, and the spatial curvature are all positive: the "Einstein static universe." The need for such a universe was soon swept away by improved astronomical observations, and the cosmological constant acquired a somewhat compromised reputation.

Later, particle physicists began to contemplate the possibility of an energy density inherent in the vacuum (defined as the state of lowest attainable energy). If the vacuum is to look Lorentz-invariant to a local observer, its energy-momentum tensor must take on the unique form

$$T_{\mu\nu}^{\text{vac}} = -\rho_{\text{vac}} g_{\mu\nu} \,, \tag{15.3}$$

where ρ_{vac} is a constant vacuum energy density. Such an energy is associated with an isotropic pressure

$$p_{\text{vac}} = -\rho_{\text{vac}} \,. \tag{15.4}$$

Comparing this kind of energy-momentum tensor to the appearance of the cosmological constant in (15.2), we find that they are formally equivalent, as can be seen by moving the $\Lambda g_{\mu\nu}$ term in (15.2) to the right-hand side and setting

$$\rho_{\text{vac}} = \rho_\Lambda \equiv \frac{\Lambda}{8\pi G} \,. \tag{15.5}$$

This equivalence is the origin of the identification of the cosmological constant with the energy of the vacuum.

From either side of Einstein's equation, the cosmological constant Λ is a completely free parameter. It has dimensions of [length]$^{-2}$ (while the energy density ρ_Λ has units [energy/volume]), and hence defines a scale, while general relativity is otherwise scale-free. Indeed, from purely classical considerations, we cannot even say whether a specific value of Λ is "large" or "small"; it is simply a constant of nature we should go out and determine through experiment.

15.2.2 Quantum Zeropoint Energy

The introduction of quantum mechanics changes this story somewhat. For one thing, Planck's constant allows us to define a gravitational length scale, the reduced Planck length

$$L_{\mathrm{P}} = (8\pi G)^{1/2} \sim 10^{-32} \text{ cm} , \qquad (15.6)$$

as well as the reduced Planck mass

$$M_{\mathrm{P}} = \left(\frac{1}{8\pi G}\right)^{1/2} \sim 10^{18} \text{ GeV} , \qquad (15.7)$$

where "reduced" means that we have included the 8π's where they really should be. (Note that, with $\hbar = 1$ and $c = 1$, we have $L = T = M^{-1} = E^{-1}$, where L represents a length scale, T a time interval, M a mass scale, and E an energy.) Hence, there is a natural expectation for the scale of the cosmological constant, namely

$$\Lambda^{(\text{guess})} \sim L_{\mathrm{P}}^{-2} , \qquad (15.8)$$

or, phrased as an energy density,

$$\rho_{\text{vac}}^{(\text{guess})} \sim M_{\mathrm{P}}^4 \sim (10^{18} \text{ GeV})^4 \sim 10^{112} \text{ erg cm}^{-3} . \qquad (15.9)$$

We can partially justify this guess by thinking about quantum fluctuations in the vacuum. At all energies probed by experiment to date, the world is accurately described as a set of quantum fields (at higher energies it may become strings or something else). If we take the Fourier transform of a free quantum field, each mode of fixed wavelength behaves like a simple harmonic oscillator. ("Free" means "noninteracting"; for our purposes this is a very good approximation.) As we know from elementary quantum mechanics, the ground-state or zeropoint energy of an harmonic oscillator with potential $V(x) = \frac{1}{2}\omega^2 x^2$ is $E_0 = \frac{1}{2}\hbar\omega$. Thus, each mode of a quantum field contributes to the vacuum energy, and the net result should be an integral over all of the modes. Unfortunately this integral diverges, so the vacuum energy appears to be infinite. However, the infinity arises from the contribution of modes with very small wavelengths; perhaps it was a mistake to include such modes, since we do not really know what might happen at such scales. To account for our ignorance, we could introduce a cutoff energy, above which we ignore any potential contributions, and hope that a more complete theory will eventually provide a physical justification for doing so. If this cutoff is at the Planck scale, we recover the estimate (15.9).

The strategy of decomposing a free field into individual modes and assigning a zeropoint energy to each one really only makes sense in a flat spacetime background. In curved space-time we can still "renormalize" the vacuum energy, relating the classical parameter to the quantum value by an infinite constant. After renormalization, the vacuum energy is completely arbitrary, just as it was in the original classical theory. But when we use general relativity we are really using an effective field theory to describe a certain limit of quantum gravity. In the context of effective field theory, if a parameter has dimensions $[\text{mass}]^n$, we expect the corresponding mass parameter to be driven up to the scale at which the effective description breaks down. Hence, if we believe classical general relativity up to the Planck scale, we would expect the vacuum energy to be given by our original guess (15.9).

However, we believe we have now measured the vacuum energy through a combination of Type Ia supernovae (Riess et al. 1998; Perlmutter et al. 1999; Knop et al. 2003; Tonry et al. 2003), microwave background anisotropies (Spergel et al. 2003), and dynamical matter measurements (Verde et al. 2002), to reveal

$$\rho_{\text{vac}}^{(\text{obs})} \sim 10^{-8} \text{ erg cm}^{-3} \sim (10^{-3} \text{ eV})^4 \,, \tag{15.10}$$

or

$$\rho_{\text{vac}}^{(\text{obs})} \sim 10^{-120} \rho_{\text{vac}}^{(\text{guess})} \,. \tag{15.11}$$

For reviews, see Sahni & Starobinski (2000), Carroll (2001), or Peebles & Ratra (2003).

Clearly, our guess was not very good. This is the famous 120-orders-of-magnitude discrepancy that makes the cosmological constant problem such a glaring embarrassment. Of course, it is somewhat unfair to emphasize the factor of 10^{120}, which depends on the fact that energy density has units of [energy]^4. We can express the vacuum energy in terms of a mass scale,

$$\rho_{\text{vac}} = M_{\text{vac}}^4 \,, \tag{15.12}$$

so our observational result is

$$M_{\text{vac}}^{(\text{obs})} \sim 10^{-3} \text{ eV} \,. \tag{15.13}$$

The discrepancy is thus

$$M_{\text{vac}}^{(\text{obs})} \sim 10^{-30} M_{\text{vac}}^{(\text{guess})} \,. \tag{15.14}$$

We should think of the cosmological constant problem as a discrepancy of 30 orders of magnitude in energy scale.

15.2.3   The Coincidence Scandal

The third issue mentioned above is the coincidence between the observed vacuum energy (15.11) and the current matter density. To understand this, we briefly review the dynamics of an expanding Robertson-Walker spacetime. The evolution of a homogeneous and isotropic universe is governed by the Friedmann equation,

$$H^2 = \frac{8\pi G}{3}\rho - \frac{\kappa}{a^2} \,, \tag{15.15}$$

where $a(t)$ is the scale factor, $H = \dot{a}/a$ is the Hubble parameter, ρ is the energy density, and κ is the spatial curvature parameter. The energy density is a sum of different components, $\rho = \sum_i \rho_i$, which will in general evolve differently as the universe expands. For matter (nonrelativistic particles) the energy density goes as $\rho_M \propto a^{-3}$, as the number density is diluted with the expansion of the universe. For radiation the energy density goes as $\rho_R \propto a^{-4}$, since each particle loses energy as it redshifts in addition to the decrease in number density. Vacuum energy, meanwhile, is constant throughout spacetime, so that $\rho_\Lambda \propto a^0$.

It is convenient to characterize the energy density of each component by its density parameter

$$\Omega_i = \frac{\rho_i}{\rho_c} \,, \tag{15.16}$$

where the critical density

$$\rho_c = \frac{3H^2}{8\pi G} \tag{15.17}$$

is that required to make the spatial geometry of the universe be flat ($\kappa = 0$). The "best-fit universe" or "concordance" model implied by numerous observations includes radiation, matter, and vacuum energy, with

$$
\begin{aligned}
\Omega_{R0} &\approx 5 \times 10^{-5} \\
\Omega_{M0} &\approx 0.3 \\
\Omega_{\Lambda 0} &\approx 0.7 \, ,
\end{aligned}
\tag{15.18}
$$

together implying a flat universe. We see that the densities in matter and vacuum are of the same order of magnitude.* But the ratio of these quantities changes rapidly as the universe expands:

$$\frac{\Omega_\Lambda}{\Omega_M} = \frac{\rho_\Lambda}{\rho_M} \propto a^3 \, . \tag{15.19}$$

As a consequence, at early times the vacuum energy was negligible in comparison to matter and radiation, while at late times matter and radiation are negligible. There is only a brief epoch of the universe's history during which it would be possible to witness the transition from domination by one type of component to another. This is illustrated in Figure 15.1, in which the various density parameters Ω_i are plotted as a function of the scale factor. At early times Ω_R is close to unity; the matter-radiation transition happens relatively gradually, while the matter-vacuum transition happens quite rapidly.

How finely tuned is it that we exist in the era when vacuum and matter are comparable? Between the Planck time and now, the universe has expanded by a factor of approximately 10^{32}. To be fair, we should consider an interval of logarithmic expansion that is centered around the present time; this would describe a total expansion by a factor of 10^{64}. If we take the transitional period between matter and vacuum to include the time from $\Omega_\Lambda/\Omega_M = 0.1$ to $\Omega_\Lambda/\Omega_M = 10$, the universe expands by a factor of $100^{1/3} \approx 10^{0.67}$. Thus, there is an approximately 1% chance that an observer living in a randomly selected logarithmic expansion interval in the history of our universe would be lucky enough to have Ω_M and Ω_Λ be the same order of magnitude. Everyone will have his own favorite way of quantifying such unnaturalness, but the calculation here gives some idea of the fine-tuning involved; it is substantial, but not completely ridiculous.

As we will discuss below, there is room to imagine that we are actually not observing the effects of an ordinary cosmological constant, but perhaps a dark energy source that varies gradually as the universe expands, or even a breakdown of general relativity on large scales. By itself, however, making dark energy dynamical does not offer a solution to the coincidence scandal; purely on the basis of observations, it seems clear that the universe has begun to accelerate recently, which implies a scale at which something new is kicking in. In particular, it is fruitless to try to explain the matter/dark energy coincidence by invoking mechanisms that make the dark energy density time dependent in such a way as to *always* be proportional to that in matter. Such a scenario would either imply that the dark energy would

---

* Of course, the "matter" contribution consists both of ordinary baryonic matter and nonbaryonic dark matter, with $\Omega_b \approx 0.04$ and $\Omega_{DM} \approx 0.25$. The similarity between these apparently independent quantities is another coincidence problem, but at least one that is independent of time; we have nothing to say about it here.

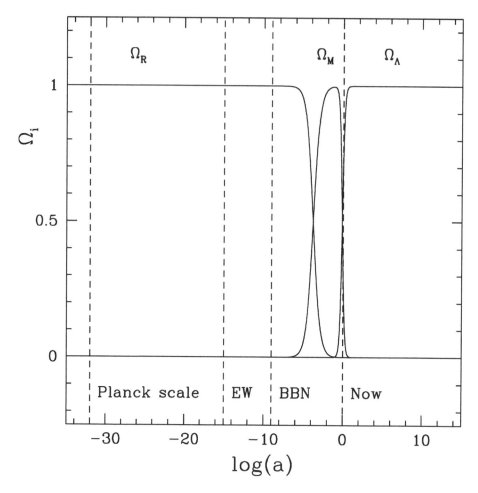

Fig. 15.1. Density parameters Ω_i for radiation (R), matter (M), and vacuum (Λ), as a function of the scale factor a, in a universe with $\Omega_{\Lambda 0} = 0.7$, $\Omega_{M0} = 0.3$, $\Omega_{R0} = 5 \times 10^{-5}$. Scale factors corresponding to the Planck era, electroweak symmetry breaking (EW), and Big Bang nucleosynthesis (BBN) are indicated, as well as the present day.

redshift away as $\rho_{dark} \propto a^{-3}$, which from (15.15) would lead to a nonaccelerating universe, or require departures from conventional general relativity of the type that (as discussed below) are excluded by other measurements.

15.3   What Might be Going on?

Observations have led us to a picture of the universe that differs dramatically from what we might have expected. In this section we discuss possible ways to come to terms with this situation; the approaches we consider include both attempts to explain a small but nonzero vacuum energy, and more dramatic ideas that move beyond a simple cosmological constant. We certainly are not close to settling on a favored explanation, neither for the low

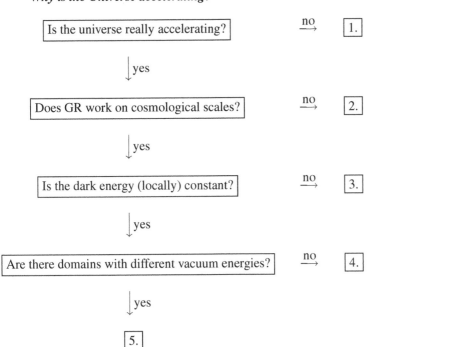

Fig. 15.2. A flowchart classifying reasons why the universe might be accelerating. The possibilities include: 1. misinterpretation of the data; 2. breakdown of general relativity; 3. dynamical dark energy; 4. unique vacuum energy; 5. environmental selection.

value of the vacuum energy nor the recent onset of universal acceleration, but we can try to categorize the different types of conceivable scenarios.

The flowchart portrayed in Figure 15.2 represents a classification of scenarios to explain our observations. Depending on the answers to various questions, we have the following possibilities to explain why the universe appears to be accelerating:

(1) Misinterpretation of the data.
(2) Breakdown of general relativity.
(3) Dynamical dark energy.
(4) Unique vacuum energy.
(5) Environmental selection.

Let us examine each possibility in turn.

15.3.1 *Are We Misinterpreting the Data?*

After the original supernova results (Riess et al. 1998; Perlmutter et al. 1999) were announced in 1998, cosmologists converted rather quickly from skepticism about universal acceleration to a tentative acceptance, which has grown substantially stronger with time. The primary reason for this sudden conversion has been the convergence of several complementary lines of evidence in favor of a concordance model; foremost among the relevant observations are the anisotropy spectrum of the cosmic microwave background (Spergel et al. 2003) and the power spectrum of large-scale structure (Verde et al. 2002), but a number of other methods have yielded consistent answers.

Nevertheless, it remains conceivable that we have dramatically misinterpreted the data, and the apparent agreement of an $\Omega_\Lambda = 0.7$, $\Omega_M = 0.3$ cosmology with a variety of observations is masking the true situation. For example, the supernova observations rely on the nature of Type Ia supernovae as "standardizable candles," an empirical fact about low-redshift supernovae that could somehow fail at high redshifts (although numerous consistency checks have confirmed basic similarities between supernovae at all redshifts). Given the many other observations, this failure would not be enough to invalidate our belief in an accelerating universe; however, we could further imagine that these other methods are conspiring to point to the wrong conclusion. This point of view has been taken by Blanchard et al. (2004), who argue that a flat matter-dominated ($\Omega_M = 1$) universe remains consistent with the data. To maintain this idea, it is necessary to discard the supernova results, to imagine that the Hubble constant is approximately 46 km s^{-1} Mpc^{-1} (in contrast to the Key Project determination of 70 ± 7 km s^{-1} Mpc^{-1}; Freedman et al. 2001), to interpret data on clusters and large-scale structure in a way consistent with $\Omega_M = 1$, to relax the conventional assumption that the power spectrum of density fluctuations can be modeled as a single power law, and to introduce some source beyond ordinary cold dark matter (such as massive neutrinos) to suppress power on small scales. To most workers in the field this conspiracy of effects seems (even) more unlikely than an accelerating universe.

A yet more drastic route is to imagine that our interpretation of the observations has been skewed by the usual assumption of an isotropic universe. It has been argued (Linde, Linde, & Mezhlumian 1995) that some versions of the anthropic principle in an eternally inflating universe lead to a prediction that most galaxies on a spacelike hypersurface are actually at the center of spherically symmetric domains with radially dependent density distributions; such a configuration could skew the distance-redshift relation at large distances even without dark energy. This picture relies heavily on a choice of measure in determining what "most" galaxies are like, an issue for which there is no obvious correct choice.

The lengths to which it seems necessary to go in order to avoid concluding that the universe is accelerating is a strong argument in favor of the concordance model.

15.3.2   *Is General Relativity Breaking Down?*

If we believe that we live in a universe that is homogeneous, isotropic, and accelerating, general relativity (GR) is unambiguous about the need for some sort of dark energy source. GR has been fantastically successful in passing classic experimental tests in the solar system, as well as at predicting the amount of gravitational radiation emitted from the binary pulsar (Will 2001). Nevertheless, the possibility remains open that gravitation might deviate from conventional GR on scales corresponding to the radius of the entire universe. For our present purposes, such deviations may either be relevant to the cosmological constant problem, or to the dark energy puzzle.

The idea behind modifying gravity to address the cosmological constant problem is to somehow allow for the vacuum energy to be large, but yet not lead to an appreciable spacetime curvature (as manifested in a rapidly expanding universe). Of course, we still need to allow ordinary matter to warp spacetime, so there has to be something special about vacuum energy. One special thing is that vacuum energy comes with a negative pressure $p_{\rm vac} = -\rho_{\rm vac}$, as in (15.4). We might therefore imagine a theory that gave rise to a modified version of the Friedmann equation, of the form

$$H^2 \sim \rho + p \,. \tag{15.20}$$

With such an equation, ordinary matter (for which p vanishes) leads to conventional expansion, while vacuum energy decouples entirely. Such a theory has been studied (Carroll & Mersini 2001), and may even arise in "self-tuning" models of extra dimensions (Arkani-Hamed et al. 2000; Kachru, Schulz, & Silverstein 2000). Unfortunately, close examination of self-tuning models reveals that there is a hidden fine-tuning, expressed as a boundary condition chosen at a naked singularity in the extra dimension. Furthermore, any alternative to the conventional Friedmann equation is also constrained by observations: any alternative must predict the right abundances of light elements from Big Bang nucleosynthesis (BBN; see Burles, Nollett, & Turner 2001), the correct evolution of a sensible spectrum of primordial density fluctuations into the observed spectrum of temperature anisotropies in the cosmic microwave background (CMB) and the power spectrum of large-scale structure (Tegmark 2002; Lue, Scoccimarro, & Starkman 2003; Zahn & Zaldarriaga 2003), and that the age of the universe is approximately 13 billion years. The most straightforward test comes from BBN (Carroll & Kaplinghat 2002; Masso & Rota 2003), since the light-element abundances depend on the expansion rate during a relatively brief period (rather than on the behavior of perturbations, or an integral of the expansion rate over a long period). Studies of BBN in alternate cosmologies indicate that it is possible for modifications of GR to remain consistent with observations, but only for a very narrow set of possibilities. It seems likely that the success of conventional BBN, including its agreement with the baryon density as determined by CMB fluctuations (Spergel et al. 2003), is not a misleading accident, but rather an indication that GR provides an accurate description of cosmology when the universe was of the order of one minute old. The idea of modifying GR to solve the cosmological constant problem is not completely dead, but is evidently not promising.

Rather than trying to solve the cosmological constant problem, we can put aside the issue of why the magnitude of the vacuum energy is small and focus instead on whether the current period of acceleration can be traced to a modification of GR. A necessary feature of any such attempt is to include a new scale in the theory, since acceleration has only begun relatively recently.* From a purely phenomenological point of view we can imagine modifying the Friedmann equation (15.15) so that acceleration kicks in when either the energy density approaches a certain value ρ_*,

$$H^2 = \frac{8\pi G}{3} \left[\rho + \left(\frac{\rho}{\rho_*} \right)^\alpha \right] \,, \tag{15.21}$$

or when the Hubble parameter approaches a certain value H_*,

---

* One way of characterizing this scale is in terms of the Hubble parameter when the universe starts accelerating, $H_0 \sim 10^{-18}$ s^{-1}. It is interesting in this context to recall the coincidence pointed out by Milgrom (1983), that dark *matter* only becomes important in galaxies when the acceleration due to gravity dips below a fixed value, $a_0/c \leq 10^{-18}$ s^{-1}. Milgrom himself has suggested that the explanation for this feature of galactic dynamics can be explained by replacing dark matter by a modified dynamics, and it is irresistible to speculate that both dark matter and dark energy could be replaced by a single (as yet undiscovered) modified theory of gravity. However, hope for this possibility seems to be gradually becoming more difficult to maintain, as different methods indicate the existence of gravitational forces that point in directions other than where ordinary matter is (Van Waerbeke et al. 2000; Dalal & Kochanek 2002; Kneib et al. 2003)—a phenomenon that is easy to explain with dark matter, but difficult with modified gravity—and explanations are offered for $a_0/c \sim H_0$ within conventional cold dark matter (Scott et al. 2001; Kaplinghat & Turner 2002).

$$H^2 + \left(\frac{H}{H_*}\right)^\beta = \frac{8\pi G}{3}\rho \, . \tag{15.22}$$

The former idea has been suggested by Freese & Lewis (2002), the latter by Dvali & Turner (2003); in both cases we can fit the data for appropriate choices of the new parameters. It is possible that equations of this type arise in brane-world models with large extra spatial dimensions; it is less clear whether the appropriate parameters can be derived. An even more dramatic mechanism also takes advantage of extra dimensions, but allows for separate gravitational dynamics on and off of our brane; in this case gravity can be four-dimensional *below* a certain length scale (which would obviously have to be very large), and appear higher-dimensional at large distances (Dvali, Gabadadze, & Porrati 2000; Arkani-Hamed et al. 2002; Deffayet, Dvali, & Gabadadze 2002). These scenarios can also make the universe accelerate at late times, and may even lead to testable deviations from GR in the solar system (Dvali, Gruzinov, & Zaldarriaga 2003).

As an alternative to extra dimensions, we may look for an ordinary four-dimensional modification of GR. This would be unusual behavior, as we are used to thinking of effective field theories as breaking down at high energies and small length scales, but being completely reliable in the opposite regime. Nevertheless, it is worth exploring whether a simple phenomenological model can easily accommodate the data. Einstein's equation can be derived by minimizing an action given by the spacetime integral of the curvature scalar R,

$$S = \int d^4x \sqrt{|g|}\, R \, . \tag{15.23}$$

A simple way to modify the theory when the curvature becomes very small (at late times in the universe) is to simply add a piece proportional to $1/R$,

$$S = \int d^4x \sqrt{|g|} \left(R - \frac{\mu^4}{R}\right) , \tag{15.24}$$

where μ is a parameter with dimensions of mass (Carroll et al. 2003a). It is straightforward to show that this theory admits accelerating solutions; unfortunately, it also brings to life a new scalar degree of freedom, which ruins the success of GR in the solar system (Chiba 2003). Investigations are still ongoing to see whether a simple modification of this idea could explain the acceleration of the universe while remaining consistent with experimental tests; in the meantime, the difficulty in finding a simple extension of GR that does away with the cosmological constant provides yet more support for the standard scenario.

15.3.3 Is Dark Energy Dynamical?

If general relativity is correct, cosmic acceleration implies there must be a dark energy density that diminishes relatively slowly as the universe expands. This can be seen directly from the Friedmann equation (15.15), which implies

$$\dot{a}^2 \propto a^2\rho + \text{constant} \, . \tag{15.25}$$

From this relation, it is clear that the only way to get acceleration ($\dot{a}$ increasing) in an expanding universe is if ρ falls off more slowly than a^{-2}; neither matter ($\rho_M \propto a^{-3}$) nor radiation ($\rho_R \propto a^{-4}$) will do the trick. Vacuum energy is, of course, strictly constant; but the data are consistent with smoothly distributed sources of dark energy that vary slowly with time.

There are good reasons to consider dynamical dark energy as an alternative to an honest cosmological constant. First, a dynamical energy density can be evolving slowly to zero, allowing for a solution to the cosmological constant problem that makes the ultimate vacuum energy vanish exactly. Second, it poses an interesting and challenging observational problem to study the evolution of the dark energy, from which we might learn something about the underlying physical mechanism. Perhaps most intriguingly, allowing the dark energy to evolve opens the possibility of finding a dynamical solution to the coincidence problem, if the dynamics are such as to trigger a recent takeover by the dark energy (independently of, or at least for a wide range of, the parameters in the theory). To date this hope has not quite been met, but dynamical mechanisms at least allow for the possibility (unlike a true cosmological constant).

The simplest possibility along these lines involves the same kind of source typically invoked in models of inflation in the very early universe: a scalar field ϕ rolling slowly in a potential, sometimes known as "quintessence" (Frieman, Hill, & Watkins 1992; Frieman et al. 1995; Caldwell, Dave, & Steinhardt 1998; Peebles & Ratra 1998; Ratra & Peebles 1998; Wetterich 1998; Huey et al. 1999). The energy density of a scalar field is a sum of kinetic, gradient, and potential energies,

$$\rho_\phi = \frac{1}{2}\dot{\phi}^2 + \frac{1}{2}(\nabla\phi)^2 + V(\phi) . \tag{15.26}$$

For a homogeneous field ($\nabla\phi \approx 0$), the equation of motion in an expanding universe is

$$\ddot{\phi} + 3H\dot{\phi} + \frac{dV}{d\phi} = 0 . \tag{15.27}$$

If the slope of the potential V is quite flat, we will have solutions for which ϕ is nearly constant throughout space and only evolving very gradually with time; the energy density in such a configuration is

$$\rho_\phi \approx V(\phi) \approx \text{constant} . \tag{15.28}$$

Thus, a slowly rolling scalar field is an appropriate candidate for dark energy.

However, introducing dynamics opens up the possibility of introducing new problems, the form and severity of which will depend on the specific kind of model being considered. Most quintessence models feature scalar fields ϕ with masses of order the current Hubble scale,

$$m_\phi \sim H_0 \sim 10^{-33} \text{ eV} . \tag{15.29}$$

(Fields with larger masses would typically have already rolled to the minimum of their potentials.) In quantum field theory, light scalar fields are unnatural; renormalization effects tend to drive scalar masses up to the scale of new physics. The well-known hierarchy problem of particle physics amounts to asking why the Higgs mass, thought to be of order 10^{11} eV, should be so much smaller than the grand unification/Planck scale, 10^{25}–10^{27} eV. Masses of 10^{-33} eV are correspondingly harder to understand.

Nevertheless, this apparent fine-tuning might be worth the price, if we were somehow able to explain the coincidence problem. To date, many investigations have considered scalar fields with potentials that asymptote gradually to zero, of the form $e^{1/\phi}$ or $1/\phi$. These can have cosmologically interesting properties, including "tracking" behavior that makes the current energy density largely independent of the initial conditions (Zlatev, Wang, &

Steinhardt 1999). They do not, however, provide a solution to the coincidence problem, as the era in which the scalar field begins to dominate is still set by finely tuned parameters in the theory. One way to address the coincidence problem is to take advantage of the fact that matter/radiation equality was a relatively recent occurrence (at least on a logarithmic scale); if a scalar field has dynamics that are sensitive to the difference between matter- and radiation-dominated universes, we might hope that its energy density becomes constant only after matter/radiation equality. An approach that takes this route is k-essence (Armendariz-Picon, Mukhanov, & Steinhardt 2000), which modifies the form of the kinetic energy for the scalar field. Instead of a conventional kinetic energy $K = \frac{1}{2}(\dot{\phi})^2$, in k-essence we posit a form

$$K = f(\phi)g(\dot{\phi}^2) ,$$ (15.30)

where f and g are functions specified by the model. For certain choices of these functions, the k-essence field naturally tracks the evolution of the total radiation energy density during radiation domination, but switches to being almost constant once matter begins to dominate. Unfortunately, it seems necessary to choose a finely tuned kinetic term to get the desired behavior (Malquarti, Copeland, & Liddle 2003).

An alternative possibility is that there is nothing special about the present era; rather, acceleration is just something that happens from time to time. This can be accomplished by oscillating dark energy (Dodelson, Kaplinghat, & Stewart 2000). In these models the potential takes the form of a decaying exponential (which by itself would give scaling behavior, so that the dark energy remained proportional to the background density) with small perturbations superimposed:

$$V(\phi) = e^{-\phi}[1 + \alpha\cos(\phi)] .$$ (15.31)

On average, the dark energy in such a model will track that of the dominant matter/radiation component; however, there will be gradual oscillations from a negligible density to a dominant density and back, on a time scale set by the Hubble parameter, leading to occasional periods of acceleration. In the previous section we mentioned the success of the conventional picture in describing primordial nucleosynthesis (when the scale factor was $a_{BBN} \sim 10^{-9}$) and temperature fluctuations imprinted on the CMB at recombination ($a_{CMB} \sim 10^{-3}$), which implies that the oscillating scalar must have had a negligible density during those periods; but explicit models are able to accommodate this constraint. Unfortunately, in neither the k-essence models nor the oscillating models do we have a compelling particle physics motivation for the chosen dynamics, and in both cases the behavior still depends sensitively on the precise form of parameters and interactions chosen. Nevertheless, these theories stand as interesting attempts to address the coincidence problem by dynamical means.

15.3.4 *Did We Just Get Lucky?*

By far the most straightforward explanation for the observed acceleration of the universe is an absolutely constant vacuum energy, or cosmological constant. Even in this case we can distinguish between two very different scenarios: one in which the vacuum energy is some fixed number that as yet we simply do not know how to calculate, and an alternative in which there are many distinct domains in the universe, with different values of the vacuum energy in each. In this section we concentrate on the first possibility. Note that such a scenario requires that we essentially give up on finding a dynamical resolution to the coincidence scandal; instead, the vacuum energy is fixed once and for all, and we are simply fortunate that it takes on a sufficiently gentle value that life has enough time and space to exist.

To date, there are not any especially promising approaches to calculating the vacuum energy and getting the right answer; it is nevertheless instructive to consider the example of supersymmetry, which relates to the cosmological constant problem in an interesting way. Supersymmetry posits that for each fermionic degree of freedom there is a matching bosonic degree of freedom, and *vice versa*. By "matching" we mean, for example, that the spin-1/2 electron must be accompanied by a spin-0 "selectron" with the same mass and charge. The good news is that, while bosonic fields contribute a positive vacuum energy, for fermions the contribution is negative. Hence, if degrees of freedom exactly match, the net vacuum energy sums to zero. Supersymmetry is thus an example of a theory, other than gravity, where the absolute zeropoint of energy is a meaningful concept. (This can be traced to the fact that supersymmetry is a spacetime symmetry, relating particles of different spins.)

We do not, however, live in a supersymmetric state; there is no selectron with the same mass and charge as an electron, or we would have noticed it long ago. If supersymmetry exists in nature, it must be broken at some scale M_{SUSY}. In a theory with broken supersymmetry, the vacuum energy is not expected to vanish, but to be of order

$$M_{\text{vac}} \sim M_{\text{SUSY}} , \qquad \text{(theory)} \qquad (15.32)$$

with $\rho_{\text{vac}} = M_{\text{vac}}^4$. What should M_{SUSY} be? One nice feature of supersymmetry is that it helps us understand the hierarchy problem—why the scale of electroweak symmetry breaking is so much smaller than the scales of quantum gravity or grand unification. For supersymmetry to be relevant to the hierarchy problem, we need the supersymmetry-breaking scale to be just above the electroweak scale, or

$$M_{\text{SUSY}} \sim 10^3 \text{ GeV} . \qquad (15.33)$$

In fact, this is very close to the experimental bound, and there is good reason to believe that supersymmetry will be discovered soon at Fermilab or CERN, if it is connected to electroweak physics.

Unfortunately, we are left with a sizable discrepancy between theory and observation:

$$M_{\text{vac}}^{(\text{obs})} \sim 10^{-15} M_{\text{SUSY}} . \qquad \text{(experiment)} \qquad (15.34)$$

Compared to (15.14), we find that supersymmetry has, in some sense, solved the problem halfway (on a logarithmic scale). This is encouraging, as it at least represents a step in the right direction. Unfortunately, it is ultimately discouraging, since (15.14) was simply a guess, while (15.34) is actually a reliable result in this context; supersymmetry renders the vacuum energy finite and calculable, but the answer is still far away from what we need. (Subtleties in supergravity and string theory allow us to add a negative contribution to the vacuum energy, with which we could conceivably tune the answer to zero or some other small number; but there is no reason for this tuning to actually happen.)

But perhaps there is something deep about supersymmetry that we do not understand, and our estimate $M_{\text{vac}} \sim M_{\text{SUSY}}$ is simply incorrect. What if instead the correct formula were

$$M_{\text{vac}} \sim \left(\frac{M_{\text{SUSY}}}{M_{\text{P}}} \right) M_{\text{SUSY}} ? \qquad (15.35)$$

In other words, we are guessing that the supersymmetry-breaking scale is actually the geometric mean of the vacuum scale and the Planck scale. Because M_P is 15 orders of magnitude larger than M_{SUSY}, and M_{SUSY} is 15 orders of magnitude larger than M_{vac}, this guess gives us the correct answer! Unfortunately this is simply optimistic numerology; there is no theory that actually yields this answer (although there are speculations in this direction; Banks 2003). Still, the simplicity with which we can write down the formula allows us to dream that an improved understanding of supersymmetry might eventually yield the correct result.

Besides supersymmetry, we do know of other phenomena that may in principle affect our understanding of vacuum energy. One example is the idea of large extra dimensions of space, which become possible if the particles of the Standard Model are confined to a three-dimensional brane (Arkani-Hamed, Dimopoulos, & Dvali 1998; Randall & Sundrum 1999). In this case gravity is not simply described by four-dimensional general relativity, as alluded to in the previous section. Furthermore, current experimental bounds on simple extra-dimensional models limit the scale characterizing the extra dimensions to less than 10^{-2} cm, which corresponds to an energy of approximately 10^{-3} eV; this is coincidentally the same as the vacuum-energy scale (15.10). As before, nobody has a solid reason why these two scales should be related, but it is worth searching for one. The fact that we are forced to take such slim hopes seriously is a measure of how difficult the cosmological constant problem really is.

15.3.5   *Are We Witnessing Environmental Selection?*

If the vacuum energy can in principle be calculated in terms of other measurable quantities, then we clearly do not yet know how to calculate it. Alternatively, however, it may be that the vacuum energy is not a fundamental quantity, but simply a feature of our local environment. We do not turn to fundamental theory for an explanation of the average temperature of the Earth's atmosphere, nor are we surprised that this temperature is noticeably larger than in most places in the universe; perhaps the cosmological constant is on the same footing.

To make this idea work, we need to imagine that there are many different regions of the universe in which the vacuum energy takes on different values; then we would expect to find ourselves in a region that was hospitable to our own existence. Although most humans do not think of the vacuum energy as playing any role in their lives, a substantially larger value than we presently observe would either have led to a rapid recollapse of the universe (if ρ_{vac} were negative) or an inability to form galaxies (if ρ_{vac} were positive). Depending on the distribution of possible values of ρ_{vac}, one can argue that the observed value is in excellent agreement with what we should expect (Efstathiou 1995; Vilenkin 1995; Martel, Shapiro, & Weinberg 1998; Garriga & Vilenkin 2000, 2003).

The idea of understanding the vacuum energy as a consequence of environmental selection often goes under the name of the "anthropic principle," and has an unsavory reputation in some circles. There are many bad reasons to be skeptical of this approach, and at least one good reason. The bad reasons generally center around the idea that it is somehow an abrogation of our scientific responsibilities to give up on calculating something as fundamental as the vacuum energy, or that the existence of many unseen domains in the universe is a metaphysical construct without any testable consequences, and hence unscientific. The problem with these objections is that they say nothing about whether environmental selection actually happens; they are only declarations that we hope it does not happen, or it would be difficult for us to prove once and for all that it does. The good reason to be skeptical is that environmental selection only works under certain special circumstances, and we are far from understanding whether those conditions hold in our universe. In particular, we need to show that there can be a huge number of different domains with slightly different values of the vacuum energy, and that the domains can be big enough that our entire observable universe is a single domain, and that the possible variation of other physical quantities from domain to domain is consistent with what we observe in ours.*

Recent work in string theory has lent some support to the idea that there are a wide variety of possible vacuum states rather than a unique one (Bousso & Polchinski 2000; Giddings, Kachru, & Polchinski 2002; Kachru et al. 2003; Susskind 2003). String theorists have been investigating novel ways to compactify extra dimensions, in which crucial roles are played by branes and gauge fields. By taking different combinations of extra-dimensional geometries, brane configurations, and gauge-field fluxes, it seems plausible that a wide variety of states may be constructed, with different local values of the vacuum energy and other physical parameters. (The set of configurations is sometimes known as the "landscape," and the discrete set of vacuum configurations is unfortunately known as the "discretuum.") An obstacle to understanding these purported solutions is the role of supersymmetry, which is an important part of string theory but needs to be broken to obtain a realistic universe. From the point of view of a four-dimensional observer, the compactifications that have small values of the cosmological constant would appear to be exactly the states alluded to in the previous section, where one begins with a supersymmetric state with a negative vacuum energy, to which supersymmetry breaking adds just the right amount of positive vacuum energy to give a small overall value. The necessary fine-tuning is accomplished simply by imagining that there are many (more than 10^{100}) such states, so that even very unlikely things will sometimes occur. We still have a long way to go before we understand this possibility; in particular, it is not clear that the many states obtained have all the desired properties (Banks, Dine, & Motl 2001; Banks, Dine, & Gorbatov 2003), or even if they are stable enough to last for the age of the universe (Hertog, Horowitz, & Maeda 2003).

Even if such states are allowed, it is necessary to imagine a universe in which a large number of them actually exist in local regions widely separated from each other. As is well known, inflation works to take a small region of space and expand it to a size larger than the observable universe; it is not much of a stretch to imagine that a multitude of different domains may be separately inflated, each with different vacuum energies. Indeed, models of inflation generally tend to be eternal, in the sense that the universe continues to inflate in some regions even after inflation has ended in others (Vilenkin 1983; Linde 1985; Goncharov, Linde, & Mukhanov 1987). Thus, our observable universe may be separated by inflating regions from other "universes" that have landed in different vacuum states; this is precisely what is needed to empower the idea of environmental selection.

Nevertheless, it seems extravagant to imagine a fantastic number of separate regions of the universe, outside the boundary of what we can ever possibly observe, just so that we may understand the value of the vacuum energy in our region. But again, this does not mean it is not true. To decide once and for all will be extremely difficult, and will at the least require a much better understanding of how both string theory (or some alternative) and inflation

---

* For example, if we have a theory that allows for any possible value of the vacuum energy, but insists that the vacuum energy scale be equal to the supersymmetry breaking scale, we have not solved any problems.

operate—an understanding that we will undoubtedly require a great deal of experimental input to achieve.

15.4   Observational Issues

From the above discussion, it is clear that theorists are in desperate need of further input from experiment—in particular, we need to know if the dark energy is constant or dynamical, and if it is dynamical what form it takes. The observational program to test these ideas has been discussed in detail elsewhere (Sahni & Starobinski 2000; Carroll 2001; Peebles & Ratra 2003); here we briefly draw attention to a couple of theoretical issues that can affect the observational strategies.

15.4.1   *Equation-of-state Parameter*

Given that the universe is accelerating, the next immediate question is whether the acceleration is caused by a strictly constant vacuum energy or something else; the obvious place to look is for some time dependence to the dark energy density. In principle any behavior is possible, but it is sensible to choose a simple parameterization that would characterize dark energy evolution in the measurable regime of relatively nearby redshifts (order unity or less). For this purpose it is common to imagine that the dark energy evolves as a power law with the scale factor:

$$\rho_{\text{dark}} \propto a^{-n} . \tag{15.36}$$

Even if ρ_{dark} is not strictly a power law, this ansatz can be a useful characterization of its effective behavior at low redshifts. We can then define an equation-of-state parameter relating the energy density to the pressure,

$$p = w\rho . \tag{15.37}$$

Using the equation of energy-momentum conservation,

$$\dot{\rho} = -3(\rho + p)\frac{\dot{a}}{a} , \tag{15.38}$$

a constant exponent n of (15.36) implies a constant w with

$$n = 3(1 + w) . \tag{15.39}$$

As n varies from 3 (matter) to 0 (cosmological constant), w varies from 0 to -1. Some limits from supernovae and large-scale structure from Melchiorri et al. (2003) are shown in Figure 15.3; see Spergel et al (2003) for limits from *WMAP* observations of the cosmic microwave background, and Tonry et al. (2003) and Knop et al. (2003) for more recent supernova limits. These constraints apply to the Ω_M-w plane, under the assumption that the universe is flat ($\Omega_M + \Omega_{\text{dark}} = 1$). We see that the observationally favored region features $\Omega_M \approx 0.3$ and an honest cosmological constant, $w = -1$. However, there is room for alternatives; one of the most important tasks of observational cosmology will be to reduce the error regions on plots such as this to pin down precise values of these parameters.

It is clear that $w = -1$ is a special value; for $w > -1$ the dark energy density slowly decreases as the universe expands, while for $w < -1$ it would actually be *increasing*. In most conventional models, unsurprisingly, we have $w \geq -1$; this is also required (for sources with positive energy densities) by the energy conditions of general relativity (Garnavich et

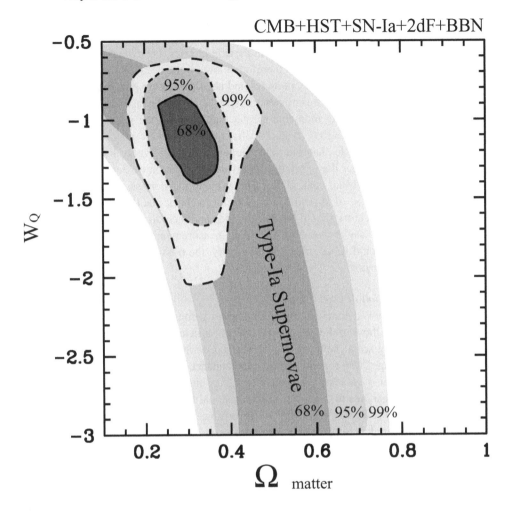

Fig. 15.3. Limits on the equation-of-state parameter w in a flat universe, where $\Omega_M + \Omega_X = 1$. (Adapted from Melchiorri et al. 2003.)

al. 1998). Nevertheless, it is interesting to ask whether we should bother to consider $w < -1$ (Parker & Raval 1999; Sahni & Starobinski 2000; Caldwell 2002; Carroll, Hoffman, & Trodden 2003b). If w is constant in such a model, the universe will expand ever faster until a future singularity is reached, the "Big Rip" (Caldwell, Kamionkowski, & Weinberg 2003); but such behavior is by no means necessary. An explicit model is given by so-called phantom fields (Caldwell 2002), scalar fields with negative kinetic and gradient energy,

$$\rho_\phi = -\frac{1}{2}\dot{\phi}^2 - \frac{1}{2}(\nabla\phi)^2 + V(\phi) , \qquad (15.40)$$

in contrast with the conventional expression (15.26). (A phantom may be thought of as a physical realization of the "ghost" fields used in some calculations in quantum field theory.) A phantom field rolls to the maximum of its potential, rather than the minimum; if there is a maximum with positive potential energy, we will have $w < -1$ while the field is rolling, but it will settle into a state with $w = -1$.

However, such fields are very dangerous in particle physics; the excitations of the phantom will be negative-mass particles, and therefore allow for the decay of empty space into a collection of positive-energy ordinary particles and negative-energy phantoms. Naively the decay rate is infinite, because there is no boundary to the allowed phase space; if we impose a cutoff by hand by disallowing momenta greater than 10^{-3} eV, the vacuum can be stable for the age of the universe (Carroll et al. 2003b). Of course, there may be other ways to get $w < -1$ other than a simple phantom field (Parker & Raval 1999; Dvali & Turner 2003), and there is a lurking danger that a rapidly time-varying equation of state might trick you into thinking that $w < -1$ (Maor et al. 2002). The moral of the story should be that theorists proposing models with $w < -1$ should be very careful to check that their theories are sufficiently stable, while observers should be open-minded and include $w < -1$ in the parameter space they constrain. To say the least, a convincing measurement that the effective value of w were less than -1 would be an important discovery, the possibility of which one would not want to exclude *a priori*.

15.4.2   *Direct Detection of Dark Energy*

If dark energy is dynamical rather than simply a constant, it is able to interact with other fields, including those of the Standard Model of particle physics. For the particular example of an ultra-light scalar field, interactions introduce the possibility of two observable phenomena: long-range "fifth forces" and time dependence of the constants of nature. Even if a dark energy scalar ϕ interacts with ordinary matter only through indirect gravitational-strength couplings, searches for these phenomena should have already enabled us to detect the quintessence field (Carroll 1998; Dvali & Zaldarriaga 2002); to avoid detection, we need to introduce dimensionless suppression factors of order 10^{-5} or less in the coupling constants. On the other hand, there has been some evidence from quasar absorption spectra that the fine-structure constant α was slightly smaller ($\Delta\alpha/\alpha \approx -10^{-5}$) at redshifts $z \approx 0.5 - 3$ (Murphy et al. 2001). On the most optimistic reading, this apparent shift might be direct evidence of a quintessence field; this would place strong constraints on the quintessence potential (Chiba & Kohri 2002). Before such an interpretation is accepted, however, it will be necessary to be certain that all possible sources of systematic error in the quasar measurements are understood, and that models can be constructed that fit the quasar data while remaining consistent with other experimental bounds (Uzan 2003).

More likely, we should work to construct particle physics models of quintessence in which both the mass and the interactions of the scalar field with ordinary matter are naturally suppressed. These requirements are met by Pseudo-Nambu-Goldstone bosons (PNGBs) (Frieman et al. 1992, 1995), which arise in models with approximate global symmetries of the form

$$\phi \rightarrow \phi + \text{constant}. \tag{15.41}$$

Clearly such a symmetry should not be exact, or the potential would be precisely flat; however, even an approximate symmetry can naturally suppress masses and couplings. PNGBs typically arise as the angular degrees of freedom in Mexican-hat potentials that are "tilted" by a small explicitly symmetry breaking, and the PNGB potential takes on a sinusoidal form:

$$V(\phi) = \mu^4 [1 + \cos(\phi)] . \tag{15.42}$$

Fields of this type are ubiquitous in string theory, and it is possible that one of them may have the right properties to be the dark energy (Choi 2000; Kim 2000; Kim & Nilles 2003). Interestingly, while the symmetry (15.41) suppresses most possible interactions with ordinary matter, it leaves open one possibility—a pseudoscalar electromagnetic interaction in which ϕ couples to $\mathbf{E} \cdot \mathbf{B}$. The effect of such an interaction would be to gradually rotate the plane of polarization of light from distant sources (Carroll 1998; Lue, Wang, & Kamionkowski 1999); current limits on such a rotation are not quite sensitive enough to tightly constrain this coupling. It is therefore very plausible that a pseudoscalar quintessence field will be directly detected by improved polarization measurements in the near future.

Even if we manage to avoid detectable interactions between dark energy and ordinary matter, we may still consider the possibility of nontrivial interactions between dark matter and dark energy. Numerous models along these lines have been proposed (Casas, Garcia-Bellido, & Quiros 1992; Wetterich 1995; Anderson & Carroll 1998; Amendola 2000; Bean 2001; for recent work and further references, see Farrar & Peebles 2003; Hoffman 2003). If these two dark components constitute 95% of the universe, the idea that they are separate and noninteracting may simply be a useful starting point. Investigations thus far seem to indicate that some sorts of interactions are possible, but constraints imposed by the cosmic microwave background and large-scale structure are actually able to exclude a wide range of possibilities. It may be that the richness of interaction we observe in the ordinary-matter sector is an exception rather than the rule.

15.5   Conclusions

The acceleration of the universe presents us with mysteries and opportunities. The fact that this behavior is so puzzling is a sign that there is something fundamental we do not understand. We do not even know whether our misunderstanding originates with gravity as described by general relativity, with some source of dynamical or constant dark energy, or with the structure of the universe on ultra-large scales. Regardless of what the answer is, we seem poised to discover something profound about how the universe works.

Acknowledgements. It is a pleasure to thank Wendy Freedman for organizing a stimulating meeting, and participants at the Seven Pines Symposium on "The Concept of the Vacuum in Physics" and the Kavli Institute for Theoretical Physics program on "String Theory and Cosmology" for numerous helpful conversations. This work was supported in part by U.S. Dept. of Energy contract DE-FG02-90ER-40560, National Science Foundation Grant PHY-0114422 (CfCP), and the David and Lucile Packard Foundation.

References

Amendola, L. 2000, Phys. Rev. D, 62, 043511
Anderson, G. W., & Carroll, S. M. 1998, in COSMO-97, First International Workshop on Particle Physics and the Early Universe, ed. L. Roszkowski (New Jersey : World Scientific), 227
Arkani-Hamed, N., Dimopoulos, S., & Dvali, G. 1998, Phys. Lett. B, 429, 263
Arkani-Hamed, N., Dimopoulos, S., Dvali, G., & Gabadadze, G. 2002, hep-th/0209227
Arkani-Hamed, N., Dimopoulos, S., Kaloper, N., & Sundrum, R. 2000, Phys. Lett. B, 480, 193
Armendariz-Picon, C., Mukhanov, V., & Steinhardt, P. J. 2000, Phys. Rev. Lett., 85, 4438
Banks, T. 2003, hep-th/0305206
Banks, T., Dine, M., & Gorbatov, E. 2003, hep-th/0309170
Banks, T., Dine, M., & Motl, L. 2003, JHEP, 0101, 031
Bean, R. 2001, Phys. Rev. D, 64, 123516
Blanchard, A., Douspis, M., Rowan-Robinson, M., & Sarkar, S. 2004, A&A, in press (astro-ph/0304237)

Bousso, R., & Polchinski, J. 2000, JHEP, 0006, 006

Burles, S., Nollett, K. M., & Turner, M. S. 2001, ApJ, 552, L1

Caldwell, R. R. 2002, Phys. Lett. B, 545, 23

Caldwell, R. R., Dave, R., & Steinhardt, P. J. 1998, Phys. Rev. Lett., 80, 1582

Caldwell, R. R., Kamionkowski, M., & Weinberg, N. N. 2003, Phys. Rev. Lett., 91, 71301

Carroll, S. M. 1998, Phys. Rev. Lett., 81, 3067

Carroll, S. M. 2001, Living Reviews in Relativity, 4, 1

Carroll, S. M., Duvvuri, V., Trodden, M., & Turner, M. S. 2003a, astro-ph/0306438

Carroll, S. M., Hoffman, M., & Trodden, M. 2003b, Phys. Rev. D, 68, 23509

Carroll, S. M., & Kaplinghat, M. 2002, Phys. Rev. D, 65, 63507

Carroll, S. M., & Mersini, L. 2001, Phys. Rev. D, 64, 124008

Casas, J. A., Garcia-Bellido, J., & Quiros, M. 1992, Class. Quant. Grav., 9, 1371

Chiba, T. 2003, astro-ph/0307338

Chiba, T., & Kohri, K. 2002, Progress of Theoretical Physics, 107, 631

Choi, K. 2000, Phys. Rev. D, 62, 043509

Dalal, N., & Kochanek, C. S. 2002, ApJ, 572, 25

Deffayet, C., Dvali, G., & Gabadadze, G. 2002, Phys. Rev. D, 65, 44023

Dodelson, S., Kaplinghat, M., & Stewart, E. 2000, Phys. Rev. Lett., 85, 5276

Dvali, G., Gabadadze, G., & Porrati, M. 2000, Phys. Lett. B, 485, 208

Dvali, G., Gruzinov, A., & Zaldarriaga, M. 2003, Phys. Rev. D, 68, 24012

Dvali, G., & Turner, M. S. 2003, astro-ph/0301510

Dvali, G., & Zaldarriaga, M. 2002, Phys. Rev. Lett. , 88, 91303

Efstathiou, G. 1995, MNRAS, 274, L73

Farrar, G. R., & Peebles, P. J. E. 2003, astro-ph/0307316

Freedman, W. L., et al. 2001, ApJ, 553, 47

Freese, K., & Lewis, M. 2002, Phys. Lett. B, 540, 1

Frieman, J. A., Hill, C. T., Stebbins, A., & Waga, I. 1995, Phys. Rev. Lett., 75, 2077

Frieman, J. A., Hill, C. T., & Watkins, R. 1992, Phys. Rev. D, 46, 1226

Garcia-Bellido, J. 1993, Int. J. Mod. Phys., D2, 85

Garnavich, P. M., et al. 1998, ApJ, 509, 74

Garriga, J., & Vilenkin, A. 2000, Phys. Rev. D, 61, 83502

———. 2003, Phys. Rev. D, 67, 43503

Giddings, S. B., Kachru, S., & Polchinski, J. 2002, Phys. Rev. D, 66, 106006

Goncharov, A. S., Linde, A. D., & Mukhanov, V. F. 1987, Int. J. Mod. Phys. A, 2, 561

Hertog, T., Horowitz, G. T., & Maeda, K. 2003, JHEP, 0305, 060

Hoffman, M. 2003, astro-ph/0307350

Huey, G., Wang, L., Dave, R., Caldwell, R. R., & Steinhardt, P. J. 1999, Phys. Rev. D, 59, 63005

Kachru, S., Kallosh, R., Linde, A., & Trivedi, S. P. 2003, Phys. Rev. D, 68, 046005

Kachru, S., Schulz, M. B., & Silverstein, E. 2000, Phys. Rev. D, 62, 045021

Kaplinghat, M., & Turner, M. S. 2002, ApJ, 569, L19

Kim, J. E. 2000, JHEP, 0006, 016

Kim, J. E., & Nilles, H. P. 2003, Phys. Lett. B, 553, 1

Kneib, J.-P., et al. 2003, ApJ, in press (astro-ph/0307299)

Knop, R. A., et al. 2003, ApJ, in press (astro-ph/0309368)

Linde, A. 1986, Phys. Lett. B, 175, 395

Linde, A. D., Linde, D. A., & Mezhlumian, A. 1995, Phys. Lett. B, 345, 203

Lue, A., Scoccimarro, R., & Starkman, G. 2003, astro-ph/0307034

Lue, A., Wang, L.-M., & Kamionkowski, M. 1999, Phys. Rev. Lett., 83, 1506

Malquarti, M., Copeland, E. J., & Liddle, A. R. 2003, Phys. Rev. D, 68, 23512

Maor, I., Brustein, R., McMahon, J., & Steinhardt, P. J. 2002, Phys. Rev. D, 65, 123003

Martel, H., Shapiro, P. R., & Weinberg, S. 1998, ApJ, 492, 29

Masso, E., & Rota, F. 2003, Phys. Rev. D, in press (astro-ph/0302554)

Milgrom, M. 1983, ApJ, 270, 365

Melchiorri, A., Mersini, L., Ödman, C. J., & Trodden, M. 2003, Phys. Rev. D, 68, 43509

Murphy, M. T., Webb, J. K., Flambaum, V. V., Dzuba, V. A., Churchill, C. W., Prochaska, J. X., Barrow, J. D., & Wolfe, A. M. 2001, MNRAS, 327, 1208

Parker, L., & Raval, A. 1999, Phys. Rev. D, 60, 063512

Peebles, P. J., & Ratra, B. 1998, ApJ, 325, L17

———. 2003, Rev. Mod. Phys., 75, 559

Perlmutter, S., et al. 1999, ApJ, 517, 565

Randall, L., & Sundrum, R. 1999, Phys. Rev. Lett., 83, 4690

Ratra. B., & Peebles, P. J. 1988 Phys. Rev. D, 37, 3406

Riess, A. G., et al. 1998, AJ, 116, 1009

Sahni, V., & Starobinsky, A. 2000, Int. J. Mod. Phys. D, 9, 373

Scott, D., White, M., Cohn, J. D., & Pierpaoli, E. 2001, astro-ph/0104435

Spergel, D. N., et al. 2003, ApJS, 148, 175

Susskind, L. 2003, hep-th/0302219

Tegmark, M. 2002, Phys. Rev. D, 66, 103507

Tonry, J. L., et al. 2003, ApJ, 594, 1

Uzan, J. 2003, Rev. Mod. Phys., 75, 403

Van Waerbeke, L., et al. 2000, A&A, 358, 30

Verde, L., et al. 2002, MNRAS, 335, 432

Vilenkin, A. 1983, Phys. Rev. D, 27, 2848

———. 1995, Phys. Rev. Lett., 74, 846

Weinberg, S. 1989, Rev. Mod. Phys., 61, 1

Wetterich, C. 1995, A&A, 301, 321

———. 1998, Nucl. Phys. B, 302, 668

Will, C. M. 2001, Living Reviews in Relativity, 4, 4

Zahn, O., & Zaldarriaga, M. 2003, Phys. Rev. D, 67, 0630002

Zlatev, I, Wang, L., & Steinhardt, P. J. 1999, Phys. Rev. Lett., 82, 896

16

Cosmology and life

MARIO LIVIO
Space Telescope Science Institute

Abstract

I examine some recent findings in cosmology and their potential implications for the emergence of life in the Universe. In particular, I discuss the requirements for carbon-based life, anthropic considerations with respect to the nature of dark energy, the possibility of time-varying constants of nature, and the question of the rarity of intelligent life.

16.1   Introduction

The progress in cosmology in the past few decades leads also to new insights into the global question of the emergence of intelligent life in the Universe. Here I am not referring to discoveries that are related to very localized regions, such as the detection of extrasolar planetary systems, but rather to properties of the Universe at large.

In order to set the stage properly for the topics to follow, I would like to start with four observations with which essentially all astronomers agree. These four observations *define* the cosmological context of our Universe.

(1) Ever since the observations of Vesto Slipher in 1912–1922 (Slipher 1917) and Hubble (1929), we know that the spectra of distant galaxies are redshifted.

(2) Observations with the *Cosmic Background Explorer (COBE)* have shown that, to a precision of better than 10^{-4}, the cosmic microwave background (CMB) is *thermal*, at a temperature of 2.73 K (Mather et al. 1994).

(3) Light elements, such as deuterium and helium, have been synthesized in a high-temperature phase in the past (e.g., Gamow 1946; Alpher, Bethe, & Gamow 1948; Hoyle & Tayler 1964; Peebles 1966; Wagoner, Fowler, & Hoyle 1967).

(4) Deep observations, such as the Hubble Deep Field, have shown that galaxies in the distant Universe look younger. Namely, their sizes are smaller (e.g., Roche et al. 1996), and there is a higher fraction of irregular morphologies (e.g., Abraham et al. 1996). This is what one would expect from a higher rate of interactions, and from "building blocks" of today's galaxies.

When the above four observational facts are combined and considered together, there is no escape from the conclusion that our Universe is *expanding and cooling*. This conclusion is entirely *consistent* with the "hot big bang" model. Sometimes the stronger statement, that these observations "prove" that there was a hot big bang, is made. However, the scientific method does not truly produce "proofs."

During the past decade, deep observations with a variety of ground-based and space-based observatories have advanced our understanding of the history of the Universe far beyond the mere statement that a big bang had occurred (see, e.g., the determination of cosmological parameters by the *Wilkinson Microwave Anisotropy Probe (WMAP)*; Spergel et al. 2003). In particular, remarkable progress has been achieved in the understanding of the cosmic star formation history.

Using different observational tracers (e.g., the UV luminosity density) of star formation in high-redshift galaxies, tentative plots for the star formation rate (SFR) as a function of redshift have been produced (e.g., Lilly et al. 1996; Madau et al. 1996; Steidel et al. 1999). There is little doubt that the SFR rises from the present to about $z \approx 1$. What happens in the redshift range $z \approx 1$–5 is still somewhat controversial. While some studies suggest that the SFR reaches a peak at $z \approx 1$–2 and then declines slightly toward higher redshifts (e.g., Steidel et al. 1999; or maybe even more than slightly, Stanway, Bunker, & McMahon 2003), or stays fairly flat up to $z \approx 5$ (e.g., Calzetti & Heckman 1999; Pei, Fall, & Hauser 1999), others claim that the SFR continues to rise to $z \approx 8$ (Lanzetta et al. 2002). The latter claim is based on the suggestion that previous studies had failed to account for surface brightness dimming effects. For my present purposes, however, it is sufficient that the history of the *global* SFR is on the verge of being determined (if it has not been determined already). A knowledge of the SFR as a function of redshift allows for the first time for meaningful constraints to be placed on the global emergence of carbon-based life.

16.2   Remarks about Carbon-based Life

The main contributors of carbon to the interstellar medium are intermediate-mass (1–8 $M_\odot$) stars (e.g., Wood 1981; Yungelson, Tutukov, & Livio 1993; Timmes, Woosley, & Weaver 1995), through the asymptotic giant branch and planetary nebulae phases. A knowledge of the cosmic SFR history, together with a knowledge of the initial mass function (presently still uncertain for high redshift), therefore allows for an approximate calculation of the rate of carbon production as a function of redshift (Livio 1999). For a peaked SFR, of the type obtained by Madau et al. (1996), for example, the peak in the carbon production rate is somewhat delayed (by $\lesssim 1$ billion years) with respect to the SFR peak. The decline in the carbon production rate is also shallower for $z \lesssim 1$ (than the decline in the SFR), owing to the buildup of a stellar reservoir in the earlier epochs.

Assuming a "principle of mediocrity," one would expect the emergence of most carbon-based life in the Universe to be perhaps not too far from the peak in the carbon production rate—around $z \approx 1$ (for a peak in the SFR at $z \approx 1$–2). Since the time scale required to develop intelligent civilizations may be within a factor of 2 of the lifetime of F5 to mid-K stars (the ones possessing continuously habitable zones; Kasting, Whitmore, & Reynolds 1993; and see § 16.5 below), it can be expected that intelligent civilizations have emerged when the Universe was $\gtrsim 10$ Gyr old. A younger emergence age may be obtained if the SFR does not decline at redshifts $1.2 \lesssim z \lesssim 8$ (e.g., Lanzetta et al. 2002).

Carbon features in most anthropic arguments. In particular, it is often argued that the existence of an excited state of the carbon nucleus (the 0_2^+ state) is a manifestation of fine-tuning of the constants of nature, which allowed for the appearance of carbon-based life.

Carbon is formed through the triple-α process in two steps. In the first, two α particles form the unstable (lifetime $\sim 10^{-16}$ s) ^{8}Be. In the second, a third α particle is captured, via ^{8}Be$(\alpha, \gamma)^{12}$C. Hoyle argued that in order for the 3α reaction to proceed at a rate sufficient to produce the observed cosmic carbon, a resonant level must exist in ^{12}C, a few hundred keV above the ^{8}Be + ^{4}He threshold. Such a level was indeed found experimentally (Dunbar et al. 1953; Hoyle, Dunbar, & Wenzel 1953; Cook, Fowler, & Lauritsen 1957).

The question of how fine-tuned this level needs to be for the existence of carbon-based life has been the subject of considerable research. The most recent work on this topic was done by Oberhummer and collaborators (e.g., Oberhummer, Csótó, & Schlattl 2000; Csótó, Oberhummer, & Schlattl 2001; Schlattl et al. 2004). These authors used a model that treats the ^{12}C nucleus as a system of 12 interacting nucleons, with the approximate resonant reaction rate

$$r_{3\alpha} = 3^{3/2} N_\alpha^3 \left(\frac{2\pi\hbar^2}{M_\alpha k_B T} \right)^3 \frac{\Gamma_\gamma}{\hbar} \exp\left(-\frac{\varepsilon}{k_B T} \right). \tag{16.1}$$

Here M_α and N_α are the mass and number density of α particles, respectively, ε is the resonance energy (in the center-of-mass frame), Γ_γ is the relative width, and all other symbols have their usual meaning. Oberhummer et al. introduced small variations in the strengths of the nucleon-nucleon interaction and in the fine structure constant (affecting ε and Γ_γ), and calculated stellar models using the modified rates. In their initial work, Oberhummer et al. (2000) concluded that a change of more than 0.5% in the strength of the strong interaction or more than 4% in the strength of the electromagnetic interaction would result in essentially no production of carbon or oxygen [considering the ^{12}C$(\alpha, \gamma)^{16}$O and ^{16}O$(\alpha, \gamma)^{20}$Ne reactions] in any star. More specifically, a decrease in the strong-interaction strength by 0.5%, coupled with an increase in the fine structure constant by 4%, resulted in a decrease in the carbon production by a factor of a few tens in 20 $M_\odot$ stars, and by a factor of ~100 in 1.3 $M_\odot$ stars. Taken at face value, this seemed to support anthropic claims for extreme fine-tuning necessary for the emergence of carbon-based life.

Earlier calculations by Livio et al. (1989) indicated less impressive fine-tuning. Livio et al. showed that shifting (artificially) the energy of the carbon resonant state by up to 0.06 MeV does not result in a significant reduction in the production of carbon. Since this 0.06 MeV should be compared to the *difference* between the resonance energy in ^{12}C and the 3α threshold (calculated with the basic nucleon-nucleon interaction), it was not obvious that a particularly fantastic fine-tuning was required. Most recently, however, Schlattl et al. (2004) reinvestigated the dependence of carbon and oxygen production in stars on the 3α rate. These authors found that following the entire stellar evolution was crucial. They concluded that in massive stars the C and O production strongly depends on the initial mass. In intermediate- and low-mass stars, Schlattl et al. found that the high carbon production during He shell flashes leads to a *lower* sensitivity of the C and O production on the 3α rate than inferred by Oberhummer et al. (2000). Schlattl et al. (2004) concluded by saying that "fine-tuning with respect to the obtained carbon and oxygen abundance is more complicated and far less spectacular" than that found by Oberhummer et al. (2000).

16.3   Dark Energy and Life

In 1998, two teams of astronomers, working independently, presented evidence that the expansion of the Universe is accelerating (Riess et al. 1998; Perlmutter et al. 1999). The evidence was based primarily on the unexpected faintness (by ~0.25 mag) of distant ($z \approx 0.5$) Type Ia supernovae, compared to their expected brightness in a universe decelerating under its own gravity. The results favored values of $\Omega_m \approx 0.3$ and $\Omega_\Lambda \approx 0.7$ for the matter and "dark energy" density parameters, respectively. Subsequent observations of the supernova SN 1997ff, at the redshift of $z \simeq 1.7$, strengthened the conclusion of an accelerating Universe (Riess et al. 2001). This supernova appeared *brighter* relative to SNe in a coasting universe, as expected from the fact that at $z \approx 1.7$ a universe with $\Omega_m \approx 0.3$ and $\Omega_\Lambda \approx 0.7$ would still be in its decelerating phase. The observations of SN 1997ff do not support any alternative interpretation (such as dust extinction or evolutionary effects) in which supernovae are expected to dim monotonically with redshift. Measurements of the power spectrum of the cosmic microwave background (e.g., Abroe et al. 2002; de Bernardis et al. 2002; and Netterfield et al. 2002; and, most recently the *WMAP* results, Bennett et al. 2003) provide strong evidence for flatness ($\Omega_m + \Omega_\Lambda = 1$). When combined with estimates of Ω_m based on mass-to-light ratios, X-ray temperatures of intracluster gas, and dynamics of clusters (all of which give $\Omega_m \lesssim 0.3$; e.g., Strauss & Willick 1995; Carlberg et al. 1996; Bahcall et al. 2000), again a value of $\Omega_\Lambda \approx 0.7$ is obtained.

Arguably the two greatest puzzles physics is facing today are:

(1) Why is the dark energy (vacuum energy) density, ρ_v, so small, but not zero? (Or, why does the vacuum energy gravitate so little?)
(2) Why *now*? Namely, why do we find at present that $\Omega_\Lambda \approx \Omega_m$?

The first question reflects the fact that taking graviton energies up to the Planck scale, M_P, would produce a dark energy density

$$\rho_v \approx M_P^4 \approx (10^{18} \text{ GeV})^4, \tag{16.2}$$

which misses the observed one, $\rho_v \approx (10^{-3} \text{ eV})^4$, by more than 120 orders of magnitude. Even if the energy density in fluctuations in the gravitational field is taken only up to the supersymmetry-breaking scale, M_{SUSY}, we still miss the mark by a factor of 60 orders of magnitude, since $\rho_v \approx M_{\text{SUSY}}^4 \approx (1 \text{ TeV})^4$. Interestingly, though, a scale $M_v \approx (M_{\text{SUSY}}/M_P)M_{\text{SUSY}}$ produces the right order of magnitude. However, while a few attempts in this direction have been made (e.g., Arkani-Hamed et al. 2000), no satisfactory model has been developed.

The second question is related to the anti-Copernican fact that Ω_Λ may be associated with a cosmological constant, while Ω_m declines continuously (and in any case, ρ_v may be expected to have a different time behavior from ρ_m), and yet the first time that we are able to measure both reliably, we find that they are of the same order.

The attempts to solve these problems fall into three general categories:

(1) The behavior of "quintessence" fields
(2) Alternative theories of gravity
(3) Anthropic considerations

The attempts of the first type have concentrated in particular on "tracker" solutions (e.g., Zlatev, Wang, & Steinhardt 1998; Albrecht & Skordis 2002), in which the smallness of Ω_Λ is a direct consequence of the Universe's old age. Generally, a uniform scalar field, ϕ, is taken to evolve according to

$$\ddot{\phi} + 3H\dot{\phi} + V'(\phi) = 0, \tag{16.3}$$

where $V'(\phi) = \frac{dV}{d\phi}$ and H is the Hubble parameter. The energy density of the scalar field is given by

$$\rho_\phi = \frac{1}{2}\dot{\phi}^2 + V(\phi),$$

(16.4)

and that of matter and radiation, ρ_m, by

$$\dot{\rho}_m = -3H(\rho_m + P_m),$$

(16.5)

where P_m is the pressure. For a potential of the form

$$V(\phi) = \phi^{-\alpha} M^{4+\alpha},$$

(16.6)

where $\alpha > 0$ and M is an adjustable constant ($M \ll M_P$), and a field that is initially much smaller than the Planck mass, one obtains a solution in which a transition occurs from an early ρ_m-dominance to a late ρ_ϕ-dominance (with no need to fine-tune the initial conditions). Nevertheless, for the condition $\rho_\phi \approx \rho_m$ to actually be satisfied at the present time requires (Weinberg 2001) that the parameter M would satisfy

$$M^{4+\alpha} \simeq (8\pi G)^{-1-\alpha/2} H_0^2,$$

(16.7)

which is not easily explicable.

In order to overcome this problem, some quintessence models choose potentials in which the Universe has periodically been accelerating in the past (e.g., Dodelson, Kaplinghat, & Stewart 2000), so that the dark energy's dominance today appears naturally.

A very different approach regards the accelerating expansion not as being propelled by dark energy, but rather as being the result of a modified gravity. For example, models have been developed (Deffayet, Dvali, & Gabadadze 2002), in which ordinary particles are localized on a three-dimensional surface (3-brane) embedded in infinite-volume extra dimensions to which gravity can spread. The model is constructed in such a way that observers on the brane discover Newtonian gravity (four dimensional) at distances that are shorter than a crossover scale, r_c, which can be of astronomical size. In one version, the Friedmann equation is replaced by

$$H^2 + \frac{k}{a^2} = \left(\sqrt{\frac{\rho}{3M_P^2} + \frac{1}{4r_c^2}} + \epsilon\frac{1}{2r_c^2} \right)^2,$$

(16.8)

where ρ is the total energy density, a is the scale factor and $\epsilon = \pm 1$.

In this case, the dynamics of gravity are governed by whether ρ/M_P^2 is larger or smaller than $1/r_c^2$. Choosing $r_c \approx H_0^{-1}$ preserves the usual cosmological results. At large cosmic distances, however, gravity spreads into extra dimensions (the force law becomes five dimensional), and becomes weaker—directly affecting the cosmic expansion. Basically, at late times, the model has a self-accelerating cosmological branch with $H = 1/r_c$ (to leading-order Equation 1.8 can be parameterized as $H^2 - H/r_c \simeq \rho/3M_P^2$). Interestingly, it has recently been suggested that the viability of these models can be tested by lunar ranging experiments (Dvali, Gruzinov, & Zaldarriaga 2003). I should also note that the *WMAP* results indicated an intriguing lack of correlated signal on angular scales greater than 60 degrees (Spergel et al. 2003), reinforcing the low quadrupole seen already in *COBE* results. One possible, although at this stage speculative, interpretation of these results is that they signal the breakdown of conventional gravity on large scales.

A third class of proposed solutions to the dark energy problems relies on anthropic selection effects, and therefore on the *existence* of intelligent life in our Universe. The basic premise of this approach is that some of the constants of nature are actually random variables, whose range of values and *a priori* probabilities are nevertheless determined by the laws of physics. The observed big bang, in this picture, is simply one member of an ensemble. It is further assumed that a "principle of mediocrity" applies; namely, we can expect to observe the most probable values (Vilenkin 1995). Using this approach, Garriga, Livio, & Vilenkin (2000; following the original idea of Weinberg 1987) were able to show that when the cosmological constant Λ is the only variable parameter, the order of magnitude coincidence $t_0 \approx t_\Lambda \approx t_G$ (where t_0 is the present time; t_Λ is the time Ω_Λ starts to dominate; t_G is the time when giant galaxies were assembled) finds a natural explanation (see also Bludman 2000).

Qualitatively, the argument works as follows.

In a geometrically flat universe with a cosmological constant, gravitational clustering can no longer occur after redshift $(1 + z_\Lambda) \approx (\rho_\Lambda / \rho_{m0})^{1/3}$ (where ρ_{m0} is the present matter density). Therefore, requiring that ρ_Λ does not dominate before redshift z_{max}, at which the earliest galaxies formed, requires (e.g., Weinberg 1987)

$$\rho_\Lambda \lesssim (1 + z_{max})^3 \rho_{m0}. \tag{16.9}$$

One can expect the *a priori* (independent of observers) probability distribution $P(\rho_\Lambda)$ to vary on some characteristic scale, $\Delta\rho_\Lambda \approx \eta^4$, determined by the underlying physics. Irrespective of whether η is determined by the Planck scale ($\sim 10^{18}$ GeV), the grand unification scale ($\sim 10^{16}$ GeV) or the electroweak scale ($\sim 10^2$ GeV), $\Delta\rho_\Lambda$ exceeds the anthropically allowed range of ρ_Λ (Eq. 1.9) by so many orders of magnitude that it looks reasonable to assume that

$$P(\rho_\Lambda) = const, \tag{16.10}$$

over the range of interest. Garriga & Vilenkin (2001) and Weinberg (2001) have shown that this assumption is satisfied by a broad class of models, even though not automatically. With a flat distribution, a value of ρ_Λ picked randomly (and which may characterize a "pocket" universe) from an interval $|\rho_\Lambda| \lesssim \rho_\Lambda^{max}$, will, with a high probability, be of the order of ρ_Λ^{max}. The principle of mediocrity, however, means that we should observe a value of ρ_Λ that maximizes the number of galaxies. This suggests that we should observe the largest value of ρ_Λ that is still consistent with a substantial fraction of matter having collapsed into galaxies—in other words, $t_\Lambda \approx t_G$, as observed. In § 16.2 I argued that the appearance of carbon-based life may be associated roughly with the peak in the star formation rate, t_{SFR}. The "present time," t_0, is not much different from that (in that it takes only a fraction of a stellar lifetime to develop intelligent life), hence $t_0 \approx t_{SFR}$. Finally, hierarchical structure formation models suggest that vigorous star formation is closely associated with the formation of galactic-size objects (e.g., Baugh et al. 1998; Fukugita, Hogan, & Peebles 1998). Therefore, $t_G \approx t_{SFR}$, and we obtain $t_0 \approx t_G \approx t_\Lambda$.

Garriga et al. (2000) further expanded their discussion to treat not just Λ, but also the density contrast at recombination, σ_{rec}, as a random variable (see also Tegmark & Rees 1998). The galaxy formation in this case is spread over a much wider time interval, and proper account has to be taken for the fact that the cooling of protogalactic clouds collapsing at very late times is too slow for efficient fragmentation and star formation (fragmentation occurs if the cooling time scale is shorter than the collapse time scale, $\tau_{cool} < \tau_{grav}$). Assuming an *a priori* probability distribution of the form

$$P(\sigma_{rec}) \sim \sigma_{rec}^{-\alpha}, \tag{16.11}$$

Garriga et al. found that "mediocre" observers will detect $\sigma_{rec} \approx 10^{-4}$, $t_0 \approx t_G \approx t_\Lambda \approx t_{cb}$, as observed, *if* $\alpha > 3$ (here the "cooling boundary" t_{cb} is the time after which fragmentation is suppressed).

Other anthropic explanations for the value of the cosmological constant and the "why now?" problem have been suggested in the context of maximally extended ($N = 8$) supergravity (Kallosh & Linde 2003; Linde 2003). In particular, the former authors found that the Universe can have a sufficiently long lifetime only if the scaler field satisfies initially $|\phi| \lesssim M_P$, and if the value of the potential $V(0)$, which plays the role of the cosmological constant, does not exceed the critical density $\rho_0 \approx 10^{-120} M_P^4$.

Personally, I feel that anthropic explanations to the dark energy problems should be regarded as the *last resort*, only after all attempts to find explanations based on first principles have been exhausted and failed. Nevertheless, the anthropic explanation may prove to be the correct one, if our understanding of what is truly *fundamental* is lacking. A historical example can help to clarify this last statement. Johannes Kepler (1571–1630) was obsessed by the following two questions:

(1) Why were there precisely six planets? (only Mercury, Venus, Earth, Mars, Jupiter and Saturn were known at his time)
(2) What was it that determined that the planetary orbits would be spaced as they are?

The first thing to realize is that these "why" and "what" questions were a novelty in the astronomical vocabulary. Astronomers before Kepler were usually satisfied with simply recording the observed positions of the planets; Kepler was seeking a theoretical explanation. Kepler finally came up with preposterously fantastic (and absolutely wrong) answers to his two questions in *Mysterium Cosmographicum*, published in 1597. He suggested that the reason for there being six planets is that there are precisely five Platonic solids. Taken as boundaries (with an outer spherical boundary corresponding to the fixed stars), the solids create six spacings. By choosing a particular order for the solids to be embedded in each other, with the Earth separating the solids that can stand upright (cube, tetrahedron, and dodecahedron) from those that "float" (octahedron and icosahedron), Kepler claimed to have explained the sizes of the orbits too (the spacings agreed with observations to within 10%).

Today we recognize what was the *main* problem with Kepler's model—Kepler did not understand that neither the number of planets nor their spacings are *fundamental* quantities that need to have an explanation from first principles. Rather, both are the result of historical accidents in the solar protoplanetary disk. Still, it is perfectly legitimate to give an anthropic "explanation" for the Earth's orbital radius. If that orbit were not in the continuously habitable zone around the sun (Kasting & Reynolds 1993), we would not be here to ask the question.

It is difficult to admit it, but our current model for the composition of the Universe: $\sim 73\%$ dark energy, $\sim 23\%$ cold dark matter, $\sim 4\%$ baryonic matter, and maybe $\sim 0.5\%$ neutrinos, appears no less preposterous than Kepler's model. While some version of string (or $M-$) theories may eventually provide a first-principles explanation for all of these values, it is also possible, in my opinion, that these individual values are in fact not fundamental, but accidental. Maybe the only fundamental property is the fact that *all the energy densities add up to produce a geometrically flat universe*, as predicted by inflation (Guth 1981; Hawking 1982; Steinhardt & Turner 1984) and confirmed by *WMAP* (Spergel et al. 2003). Clearly, for any anthropic explanation of the value of Ω_Λ to be meaningful at all, even in principle, one requires the existence of a large ensemble of universes, with different values of Ω_Λ. That this requirement may actually be fulfilled is precisely one of the consequences of the concept of "eternal inflation" (Steinhardt 1983; Vilenkin 1983; Linde 1986; Goncharov, Linde, & Mukhanov 1987; Linde 2003). In most inflationary models the time scale associated with the expansion is much shorter than the decay time scale of the false vacuum phase, $\tau_{\text{exp}} \ll \tau_{\text{dec}}$. Consequently, the emergence of a fractal structure of "pocket universes" surrounded by false vacuum material is almost inevitable (Garcia-Bellido & Linde 1995; Guth 2001; for a different view, see, e.g., Bucher, Goldhaber, & Turok 1995; Turok 2001).

This ensemble of pocket universes may serve as the basis on which anthropic argumentation can be constructed (even though the definition of probabilities on this infinite set is nontrivial; see, e.g., Linde, Linde, & Mezhlumian 1995; Vilenkin 1998).

16.4   Varying Constants of Nature?

Another recent finding, which, *if confirmed*, may have implications for the emergence of life in the Universe, is that of cosmological evolution of the fine structure constant $\alpha \equiv e^2/\hbar c$ (Webb et al. 1999, 2001, and references therein). Needless to say, life as we know it places significant anthropic constraints on the range of values allowed for α. For example, the requirement that the lifetime of the proton would be longer than the main sequence lifetime of stars results in an upper bound $\alpha \lesssim 1/80$ (Ellis & Nanopoulos 1981; Barrow, Sandvik, & Magueijo 2002a). The claimed detection of time variability was based on shifts in the rest wavelengths of redshifted UV resonance transitions observed in quasar absorption systems. Basically, the dependence of observed wave number at redshift z, w_z, on α can be expressed as

$$w_z = w_0 + a_1 w_1 + a_2 w_2, \tag{16.12}$$

where a_1 and a_2 represent relativistic corrections for particular atomic masses and electron configurations, and

$$w_1 = \left(\frac{\alpha_z}{\alpha_0}\right)^2 - 1 \tag{16.13}$$

$$w_2 = \left(\frac{\alpha_z}{\alpha_0}\right)^4 - 1. \tag{16.14}$$

Here α_0 and α_z represent the present day and redshift z values of α, respectively. By analyzing a multitude of absorption lines from many multiplets in different ions, such as Fe II and Mg II transitions in 28 absorption systems (in the redshift range $0.5 \lesssim z \lesssim 1.8$), and Ni II, Cr II, Zn II, and Si IV transitions in some 40 absorption systems (in the redshift range $1.8 \lesssim z \lesssim 3.5$), Webb et al. (2001) concluded that α was *smaller* in the past. Their data suggest a 4σ deviation

$$\frac{\Delta\alpha}{\alpha} = -0.72 \pm 0.18 \times 10^{-5} \tag{16.15}$$

over the redshift range $0.5 \lesssim z \lesssim 3.5$ (where $\Delta\alpha/\alpha = \frac{\alpha_z - \alpha_0}{\alpha_0}$). It should be noted, however, that the data are consistent with *no* variation for $z \lesssim 1$, in agreement with many previous studies (e.g., Bahcall, Sargent, & Schmidt 1967; Wolfe, Brown, & Roberts 1976; Cowie & Songaila 1995).

Murphy et al. (2001) conducted a comprehensive search for systematic effects that could potentially be responsible for the result (e.g., laboratory wavelength errors, isotopic abundance effects, heliocentric corrections during the quasar integration, line blending, and atmospheric dispersion). While they concluded that isotopic abundance evolution and atmospheric dispersion could have an effect, this was in the direction of actually amplifying the variation in α [to $\Delta\alpha/\alpha = (-1.19 \pm 0.17) \times 10^{-5}$]. The most recent results of Webb et al. are not inconsistent with limits on α from the Oklo natural uranium fission reactor (which was active 1.8×10^9 years ago, corresponding to $z \approx 0.1$) and with constraints from experimental tests of the equivalence principle. The former suggests $\Delta\alpha/\alpha \simeq (-0.4 \pm 1.4) \times 10^{-8}$ (Fuji et al. 2000), and the latter *allows* for a variation of the magnitude observed in the context of a general dynamical theory relating variations of α to the electromagnetic fraction of the mass density in the Universe (Bekenstein 1982; Livio & Stiavelli 1998).

Before going any further, I would like to note that what is desperately needed right now is an independent confirmation (or refutation) of the results of Webb et al. by other groups, both through additional (and preferably different) observations and via independent analysis of the data. In this respect it is important to realize that the reliability of the SNe Ia results (concerning the accelerating Universe) was enormously enhanced by the fact that two separate teams (the Supernova Cosmology Project and the High-z Supernova Team) reached the same conclusion independently, using different samples and different data analysis techniques. A first small step in the direction of testing the variable α result came from measurements of the CMB. A likelihood analysis of BOOMERanG and MAXIMA data, allowing for the possibility of a time-varying α (which, in turn, affects the recombination time) found that in general the data may prefer a smaller α in the past (although the conclusion is not free of degeneracies; Avelino et al. 2000; Battye, Crittenden, & Weller 2001). A second, much more important step, came through an extensive analysis using the nebular emission lines of [O III] $\lambda\lambda4959, 5007$ Å (Bahcall, Steinhardt, & Schlegel 2004). Bahcall et al. found $\Delta\alpha/\alpha = (-2 \pm 1.2) \times 10^{-4}$ (corresponding to $|\alpha^{-1}d\alpha/dt| < 10^{-13}$ yr^{-1}, which they consider to be a null result, given the precision of their method) for quasars in the redshift range $0.16 < z < 0.8$. While this result is not formally inconsistent with the variation claimed by Webb et al., the careful analysis of Bahcall et al. has cast some serious doubts on the ability of the "many-multiplet" method employed by Webb and his collaborators to actually reach the accuracy required to measure fractional variations in α at the 10^{-5} level. For example, Bahcall et al. have shown that to achieve that precision, one needs to assume that the velocity profiles of different ions in different clouds are essentially the same to within 1 km s^{-1}. Clearly, much more work on this topic is needed. I should also note right away that, in order not to be in conflict with the yield of ^{4}He, $|\Delta\alpha/\alpha|$ cannot exceed $\sim 2 \times 10^{-2}$ at the time of nucleosynthesis (e.g., Bergström, Igury, & Rubinstein 1999).

On the theoretical side, simple cosmological models with a varying fine structure constant have now been developed (e.g., Sandvik, Barrow, & Magueijo 2003; Barrow, Sandvik, & Magueijo 2002b). They share some properties with Kaluza-Klein-type models in which α varies at the same rate as the extra dimensions of space (e.g., Damour & Polyakov 1994), and with varying-speed-of-light theories (e.g., Albrecht & Magueijo 1999; Barrow & Magueijo 2000).

The general equations describing a geometrically flat, homogeneous, isotropic, variable-α universe are (Beckenstein 1982; Livio & Stiavelli 1998; Sandvik et al. 2002) the Friedmann equation (with $G = c \equiv 1$)

$$\left(\frac{\dot{a}}{a}\right)^2 = \frac{8\pi}{3}\left[\rho_m\left(1+|\zeta_m|e^{-2\psi}\right)+\rho_r e^{-2\psi}+\rho_\psi+\rho_\Lambda\right], \tag{16.16}$$

the evolution of the scalar field varying α ($\alpha = \exp(2\psi)e_0^2/\hbar c$)

$$\ddot{\psi}+3H\dot{\psi} = -\frac{2}{w}e^{-2\psi}\zeta_m\rho_m, \tag{16.17}$$

and the conservation equations for matter and radiation

$$\dot{\rho}_m+3H\rho_m = 0 \tag{16.18}$$
$$\dot{\rho}_r+4H\rho_r = 2\dot{\psi}\rho_r. \tag{16.19}$$

Here, ρ_m, ρ_r, ρ_ψ, ρ_Λ are the densities of matter, radiation, scalar field ($\frac{w}{2}\dot{\psi}^2$), and vacuum, respectively, $a(t)$ is the scale factor ($H \equiv \dot{a}/a$), $w = \hbar c/l^2$ is the coupling constant of the dynamic Langrangian (l is a length scale of the theory), and ξ_m is a dimensionless parameter that represents the fraction of mass in Coulomb energy of an average nucleon compared to the free proton mass.

Equations 1.16–1.19 were solved numerically by Sandvik et al. (2002) and Barrow et al. (2002), assuming a negative value of the parameter ξ_m/w, and the results are interesting both from a purely cosmological point of view and from the perspective of the emergence of life. First, the results are consistent with both the claims of a varying α of Webb et al. (which, as I noted, badly need further confirmation) and with the more secure, by now, observations of an accelerating Universe (Riess et al. 1998; Perlmutter et al. 1999; Spergel et al. 2003), while complying with the geological and nucleosynthetic constraints. Second, Barrow et al. find that α remains almost constant in the radiation-dominated era, experiences a small logarithmic time increase during the matter-dominated era, but approaches a constant value again in the Λ-dominated era. This behavior has interesting anthropic consequences. The existence of a nonzero vacuum energy contribution is now *required* in this picture to dynamically stabilize the fine structure constant. In a universe with zero Λ, α would continue to grow in the matter-dominate era to values that would make the emergence of life impossible (Barrow et al. 2001).

Clearly, the viability of all of the speculative ideas above relies at this point on the confirmation or refutation of time-varying constants of nature.

16.5 Is Intelligent Life Extremely Rare?

With the discovery of ~ 100 massive extrasolar planets (Mayor & Queloz 1995; Marcy & Butler 1996, 2000), the question of the potential existence of extraterrestrial, Galactic, intelligent life has certainly become more intriguing than ever. This topic has attracted much attention and generated many speculative (by necessity) probability estimates. Nevertheless, in a quite remarkable paper, Carter (1983) concluded on the basis of the near-equality between the lifetime of the sun, $t_\odot$, and the time scale of biological evolution on Earth, t_ℓ, that extraterrestrial intelligent civilizations are exceedingly rare in the Galaxy. Most significantly, Carter's conclusion is supposed to hold even if the conditions optimal for the emergence of life are relatively common.

Let me reproduce here, very briefly, Carter's argument. The basic, and very crucial assumption on which the argument is based is that the lifetime of a star, t_*, and the time scale of biological evolution on a planet around that star, t_ℓ (taken here, for definiteness, to be the time scale for the appearance of complex land life), are *a priori entirely independent*. In other words, the assumption is that land life appears at some *random* time with respect to the main sequence lifetime of the star. Under this assumption, one expects that generally one of the two relations $t_\ell \gg t_*$ or $t_\ell \ll t_*$ applies (the set where $t_\ell \approx t_*$ is of negligible measure for two independent quantities). Let us examine each one of these possibilities. If *generally* $t_\ell \ll t_*$, it is very difficult to understand why in the first system found to contain complex land life, the Earth-Sun system, the two time scales are nearly equal, $t_\ell \approx t_*$. If, on the other hand, *generally* $t_\ell \gg t_*$, then clearly the first system we find must exhibit $t_\ell \approx t_*$ (since for $t_\ell \gg t_*$ complex land life would not have developed). Therefore, one has to conclude that *typically* $t_\ell \gg t_*$, and that consequently, complex land life will generally not develop—the Earth is an extremely rare exception.

Carter's argument is quite powerful and not easily refutable. Its basic assumption (the independence of t_ℓ and t_*) appears on the face of it to be solid, since t_* is determined primarily by nuclear burning reactions, while t_ℓ is determined by biochemical reactions and the evolution of species. Nevertheless, the fact that the star is the main energy source for biological evolution (light energy exceeds the other sources by 2–3 orders of magnitude; e.g., Deamer 1997), already implies that the two quantities are not completely independent.

Let me first take a purely mathematical approach and examine what would it take for the condition $t_\ell \approx t_*$ to be satisfied in the Earth-Sun system *without* implying that extraterrestrial intelligent life is extremely rare. Imagine that t_ℓ and t_* are not independent, but rather that

$$t_\ell/t_* = f(t_*), \tag{16.20}$$

where $f(t_*)$ is some *monotonically increasing* function in the narrow range $t_*^{\min} \lesssim t_* \lesssim t_*^{\max}$ that allows the emergence of complex land life through the existence of continuously habitable zones (corresponding to stellar spectral types F5 to mid-K; Kasting et al. 1993). Note that, for a Salpeter (1955) initial mass function, the distribution of stellar lifetimes behaves as

$$\psi(t_*) \approx t_*. \tag{16.21}$$

Consequently, if relation 1.20 were to hold, it would in fact be the *most probable* that in the first place where we encounter an intelligent civilization we would find that $t_\ell/t_* \approx 1$, as in the Earth-Sun system. In other words, if we could identify some processes that are likely to produce a monotonically increasing $t_* - t_\ell/t_*$ relation, then the near equality of t_ℓ and t_* in the Earth-Sun system would find a natural explanation, with no implications whatsoever for the frequency of intelligent civilizations. A few years ago, I proposed a simple toy-model for how such a relation might arise (Livio 1999). The toy-model was based on the assumption that the appearance of land life has to await the build-up of a sufficient layer of protective ozone (Berkner & Marshall 1965; Hart 1978), and on the fact that oxygen in a planet's atmosphere is released in the first phase from the dissociation of water (Hart 1978; Levine, Hayes, & Walker 1979). Given that the duration of this phase is inversely proportioned to the intensity of radiation in the 1000–2000 Å range, a relation between t_ℓ and t_* can be established. In fact, a simple calculation gave

$$t_\ell/t_* \simeq 0.4(t_*/t_\odot)^{1.7}, \tag{16.22}$$

precisely the type of monotonic relation needed.

I should be the first to point out that the toy-model above is nothing more than that—a toy model. It does point out, however, that at the very least, establishing a link between the biochemical and astrophysical time scales may not be impossible. Clearly, the emergence of complex life on Earth required many factors operating together. These include processes that appear entirely accidental, such as the stabilization of the Earth's tilt against chaotic evolution by the Moon (e.g., Laskar, Joutel, & Boudin 1993). Nevertheless, we should not be so arrogant as to conclude everything from the one example we know. The discovery of many "hot Jupiters" (giant planets with orbital radii $\lesssim 0.05$ AU) has already demonstrated that the solar system may not be typical. We should keep an open mind to the possibility that biological complexity may find other paths to emerge, making various "accidents," coincidences, and fine-tuning unnecessary. In any case, the final scientific assessment on life in the Universe will probably come from biologists and observers—not from speculating theorists like myself.

Acknowledgements. This work has been supported by Grant 938-COS191 from the Templeton Foundation. I am grateful to Andrei Linde and Heinz Oberhummer for helpful comments.

References

Abraham, R. G., van den Bergh, S., Glazebrook, K., Ellis, R. S., & Santiago, B. X. 1996, ApJS, 101, 1

Abroe, M. E., et al. 2002, MNRAS, 334, 11

Albrecht, A., & Magueijo, J. 1999, Phys. Rev. D, 59, 043516

Albrecht, A., & Skordis, C. 2002, Phys. Rev. Lett., 84, 2076

Alpher, R. A., Bethe, H., & Gamow, G. 1948, Phys. Rev., 73, 803

Arkani-Hamed, N., Hall, L. J., Colda, C., & Murayama, H. 2000, Phys. Rev. Lett., 85, 4434

Avelino, P. P., Martins, C. J. A. P., Rocha, G., & Viana, P. 2000, Phys. Rev. D, 62, 123508

Bahcall, J. N., Sargent, W. L. W, & Schmidt, M. 1967, ApJ, 149, L11

Bahcall, J. N., Steinhardt, C. L., & Schlegel, D. 2004, ApJ, in press (astro-ph/0301507)

Bahcall, N. A., Cen, R., Davé, R., Ostriker, J. P., & Yu, O. 2000, ApJ, 541, 1

Barrow, J. D., & Magueijo, J. 2000, ApJ, 532, L87

Barrow, J. D., Sandvik, H. B., & Magueijo, J. 2002a, Phys. Rev. D, 65, 123501

——. 2002b, Phys. Rev. D, 65, 063504

Battye, R. A., Crittenden, R., & Weller, J. 2001, Phys. Rev. D, 63, 043505

Baugh, C. M., Cole, S., Frenk, C. S., & Lacey, C. G. 1998, ApJ, 498, 504

Bekenstein, J. D. 1982, Phys. Rev. D, 25, 1527

Bennett, C. L., et al. 2003, ApJS, 148, 1

Bergström, L., Iguri, S., & Rubinstein, H. 1999, Phys. Rev. D, 60, 045005

Berkner, L. V., & Marshall, K. C. 1965, J. Atmos. Sci., 22, 225

Bludman, S. 2000, Nucl. Phys. A, 663–664, 865

Bucher, M., Goldhaber, A., & Turok, N. 1995, Phys. Rev. D, 52, 3314

Calzetti, D., & Heckman, T. M. 1999, ApJ, 519, 27

Carlberg, R. G., Yee, H. K. C., Ellingson, E., Abraham, R., Grabel, P., Morris, S., & Pritchet, C. J. 1996, ApJ, 462, 32

Carter, B. 1983, Phil. Trans. R. Soc. London A, 310, 347

Cook, C. W., Fowler, W. A., & Lauritsen, T. 1957, Phys. Rev., 107, 508

Cowie, L. L., & Songaila, A. 1995, ApJ, 453, 596

Csótó, A., Oberhummer, H., & Schlattl, H. 2001, Nucl. Phys. A, 688, 560

Damour, T., & Polyakov, A. M. 1994, Nucl. Phys. B, 423, 532

Deamer, D. W. 1997, Microb. and Molec. Bio. Rev., 61, 239

de Bernardis, P., et al. 2002, ApJ, 564, 559

Deffayet, C., Dvali, G., & Gabadadze, G. 2002, Phys. Rev. D, 65, 044023

Dodelson, S., Kaplinghat, M., & Stewart, E. 2000, Phys. Rev. Lett., 85, 5276

Dunbar, D. N. F., Pixley, R. E., Wenzel, W. A., & Whaling, W. 1953, Phys. Rev., 92, 649

Dvali, G., Gruzinov, A., & Zaldarriaga, M. 2003, Phys. Rev. D, 68, 024012

Ellis, J. D., & Nanopoulos, D. V. 1981, Nature, 292, 436

Fuji, V., et al. 2000, Nucl. Phys. B, 573, 377

Fukugita, M., Hogan, C. J., & Peebles, P. J. E. 1998, ApJ, 503, 518

Gamow, G. 1946, Phys. Rev., 70, 527

Garcia-Bellido, J., & Linde, A. O. 1995, Phys. Rev. D, 51, 429

Garriga, J., Livio, M., & Vilenkin, A. 2000, Phys. Rev. D., 61, 023503

Garriga, J., & Vilenkin, A. 2001, Phys. Rev. D., 64, 023517

Goncharov, A. S., Linde, A. D., & Mukhanov, V. F. 1987, Int. J. Mod. Phys. A, 2, 561

Guth, A. H. 1981, Phys. Rev. D, 23, 347

Guth, A. H. 2001, in Astrophysical Ages and Time Scales, ed. T. von Hippel, C. Simpson, & N. Mansit (San Francisco: ASP), 3

Hart, M. H. 1978, Icarus, 33, 23

Hawking, S. W. 1982, Phys. Lett. B, 115, 295

Hoyle, F., Dunbar, D. N. F., & Wenzel, W. A. 1953, Phys. Rev., 92, 1095

Hoyle, F., & Tayler, R. J. 1964, Nature, 203, 1108

Hubble, E. 1929, Proc. Nat. Acad. Sci., 15, 168

Kallosh, R., & Linde, A. 2003, Phys. Rev. D., 67, 023510

Kasting, J. F., & Reynolds, R. T. 1993, Icarus, 101, 108

Kasting, J. F., Whitmore, D. P., & Reynolds, R. T. 1993, Icarus, 101, 108

Lanzetta, K. M., Yahata, N., Pascarelle, S., Chen, H.-W., & Fernández-Soto, A. 2002, ApJ, 570, 492

Laskar, J., Joutel, F., & Boudin, F. 1993, Nature, 361, 615

Levine, J. S., Hayes, P. B., & Walker, J. C. G. 1979, Icarus, 39, 295

Lilly, S. J., Le Fèvre, O., Hammer, F., & Crampton, D. 1996, ApJ, 460, L1

Linde, A. 2003, in Science and Ultimate Reality: From Quantum to Cosmos, ed. J. D. Barrow, P. C. W. Davies, & C. L. Harper (Cambridge: Cambridge Univ. Press), in press

Linde, A., Linde, D., & Mezhlumian, A. 1995, Phys. Lett. B, 345, 203

Linde, A. D. 1986, Mod. Phys. Lett. A, 1, 81

Livio, M. 1999, ApJ, 511, 429

Livio, M., Hollowell, D., Weiss, A., & Truran, J. W. 1989, Nature, 340, 281

Livio, M., & Stiavelli, M. 1998, ApJ, 507, L13

Madau, P., Ferguson, H. C., Dickinson, M., Giavalisco, M., Steidel, C. C., & Fruchter, A. 1996, MNRAS, 283, 1388

Marcy, G. W., & Butler, R. P. 1996, ApJ, 464, L147

——. 2000, PASP, 112, 137

Mather, J. C., et al. 1994, ApJ, 420, 439

Mayor, M., & Queloz, D. 1995, Nature, 378, 355

Murphy, M. T., Webb, J. K., Flambaum, V. V., Churchill, C. W., & Prochaska, J. X. 2001, MNRAS, 327, 1223

Netterfield, C. B., et al. 2002, ApJ, 571, 604

Oberhummer, H., Csótó, A., & Schlattl, H. 2000, Science, 289, 88

Peebles, P. J. E. 1966, ApJ, 146, 542

Pei, Y., Fall, S. M., & Hauser, M. G. 1999, ApJ, 522, 604

Perlmutter, S., et al. 1999, ApJ, 517, 565

Riess, A. G., et al. 1998, AJ, 116, 1009

——. 2001, ApJ, 560, 49

Roche, N., Ratnatunga, K. U., Griffiths, R. E., Im, M., & Neuschaefer, L. W. 1996, MNRAS, 282, 1247

Salpeter, E. E. 1955, ApJ, 121, 161

Sandvik, H. B., Barrow, J. D., & Magueijo, J. 2002, Phys. Rev. Lett., 88, 031302

Schlattl, H., Heger, A., Oberhummer, H., Rauscher, T., & Csótó, A. 2004, MNRAS, in press

Slipher, V. M. 1917, Proc. Amer. Phil. Soc., 56, 403

Spergel, D. N., et al. 2003, ApJS, 148, 175

Stanway, E. R., Bunker, A. J., & McMahon, R. G. 2003, MNRAS, 342, 439

Steidel, C. C., Adelberger, K. L., Giavalisco, M., Dickinson, M., & Pettini, M. 1999, ApJ, 519, 1

Steinhardt, P. J. 1983, in The Very Early Universe, ed. G. W. Gibbons, S. Hawking, & S. T. C. Siklos (Cambridge: Cambridge Univ. Press), 251

Steinhardt, P. J., & Turner, M. S. 1984, Phys. Rev. D, 29, 2162

Strauss, M. A., & Willick, J. A. 1995, Phys. Rep., 261, 271

Tegmark, M., & Rees, M. J. 1998, ApJ, 499, 526

Timmes, F. X., Woosley, S. E., & Weaver, T. A. 1995, ApJS, 98, 617

Turok, N. 2001, in Birth and Evolution of the Universe, ed. K. Sato & M. Kawasaki (Tokyo: Universal Academy Press), 1

Vilenkin, A. 1983, Phys. Rev. D, 27, 2848

——. 1995, Phys. Rev. Lett., 74, 846

——. 1998, Phys. Rev. Lett., 81, 5501

Wagoner, R. V., Fowler, W. A., & Hoyle, F. 1967, ApJ, 148, 3

Webb, J. K., Flambaum, V. V., Churchill, C. W., Drinkwater, M. J., & Barrow, J. D. 1999, Phys. Rev. Lett., 82, 884

Webb, J. K., Murphy, M. T., Flambaum, V. V., Dzuba, V. A., Barrow, J. D., Churchill, C. W. Prochanska, J. X., & Wolfe, A. M. 2001, Phys. Rev. Lett., 87, 091301

Weinberg, S. 1987, Phys. Rev. Lett., 59, 2607

——. 2001, in Sources and Detection of Dark Matter and Energy in the Universe, ed. D. B. Cline (Berlin: Springer), 18

Wolfe, A. M., Brown, R. L., & Roberts, M. S. 1976, Phys. Rev. Lett., 37, 179

Wood, P. R. 1981, in Physical Processes in Red Giants, ed. I. Iben, Jr. & A. Renzini (Dordrecht: Reidel), 205

Yungelson, L., Tutukov, A. V., & Livio, M. 1993, ApJ, 418, 794

Zlatev, I., Wang, L., & Steinhardt, P. J. 1998, Phys. Rev. Lett., 82, 896

17

Evidence from Type Ia supernovae
for an accelerating Universe and dark energy

ALEXEI V. FILIPPENKO
Department of Astronomy, University of California, Berkeley

Abstract

I review the use of Type Ia supernovae (SNe Ia) for cosmological distance determinations. Low-redshift SNe Ia ($z \lesssim 0.1$) demonstrate that the Hubble expansion is linear, that $H_0 = 65 \pm 2$ (statistical) km s^{-1} Mpc^{-1}, and that the properties of dust in other galaxies are similar to those of dust in the Milky Way. The light curves of high-redshift ($z = 0.3-1$) SNe Ia are stretched in a manner consistent with the expansion of space; similarly, their spectra exhibit slower temporal evolution (by a factor of $1+z$) than those of nearby SNe Ia. The measured luminosity distances of SNe Ia as a function of redshift have shown that the expansion of the Universe is currently accelerating, probably due to the presence of repulsive dark energy such as Einstein's cosmological constant (Λ). Combining our data with existing measurements of the cosmic microwave background (CMB) radiation and with the results of large-scale structure surveys, we find a best fit for Ω_m and Ω_Λ of about 0.3 and 0.7, respectively. Other studies (e.g., masses of clusters of galaxies) also suggest that $\Omega_m \approx 0.3$. The sum of the densities, ~ 1.0, agrees with the value predicted by most inflationary models for the early Universe: the Universe is flat on large scales. A number of possible systematic effects (dust, supernova evolution) thus far do not seem to eliminate the need for $\Omega_\Lambda > 0$. Most recently, analyses of SNe Ia at $z = 1.0 - 1.7$ provide further support for current acceleration, and give tentative evidence for an early epoch of deceleration. Current projects include the search for additional SNe Ia at $z > 1$ to confirm the early deceleration, and the measurement of a few hundred SNe Ia at $z = 0.2 - 0.8$ to determine the equation of state of the dark energy, $w = P/(\rho c^2)$.

17.1   Introduction

Supernovae (SNe) come in two main varieties (see Filippenko 1997b for a review). Those whose optical spectra exhibit hydrogen are classified as Type II, while hydrogen-deficient SNe are designated Type I. SNe I are further subdivided according to the appearance of the early-time spectrum: SNe Ia are characterized by strong absorption near 6150 Å (now attributed to Si II), SNe Ib lack this feature but instead show prominent He I lines, and SNe Ic have neither the Si II nor the He I lines. SNe Ia are believed to result from the thermonuclear disruption of carbon-oxygen white dwarfs, while SNe II come from core collapse in massive supergiant stars. The latter mechanism probably produces most SNe Ib/Ic as well, but the progenitor stars previously lost their outer layers of hydrogen or even helium.

It has long been recognized that SNe Ia may be very useful distance indicators for a number of reasons; see Branch & Tammann (1992), Branch (1998), and references therein.

(1) They are exceedingly luminous, with peak M_B averaging -19.2 mag if $H_0 = 65$ km s^{-1} Mpc^{-1}. (2) "Normal" SNe Ia have small dispersion among their peak absolute magnitudes ($\sigma \lesssim 0.3$ mag). (3) Our understanding of the progenitors and explosion mechanism of SNe Ia is on a reasonably firm physical basis. (4) Little cosmic evolution is expected in the peak luminosities of SNe Ia, and it can be modeled. This makes SNe Ia superior to galaxies as distance indicators. (5) One can perform *local* tests of various possible complications and evolutionary effects by comparing nearby SNe Ia in different environments.

Research on SNe Ia in the 1990s has demonstrated their enormous potential as cosmological distance indicators. Although there are subtle effects that must indeed be taken into account, it appears that SNe Ia provide among the most accurate values of H_0, q_0 (the deceleration parameter), Ω_m (the matter density), and Ω_Λ [the cosmological constant, $\Lambda c^2/(3H_0^2)$].

There have been two major teams involved in the systematic investigation of high-redshift SNe Ia for cosmological purposes. The "Supernova Cosmology Project" (SCP) is led by Saul Perlmutter of the Lawrence Berkeley Laboratory, while the "High-Z Supernova Search Team" (HZT) is led by Brian Schmidt of the Mt. Stromlo and Siding Springs Observatories. I have been privileged to work with both teams (see Filippenko 2001 for a personal account), but my primary allegiance is now with the HZT.

17.2   Homogeneity and Heterogeneity

Until the mid-1990s, the traditional way in which SNe Ia were used for cosmological distance determinations was to assume that they are perfect "standard candles" and to compare their observed peak brightness with those of SNe Ia in galaxies whose distances had been independently determined (e.g., with Cepheid variables). The rationale was that SNe Ia exhibit relatively little scatter in their peak blue luminosity ($\sigma_B \approx 0.4$–0.5 mag; Branch & Miller 1993), and even less if "peculiar" or highly reddened objects were eliminated from consideration by using a color cut. Moreover, the optical spectra of SNe Ia are usually rather homogeneous, if care is taken to compare objects at similar times relative to maximum brightness (Riess et al. 1997, and references therein). Over 80% of all SNe Ia discovered through the early 1990s were "normal" (Branch, Fisher, & Nugent 1993).

From a Hubble diagram constructed with unreddened, moderately distant SNe Ia ($z \lesssim 0.1$) for which peculiar motions should be small and relative distances (as given by ratios of redshifts) are accurate, Vaughan et al. (1995) find that

$$\langle M_B(\max) \rangle = (-19.74 \pm 0.06) + 5 \log(H_0/50) \text{ mag.}$$

In a series of papers, Sandage et al. (1996) and Saha et al. (1997) combine similar relations with *Hubble Space Telescope (HST)* Cepheid distances to the host galaxies of seven SNe Ia to derive $H_0 = 57 \pm 4$ km s^{-1} Mpc^{-1}.

Over the past two decades it has become clear, however, that SNe Ia do *not* constitute a perfectly homogeneous subclass (e.g., Filippenko 1997a,b). In retrospect this should have been obvious: the Hubble diagram for SNe Ia exhibits scatter larger than the photometric errors, the dispersion actually *rises* when reddening corrections are applied (under the assumption that all SNe Ia have uniform, very blue intrinsic colors at maximum; van den Bergh & Pazder 1992; Sandage & Tammann 1993), and there are some significant outliers whose anomalous magnitudes cannot possibly be explained by extinction alone.

Spectroscopic and photometric peculiarities have been noted with increasing frequency in well-observed SNe Ia. A striking case is SN 1991T; its pre-maximum spectrum did not exhibit Si II or Ca II absorption lines, yet two months past maximum brightness the spectrum was nearly indistinguishable from that of a classical SN Ia (Filippenko et al. 1992b; Phillips et al. 1993). The light curves of SN 1991T were slightly broader than the SN Ia template curves, and the object was probably somewhat more luminous than average at maximum. Another well-observed, peculiar SNe Ia is SN 1991bg (Filippenko et al. 1992a; Leibundgut et al. 1993; Turatto et al. 1996). At maximum brightness it was subluminous by 1.6 mag in *V* and 2.5 mag in *B*, its colors were intrinsically red, and its spectrum was peculiar (with a deep absorption trough due to Ti II). Moreover, the decline from maximum was very steep, the *I*-band light curve did not exhibit a secondary maximum like normal SNe Ia, and the velocity of the ejecta was unusually low. The photometric heterogeneity among SNe Ia is well demonstrated by Suntzeff (1996) with objects having excellent *BVRI* light curves.

17.3   Cosmological Uses: Low Redshifts

Although SNe Ia can no longer be considered perfect "standard candles," they are still exceptionally useful for cosmological distance determinations. Excluding those of low luminosity (which are hard to find, especially at large distances), most of the nearby SNe Ia that had been discovered through the early 1990s were *nearly* standard (Branch et al. 1993; but see Li et al. 2001b for more recent evidence of a higher intrinsic peculiarity rate). Also, after many tenuous suggestions (e.g., Pskovskii 1977, 1984; Branch 1981), Phillips (1993) found convincing evidence for a correlation between light curve shape and the luminosity at maximum brightness by quantifying the photometric differences among a set of nine well-observed SNe Ia, using a parameter [$\Delta m_{15}(B)$] that measures the total drop (in *B* magnitudes) from *B*-band maximum to $t = 15$ days after *B* maximum. In all cases the host galaxies of his SNe Ia have accurate relative distances from surface brightness fluctuations or from the Tully-Fisher relation. The intrinsically bright SNe Ia clearly decline more slowly than dim ones, but the correlation is stronger in *B* than in *V* or *I*.

Using SNe Ia discovered during the Calán/Tololo survey ($z \lesssim 0.1$), Hamuy et al. (1995, 1996b) confirm and refine the Phillips (1993) correlation between peak luminosity and $\Delta m_{15}(B)$. Apparently the slope is steep only at low luminosities; thus, objects such as SN 1991bg skew the slope of the best-fitting single straight line. Hamuy et al. reduce the scatter in the Hubble diagram of normal, unreddened SNe Ia to only 0.17 mag in *B* and 0.14 mag in *V*; see also Tripp (1997). Yet another parameterization is the "stretch" method of Perlmutter et al. (1997) and Goldhaber et al. (2001): the *B*-band light curves of SNe Ia appear nearly identical when expanded or contracted temporally by a factor $(1+s)$, where the value of *s* varies among objects. In a similar but distinct effort, Riess, Press, & Kirshner (1995) show that the luminosity of SNe Ia correlates with the detailed *shape* of the overall light curve.

By using light curve shapes measured through several different filters, Riess, Press, & Kirshner (1996a) extend their analysis and objectively eliminate the effects of interstellar extinction: a SN Ia that has an unusually red *B−V* color at maximum brightness is assumed to be *intrinsically* subluminous if its light curves rise and decline quickly, or of normal luminosity but significantly *reddened* if its light curves rise and decline more slowly. With a set of 20 SNe Ia consisting of the Calán/Tololo sample and their own objects, they show that the dispersion decreases from 0.52 mag to 0.12 mag after application of this "multi-color light curve shape" (MLCS) method. The results from a recent, expanded set of nearly 50 SNe Ia indicate that the dispersion decreases from 0.44 mag to 0.15 mag (Fig. 17.1). The resulting Hubble constant is 65 ± 2 (statistical) km s^{-1} Mpc^{-1}, with an additional systematic

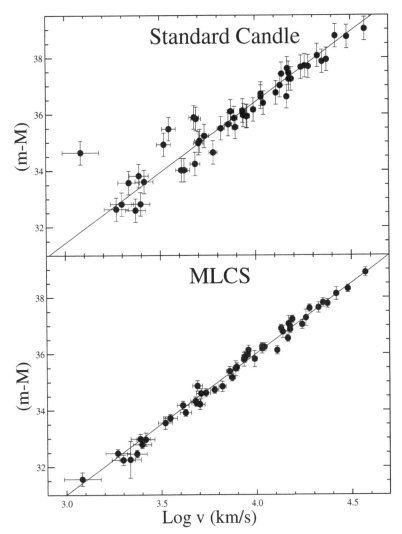

Fig. 17.1. Hubble diagrams for SNe Ia (A. G. Riess 2002, private communication) with velocities (km s^{-1}) in the *COBE* rest frame on the Cepheid distance scale. *Top:* The objects are assumed to be *standard candles* and there is no correction for extinction; the result is $\sigma = 0.42$ mag and $H_0 = 58 \pm 8$ km s^{-1} Mpc^{-1}. *Bottom:* The same objects, after correction for reddening and intrinsic differences in luminosity. The result is $\sigma = 0.15$ mag and $H_0 = 65 \pm 2$ (statistical) km s^{-1} Mpc^{-1}.

and zeropoint uncertainty of ± 5 km s^{-1} Mpc^{-1}. Riess et al. (1996a) also show that the Hubble flow is remarkably linear; indeed, SNe Ia now constitute the best evidence for linearity. Finally, they argue that the dust affecting SNe Ia is *not* of circumstellar origin, and show quantitatively that the extinction curve in external galaxies typically does not differ from that in the Milky Way (cf. Branch & Tammann 1992, but see Tripp 1998).

The advantage of systematically correcting the luminosities of SNe Ia at high redshifts rather than trying to isolate "normal" ones seems clear in view of evidence that the luminosity of SNe Ia may be a function of stellar population. If the most luminous SNe Ia occur in young stellar populations (e.g., Hamuy et al. 1996a, 2000; Branch, Baron, & Roman-ishin 1996; Ivanov, Hamuy, & Pinto 2000), then we might expect the mean peak luminosity of high-z SNe Ia to differ from that of a local sample. Alternatively, the use of Cepheids (Population I objects) to calibrate local SNe Ia can lead to a zeropoint that is too luminous. On the other hand, as long as the physics of SNe Ia is essentially the same in young stellar populations locally and at high redshift, we should be able to adopt the luminosity correction methods (photometric and spectroscopic) found from detailed studies of low-z SNe Ia.

Large numbers of nearby SNe Ia are now being found by my team's Lick Observatory Su-pernova Search (LOSS) conducted with the 0.76-m Katzman Automatic Imaging Telescope (KAIT; Li et al. 2000; Filippenko et al. 2001; see http://astro.berkeley.edu/~bait/kait.html). CCD images are taken of ~ 1000 galaxies per night and compared with KAIT "template im-ages" obtained earlier; the templates are automatically subtracted from the new images and analyzed with computer software. The system reobserves the best candidates the same night, to eliminate star-like cosmic rays, asteroids, and other sources of false alarms. The next day, undergraduate students at UC Berkeley examine all candidates, including weak ones, and they glance at all subtracted images to locate SNe that might be near bright, poorly sub-tracted stars or galactic nuclei. LOSS discovered 20 SNe (of all types) in 1998, 40 in 1999, 38 in 2000, 69 in 2001, and 82 in 2002, making it by far the world's most successful search for nearby SNe. The most important objects were photometrically monitored through *BVRI* (and sometimes *U*) filters (e.g., Li et al. 2001a, 2003; Modjaz et al. 2001; Leonard et al. 2002a,b), and unfiltered follow-up observations (e.g., Matheson et al. 2001) were made of most of them during the course of the SN search. This growing sample of well-observed SNe Ia should allow us to more precisely calibrate the MLCS method, as well as to look for correlations between the observed properties of the SNe and their environment (Hubble type of host galaxy, metallicity, stellar population, etc.).

17.4   Cosmological Uses: High Redshifts

These same techniques can be applied to construct a Hubble diagram with high-redshift SNe Ia, from which the value of $q_0 = (\Omega_m/2) - \Omega_\Lambda$ can be determined. With enough objects spanning a range of redshifts, we can determine Ω_m and Ω_Λ independently (e.g., Goobar & Perlmutter 1995). Contours of peak apparent R-band magnitude for SNe Ia at two redshifts have different slopes in the Ω_m–Ω_Λ plane, and the regions of intersection provide the answers we seek.

17.4.1   *The Search*

Based on the pioneering work of Norgaard-Nielsen et al. (1989), whose goal was to find SNe in moderate-redshift clusters of galaxies, the SCP (Perlmutter et al. 1995a, 1997) and the HZT (Schmidt et al. 1998) devised a strategy that almost guarantees the discov-ery of many faint, distant SNe Ia "on demand," during a predetermined set of nights. This "batch" approach to studying distant SNe allows follow-up spectroscopy and photometry to be scheduled in advance, resulting in a systematic study not possible with random discover-ies. Most of the searched fields are equatorial, permitting follow-up from both hemispheres. The SCP was the first group to convincingly demonstrate the ability to find SNe in batches.

Our approach is simple in principle. Pairs of first-epoch images are obtained with wide-field cameras on large telescopes (e.g., the Big Throughput Camera on the CTIO 4-m Blanco

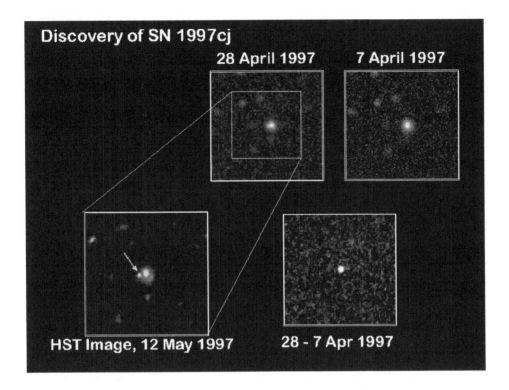

Fig. 17.2. Discovery image of SN 1997cj (28 April 1997), along with the template image and an *HST* image obtained subsequently. The net (subtracted) image is also shown.

telescope) during the nights around new moon, followed by second-epoch images 3–4 weeks later. (Pairs of images permit removal of cosmic rays, asteroids, and distant Kuiper-belt objects.) These are compared immediately using well-tested software, and new SN candidates are identified in the second-epoch images (Fig. 17.2). Spectra are obtained as soon as possible after discovery to verify that the objects are SNe Ia and determine their redshifts. Each team has already found over 150 SNe in concentrated batches, as reported in numerous *IAU Circulars* (e.g., Perlmutter et al. 1995b, 11 SNe with $0.16 \lesssim z \lesssim 0.65$; Suntzeff et al. 1996, 17 SNe with $0.09 \lesssim z \lesssim 0.84$). The observed SN Ia rate at $z \approx 0.5$ is consistent with the low-z SN Ia rate together with plausible star-formation histories (Pain et al. 2002; Tonry et al. 2003), but the error bars on the high-z rate are still quite large.

Intensive photometry of the SNe Ia commences within a few days after procurement of the second-epoch images; it is continued throughout the ensuing and subsequent dark runs. In a few cases *HST* images are obtained. As expected, most of the discoveries are *on the rise or near maximum brightness*. When possible, the SNe are observed in filters that closely match the redshifted B and V bands; this way, the K-corrections become only a second-order effect (Kim, Goobar, & Perlmutter 1996; Nugent, Kim, & Perlmutter 2002). We try to obtain excellent multi-color light curves, so that reddening and luminosity corrections can be applied (Riess et al. 1996a; Hamuy et al. 1996a,b).

Although SNe in the magnitude range 22–22.5 can sometimes be spectroscopically confirmed with 4-m class telescopes, the signal-to-noise ratios are low, even after several hours of integration. Certainly Keck or the VLT are required for the fainter objects (22.5–24.5 mag). With the largest telescopes, not only can we rapidly confirm a substantial number of candidate SNe, but we can search for peculiarities in the spectra that might indicate evolution of SNe Ia with redshift. Moreover, high-quality spectra allow us to measure the age of a SN: we have developed a method for automatically comparing the spectrum of a SN Ia with a library of spectra corresponding to many different epochs in the development of SNe Ia (Riess et al. 1997). Our technique also has great practical utility at the telescope: we can determine the age of a SN "on the fly," within half an hour after obtaining its spectrum. This allows us to decide rapidly which SNe are best for subsequent photometric follow-up, and we immediately alert our collaborators on other telescopes.

17.4.2   Results

First, we note that the light curves of high-redshift SNe Ia are broader than those of nearby SNe Ia; the initial indications (Leibundgut et al. 1996; Goldhaber et al. 1997), based on small numbers of SNe Ia, are amply confirmed with the larger samples (Goldhaber et al. 2001). Quantitatively, the amount by which the light curves are "stretched" is consistent with a factor of $1 + z$, as expected if redshifts are produced by the expansion of space rather than by "tired light" and other non-expansion hypotheses for the redshifts of objects at cosmological distances. [For non-standard cosmological interpretations of the SN Ia data, see Narlikar & Arp (1997) and Hoyle, Burbidge, & Narlikar (2000).] We also demonstrate this *spectroscopically* at the 2σ confidence level for a single object: the spectrum of SN 1996bj ($z = 0.57$) evolved more slowly than those of nearby SNe Ia, by a factor consistent with $1 + z$ (Riess et al. 1997). Although one might be able to argue that something other than universal expansion could be the cause of the apparent stretching of SN Ia light curves at high redshifts, it is much more difficult to attribute apparently slower evolution of spectral details to an unknown effect.

The formal value of Ω_m derived from SNe Ia has changed with time. The SCP published the first result (Perlmutter et al. 1995a), based on a single object, SN 1992bi at $z = 0.458$: $\Omega_m = 0.2 \pm 0.6 \pm 1.1$ (assuming that $\Omega_\Lambda = 0$). The SCP's analysis of their first seven objects (Perlmutter et al. 1997) suggested a much larger value of $\Omega_m = 0.88 \pm 0.6$ (if $\Omega_\Lambda = 0$) or $\Omega_m = 0.94 \pm 0.3$ (if $\Omega_{\text{total}} = 1$). Such a high-density universe seemed at odds with other, independent measurements of Ω_m. However, with the subsequent inclusion of just one more object, SN 1997ap at $z = 0.83$ (the highest known for a SN Ia at the time; Perlmutter et al. 1998), their estimates were revised back down to $\Omega_m = 0.2 \pm 0.4$ if $\Omega_\Lambda = 0$, and $\Omega_m = 0.6 \pm 0.2$ if $\Omega_{\text{total}} = 1$; the apparent brightness of SN 1997ap had been precisely measured with *HST*, so it substantially affected the best fits.

Meanwhile, the HZT published (Garnavich et al. 1998a) an analysis of four objects (three of them observed with *HST*), including SN 1997ck at $z = 0.97$, at that time a redshift record, although they cannot be absolutely certain that the object was a SN Ia because the spectrum is too poor. From these data, the HZT derived that $\Omega_m = -0.1 \pm 0.5$ (assuming $\Omega_\Lambda = 0$) and $\Omega_m = 0.35 \pm 0.3$ (assuming $\Omega_{\text{total}} = 1$), inconsistent with the high Ω_m initially found by Perlmutter et al. (1997) but consistent with the revised estimate in Perlmutter et al. (1998). An independent analysis of 10 SNe Ia using the "snapshot" distance method (with which conclusions are drawn from sparsely observed SNe Ia) gave quantitatively similar

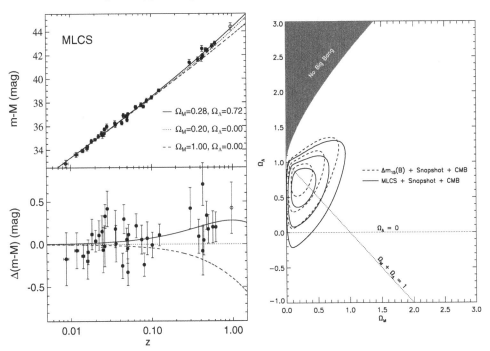

Fig. 17.3. *Left:* The upper panel shows the Hubble diagram for the low-z and high-z HZT SN Ia sample with MLCS distances; see Riess et al. (1998b). Overplotted are three world models: "low" and "high" Ω_m with $\Omega_\Lambda = 0$, and the best fit for a flat universe ($\Omega_m = 0.28$, $\Omega_\Lambda = 0.72$). The bottom panel shows the difference between data and models from the $\Omega_m = 0.20$, $\Omega_\Lambda = 0$ prediction. Only the 10 best-observed high-z SNe Ia are shown. The average difference between the data and the $\Omega_m = 0.20$, $\Omega_\Lambda = 0$ prediction is ~ 0.25 mag. *Right:* The HZT's combined constraints from SNe Ia (left) and the position of the first acoustic peak of the CMB angular power spectrum, based on data available in mid-1998; see Garnavich et al. (1998b). The contours mark the 68.3%, 95.4%, and 99.7% enclosed probability regions. Solid curves correspond to results from the MLCS method; dotted ones are from the $\Delta m_{15}(B)$ method; all 16 SNe Ia in Riess et al. (1998b) were used.

conclusions (Riess et al. 1998a). However, none of these early data sets carried the statistical discriminating power to detect cosmic acceleration.

The SCP's next results were announced at the 1998 January AAS meeting in Washington, DC. A press conference was scheduled, with the stated purpose of presenting and discussing the then-current evidence for a low-Ω_m universe as published by Perlmutter et al. (1998; SCP) and Garnavich et al. (1998a; HZT). When showing the SCP's Hubble diagram for SNe Ia, however, Perlmutter also pointed out tentative evidence for *acceleration*! He stated that the conclusion was uncertain, and that the data were equally consistent with no acceleration; the systematic errors had not yet been adequately assessed. Essentially the same conclusion was given by the SCP in their talks at a conference on dark matter, near Los Angeles, in February 1998 (Goldhaber & Perlmutter 1998).

Although it chose not to reveal them at the same 1998 January AAS meeting, the HZT already had similar, tentative evidence for acceleration in their own SN Ia data set. The HZT continued to perform numerous checks of their data analysis and interpretation, including fairly thorough consideration of various possible systematic effects. Unable to find any significant problems, even with the possible systematic effects, the HZT reported detection of a *nonzero* value for Ω_Λ (based on 16 high-z SNe Ia) at the Los Angeles dark matter conference in February 1998 (Filippenko & Riess 1998), and soon thereafter submitted a formal paper that was published in September 1998 (Riess et al. 1998b). Their original Hubble diagram for the 10 best-observed high-z SNe Ia is given in Figure 17.3 (*left*). With the MLCS method applied to the full set of 16 SNe Ia, the HZT's formal results were $\Omega_m = 0.24 \pm 0.10$ if $\Omega_{\text{total}} = 1$, or $\Omega_m = -0.35 \pm 0.18$ (unphysical) if $\Omega_\Lambda = 0$. If one demanded that $\Omega_m = 0.2$, then the best value for Ω_Λ was 0.66 ± 0.21. These conclusions did not change significantly when only the 10 best-observed SNe Ia were used (Fig. 17.3, *left*; $\Omega_m = 0.28 \pm 0.10$ if $\Omega_{\text{total}} = 1$).

Another important constraint on the cosmological parameters could be obtained from measurements of the angular scale of the first acoustic peak of the CMB (e.g., Zaldarriaga, Spergel, & Seljak 1997; Eisenstein, Hu, & Tegmark 1998); the SN Ia and CMB techniques provide nearly complementary information. A stunning result was already available by mid-1998 from existing measurements (e.g., Hancock et al. 1998; Lineweaver & Barbosa 1998): the HZT's analysis of the SN Ia data in Riess et al. (1998b) demonstrated that $\Omega_m + \Omega_\Lambda = 0.94 \pm 0.26$ (Fig. 17.3, *right*), when the SN and CMB constraints were combined (Garnavich et al. 1998b; see also Lineweaver 1998, Efstathiou et al. 1999, and others).

Somewhat later (June 1999), the SCP published almost identical results, implying an accelerating expansion of the Universe, based on an essentially independent set of 42 high-z SNe Ia (Perlmutter et al. 1999). Their data, together with those of the HZT, are shown in Figure 17.4 (*left*), and the corresponding confidence contours in the Ω_Λ vs. Ω_m plane are given in Figure 17.4 (*right*). This incredible agreement suggested that neither group had made a large, simple blunder; if the result was wrong, the reason must be subtle. Had there been only one team working in this area, it is likely that far fewer astronomers and physicists throughout the world would have taken the result seriously.

Moreover, already in 1998–1999 there was tentative evidence that the "dark energy" driving the accelerated expansion was indeed consistent with the cosmological constant, Λ. If Λ dominates, then the equation of state of the dark energy should have an index $w = -1$, where the pressure (P) and density (ρ) are related according to $w = P/(\rho c^2)$. Garnavich et al. (1998b) and Perlmutter et al. (1999) already set an interesting limit, $w \lesssim -0.60$ at the 95% confidence level. However, more high-quality data at $z \approx 0.5$ are needed to narrow the allowed range, in order to test other proposed candidates for dark energy such as various forms of "quintessence" (e.g., Caldwell, Davé, & Steinhardt 1998).

Although the CMB results appeared reasonably persuasive in 1998–1999, one could argue that fluctuations on different scales had been measured with different instruments, and that suble systematic effects might lead to erroneous conclusions. These fears were dispelled only 1–2 years later, when the more accurate and precise results of the BOOMERANG collaboration were announced (de Bernardis et al. 2000, 2002). Shortly thereafter the MAXIMA collaboration distributed their very similar findings (Hanany et al. 2000; Balbi et al. 2000; Netterfield et al. 2002; see also the TOCO, DASI, and many other measurements). Figure 17.4 (*right*) illustrates that the CMB measurements tightly constrain Ω_{total} to be close to unity; we appear to live in a flat universe, in agreement with most inflationary models for the early Universe! Combined with the SN Ia results, the evidence for nonzero Ω_Λ was fairly strong. Making the argument even more compelling was the fact that various studies

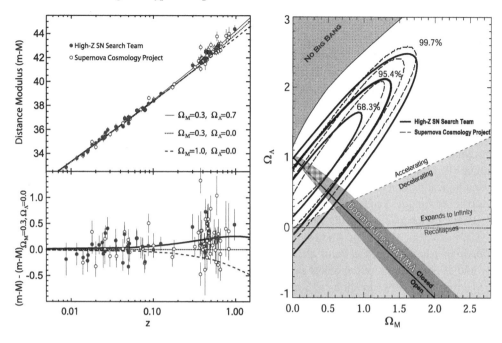

Fig. 17.4. *Left:* As in Fig. 17.3, but this time including both the HZT (Riess et al. 1998b) and SCP (Perlmutter et al. 1999) samples of low-redshift and high-redshift SNe Ia. Over-plotted are three world models: $\Omega_m = 0.3$ and 1.0 with $\Omega_\Lambda = 0$, and a flat universe ($\Omega_{total} = 1.0$) with $\Omega_\Lambda = 0.7$. The bottom panel shows the difference between data and models from the $\Omega_m = 0.3$, $\Omega_\Lambda = 0$ prediction. *Right:* The combined constraints from SNe Ia (left) and the position of thefirst acoustic peak of the CMB angular power spectrum, based on BOOMERANG and MAXIMA data. The contours mark the 68.3%, 95.4%, and 99.7% enclosed probability regions determined from the SNe Ia. According to the CMB, $\Omega_{total} \approx 1.0$.

of clusters of galaxies (see summary by Bahcall et al. 1999) showed that $\Omega_m \approx 0.3$, consistent with the results in Figures 17.3 and 17.4. Thus, a "concordance cosmology" had emerged: $\Omega_m \approx 0.3$, $\Omega_\Lambda \approx 0.7$—consistent with what had been suspected some years earlier by Ostriker & Steinhardt (1995; see also Carroll, Press, & Turner 1992).

Yet another piece of evidence for a nonzero value of Λ was provided by the Two-Degree Field Galaxy Redshift Survey (2dFGRS; Peacock et al. 2001; Percival et al. 2001; Efstathiou et al. 2002). Combined with the CMB maps, their results are inconsistent with a universe dominated by gravitating dark matter. Again, the implication is that about 70% of the mass-energy density of the Universe consists of some sort of dark energy whose gravitational effect is repulsive. Just as this review was going to press, results from the *Wilkinson Microwave Anisotropy Prove (WMAP)* appeared; together with the 2dFGRS constraints, they confirm and refine the concordance cosmology ($\Omega_m = 0.27$, $\Omega_\Lambda = 0.73$, $\Omega_{baryon} = 0.044$, $H_0 = 71 \pm 4$ km s^{-1} Mpc^{-1}; Spergel et al. 2003).

The dynamical age of the Universe can be calculated from the cosmological parameters. In an empty Universe with no cosmological constant, the dynamical age is simply the "Hubble time" (i.e., the inverse of the Hubble constant); there is no deceleration. SNe Ia yield

$H_0 = 65 \pm 2$ km s^{-1} Mpc^{-1} (statistical uncertainty only), and a Hubble time of 15.1 ± 0.5 Gyr. For a more complex cosmology, integrating the velocity of the expansion from the current epoch ($z = 0$) to the beginning ($z = \infty$) yields an expression for the dynamical age. As shown in detail by Riess et al. (1998b), by mid-1998 the HZT had obtained a value of $14.2^{+1.0}_{-0.8}$ Gyr using the likely range for (Ω_m, Ω_Λ) that they measured. (The precision was so high because their experiment was sensitive to roughly the *difference* between Ω_m and Ω_Λ, and the dynamical age also varies in approximately this way.) Including the *systematic* uncertainty of the Cepheid distance scale, which may be up to 10%, a reasonable estimate of the dynamical age was 14.2 ± 1.7 Gyr (Riess et al. 1998b). Again, the SCP's result was very similar (Perlmutter et al. 1999), since it was based on nearly the same derived values for the cosmological parameters. This expansion age is consistent with ages determined from various other techniques such as the cooling of white dwarfs (Galactic disk > 9.5 Gyr; Oswalt et al. 1996), radioactive dating of stars via the thorium and europium abundances (15.2 ± 3.7 Gyr; Cowan et al. 1997), and studies of globular clusters (10–15 Gyr, depending on whether *Hipparcos* parallaxes of Cepheids are adopted; Gratton et al. 1997; Chaboyer et al. 1998). By mid-1998, the ages of the oldest stars no longer seemed to exceed the expansion age of the Universe; the long-standing "age crisis" had evidently been resolved.

17.5   Discussion

Although the convergence of different methods on the same answer is reassuring, and suggests that the concordance cosmology is correct, it is important to vigorously test each method to make sure it is not leading us astray. Moreover, only through such detailed studies will the accuracy and precision of the methods improve, allowing us to eventually set better constraints on the equation of state parameter, w. Here I discuss the systematic effects that could adversely affect the SN Ia results.

High-redshift SNe Ia are observed to be dimmer than expected in an empty Universe (i.e., $\Omega_m = 0$) with no cosmological constant. At $z \approx 0.5$, where the SN Ia observations have their greatest leverage on Λ, the difference in apparent magnitude between an $\Omega_m = 0.3$ ($\Omega_\Lambda = 0$) universe and a flat universe with $\Omega_\Lambda = 0.7$ is only about 0.25 mag. Thus, we need to find out if chemical abundances, stellar populations, selection bias, gravitational lensing, or grey dust can have an effect this large. Although both the HZT and SCP had considered many of these potential systematic effects in their original discovery papers (Riess et al. 1998b; Perlmutter et al. 1999), and had shown with reasonable confidence that obvious ones were not greatly affecting their conclusions, if was of course possible that they were wrong, and that the data were being misinterpreted.

17.5.1   Evolution

Perhaps the most obvious possible culprit is *evolution* of SNe Ia over cosmic time, due to changes in metallicity, progenitor mass, or some other factor. If the peak luminosity of SNe Ia were lower at high redshift, then the case for $\Omega_\Lambda > 0$ would weaken. Conversely, if the distant explosions are more powerful, then the case for acceleration strengthens. Theorists are not yet sure what the sign of the effect will be, if it is present at all; different assumptions lead to different conclusions (Höflich et al. 1998; Umeda et al. 1999; Nomoto et al. 2000; Yungelson & Livio 2000).

Of course, it is extremely difficult, if not effectively impossible, to obtain an accurate, independent measure of the peak luminosity of high-z SNe Ia, and hence to directly test for luminosity evolution. However, we can more easily determine whether *other* observable properties of low-z and high-z SNe Ia differ. If they are all the same, it is more probable that the peak luminosity is constant as well—but if they differ, then the peak luminosity might also be affected (e.g., Höflich et al. 1998). Drell, Loredo, & Wasserman (2000), for example, argue that there are reasons to suspect evolution, because the average properties of existing samples of high-z and low-z SNe Ia seem to differ (e.g., the high-z SNe Ia are more uniform).

The local sample of SNe Ia displays a weak correlation between light curve shape (or peak luminosity) and host galaxy type, in the sense that the most luminous SNe Ia with the broadest light curves only occur in late-type galaxies. Both early-type and late-type galaxies provide hosts for dimmer SNe Ia with narrower light curves (Hamuy et al. 1996a). The mean luminosity difference for SNe Ia in late-type and early-type galaxies is ~ 0.3 mag. In addition, the SN Ia rate per unit luminosity is almost twice as high in late-type galaxies as in early-type galaxies at the present epoch (Cappellaro et al. 1997). These results may indicate an evolution of SNe Ia with progenitor age. Possibly relevant physical parameters are the mass, metallicity, and C/O ratio of the progenitor (Höflich et al. 1998).

We expect that the relation between light curve shape and peak luminosity that applies to the range of stellar populations and progenitor ages encountered in the late-type and early-type hosts in our nearby sample should also be applicable to the range we encounter in our distant sample. In fact, the range of age for SN Ia progenitors in the nearby sample is likely to be *larger* than the change in mean progenitor age over the 4–6 Gyr lookback time to the high-z sample. Thus, to first order at least, our local sample should correct the distances for progenitor or age effects.

We can place empirical constraints on the effect that a change in the progenitor age would have on our SN Ia distances by comparing subsamples of low-redshift SNe Ia believed to arise from old and young progenitors. In the nearby sample, the mean difference between the distances for the early-type hosts (8 SNe Ia) and late-type hosts (19 SNe Ia), at a given redshift, is 0.04 ± 0.07 mag from the MLCS method. This difference is consistent with zero. Even if the SN Ia progenitors evolved from one population at low redshift to the other at high redshift, we still would not explain the surplus in mean distance of 0.25 mag over the $\Omega_\Lambda = 0$ prediction. Moreover, in a major study of high-redshift SNe Ia as a function of galaxy morphology, the SCP found no clear differences (except for the amount of scatter; see §17.5.2) between the cosmological results obtained with SNe Ia in late-type and early-type galaxies (Sullivan et al. 2004).

It is also reassuring that initial comparisons of high-z SN Ia spectra appear remarkably similar to those observed at low redshift. For example, the spectral characteristics of SN 1998ai ($z = 0.49$) appear to be essentially indistinguishable from those of normal low-z SNe Ia; see Figure 17.5 (*left*). In fact, the most obviously discrepant spectrum in this figure is the second one from the top, that of SN 1994B ($z = 0.09$); it is intentionally included as a "decoy" that illustrates the degree to which even the spectra of nearby, relatively normal SNe Ia can vary. Nevertheless, it is important to note that a dispersion in luminosity (perhaps 0.2 mag) exists even among the other, more normal SNe Ia shown in Figure 17.5 (*left*); thus, our spectra of SN 1998ai and other high-z SNe Ia are not yet sufficiently good for independent, *precise* determinations of peak luminosity from spectral features (Nugent et al. 1995). Many of them, however, are sufficient for ruling out other SN types (Fig. 17.5, *right*), or for

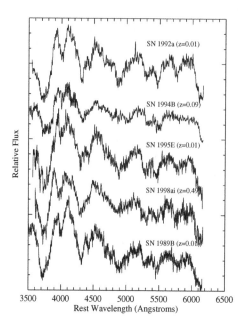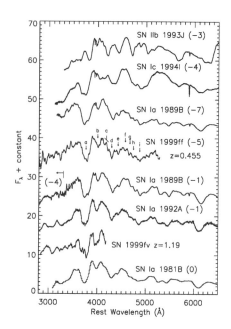

Fig. 17.5. *Left:* Spectral comparison (in f_λ) of SN 1998ai ($z = 0.49$; Keck spectrum) with low-redshift ($z < 0.1$) SNe Ia at a similar age (~ 5 days before maximum brightness), from Riess et al. (1998b). The spectra of the low-redshift SNe Ia were resampled and convolved with Gaussian noise to match the quality of the spectrum of SN 1998ai. Overall, the agreement in the spectra is excellent, tentatively suggesting that distant SNe Ia are physically similar to nearby SNe Ia. SN 1994B ($z = 0.09$) differs the most from the others, and was included as a "decoy." *Right:* Heavily smoothed spectra of two high-z SNe (SN 1999ff at $z = 0.455$ and SN 1999fv at $z = 1.19$; quite noisy below ~ 3500 Å) are presented along with several low-z SN Ia spectra (SNe 1989B, 1992A, and 1981B), a SN Ib spectrum (SN 1993J), and a SN Ic spectrum (SN 1994I); see Filippenko (1997) for a discussion of spectra of various types of SNe. The date of the spectra relative to *B*-band maximum is shown in parentheses after each object's name. Specific features seen in SN 1999ff and labeled with a letter are discussed by Coil et al. (2000). This comparison shows that the two high-z SNe are most likely SNe Ia.

identifying gross peculiarities such as those shown by SNe 1991T and 1991bg; see Coil et al. (2000).

We can help verify that the SNe at $z \approx 0.5$ being used for cosmology do not belong to a subluminous population of SNe Ia by examining restframe *I*-band light curves. Normal, nearby SNe Ia show a pronounced second maximum in the *I* band about a month after the first maximum and typically about 0.5 mag fainter (e.g., Ford et al. 1993; Suntzeff 1996). Subluminous SNe Ia, in contrast, do not show this second maximum, but rather follow a linear decline or show a muted second maximum (Filippenko et al. 1992a). As discussed by Riess et al. (2000), tentative evidence for the second maximum is seen from the HZT's existing *J*-band (restframe *I*-band) data on SN 1999Q ($z = 0.46$); see Figure 17.6 (*left*). Additional tests with spectra and near-infrared light curves are currently being conducted.

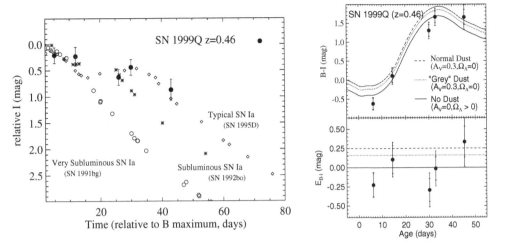

Fig. 17.6. *Left:* Restframe *I*-band (observed *J*-band) light curve of SN 1999Q ($z = 0.46$, 5 solid points; Riess et al. 2000), and the *I*-band light curves of several nearby SNe Ia. Subluminous SNe Ia exhibit a less prominent second maximum than do normal SNe Ia. *Right:* Color excess, E_{B-I}, for SN 1999Q and different dust models (Riess et al. 2000). The data are most consistent with no dust and $\Omega_\Lambda > 0$.

Another way of using light curves to test for possible evolution of SNe Ia is to see whether the rise time (from explosion to maximum brightness) is the same for high-redshift and low-redshift SNe Ia; a difference might indicate that the peak luminosities are also different (Höflich et al. 1998). Riess et al. (1999c) measured the risetime of nearby SNe Ia, using data from KAIT, the Beijing Astronomical Observatory (BAO) SN search, and a few amateur astronomers. Though the exact value of the risetime is a function of peak luminosity, for typical low-redshift SNe Ia it is 20.0 ± 0.2 days. Riess et al. (1999b) pointed out that this differs by 5.8σ from the *preliminary* risetime of 17.5 ± 0.4 days reported in conferences by the SCP (Goldhaber et al. 1998a,b; Groom 1998). However, more thorough analyses of the SCP data (Aldering, Knop, & Nugent 2000; Goldhaber et al. 2001) show that the high-redshift uncertainty of ± 0.4 days that the SCP originally reported was much too small because it did not account for systematic effects. The revised discrepancy with the low-redshift risetime is about 2σ or less. Thus, the apparent difference in risetimes might be insignificant. Even if the difference is real, however, its relevance to the peak luminosity is unclear; the light curves may differ only in the first few days after the explosion, and this could be caused by small variations in conditions near the outer part of the exploding white dwarf that are inconsequential at the peak.

17.5.2 *Extinction*

Our SN Ia distances have the important advantage of including corrections for interstellar extinction occurring in the host galaxy and the Milky Way. Extinction corrections based on the relation between SN Ia colors and luminosity improve distance precision for a sample of nearby SNe Ia that includes objects with substantial extinction (Riess et al. 1996a); the scatter in the Hubble diagram is much reduced. Moreover, the consistency of the measured Hubble flow from SNe Ia with late-type and early-type hosts (see §17.5.1) shows that the extinction corrections applied to dusty SNe Ia at low redshift do not alter the expansion rate from its value measured from SNe Ia in low-dust environments.

In practice, the high-redshift SNe Ia generally appear to suffer very little extinction; their $B-V$ colors at maximum brightness are normal, suggesting little color excess due to reddening. The most detailed available study is that of the SCP (Sullivan et al. 2004): they found that the scatter in the Hubble diagram is minimal for SNe Ia in early-type host galaxies, but increases for SNe Ia in late-type galaxies. Moreover, on average the SNe in late-type galaxies are slightly fainter (by 0.14 ± 0.09 mag) than those in early-type galaxies. Finally, at peak brightness the colors of SNe Ia in late-type galaxies are marginally redder than those in early-type galaxies. Sullivan et al. (2004) conclude that extinction by dust in the host galaxies of SNe Ia is one of the major sources of scatter in the high-redshift Hubble diagram. By restricting their sample to SNe Ia in early-type host galaxies (presumably with minimal extinction), they obtain a very tight Hubble diagram that suggests a nonzero value for Ω_Λ at the 5σ confidence level, under the assumption that $\Omega_{total} = 1$. In the absence of this assumption, SNe Ia in early-type hosts still imply that $\Omega_\Lambda > 0$ at nearly the 98% confidence level. The results for Ω_Λ with SNe Ia in late-type galaxies are quantitatively similar, but statistically less secure because of the larger scatter.

Riess, Press, & Kirshner (1996b) found indications that the Galactic ratios between selective absorption and color excess are similar for host galaxies in the nearby ($z \le 0.1$) Hubble flow. Yet, what if these ratios changed with lookback time (e.g., Aguirre 1999a)? Could an evolution in dust-grain size descending from ancestral interstellar "pebbles" at higher redshifts cause us to underestimate the extinction? Large dust grains would not imprint the reddening signature of typical interstellar extinction upon which our corrections rely.

However, viewing our SNe through such gray interstellar grains would also induce a *dispersion* in the derived distances. Using the results of Hatano, Branch, & Deaton (1998), Riess et al. (1998b) estimate that the expected dispersion would be 0.40 mag if the mean gray extinction were 0.25 mag (the value required to explain the measured MLCS distances without a cosmological constant). This is significantly larger than the 0.21 mag dispersion observed in the high-redshift MLCS distances. Furthermore, most of the observed scatter is already consistent with the estimated *statistical* errors, leaving little to be caused by gray extinction. Nevertheless, if we assumed that *all* of the observed scatter were due to gray extinction, the mean shift in the SN Ia distances would be only 0.05 mag. With the existing observations, it is difficult to rule out this modest amount of gray interstellar extinction.

Gray *intergalactic* extinction could dim the SNe without either telltale reddening or dispersion, if all lines of sight to a given redshift had a similar column density of absorbing material. The component of the intergalactic medium with such uniform coverage corresponds to the gas clouds producing Lyman-α forest absorption at low redshifts. These clouds have individual H I column densities less than about 10^{15} cm^{-2} (Bahcall et al. 1996). However, they display low metallicities, typically less than 10% of solar. Gray extinction would require larger dust grains which would need a larger mass in heavy elements than typical interstellar grain size distributions to achieve a given extinction. It is possible that large dust grains are blown out of galaxies by radiation pressure, and are therefore not associated with Lyman-α clouds (Aguirre 1999b).

But even the dust postulated by Aguirre (1999a,b) and Aguirre & Haiman (1999) is not *completely* gray, having a size of about 0.1 μm. We can test for such nearly gray dust by

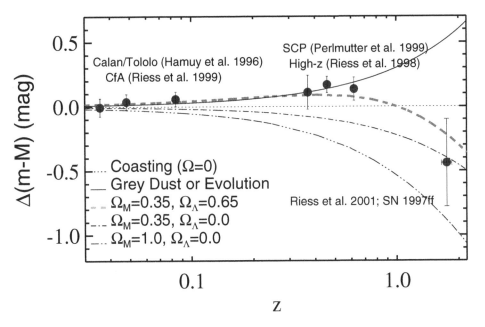

Fig. 17.7. Hubble diagram for SNe Ia relative to an empty universe ($\Omega = 0$) compared with cosmological and astrophysical models (Riess et al. 2001). Low-redshift SNe Ia are from Hamuy et al. (1996a) and Riess et al. (1999a). The magnitude of SN 1997ff at $z = 1.7$ has been corrected for gravitational lensing (Benítez et al. 2002). The measurements of SN 1997ff are inconsistent with astrophysical effects that could mimic previous evidence for an accelerating universe from SNe Ia at $z \approx 0.5$.

observing high-redshift SNe Ia over a wide wavelength range to measure the color excess it would introduce. If $A_V = 0.25$ mag, then $E(U-I)$ and $E(B-I)$ should be 0.12–0.16 mag (Aguirre 1999a,b). If, on the other hand, the 0.25 mag faintness is due to Λ, then no such reddening should be seen. This effect is measurable using proven techniques; so far, with just one SN Ia (SN 1999Q; Fig. 17.6, *right*), our results favor the no-dust hypothesis to better than 2σ (Riess et al. 2000). More work along these lines is in progress.

17.5.3 *The Smoking Gun*

Suppose, however, that for some reason the dust is *very* gray, or our color measurements are not sufficiently precise to rule out Aguirre's (or other) dust. Or, perhaps some other astrophysical systematic effect is fooling us, such as possible evolution of the white dwarf progenitors (e.g., Höflich et al. 1998; Umeda et al. 1999), or gravitational lensing (Wambsganss, Cen, & Ostriker 1998). The most decisive test to distinguish between Λ and cumulative systematic effects is to examine the *deviation* of the observed peak magnitude of SNe Ia from the magnitude expected in the low-Ω_m, zero-Λ model. If Λ is positive, the deviation should actually begin to *decrease* at $z \approx 1$; we will be looking so far back in time that the Λ effect becomes small compared with Ω_m, and the Universe is decelerating at that epoch. If, on the other hand, a systematic bias such as gray dust or evolution of the white dwarf progenitors is the culprit, we expect that the deviation of the apparent magnitude will continue growing, unless the systematic bias is set up in such an unlikely way as to mimic the effects of Λ (Drell et al. 2000). A turnover, or decrease of the deviation of apparent magnitude at high redshift, can be considered the "smoking gun" of Λ.

In a wonderful demonstration of good luck and hard work, Riess et al. (2001) report on *HST* observations of a probable SN Ia at $z \approx 1.7$ (SN 1997ff, the most distant SN ever observed) that suggest the expected turnover is indeed present, providing a tantalizing glimpse of the epoch of deceleration. (See also Benítez et al. 2002, which corrects the observed magnitude of SN 1997ff for gravitational lensing.) SN 1997ff was discovered by Gilliland & Phillips (1998) in a repeat *HST* observation of the Hubble Deep Field–North, and serendipitously monitored in the infrared with *HST*/NICMOS. The peak apparent SN brightness is consistent with that expected in the decelerating phase of the concordance cosmological model, $\Omega_m \approx 0.3$, $\Omega_\Lambda \approx 0.7$ (Fig. 17.7). It is inconsistent with gray dust or simple luminosity evolution, when combined with the data for SNe Ia at $z \approx 0.5$. On the other hand, it is wise to remain cautious: the error bars are large, and it is always possible that we are being fooled by this one object. The HZT and SCP currently have programs to find and measure more SNe Ia at such high redshifts. For example, SN candidates at very high redshifts (e.g., Giavalisco et al. 2002) have been found by "piggybacking" on the Great Observatories Origins Deep Survey (GOODS) being conducted with the Advanced Camera for Surveys aboard *HST*.

Less ambitious programs, concentrating on SNe Ia at $z \gtrsim 0.8$, have already been completed (HZT; Tonry et al. 2003) or are nearing completion (SCP). Tonry et al. (2003) measured several SNe Ia at $z \approx 1$, and their deviation of apparent magnitude from the low-Ω_m, zero-Λ model is roughly the same as that at $z \approx 0.5$, in agreement with expectations based on the results of Riess et al. (2001). Moreover, the new sample of high-redshift SNe Ia presented by Tonry et al., analyzed with methods distinct from (but similar to) those used previously, confirm the result of Riess et al. (1998b) and Perlmutter et al. (1999) that the expansion of the Universe is accelerating. By combining all of the available data sets, Tonry et al. are able to use 230 SNe Ia, and they place the following constraints on cosmological quantities. (1) If the equation of state parameter of the dark energy is $w = -1$, then $H_0 t_0 = 0.96 \pm 0.04$, and $\Omega_\Lambda - 1.4\Omega_m = 0.35 \pm 0.14$. (2) Including the constraint of a flat universe, they find that $\Omega_m = 0.28 \pm 0.05$, independent of any large-scale structure measurements. (3) Adopting a prior based on the 2dFGRS constraint on Ω_m (Percival et al. 2001) and assuming a flat universe, they derive that $-1.48 < w < -0.72$ at 95% confidence. These constraints are similar in precision and in value to very recent conclusions reported using *WMAP* (Spergel et al. 2003), also in combination with the 2dFGRS. Complete details on the SN Ia results, as well as figures, can be found in Tonry et al. (2003).

17.5.4   *Measuring the Dark Energy Equation of State*

Every energy component in the Universe can be parameterized by the way its density varies as the Universe expands (scale factor a), with $\rho \propto a^{-3(1+w)}$, and w reflects the component's equation of state, $w = P/(\rho c^2)$, where P is the pressure exerted by the component. So for matter, $w = 0$, while an energy component that does not vary with scale factor has $w = -1$, as in the cosmological constant Λ. Some really strange energies may have

$w < -1$: their density increases with time (Carroll, Hoffman, & Trodden 2003)! Clearly, a good estimate of w becomes the key to differentiating between models.

The CMB observations imply that the geometry of the universe is close to flat, so the energy density of the dark component is simply related to the matter density by $\Omega_x = 1 - \Omega_m$. This allows the luminosity distance as a function of redshift to be written as

$$D_L(z) = \frac{c(1+z)}{H_0} \int_0^z \frac{[1+\Omega_x((1+z)^{3w}-1)]^{-1/2}}{(1+z)^{3/2}} \, dz \,,$$

showing that the dark energy density and equation of state directly influence the apparent brightness of standard candles. As demonstrated graphically in Figure 17.8 (*left*), SNe Ia observed over a wide range of redshifts can constrain the dark energy parameters to a cosmologically interesting accuracy.

But there are two major problems with using SNe Ia to measure w. First, systematic uncertainties in SN Ia peak luminosity limit how well $D_L(z)$ can be measured. While statistical uncertainty can be arbitrarily reduced by finding thousands of SNe Ia, intrinsic SN properties such as evolution and progenitor metallicity, and observational limits like photometric calibrations and K-corrections, create a systematic floor that cannot be decreased by sheer force of numbers. We expect that systematics can be controlled to at best 3%.

Second, SNe at $z > 1.0$ are very hard to discover and study from the ground. As discussed above, both the HZT and the SCP have found a few SNe Ia at $z > 1.0$, but the numbers and quality of these light curves are insufficient for a w measurement. Large numbers of SNe Ia at $z > 1.0$ are best left to a wide-field optical/infrared imager in space, such as the proposed *Supernova/ Acceleration Probe* (*SNAP*; Nugent et al. 2000) satellite.

Fortunately, an interesting measurement of w can be made at present. The current values of Ω_m from many methods (most recently *WMAP*: 0.27; Spergel et al. 2003) make an excellent substitute for those expensive SNe at $z > 1.0$. Figure 17.8 (*left*) shows that a SN Ia sample with a maximum redshift of $z = 0.8$, combined with the current 10% error on Ω_m, will do as well as a SN Ia sample at much higher redshifts. Within a few years, the Sloan Digital Sky Survey and *WMAP* will solidify the estimate of Ω_m and sharpen w further.

Both the SCP and the HZT are involved in multi-year programs to discover and monitor hundreds of SNe Ia for the purpose of measuring w. For example, the HZT's project, ESSENCE (Equation of State: SupErNovae trace Cosmic Expansion), is designed to discover 200 SNe Ia evenly distributed in the $0.2 < z < 0.8$ range. The CTIO 4-m telescope and mosaic camera are being used to find and follow the SNe by imaging on every other dark night for several consecutive months of the year. Keck and other large telescopes are being used to get the SN spectra and redshifts. Project ESSENCE will eventually provide an estimate of w to an accuracy of $\sim 10\%$ (Fig. 17.8, *right*).

Farther in the future, large numbers of SNe Ia to be found by the *SNAP* satellite and the Large-area Synoptic Survey Telescope (the "Dark Matter Telescope"; Tyson & Angel 2001) could reveal whether the value of w depends on redshift, and hence should give additional constraints on the nature of the dark energy. High-redshift surveys of galaxies such as DEEP2 (Davis et al. 2001), as well as space-based missions to map the CMB (*Planck*), should provide additional evidence for (or against) Λ. Observational cosmology promises to remain exciting for quite some time!

Acknowledgements. I thank all of my HZT collaborators for their contributions to our

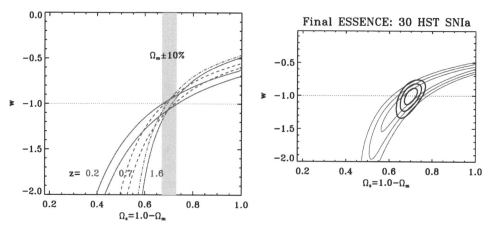

Fig. 17.8. *Left:* Constraints on Ω_x and w from SN data sets collected at $z = 0.2$ (solid lines), $z = 0.7$ (dashed lines), and $z = 1.6$ (dash-dot lines). The shaded area indicates how an independent estimate of Ω_m with a 10% error can help constrain w. *Right:* Expected constraints on w with the desired final ESSENCE data set of 200 SNe Ia, 30 of which (in the redshift range $0.6 < z < 0.8$) are to be observed with *HST*. The thin lines are for SNe alone while the thick lines assume an uncertainty in Ω_m of 7%. The final ESSENCE data will constrain the value of w to $\sim 10\%$.

team's research, and members of the SCP for their seminal complementary work on the accelerating Universe. My group's work at U.C. Berkeley has been supported by NSF grant AST–9987438, as well as by grants GO–7505, GO/DD–7588, GO–8177, GO–8641, GO–9118, and GO–9352 from the Space Telescope Science Institute, which is operated by the Association of Universities for Research in Astronomy, Inc., under NASA contract NAS 5–26555. KAIT was made possible by generous donations from Sun Microsystems, Inc., the Hewlett-Packard Company, AutoScope Corporation, Lick Observatory, the National Science Foundation, the University of California, and the Sylvia and Jim Katzman Foundation.

References

Aguirre, A. N. 1999a, ApJ, 512, L19

——. 1999b, ApJ, 525, 583

Aguirre, A. N., & Haiman, Z. 1999, ApJ, 525, 583

Aldering, G., Knop, R., & Nugent, P. 2000, AJ, 119, 2110

Bahcall, J. N., et al. 1996, ApJ, 457, 19

Bahcall, N. A., Ostriker, J. P., Perlmutter, S., & Steinhardt, P. J. 1999, Science, 284, 1481

Balbi, A., et al. 2000, ApJ, 545, L1

Benítez, N., Riess, A., Nugent, P., Dickinson, M., Chornock, R., & Filippenko, A. V. 2002, ApJ, 577, L1

Branch, D. 1981, ApJ, 248, 1076

——. 1998, ARA&A, 36, 17

Branch, D., Fisher, A., & Nugent, P. 1993, AJ, 106, 2383

Branch, D., & Miller, D. L. 1993, ApJ, 405, L5

Branch, D., Romanishin, W., & Baron, E. 1996, ApJ, 465, 73 (erratum: 467, 473)

Branch, D., & Tammann, G. A. 1992, ARA&A, 30, 359

Caldwell, R. R., Davé, R., & Steinhardt, P. J. 1998, Ap&SS, 261, 303

Cappellaro, E., Turatto, M., Tsvetkov, D. Yu., Bartunov, O. S., Pollas, C., Evans, R., & Hamuy, M. 1997, A&A, 322, 431

Carroll, S. M., Hoffman, M., & Trodden, M. 2003, Phys. Rev. D, 68, 023509

Carroll, S. M., Press, W. H., & Turner, E. L. 1992, ARA&A, 30, 499

Chaboyer, B., Demarque, P., Kernan, P. J., & Krauss, L. M. 1998, ApJ, 494, 96

Coil, A. L., et al. 2000, ApJ, 544, L111

Cowan, J. J., McWilliam, A., Sneden, C., & Burris, D. L. 1997, ApJ, 480, 246

Davis, M., Newman, J. A., Faber, S. M., & Phillips, A. C. 2001, in Deep Fields, ed. S. Cristiani, A. Renzini, & R. E. Williams (Berlin: Springer), 241

de Bernardis, P., et al. 2000, Nature, 404, 955

———. 2002, ApJ, 564, 559

Drell, P. S., Loredo, T. J., & Wasserman, I. 2000, ApJ, 530, 593

Efstathiou, G., et al. 1999, MNRAS, 303, L47

———. 2002, MNRAS, 330, L29

Eisenstein, D. J., Hu, W., & Tegmark, M. 1998, ApJ, 504, L57

Filippenko, A. V. 1997a, in Thermonuclear Supernovae, ed. P. Ruiz-Lapuente et al. (Dordrecht: Kluwer), 1

———. 1997b, ARA&A, 35, 309

———. 2001, PASP, 113, 1441

Filippenko, A. V., et al. 1992a, AJ, 104, 1543

———. 1992b, ApJ, 384, L15

Filippenko, A. V., Li, W. D., Treffers, R. R., & Modjaz, M. 2001, in Small-Telescope Astronomy on Global Scales, ed. W. P. Chen, C. Lemme, & B. Paczyński (San Francisco: ASP), 121

Filippenko, A. V., & Riess, A. G. 1998, Phys. Rep., 307, 31

Ford, C. H., et al. 1993, AJ, 106, 1101

Garnavich, P., et al. 1998a, ApJ, 493, L53

———. 1998b, ApJ, 509, 74

Giavalisco, M., et al. 2002, IAUC 7981

Gilliland, R. L., & Phillips, M. M. 1998, IAUC 6810

Goldhaber, G., et al. 1997, in Thermonuclear Supernovae, ed. P. Ruiz-Lapuente et al. (Dordrecht: Kluwer), 777

———. 1998a, BAAS, 30, 1325

———. 1998b, in Gravity: From the Hubble Length to the Planck Length, SLAC Summer Institute (Stanford, CA: SLAC)

———. 2001, ApJ, 558, 359

Goldhaber, G., & Perlmutter, S. 1998, Phys. Rep., 307, 325

Goobar, A., & Perlmutter, S. 1995, ApJ, 450, 14

Gratton, R. G., Fusi Pecci, F., Carretta, E., Clementini, G., Corsi, C. E., & Lattanzi, M. 1997, ApJ, 491, 749

Groom, D. E. 1998, BAAS, 30, 1419

Hamuy, M., Phillips, M. M., Maza, J., Suntzeff, N. B., Schommer, R. A., & Aviles, R. 1995, AJ, 109, 1

———. 1996a, AJ, 112, 2391

———. 1996b, AJ, 112, 2398

Hamuy, M., Trager, S. C., Pinto, P. A., Phillips, M. M., Schommer, R. A., Ivanov, V., & Suntzeff, N. B. 2000, AJ, 120, 1479

Hanany, S., et al. 2000, ApJ, 545, L5

Hancock, S., Rocha, G., Lazenby, A. N., & Gutiérrez, C. M. 1998, MNRAS, 294, L1

Hatano, K., Branch, D., & Deaton, J. 1998, ApJ, 502, 177

Höflich, P., Wheeler, J. C., & Thielemann, F. K. 1998, ApJ, 495, 617

Hoyle, F., Burbidge, G., & Narlikar, J. V. 2000, A Different Approach to Cosmology (Cambridge: Cambridge Univ. Press)

Ivanov, V. D., Hamuy, M., & Pinto, P. A. 2000, ApJ, 542, 588

Kim, A., Goobar, A., & Perlmutter, S. 1996, PASP, 108, 190

Leibundgut, B., et al. 1993, AJ, 105, 301

———. 1996, ApJ, 466, L21

Leonard, D. C., et al. 2002a, PASP, 114, 35 (erratum: 114, 1291)

———. 2002b, AJ, 124, 2490

Li, W., et al. 2000, in Cosmic Explosions, ed. S. S. Holt & W. W. Zhang (New York: AIP), 103

———. 2001a, PASP, 113, 1178

———. 2003, PASP, 115, 453

Li, W., Filippenko, A. V., Treffers, R. R., Riess, A. G., Hu, J., & Qiu, Y. 2001b, ApJ, 546, 734

Lineweaver, C. H. 1998, ApJ, 505, L69

Lineweaver, C. H., & Barbosa, D. 1998, ApJ, 496, 624

Matheson, T., Filippenko, A. V., Li, W., Leonard, D. C., & Shields, J. C. 2001, AJ, 121, 1648

Modjaz, M., Li, W., Filippenko, A. V., King, J. Y., Leonard, D. C., Matheson, T., Treffers, R. R., & Riess, A. G. 2001, PASP, 113, 308

Narlikar, J. V., & Arp, H. C. 1997, ApJ, 482, L119

Netterfield, C. B., et al. 2002, ApJ, 571, 604

Nomoto, K., Umeda, H., Hachisu, I., Kato, M., Kobayashi, C., & Tsujimoto, T. 2000, in Type Ia Supernovae: Theory and Cosmology, ed. J. C. Niemeyer & J. W. Truran (Cambridge: Cambridge Univ. Press), 63

Norgaard-Nielsen, H., et al. 1989, Nature, 339, 523

Nugent, P., 2000, in Particle Physics and Cosmology: Second Tropical Workshop, ed. J. F. Nieves (New York: AIP), 263

Nugent, P., Kim, A., & Perlmutter, S. 2002, PASP, 114, 803

Nugent, P., Phillips, M., Baron, E., Branch, D., & Hauschildt, P. 1995, ApJ, 455, L147

Ostriker, J. P., & Steinhardt, P. J. 1995, Nature, 377, 600

Oswalt, T. D., Smith, J. A., Wood, M. A., & Hintzen, P. 1996, Nature, 382, 692

Pain, R., et al. 2002, ApJ, 577, 120

Peacock, J. A., et al. 2001, Nature, 410, 169

Percival, W., et al. 2001, MNRAS, 327, 1297

Perlmutter, S., et al. 1995a, ApJ, 440, L41

——. 1995b, IAUC 6270

——. 1997, ApJ, 483, 565

——. 1998, Nature, 391, 51

——. 1999, ApJ, 517, 565

Phillips, M. M. 1993, ApJ, 413, L105

Phillips, M. M., et al. 1992, AJ, 103, 1632

Pskovskii, Yu. P. 1977, Sov. Astron., 21, 675

——. 1984, Sov. Astron., 28, 658

Riess, A. G., et al. 1997, AJ, 114, 722

——. 1998b, AJ, 116, 1009

——. 1999a, AJ, 117, 707

——. 1999c, AJ, 118, 2675

——. 2000, ApJ, 536, 62

——. 2001, ApJ, 560, 49

Riess, A. G., Filippenko, A. V., Li, W. D., & Schmidt, B. P. 1999b, AJ, 118, 2668

Riess, A. G., Nugent, P. E., Filippenko, A. V., Kirshner, R. P., & Perlmutter, S. 1998a, ApJ, 504, 935

Riess, A. G., Press, W. H., & Kirshner, R. P. 1995, ApJ, 438, L17

——. 1996a, ApJ, 473, 88

——. 1996b ApJ, 473, 588.

Saha, A., et al. 1997, ApJ, 486, 1

Sandage, A., et al. 1996, ApJ, 460, L15

Sandage, A., & Tammann, G. A. 1993, ApJ, 415, 1

Schmidt, B. P., et al. 1998, ApJ, 507, 46

Spergel, D. N., et al. 2003, ApJS, 148, 175

Sullivan, M., et al. 2004, MNRAS, in press (astro-ph/0211444)

Suntzeff, N. 1996, in Supernovae and Supernova Remnants, ed. R. McCray & Z. Wang (Cambridge: Cambridge Univ. Press), 41

Suntzeff, N., et al. 1996, IAUC 6490

Tonry, J. L., et al. 2003, ApJ, 594, 1

Tripp, R. 1997, A&A, 325, 871

——. 1998, A&A, 331, 815

Turatto, M., et al. 1996, MNRAS, 283, 1

Tyson, J. A., & Angel, R. 2001, in The New Era of Wide Field Astronomy, ed. R. Clowes et al. (San Francisco: ASP), 347

Umeda, H., et al. 1999, ApJ, 522, L43

van den Bergh, S., & Pazder, J. 1992, ApJ, 390, 34

Vaughan, T. E., Branch, D., Miller, D. L., & Perlmutter, S. 1995, ApJ, 439, 558

Wambsganss, J., Cen, R., & Ostriker, J. P. 1998, ApJ, 494, 29

Yungelson, L. R., & Livio, M. 2000, ApJ, 528, 108

Zaldarriaga, M., Spergel, D. N., & Seljak, U. 1997, ApJ, 488, 1

18

Theoretical overview of cosmic microwave background anisotropy

EDWARD L. WRIGHT

Department of Astronomy, University of California, Los Angeles

Abstract

The theoretical basis for the prediction of anisotropies in the cosmic microwave background is very well developed. Very low amplitude density and temperature perturbations produce small gravitational effects, leading to an anisotropy that is a combination of temperature fluctuations at the surface of last scattering and gravitational redshifts both at last scattering and along the path to the observer. All of the primary anisotropy can be handled by linear perturbation theory, which allows a very accurate calculation of the predicted anisotropy from different models of the Universe.

18.1   Introduction

The first predictions of the anisotropy of the cosmic microwave background (CMB) were published shortly after the CMB was discovered by Penzias & Wilson (1965). Sachs & Wolfe (1967) calculated the anisotropies due to gravitational potential fluctuations produced by density perturbations (Figure 18.1). Because the density perturbations are given by the second derivative of the gravitational potential fluctuation in Poisson's equation, the Sachs-Wolfe effect dominates the temperature fluctuations at large scales or low spherical harmonic index ℓ. Sachs & Wolfe predicted $\Delta T/T \approx 10^{-2}$ at large scales. This prediction, which failed by a factor of 10^3, is based on correct physics but incorrect input assumptions: prior to the discovery of the CMB no one knew how uniform the Universe was on large scales.

Silk (1968) computed the density perturbations needed at the recombination epoch at $z \approx 10^3$ in order to produce galaxies, and predicted $\Delta T/T \approx 3 \times 10^{-4}$ on arcminute scales. Silk (1967) calculated the damping of waves that were partially optically thick during recombination. This process, known as "Silk damping," greatly reduces the CMB anisotropy for small angular scales.

Observations by Conklin (1969) and then Henry (1971) showed that there was a dipole anisotropy in the CMB corresponding to the motion of the Solar System with respect to the average velocity of the observable Universe. There is a discussion of the dipole observations and their interpretation in Peebles (1971) that is still valid today, except that what was then a "tentative" dipole is now known to better than 1% accuracy, after a string of improved measurements starting with Corey & Wilkinson (1976) and ending with the *COBE* DMR (Bennett et al. 1996).

Peebles & Yu (1970) calculated the baryonic oscillations resulting from interactions between photons and hydrogen in the early Universe, and also independently introduced the

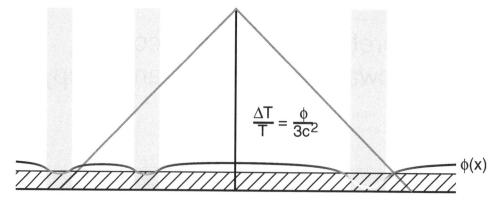

$$\frac{\Delta T}{T} = \frac{\phi}{3c^2}$$

Fig. 18.1. Sachs & Wolfe (1967) predicted that density enhancements would be cold spots in the CMB, as shown in this conformal spacetime diagram.

Harrison-Zel'dovich spectrum (Harrison 1970; Zel'dovich 1972). Pebbles & Yu predicted $\Delta T/T \approx 1.5 \times 10^{-4}$ on $1'$ scales and $\Delta T/T \approx 1.7 \times 10^{-3}$ on $7'$ scales.

Wilson & Silk (1981) further developed the theory of photon and matter interaction by scattering and gravity, and predicted $\Delta T/T = 100\ \mu K$ for a single subtracted experiment with a $7°$ throw and with a $7°$ beam like *COBE*. Of course, when *COBE* was launched in 1989 it actually observed a much smaller anisotropy.

These early predictions of a large anisotropy were greatly modified by the addition of dark matter to the recipe for the cosmos. Observational upper limits on small-scale anisotropies had reached $\Delta T/T < 4 \times 10^{-5}$ on $1'\!.5$ scales (Uson & Wilkinson 1982), which was considerably less than the predictions from universes with just baryons and photons. Peebles (1982) computed the anisotropy expected in a universe "dominated by massive, weakly interacting particles"—in other words cold dark matter (CDM), although this paper predated the use of "cold dark matter."

Bond & Efstathiou (1987) calculated the correlation function of the CMB anisotropy, $C(\theta)$, and also the angular power spectrum, C_ℓ, in the CDM cosmology. This paper contains one of the first plots showing $\ell(\ell+1)C_\ell$ vs. ℓ, with peaks originally called the "Doppler" peaks but more properly called "acoustic peaks." This paper solved the Boltzmann equation describing the evolution of the photon distribution functions. Several authors developed these "Boltzmann codes," but the calculation of the angular power spectrum up to high ℓ was very slow. These codes described the conversion of inhomogeneity at the last-scattering surface into anisotropy on the observed sky by a set of differential equations evolving the coefficients of a Legendre polynomial expansion of the radiation intensity. Since the Universe is almost completely transparent after recombination, a ray-tracing approach is much more efficient. This great step in efficiency was implemented in the CMBFAST code by Seljak & Zaldarriaga (1996).

Hu & Dodelson (2002) give a recent review of CMB anisotropies, which includes a very good tutorial on the theory of $\Delta T/T$.

18.2   Results

A simple analysis of cosmological perturbations can be obtained from the considerations in the Newtonian approximation to a homogeneous and isotropic universe. Consider a test particle at radius R from an arbitrary center. Because the model is homogeneous the choice of center does not matter. The evolution of the velocity of the test particle is given by the energy equation

$$\frac{v^2}{2} = E_{tot} + \frac{GM}{R}. \tag{18.1}$$

If the total energy E_{tot} is positive, the Universe will expand forever since M, the mass (plus energy) enclosed within R, is positive, G is positive, and R is positive. In the absence of a cosmological constant or "dark energy," the expansion of the Universe will stop, leading to a recollapse if E_{tot} is negative. But this simple connection between E_{tot} and the fate of the Universe is broken in the presence of a vacuum energy density. The mass M is proportional to R^3 because the Universe is homogeneous and the Hubble velocity v is given by $v = HR$. Thus $E_{tot} \propto R^2$.

We can find the total energy by plugging in the velocity $v_0 = H_0 R_0$ and the density ρ_0 in the Universe now. This gives

$$E_{tot} = \frac{(H_0 R_0 c)^2}{2} - \frac{4\pi G \rho_0 R_0^2}{3} = \frac{(H_0 R_0)^2}{2}\left(1 - \frac{\rho_0}{\rho_{crit}}\right), \tag{18.2}$$

with the critical density at time t_0 being $\rho_{crit} = 3H_0^2/(8\pi G)$. We define the ratio of density to critical density as $\Omega = \rho/\rho_{crit}$. This Ω includes all forms of matter and energy. Ω_m will be used to refer to the matter density.

From Equation 18.1 we can compute the time variation of Ω. Let

$$2E_{tot} = v^2 - \frac{2GM}{R} = H^2 R^2 - \frac{8\pi G \rho R^2}{3} = \text{const.} \tag{18.3}$$

If we divide this equation by $8\pi G \rho R^2/3$ we get

$$\frac{3H^2}{8\pi G\rho} - 1 = \frac{\text{const}'}{\rho R^2} = \Omega^{-1} - 1. \tag{18.4}$$

Thus $\Omega^{-1} - 1 \propto (\rho R^2)^{-1}$. When ρ declines with expansion at a rate faster than R^{-2} then the deviation of Ω from unity grows with expansion. This is the situation during the matter-dominated epoch with $\rho \propto R^{-3}$, so $\Omega^{-1} - 1 \propto R$. During the radiation-dominated epoch $\rho \propto R^{-4}$, so $\Omega^{-1} - 1 \propto R^2$. For Ω_0 to be within 0.9 and 1.1, Ω needed to be between 0.999 and 1.001 at the epoch of recombination, and within 10^{-15} of unity during nucleosynthesis. This fine-tuning problem is an aspect of the "flatness-oldness" problem in cosmology.

Inflation produces such a huge expansion that quantum fluctuations on the microscopic scale can grow to be larger than the observable Universe. These perturbations can be the seeds of structure formation and also will create the anisotropies seen by *COBE* for spherical harmonic indices $\ell \geq 2$. For perturbations that are larger than $\sim c_s t$ (or $\sim c_s/H$) we can ignore pressure gradients, since pressure gradients produce sound waves that are not able to cross the perturbation in a Hubble time. In the absence of pressure gradients, the density perturbation will evolve in the same way that a homogeneous universe does, and we can use the equation

$$\rho a^2 \left(\frac{1}{\Omega} - 1 \right) = \text{const}, \tag{18.5}$$

the assumption that $\Omega \approx 1$ for early times, and $\Delta\rho \ll \rho$ as indicated by the smallness of the ΔT's seen by *COBE*, to derive

$$-\rho a^2 \left(\frac{1}{\Omega} - 1 \right) \approx \rho_{crit} a^2 \Delta\Omega \approx \Delta\rho a^2 = \text{const}. \tag{18.6}$$

Hence,

$$\Delta\phi = \frac{G\Delta M}{R} = \frac{4\pi}{3} \frac{G\Delta\rho_0 (aL)^3}{aL} = \frac{1}{2} \frac{\Delta\rho_0}{\rho_{crit}} (H_0 L)^2, \tag{18.7}$$

where L is the comoving size of the perturbation. This is independent of the scale factor, so it does not change due to the expansion of the Universe.

During inflation (Guth 2003), the Universe is approximately in a steady state with constant H. Thus, the magnitude of $\Delta\phi$ for perturbations with physical scale c/H will be the same for all times during the inflationary epoch. But since this constant physical scale is aL, and the scale factor a changes by more than 30 orders of magnitude during inflation, this means that the magnitude of $\Delta\phi$ will be the same over 30 decades of comoving scale L. Thus, we get a strong prediction that $\Delta\phi$ will be the same on all observable scales from c/H_0 down to the scale that is no longer always larger than the sound speed horizon. This means that

$$\frac{\Delta\rho}{\rho} \propto L^{-2}, \tag{18.8}$$

so the Universe becomes extremely homogeneous on large scales even though it is quite inhomogeneous on small scales.

This behavior of $\Delta\phi$ being independent of scale is called *equal power on all scales*. It was originally predicted by Harrison (1970), Zel'dovich (1972), and Peebles & Yu (1970) based on a very simple argument: there is no scale length provided by the early Universe, and thus the perturbations should be *scale-free*—a power law. Therefore $\Delta\phi \propto L^m$. The gravitational potential divided by c^2 is a component of the metric, and if it gets comparable to unity then wild things happen. If $m < 0$ then $\Delta\phi$ gets large for small L, and many black holes would form. But we observe that this did not happen. Therefore $m \geq 0$. But if $m > 0$ then $\Delta\phi$ gets large on large scales, and the Universe would be grossly inhomogeneous. But we observe that this is not the case, so $m \leq 0$. Combining both results requires that $m = 0$, which is a *scale-invariant* perturbation power spectrum. This particular power-law power spectrum is called the Harrison-Zel'dovich spectrum. It was expected that the primordial perturbations should follow a Harrison-Zel'dovich spectrum because all other answers were wrong, but the inflationary scenario provides a good mechanism for producing a Harrison-Zel'dovich spectrum.

Sachs & Wolfe (1967) show that a gravitational potential perturbation produces an anisotropy of the CMB with magnitude

$$\frac{\Delta T}{T} = \frac{1}{3} \frac{\Delta\phi}{c^2}, \tag{18.9}$$

where $\Delta\phi$ is evaluated at the intersection of the line-of-sight and the surface of last scattering (or recombination at $z \approx 1100$). The $(1/3)$ factor arises because clocks run faster by a factor

Fig. 18.2. *Left:* A plane wave on the last-scattering hypersurface. *Right:* The spherical intersection with our past light cone is shown.

$(1 + \phi/c^2)$ in a gravitational potential, and we can consider the expansion of the Universe to be a clock. Since the scale factor is varying as $a \propto t^{2/3}$ at recombination, the faster expansion leads to a decreased temperature by $\Delta T/T = -(2/3)\Delta\phi/c^2$, which, when added to the normal gravitational redshift $\Delta T/T = \Delta\phi/c^2$ yields the $(1/3)$ factor above. This is an illustration of the "gauge" problem in calculating perturbations in general relativity. The expected variation of the density contrast as the square of the scale factor for scales larger than the horizon in the radiation-dominated epoch is only obtained after allowance is made for the effect of the potential on the time. For a plane wave with wavenumber k we have $-k^2\Delta\phi = 4\pi G\Delta\rho$, or

$$\frac{\Delta\phi}{c^2} = -\frac{3}{2}(H/ck)^2 \frac{\Delta\rho}{\rho_{crit}}, \tag{18.10}$$

so when $\rho \approx \rho_{crit}$ at recombination, the Sachs-Wolfe effect exceeds the physical temperature fluctuation $\Delta T/T = (1/4)\Delta\rho/\rho$ by a factor of $2(H/ck)^2$ if fluctuations are adiabatic (all component number densities varying by the same factor).

In addition to the physical temperature fluctuation and the gravitational potential fluctuation, these is a Doppler shift term. When the baryon fluid has a density contrast given by

$$\delta_b(x,t) = \frac{\Delta\rho_b}{\rho_b} = \delta_b \exp[ik(x - c_s t)], \tag{18.11}$$

where c_s is the sound speed, then

$$\frac{\partial \Delta\rho_b}{\partial t} = -ikc_s\delta_b\rho_b = -\rho_b\vec{\nabla}\cdot\vec{v} = ikv\rho_b. \tag{18.12}$$

As a result the velocity perturbation is given by $v = -c_s\delta_b$. But the sound speed is given by $c_s = \sqrt{\partial P/\partial\rho} = \sqrt{(4/3)\rho_\gamma c^2/(3\rho_b + 4\rho_\gamma)} \approx c/\sqrt{3}$, since $\rho_\gamma > \rho_b$ at recombination ($z = 1100$). But the photon density is only slightly higher than the baryon density at recombination so the sound speed is about 20% smaller than $c/\sqrt{3}$. The Doppler shift term in the anisotropy is given by $\Delta T/T = v/c$, as expected. This results in $\Delta T/T$ slightly less than $\delta_b/\sqrt{3}$, which is nearly $\sqrt{3}$ larger than the physical temperature fluctuation given by $\Delta T/T = \delta_b/3$.

Single k=50 plane wave   k=50 velocity term

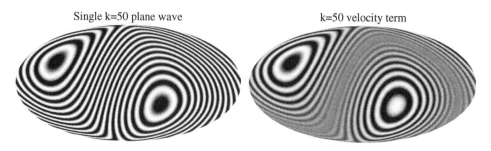

Fig. 18.3. *Left:* The scalar density and potential perturbation. *Right:* The vector velocity perturbation.

These plane-wave calculations need to be projected onto the sphere that is the intersection of our past light cone and the hypersurface corresponding to the time of recombination. Figure 18.2 shows a plane wave on these surfaces. The scalar density and potential perturbations produce a different pattern on the observed sky than the vector velocity perturbation. Figure 18.3 shows these patterns on the sky for a plane wave with $kR_{LS} = 50$, where R_{LS} is the radius of the last-scattering surface. The contribution of the velocity term is multiplied by $\cos\theta$, and since the RMS of this over the sphere is $\sqrt{1/3}$, the RMS contribution of the velocity term almost equals the RMS contribution from the density term since the speed of sound is almost $c/\sqrt{3}$.

The anisotropy is usually expanded in spherical harmonics:

$$\frac{\Delta T(\hat{n})}{T_0} = \sum_{\ell} \sum_{m=-\ell}^{\ell} a_{\ell m} Y_{\ell m}(\hat{n}). \tag{18.13}$$

Because the Universe is approximately isotropic the probability densities for all the different m's at a given ℓ are identical. Furthermore, the expected value of $\Delta T(\hat{n})$ is obviously zero, and thus the expected values of the $a_{\ell m}$'s is zero. But the variance of the $a_{\ell m}$'s is a measurable function of ℓ, defined as

$$C_{\ell} = \langle |a_{\ell m}|^2 \rangle. \tag{18.14}$$

Note that in this normalization C_{ℓ} and $a_{\ell m}$ are dimensionless. The harmonic index ℓ associated with an angular scale θ is given by $\ell \approx 180°/\theta$, but the total number of spherical harmonics contributing to the anisotropy power at angular scale θ is given by $\Delta\ell \approx \ell$ times $2\ell+1$. Thus to have equal power on all scales one needs to have approximately $C_{\ell} \propto \ell^{-2}$. Given that the square of the angular momentum operator is actually $\ell(\ell+1)$, it is not surprising that the actual angular power spectrum of the CMB predicted by "equal power on all scales" is

$$C_{\ell} = \frac{4\pi\langle Q^2 \rangle}{5T_0^2} \frac{6}{\ell(\ell+1)}, \tag{18.15}$$

where $\langle Q^2 \rangle$ or Q_{rms-PS}^2 is the expected variance of the $\ell=2$ component of the sky, which must be divided by T_0^2 because the $a_{\ell m}$'s are defined to be dimensionless. The "4π" term arises because the mean of $|Y_{\ell m}|^2$ is $1/(4\pi)$, so the $|a_{\ell m}|^2$'s must be 4π times larger to compensate. Finally, the quadrupole has 5 components, while C_{ℓ} is the variance of a single component,

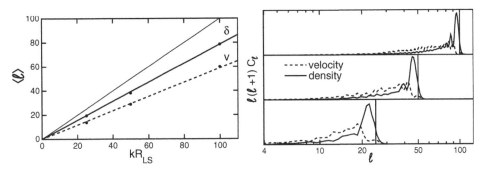

Fig. 18.4. *Left:* Mean ℓ is plotted vs. the wavenumber kR_{LS}. The light solid line shows $\ell = kR_{LS}$, while the solid line shows $\ell = (\pi/4)kR_{LS}$, and the dashed line shows $\ell = (3\pi/16)kR_{LS}$. *Right:* The angular power spectra for single k skies is plotted for $kR_{LS} = 25$, 50, and 100 (from bottom to top).

giving the "5" in the denominator. The *COBE* DMR experiment determined $\sqrt{\langle Q^2 \rangle} = 18\ \mu K$, and that the C_ℓ's from $\ell = 2$ to $\ell = 20$ were consistent with Equation 18.15.

The other common way of describing the anisotropy is in terms of

$$\Delta T_\ell^2 = \frac{T_0^2 \ell(\ell+1)C_\ell}{2\pi}. \qquad (18.16)$$

Note these definitions give $\Delta T_2^2 = 2.4\langle Q^2 \rangle$. Therefore, the *COBE* normalized Harrison-Zel'dovich spectrum has $\Delta T_\ell^2 = 2.4 \times 18^2 = 778\ \mu K^2$ for $\ell \le 20$.

It is important to realize that the relationship between the wavenumber k and the spherical harmonic index ℓ is not a simple $\ell = kR_{LS}$. Figure 18.3 shows that while $\ell = kR_{LS}$ at the "equator" the poles have lower ℓ. In fact, if $\mu = \cos\theta$, where θ is the angle between the wave vector and the line-of-sight, then the "local ℓ" is given by $kR_{LS}\sqrt{1-\mu^2}$. The average of this over the sphere is $\langle \ell \rangle = (\pi/4)kR_{LS}$. For the velocity term the power goes to zero when $\mu = 0$ on the equator, so the average ℓ is smaller, $\langle \ell \rangle = (3\pi/16)kR_{LS}$, and the distribution of power over ℓ lacks the sharp cusp at $\ell = kR_{LS}$. As a result the velocity term, while contributing about 60% as much to the RMS anisotropy as the density term, does not contribute this much to the peak structure in the angular power spectrum. Thus the old nomenclature of "Doppler" peaks was not appropriate, and the new usage of "acoustic" peaks is more correct. Figure 18.4 shows the angular power spectrum from single k skies for both the density and velocity terms for several values of k, and a graph of the variance-weighted mean ℓ vs. kR_{LS}. These curves were computed numerically but have the expected forms given by the spherical Bessel function j_ℓ for the density term and j'_ℓ for the velocity term.

Seljak (1994) considered a simple model in which the photons and baryons are locked together before recombination, and completely noninteracting after recombination. Thus the opacity went from infinity to zero instantaneously. Prior to recombination there were two fluids, the photon-baryon fluid and the CDM fluid, which interacted only gravitationally. The baryon-photon fluid has a sound speed of about $c/\sqrt{3}$ while the dark matter fluid has a sound speed of zero. Figure 18.5 shows a conformal spacetime diagrams with a traveling wave in the baryon-photon fluid and the stationary wave in the CDM. The CDM dominates the potential, so the large-scale structure (LSS) forms in the potential wells defined by the CDM.

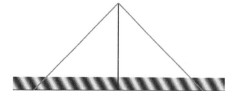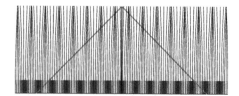

Fig. 18.5. On the left a conformal spacetime diagram showing a traveling wave in the baryon-photon fluid. On the right, the stationary CDM wave and the world lines of matter falling into the potential wells. For this wavenumber the density contrast in the baryon-photon fluid has undergone one-half cycle of its oscillation and is thus in phase with the Sachs-Wolfe effect from the CDM. This condition defines the first acoustic peak.

In Seljak's simple two-fluid model, there are five variables to follow: the density contrast in the CDM and baryons, δ_c and δ_b, the velocities of these fluids v_c and v_b, and the potential ϕ. The photon density contrast is $(4/3)\delta_b$. In Figure 18.6 the density contrasts are plotted vs. the scale factor for several values of the wavenumber. To make this plot the density contrasts were adjusted for the effect of the potential on the time, with

$$\Delta_c = \delta_c + 3H \int (\phi/c^2)dt \tag{18.17}$$

and

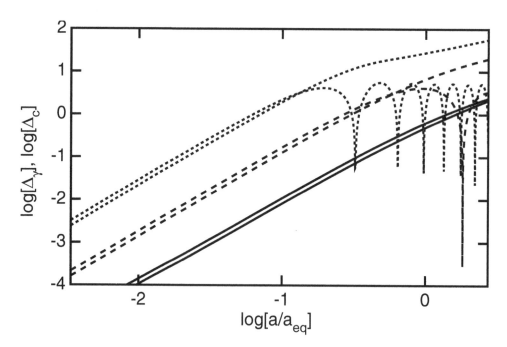

Fig. 18.6. Density contrasts in the CDM and the photons for wavenumbers $\kappa = 5, 20,$ and 80 (see Fig. 18.7) as a function of the scale factor relative to the scale factor and when matter and radiation densities were equal. The photon density contrast starts out slightly larger than the CDM density contrast but oscillates.

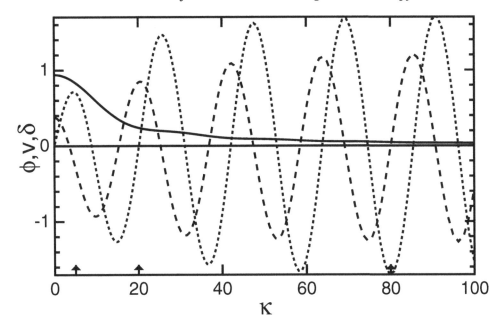

Fig. 18.7. Density contrasts at recombination as a function of wavenumber κ. The arrows on the x-axis indicate the values of κ for δ vs. a, as plotted in Figure 18.6. The solid curve shows the potential (the initial potential is always $\phi = 1$), the long dashed curve curve shows the combined potential plus density effect on the CMB temperature, while the short dashed curve shows the velocity of the baryon-photon fluid.

$$\Delta_\gamma = \delta_\gamma + 4H \int (\phi/c^2) dt. \tag{18.18}$$

Remembering that ϕ is negative when δ is positive, the two terms on the right-hand side of the above equations cancel almost entirely at early times, leaving a small residual growing like a^2 prior to a_{eq}, the scale factor when the matter density and the radiation density were equal. Thus these adjusted density contrasts evolve like $\Omega^{-1} - 1$ in homogeneous universes.

Figure 18.7 shows the potential that survives to recombination and produces LSS, the potential plus density effect on the CMB temperature, and the velocity of the baryons as function of wavenumber. A careful examination of the potential curve in the plot shows the baryonic wiggles in the LSS that may be detectable in the large redshift surveys by the 2dF and SDSS groups.

A careful examination of the angular power spectrum allows several cosmological parameters to be derived. The baryon to photon ratio and the dark matter to baryon density ratio can both be derived from the amplitudes of the first two acoustic peaks. Since the photon density is known precisely, the peak amplitudes determine the baryon density $w_b = \Omega_b h^2$ and the cold dark matter density $w_c = \Omega_{CDM} h^2$. The matter density is given by $\Omega_m = \Omega_b + \Omega_{CDM}$. The amplitude $\langle Q^2 \rangle$ and spectral index n of the primordial density perturbations are also easily observed. Finally the angular scale of the peaks depends on the ratio of the angular size distance at recombination to the distance sound can travel before recombination. Since the speed of sound is close to $c/\sqrt{3}$, this sound travel distance is primarily affected by the

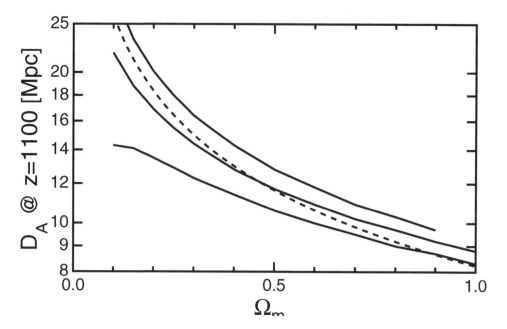

Fig. 18.8. The angular size distance vs. Ω_m for $H_0 = 60$ km s^{-1} Mpc^{-1} and three different values of Ω_{tot}(0.9, 1, and 1.1, from top to bottom). The dashed curve shows $\Omega_m^{-1/2}$.

age of the Universe at $z = 1100$. The age of the Universe goes like $t \propto \rho^{-1/2} \propto \Omega_m^{-1/2} h^{-1}$. The angular size distance is proportional to h^{-1} as well, so the Hubble constant cancels out. The angular size distance is almost proportional to $\Omega_m^{-1/2}$, but this relation is not quite exact. Figure 18.8 compares the angular size distance to $\Omega_m^{-1/2}$. One sees that a peak position that corresponds to $\Omega_{tot} = 0.95$ if $\Omega_m = 0.2$ can also be fit by $\Omega_{tot} = 1.1$ if $\Omega_m = 1$. Thus, to first order the peak position is a good measure of Ω_{tot}.

The CMBFAST code by Seljak & Zaldarriaga (1996) provides the ability to quickly compute the angular power spectrum C_ℓ. Typically CMBFAST runs in about 1 minute for a given set of cosmological parameters. However, two different groups have developed even faster methods to evaluate C_ℓ. Kaplinghat, Knox, & Skordis (2002) have published the Davis Anisotropy Shortcut (DASh), with code available for download. This program interpolates among precomputed C_ℓ's. Kosowsky, Milosavljević, & Jiminez (2002) discuss combinations of the parameters that produce simple changes in the power spectrum, and also allow accurate and fast interpolation between C_ℓ's. These shortcuts allow the computation of a C_ℓ from model parameters in about 1 second. This allows the rapid computation of the likelihood of a given data set D for a set of model parameters M, $L(D|M)$. When computing the likelihood for high signal-to-noise ratio observations of a small area of the sky, biases due to the non-Gaussian shape of the likelihood are common. This can be avoided using the offset log-normal form for the likelihood $L(C_\ell)$ advocated by Bond, Jaffe, & Knox (2000).

The likelihood is a probability distribution over the data, so $\int L(D|M)dD = 1$ for any M. It is not a probability distribution over the models, so one should never attempt to evaluate $\int LdM$. For example, one could consider the likelihood as a function of the model param-

Table 18.1. *Beam Size and Calibration Corrections*

Experiment	θ_B (')	$100(\Delta\theta)/\theta$	$100(\Delta dT)/dT$
COBE	420.0	−0.3	...
ARCHEOPS	15.0	−0.2	2.7
BOOMERanG	12.9	10.5	−7.6
MAXIMA	10.0	−0.9	0.2
DASI	5.0	−0.4	0.7
VSA	3.0	−0.5	−1.7
CBI	1.5	−1.1	−0.2

eters H_0 in km s^{-1} Mpc^{-1} and Ω_m for flat ΛCDM models, or one could use the parameters t_0 in seconds and Ω_m. For any (H_0, Ω_m) there is a corresponding (t_0, Ω_m) that makes exactly the same predictions, and therefore gives the same likelihood. But the integral of the likelihood over $dt_0 d\Omega_m$ will be much larger than the integral of the likelihood over $dH_0 d\Omega_m$ just because of the Jacobian of the transformation between the different parameter sets.

Wright (1994) gave the example of determining the primordial power spectrum power-law index n, $P(k) = A(k/k_0)^n$. Marginalizing over the amplitude by integrating the likelihood over A gives very different results for different values of k_0. Thus, it is very unfortunate that Hu & Dodelson (2002) still accept integration over the likelihood.

Instead of integrating over the likelihood one needs to define the *a posteriori* probability of the models $p_f(M)$ based on an *a priori* distribution $p_i(M)$ and Bayes' theorem:

$$p_f(M) \propto p_i(M)L(D|M). \tag{18.19}$$

It is allowable to integrate p_f over the space of models because the prior will transform when changing variables so as to keep the integral invariant.

In the modeling reported here, the *a priori* distribution is chosen to be uniform in ω_b, ω_c, n, Ω_v, Ω_{tot}, and z_{ri}. In doing the fits, the model C_ℓ's are adjusted by a factor of $\exp[a + b\ell(\ell+1)]$ before comparison with the data. Here a is a calibration adjustment, and b is a beam size correction that assumes a Gaussian beam. For *COBE*, a is the overall amplitude scaling parameter instead of a calibration correction. Marginalization over the calibration and beam size corrections for each experiment, and the overall spectral amplitude, is done by maximizing the likelihood, not by integrating the likelihood. Table 18.1 gives these beam and calibration corrections for each experiments. All of these corrections are less than the quoted uncertainties for these experiments. BOOMERanG stands out in the table for having honestly reported its uncertainties: $\pm 11\%$ for the beam size and $\pm 10\%$ for the gain. The likelihood is given by

$$
\begin{aligned}
-2\ln L = \chi^2 &= \sum_j \left\{ f(a_j/\sigma[a_j]) + f(b_j/\sigma[b_j]) \right. \\
&+ \left. \sum_i f([Z_{ij}^o - Z_{ij}^c]/\sigma[Z_{ij}]) \right\},
\end{aligned}
\tag{18.20}
$$

where j indexes over experiments, i indexes over points within each experiment, $Z = \ln(C_\ell + N_\ell)$ in the offset log normal approach of Bond et al. (2000), and N_ℓ is the noise bias. Since for

Table 18.2. *Cosmic Parameters from pre-WMAP CMB Data only*

Parameter	Mean	σ	Units
ω_b	0.0206	0.0020	
ω_c	0.1343	0.0221	
Ω_V	0.3947	0.2083	
Ω_k	-0.0506	0.0560	
z_{ri}	7.58	3.97	
n	0.9409	0.0379	
H_0	51.78	12.26	km s^{-1} Mpc^{-1}
t_0	15.34	1.60	Gyr
Γ	0.2600	0.0498	

COBE a is the overall normalization, $\sigma(a)$ is set to infinity for this term to eliminate it from the likelihood. The function $f(x)$ is x^2 for small $|x|$ but switches to $4(|x|-1)$ when $|x| > 2$. This downweighs outliers in the data. Most of the experiments have double tabulated their data. I have used both the even and odd points in my fits, but I have multiplied the σ's by $\sqrt{2}$ to compensate. Thus, I expect to get χ^2 per degree of freedom close to 0.5 but should have the correct sensitivity to cosmic parameters.

The scientific results such as the mean values and the covariance matrix of the parameters can be determined by integrations over parameter space weighted by p_f. Table 18.2 shows the mean and standard deviation of the parameters determined by integrating over the *a posteriori* probability distribution of the models. The evaluation of integrals over multi-dimensional spaces can require a large number of function evaluations when the dimensionality of the model space is large, so a Monte Carlo approach can be used. To achieve an accuracy of $\mathcal{O}(\epsilon)$ in a Monte Carlo integration requires $\mathcal{O}(\epsilon^{-2})$ function evaluations, while achieving the same accuracy with a gridding approach requires $\mathcal{O}(\epsilon^{-n/2})$ evaluations when second-order methods are applied on each axis. The Monte Carlo approach is more efficient for more than four dimensions. When the CMB data get better, the likelihood gets more and more sharply peaked as a function of the parameters, so a Gaussian approximation to $L(M)$ becomes more accurate, and concerns about banana-shaped confidence intervals and long tails in the likelihood are reduced. The Monte Carlo Markov Chain (MCMC) approach using the Metropolis-Hastings algorithm to generate models drawn from p_f is a relatively fast way to evaluate these integrals (Lewis & Bridle 2002). In the MCMC, a "trial" set of parameters is sampled from the proposal density $p_t(P';P)$, where P is the current location in parameter space, and P' is the new location. Then the trial location is accepted with a probability given by

$$\lambda = \frac{p_f(P')}{p_f(P)} \frac{p_t(P;P')}{p_t(P';P)}. \tag{18.21}$$

When a trial is accepted the Markov chain one sets $P = P'$. This algorithm guarantees that the accepted points in parameter space are sampled from the *a posteriori* probability distribution.

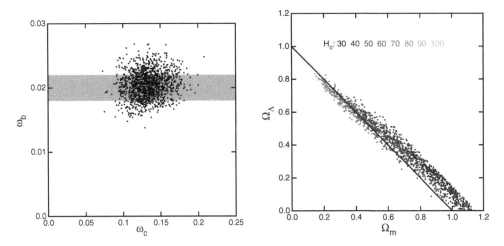

Fig. 18.9. *Left:* Clouds of models drawn from the *a posteriori* distribution based on the CMB data set as of 19 November 2002. The gray band shows the Big Bang nucleosynthesis determination of ω_b ($\pm 2\sigma$) from Burles, Nollett, & Turner (2001). *Right:* the same set of models in the Ω_m, Ω_Λ plane.

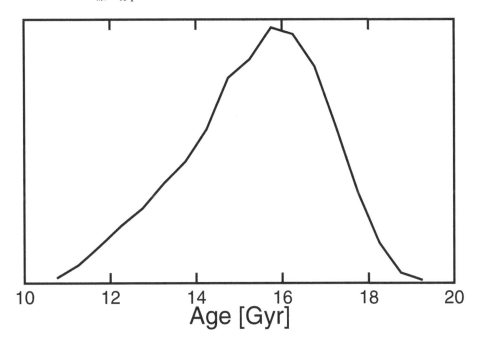

Fig. 18.10. Distribution of the age of the Universe based only on the pre-*WMAP* CMB data.

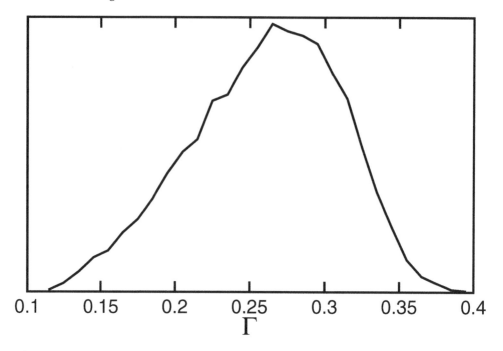

Fig. 18.11. Distribution of the LSS power spectrum shape parameter $\Gamma = \Omega_m h \exp(-2\Omega_b)$ from the pre-*WMAP* CMB data.

The most common choice for the proposal density is one that depends only on the parameter change $P'-P$. If the proposal density is a symmetric function then the ratio $p_t(P;P')/p_t P';P)$ $= 1$ and λ is then just the ratio of *a posteriori* probabilities. But the most efficient choice for the proposal density is $p_f(P)$ which is not a function of the parameter change, because this choice makes $\lambda = 1$ and all trials are accepted. However, if one knew how to sample models from p_f, why waste time calculating the likelihoods?

Just plotting the cloud of points from MCMC gives a useful indication of the allowable parameter ranges that are consistent with the data. I have done some MCMC calculations using the DASh (Kaplinghat et al. 2002) to find the C_ℓ's. I found DASh to be user unfriendly and too likely to terminate instead of reporting an error for out-of-bounds parameter sets, but it was fast. Figure 18.9 shows the range of baryon and CDM densities consistent with the CMB data set from *COBE* (Bennett et al. 1996), ARCHEOPS, BOOMERanG (Netterfield et al. 2002), MAXIMA (Lee et al. 2001), DASI (Halverson et al. 2002), VSA (Scott et al. 2003), and CBI (Pearson et al. 2003), and the range of matter and vacuum densities consistent with these data. The Hubble constant is strongly correlated with position on this diagram. Figure 18.10 shows the distribution of t_0 for models consistent with this pre-*WMAP* CMB data set. The relative uncertainty in t_0 is much smaller than the relative uncertainty in H_0 because the low-H_0 models have low vacuum energy density (Ω_V), and thus low values of the product $H_0 t_0$. The CMB data are giving a reasonable value for t_0 without using information on the distances or ages of objects, which is an interesting confirmation of the Big Bang model.

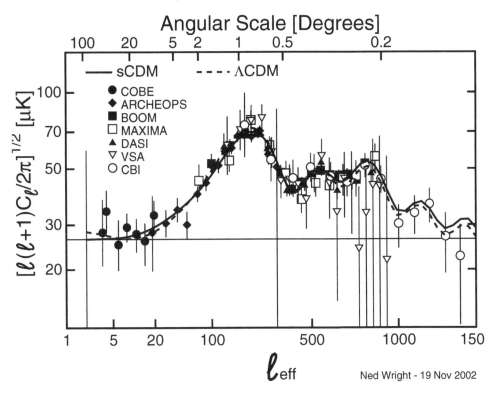

Fig. 18.12. Two flat $n = 1$ models. One shows ΛCDM with $\Omega_\Lambda = 2/3$. The best fit gives $\omega_b = 0.022$ and $\omega_c = 0.132$, implying $H_0 = 68$ km s^{-1} Mpc^{-1}. The other fit shows $\Omega_\Lambda = 0$ with $\omega_b = 0.021$ and $\omega_c = 0.196$, implying $H_0 = 47$ km s^{-1} Mpc^{-1}.

Peacock & Dodds (1994) define a shape parameter for the observed LSS power spectrum, $\Gamma = \Omega_m h \exp(-2\Omega_b)$. There are other slightly different definitions of Γ in use, but this will be used consistently here. Peacock & Dodds determine $\Gamma = 0.255 \pm 0.017 + 0.32(n^{-1} - 1)$. The CMB data specify n, so the slope correction in the last term is only 0.020 ± 0.013. Hence, the LSS power spectrum wants $\Gamma = 0.275 \pm 0.02$. The models based only on the pre-*WMAP* CMB data give the distribution in Γ shown in Figure 18.11, which is clearly consistent with the LSS data.

Two examples of flat ($\Omega_{tot} = 1$) models with equal power on all scales ($n = 1$), plotted on the pre-*WMAP* data set, are shown in Figure 18.12. Both these models are acceptable fits, but the ΛCDM model is somewhat favored based on the positions of the peaks. The rise in C_ℓ at low ℓ for the ΛCDM model is caused by the late integrated Sachs-Wolfe effect, which is due to the changing potential that occurs for $z < 1$ in this model. The potential changes because the density contrast stops growing when Λ dominates while the Universe continues to expand at an accelerating rate. The potential change during a photon's passage through a structure produces a temperature change given by $\Delta T/T = 2\Delta\phi/c^2$ (Fig. 18.13). The factor of 2 is the same factor of 2 that enters into the gravitational deflection of starlight by the Sun. The effect should be correlated with LSS that we can see at $z \approx 0.6$. Boughn & Crittenden (2002) have looked for this correlation using *COBE* maps compared to radio source count maps from the

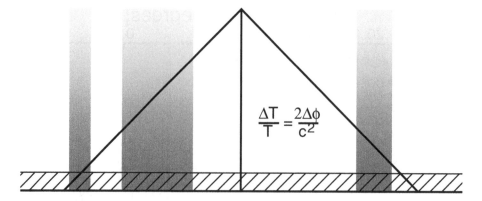

Fig. 18.13. Fading potentials cause large-scale anisotropy correlated with LSS due to the late integrated Sachs-Wolfe effect.

NVSS, and Boughn, Crittenden, & Koehrsen (2002) have looked at the correlation of *COBE* and the X-ray background. As of now the correlation has not been seen, which is an area of concern for ΛCDM, since the (non)correlation implies $\Omega_\Lambda = 0 \pm 0.33$ with roughly Gaussian errors. This correlation should arise primarily from redshifts near $z = 0.6$, as shown in Figure 18.14. The coming availability of LSS maps based on deep all-sky infrared surveys should allow a better search for this correlation.

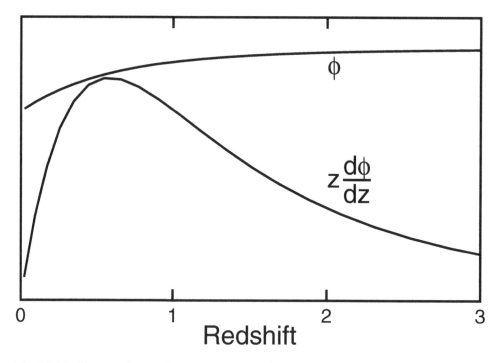

Fig. 18.14. Change of potential vs. redshift in a ΛCDM model. Note that the most significant changes occur near $z = 0.6$.

In addition to the late integrated Sachs-Wolfe effect from Λ, reionization should also enhance C_ℓ at low ℓ, as would an admixture of tensor waves. Since Λ, τ_{ri} and T/S all increase C_ℓ at low ℓ, and this increase is not seen, one has an upper limit on a weighted sum of all these parameters. If Λ is finally detected by the correlation between improved CMB and LSS maps, or if a substantial τ_{ri}, such as the $\tau = 0.1$ predicted by Cen (2003), is detected by the correlation between the E-mode polarization and the anisotropy (Zaldarriaga 2004), then one gets a greatly strengthened limit on tensor waves.

18.3   Discussion

The observed anisotropy of the CMB has an angular power spectrum that is in excellent agreement with the predictions of the ΛCDM model. But the CMB angular power spectrum is also consistent with an Einstein-de Sitter model having $\Omega_m = 1$ and a low value of $H_0 \approx 40$ km s^{-1} Mpc^{-1}. The observed lack of the expected correlation between the CMB and LSS due to the late integrated Sachs-Wolfe effect in ΛCDM slightly favors the $\Omega_m = 1$ "super Sandage" CDM model (sSCDM), which, like ΛCDM, is also consistent with the shape of the matter power spectrum $P(k)$ and the baryon fraction in clusters of galaxies. But sSCDM disagrees with the actual measurements of H_0 and with the supernova data for an accelerating Universe. Thus, ΛCDM is the overall best fit, but further efforts to confirm the CMB-LSS correlation should be encouraged.

18.4   Conclusions

The pre-*WMAP* CMB angular power spectrum assembled from multiple experiments is very well fit by a six-parameter model. Of these six parameters, the vacuum energy density Ω_V and the redshift of reionization z_{ri} are still poorly determined from CMB data alone. However, the well-determined parameters either match independent determinations or the expectations from inflation:

- The baryon density ω_b is determined to 10% and agrees with the value from Big Bang nucleosynthesis.
- The age of the Universe is determined to 11% and agrees with determinations from white dwarf cooling, main sequence turnoffs, and radioactive decay.
- The predicted shape of the LSS power spectrum $P(k)$ agrees with the observed shape.
- The curvature Ω_k is determined to 4% and agrees with the expected value from inflation.
- The spectral index n is determined to 4% and agrees with the expected value from inflation.

The angular power spectrum of the CMB can be computed using well-understood physics and linear perturbation theory. The current data set agrees with the predictions of inflation happening less than 1 picosecond after the Big Bang, the observations of light isotope abundances from the first three minutes after the Big Bang, and the observations of LSS in the current Universe. The inflationary scenario and the hot Big Bang model appear to be solidly based on confirmed quantitative predictions.

The greatly improved CMB data expected from *WMAP*, and later *Planck*, should dramatically improve our knowledge of the Universe.

References

Bennett, C. L., et al. 1996, ApJ, 464, L1
Bond, J. R., & Efstathiou, G. 1987, MNRAS, 226, 655
Bond, J. R., Jaffe, A. H., & Knox, L. 2000, ApJ, 533, 19

Boughn, S. P., & Crittenden, R. G. 2002, Phys. Rev. Lett., 88, 021302

Boughn, S. P., Crittenden, R. G., & Koehrsen, G. P. 2002, ApJ, 580, 672

Burles, S., Nollett, K. M., & Turner, M. S. 2001, Phys. Rev. D., 63, 063512

Cen, R. 2003, ApJ, 591, 12

Conklin, E. K. 1969, Nature, 222, 971

Corey, B. E., & Wilkinson, D. T. 1976, BAAS, 8, 351

Guth, A. H. 2004, in Carnegie Observatories Astrophysics Series, Vol. 2: Measuring and Modeling the Universe, ed. W. L. Freedman (Cambridge: Cambridge Univ. Press), in press

Halverson, N. W., et al. 2002, ApJ, 568, 38

Harrison, E. R. 1970, Phys Rev D, 1, 2726

Henry, P. S. 1971, Nature, 231, 516

Hu, W., & Dodelson, S. 2002, ARA&A, 40, 171

Kaplinghat, M., Knox, L., & Skordis, C. 2002, ApJ, 575, 665

Kosowsky, A., Milosavljević, M., & Jiminez, R. 2002, Phys. Rev. D, 66, 063007

Lee, A. T., et al. 2001, ApJ, 561, L1

Lewis, A., & Bridle, S. 2002, Phys. Rev. D, 66, 103511

Netterfield, C. B., et al. 2002, ApJ, 571, 604

Peacock, J. A., & Dodds, S. J. 1994, MNRAS, 267, 1020

Pearson, T. J., et al. 2003, ApJ, 591, 556

Peebles, P. J. E. 1971, Physical Cosmology (Princeton: Princeton Univ. Press)

——. 1982, ApJ, 263, L1

Peebles, P. J. E., & Yu, J. T. 1970, ApJ, 162, 815

Penzias, A. A., & Wilson, R. W. 1965, ApJ, 142, 419

Sachs, R. K., & Wolfe, A. M. 1967, ApJ, 147, 73

Scott, P. F., et al. 2003, MNRAS, 341, 1076

Seljak, U. 1994, ApJ, 435, L87

Seljak, U., & Zaldarriaga, M. 1996, ApJ, 469, 437

Silk, J. 1967, Nature, 215, 1155

——. 1968, ApJ, 151, 459

Uson, J. M., & Wilkinson, D. T. 1982, Phys. Rev. Lett., 49, 1463

Wilson, M. L., & Silk, J. 1981, ApJ, 243, 14

Wright. E. L. 1994, in The Proceedings of the 1994 CWRU Workshop on CMB Anisotropies Two Years After Cobe: Observations, Theory and the Future, ed. L. M. Krauss (World Scientific: Singapore), 21

Zaldarriaga, M. 2004, in Carnegie Observatories Astrophysics Series, Vol. 2: Measuring and Modeling the Universe, ed. W. L. Freedman (Cambridge: Cambridge Univ. Press), in press

Zel'dovich, Ya. B. 1972, MNRAS, 160, 1P

19

The polarization of the cosmic microwave background

MATIAS ZALDARRIAGA
Astronomy & Physics Departments, Harvard University

Abstract
We summarize the physical mechanism by which the cosmic microwave background acquires a small degree of polarization. We discuss the imprint left by gravitational waves and the use of polarization as a test of the inflationary paradigm. We discuss some physical processes that affect the cosmic microwave background polarization after recombination, such as gravitational lensing and the reionization of the Universe.

19.1   Introduction

Since its discovery in 1965 the cosmic microwave background (CMB) has been one of the pillars of the Big Bang model (Penzias & Wilson 1965). The various measurements of its spectrum, in particular by the *COBE*/FIRAS instruments firmly established the hot big bang model as the basis of our understanding of cosmology (Mather et al. 1994).

The study of the spectrum was followed by the detection of tiny anisotropies in the CMB temperature, first by *COBE*/DMR (Smoot et al. 1992), and then by a variety of more sensitive experiments with better angular resolution[*]. The anisotropies, a natural consequence of the structure formation process, contain a wealth of information about the cosmological model. They depend on the matter content of the Universe and on the physical process that created the tiny seeds that grew under gravity to form the structure in the Universe around us. Moreover, the structure formation process leaves its imprint on the CMB through secondary effects such as gravitational lensing, the Sunyaev-Zel'dovich effect (Sunyaev & Zel'dovich 1972), or the Ostriker-Vishniak effect (Ostriker & Vishniak 1986).

The detailed study of the temperature anisotropies has taken the field into the era of "precision cosmology." Ever more sensitive temperature experiments have so far confirmed our theoretical picture. These studies have revealed the presence of acoustic peaks in the power spectrum and of a damping tail on small scales. The comparison of measurements and theory, most recently the results of *WMAP* (Hinshaw et al. 2003), have led to very narrow constraints on several of the cosmological parameters. Experiments already under way or being constructed, such as the *Planck* satellite, will tighten constraints even further. We warn the reader that this article was written prior to the release of the *WMAP* results; references to them were added just before publication.

The next big goal for CMB experimentalists was the detection of the even smaller CMB polarization anisotropies. This was accomplished for the first time by the beautiful DASI

---

[*] A compilation of up to date results can be found at http://www.hep.upenn.edu/ max/ . We apologize for not giving an up to date list of all experiments but that made us exceed the page limit.

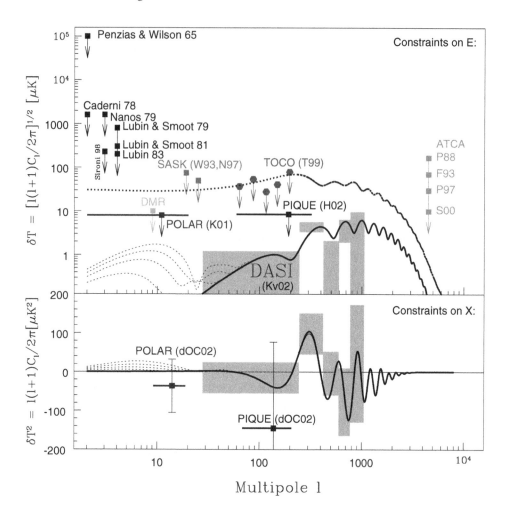

Fig. 19.1. Summary of polarization measurements and upper limits before the release of *WMAP* results, kindly provided by Angelica de Oliveira-Costa (adapted from de Oliveira-Costa et al. 2004). The top (bottom) panel shows results for the E (cross-correlation) power spectra.

experiments (Kovac et al. 2002; Leitch at al. 2002). Since the first detection of the CMB, there have been many theoretical studies of the expected polarization properties (e.g., Rees 1968; Polnarev 1985) and numerous attempts to measure or put upper limits on it. Figure 19.1 provides a summary of experimental results up to the time of writing. "Modern" experiments include POLAR (Keating et al. 2001) and PIQUE (Hedman et al. 2002), which set the most stringent upper limits before the DASI detection, roughly around 10 μK. The correlation between temperature and polarization was searched for by comparing PIQUE with Saskatoon (de Oliveira-Costa et al. 2003, 2004), was detected by DASI, and has now been measured with exquisite signal-to-noise ratio over a wide range of scales by *WMAP* (Kogut et al. 2003).

DASI opened a new window into the early Universe. Polarization is sensitive to most of the parameters in the cosmological model. The way it depends on them is often different than the way temperature anisotropies do. As a result, accurate measurements of the CMB polarization will improve the determination of many cosmological parameters (Zaldarriaga, Spergel, & Seljak 1997; Eisenstein, Hu, & Tegmark 1999). Moreover, polarization will provide an excellent test bed for consistency check on parts of the model that at present we take for granted.

The big push toward polarization, however, comes from its potential as a detector of a stochastic background of gravitational waves (GW). It has been shown that the pattern of polarization "vectors" on the sky will be different—it will have a curl-like component—if there is a stochastic background of GW (Kamionkowski, Kosowsky & Stebbins 1997a; Seljak & Zaldarriaga 1997). Inflation models predict the existence of a stochastic background of GW. A detection would provide the "smoking gun" for inflation. The amplitude of the GW in this background is directly proportional to the square of the energy scale of inflation, so that a detection of polarization would nail down key parameters of the inflationary model (e.g., Dodelson, Kinney, & Kolb 1997; Kinney 1998).

In this article we will review the mechanism by which polarization is generated and discuss what can be learned from measurements of polarization. In § 19.2 we will briefly review the physics of temperature anisotropies, although the reader is encouraged to go to one of the many reviews in the literature for further details. In § 19.3 we will discuss how polarization gets generated during recombination and to what it is sensitive. In § 19.4 we will discuss the imprint of GW. In § 19.5 we will review some of the physical processes that can affect the polarization signal after recombination. We conclude in § 19.6.

19.2 Temperature Anisotropies

In this section we will summarize the relevant facts about hydrogen recombination and temperature anisotropies. It is not our intention to provide an extensive review of these topics. The interested reader should consult other reviews (e.g., Hu & Dodelson 2002; Hu 2003). A discussion on the relevance of CMB studies for particle physics can be found in Kamionkowski & Kosowsky (1999).

19.2.1 Hydrogen Recombination and Thomson Scattering

The most abundant element in our Universe is hydrogen, and its ionization state has profound consequences on the CMB. The temperature of the CMB increased linearly with redshift, $T \propto (1+z)$. The interaction between the CMB photons and the hydrogen atoms kept hydrogen ionized until a redshift of $z \approx 1000$. At this time, corresponding to a conformal time of $\tau_R \approx 110 \, (\Omega_m h^2)^{-1/2}$ Mpc, there are not enough energetic photons in the CMB to keep hydrogen ionized, so it recombines. Conformal time is defined by $a \, d\tau = dt$, with a the expansion factor and t the physical time. It is useful because a null geodesic is simply given by $d\tau = dx$, where x are comoving coordinates ($a \, dx = dr$, where r is the physical distance). For example the fact that $\tau_R \approx 110 \, (\Omega_m h^2)^{-1/2}$ Mpc means a photon traveling on a straight line since the Big Bang until recombination would have traveled between two points that *today* are separated by a physical distance of $110 \, (\Omega_m h^2)^{-1/2}$ Mpc.

The fraction of free electrons coming from hydrogen and helium is shown in Figure 19.2 (it was calculated using RECFAST of Seager, Sasselov, & Scott 1999). One of the main

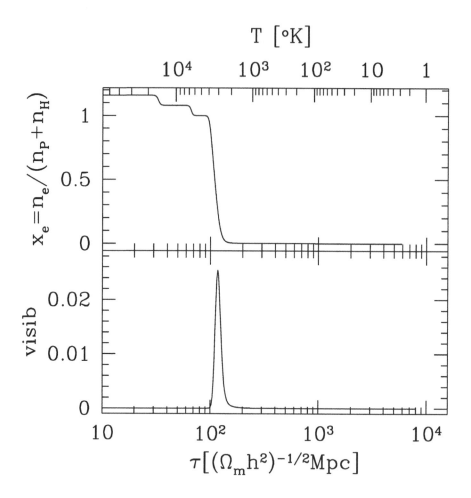

Fig. 19.2. In the top panel we show the ratio of number densities of free electrons and baryons. At early times the fraction is bigger than one because it includes electrons coming from helium. The bottom panel shows the visibility function that gives the probability a photon we observe last scattered at a particular position along the line of sight. The x-axis is labeled using conformal coordinates on the bottom and the temperature of the CMB on the top.

points to take away from the figure is that recombination happens rather fast. The width of recombination is $\delta\tau_R/\tau_R \approx 0.1-0.2$.

The other process we need to consider to understand how anisotropies are generated is Thomson scattering. When hydrogen is ionized the CMB photons can scatter with free electrons. This process conserves the number of photons and does not change their energy (because we are working in the limit of infinite electron mass). The mean free path for Thomson scattering in comoving Mpc is, $\lambda_T = (a n_e \sigma_T)^{-1} = 4$ Mpc $\frac{1}{x_e}(\frac{0.0125}{\Omega_b h^2})(\frac{0.9}{1-Y_p/2})(\frac{1000}{(1+z)})^2$, where n_e, x_e, and Y_p are the electron density, fraction of free electrons, and the primordial helium mass fraction, respectively.

Before recombination ($x_e \approx 1$) the mean free path was much smaller than the horizon at the time [$\tau_R \approx 110\,(\Omega_m h^2)^{-1/2}\,\text{Mpc}$]. Photons were hardly able to travel without scattering, a regime called "tight coupling." After recombination ($x_e \approx 3 \times 10^{-4}$) the mean free path becomes much larger than the horizon, so photons can travel in a straight line to our detectors. The Universe has become transparent for CMB photons, a process also called decoupling.

In the bottom panel of Figure 19.2 we show the visibility function, which gives the probability a photon we observe last scattered at a particular position along the line of sight. The function is strongly peaked. After recombination the photons no longer scatter, so the visibility function is almost zero. Before recombination the scatterings were so frequent that for most photons the last scattering occurs right around recombination. Thus, the CMB photons are coming from a very narrow range in distance, given by the width of the last-scattering surface, $\delta\tau_R/\tau_R \approx 0.1$. The collection of these regions in all directions on the sky form a very thin sphere of radius $D_{LSS} = \tau_0 - \tau_R \approx 6000(\Omega h^2)^{-1/2}\text{Mpc}$, called the last-scattering surface (LSS).

19.2.2 Review of Temperature Anisotropies

Now we are in a position to understand the origin of the CMB anisotropies. There are several reasons why the temperature we observe in different directions is not exactly the same. Differences in the density of photons across the last-scattering surface will lead to differences in the observed intensity as a function of position on the sky. If the density of photons is higher in a particular region, we see this as a hot spot. These same density differences across the Universe lead to gravitational potential differences. If photons climb (go down) a potential well to get to us, they get redshifted (blueshifted), and this decreases (increases) the observed temperature. The gravitational potential differences create forces that cause motions, and these motions create shifts in frequency due to the Doppler effect. Thus, if photons come to us from a region that is moving toward (away from) us, they get blueshifted (redshifted), which we observe as a higher (lower) temperature. Finally, if the gravitational potentials are changing with time, the energy of photons changes along the way, leading to the so called integrated Sachs-Wolfe (ISW) effect (Sachs & Wolfe 1967). For a detailed discussion of all the relevant equations in both synchronous and conformal gauges, see Ma & Bertschinger (1995), and for more details as to how to treat polarization and GW, see Hu et al. (1998).

In the Newtonian gauge, the temperature fluctuations $\Delta_T = \delta T/T$ observed in direction $\hat{n}$ can be written as

$$(\Delta_T + \psi)(\hat{n}) = (\frac{\delta_\gamma}{4} + \psi)|_{\tau_R} + \hat{n} \cdot v_b|_{\tau_R} + \int d\tau' e^{-\kappa(\tau_0,\tau')}(\dot{\phi} + \dot{\psi}). \tag{19.1}$$

We have introduced $\delta_\gamma = \delta\rho_\gamma/\rho_\gamma$, the fractional energy density fluctuations in the CMB, v_b the baryon velocity, ϕ and ψ the two gravitational potentials in terms of which the metric fluctuations are given by $ds^2 = a^2(\tau)[-(1+2\psi)d\tau^2 + (1-2\phi)dx^2]$, and the optical depth $\kappa(\tau_2,\tau_1) = \int_{\tau_2}^{\tau_1} an_e x_e d\tau$. The first term is usually referred to as the monopole contribution, while the second is the Doppler or dipole contribution.

In Equation 19.1 we have not included the terms coming from polarization because they are sub-dominant. We have approximated the visibility function by a δ-function; that is why the first two terms are evaluated at recombination. The different terms describe the physical effects mentioned above. From left to right, they correspond to photon energy density fluctuations, gravitational potential redshifts, Doppler shifts, and the ISW effect. Note that the ISW effect is an integral along the line of sight not constrained to recombination.

To be able to calculate the power spectrum of the anisotropies, we need to evaluate the different terms in Equation 19.1. Because we are dealing with small perturbations and linear theory, perturbations are usually expanded in Fourier modes, and the anisotropies are calculated for each Fourier mode individually. The contributions of individual modes are then added to calculate the power spectrum.

As discussed in the previous section, before recombination Thomson scattering was very efficient. As a result it is a good approximation to treat photons and baryons as a single fluid. This treatment is called the tight-coupling approximation and will allow us to evolve the perturbations until recombination to calculate the different terms in Equation 19.1.

The equation for the photon density perturbations for one Fourier mode of wavenumber k is that of a forced and damped harmonic oscillator (e.g., Seljak 1994; Hu & Sugiyama 1995):

$$\ddot{\delta}_\gamma + \frac{\dot{R}}{(1+R)}\dot{\delta}_\gamma + k^2 c_s^2 \delta_\gamma \;\; = \;\; F$$

$$F \;\; = \;\; 4[\ddot{\phi} + \frac{\dot{R}}{(1+R)}\dot{\phi} - \frac{1}{3}k^2\psi]$$

$$\dot{\delta}_\gamma \;\; = \;\; -\frac{4}{3}kv_\gamma + 4\dot{\phi}. \tag{19.2}$$

The photon-baryon fluid can sustain acoustic oscillations. The inertia is provided by the baryons, while the pressure is provided by the photons. The sound speed is $c_s^2 = 1/3(1+R)$, with $R = 3\rho_b/4\rho_\gamma = 31.5\,(\Omega_b h^2)(T/2.7)^{-4}[(1+z)/10^3]^{-1}$. As the baryon fraction goes down, the sound speed approaches $c_s^2 \to 1/3$. The third equation above is the continuity equation.

As a toy problem, we will solve Equation 19.2 under some simplifying assumptions. If we consider a matter-dominated universe, the driving force becomes a constant, $F = -4/3k^2\psi$, because the gravitational potential remains constant in time. We neglect anisotropic stresses so that $\psi = \phi$, and, furthermore, we neglect the time dependence of R. Equation 19.2 becomes that of a harmonic oscillator that can be trivially solved. This is a very simplified picture, but it captures most of the relevant physics we want to discuss. More elaborate approximation schemes can be found in the literature. They allow the calculation of the power spectrum with an accuracy of roughly 10% (Seljak 1994; Hu & Sugiyama 1995; Weinberg 2001a,b; Mukhanov 2003).

To obtain the final solution we need to specify the initial conditions. We will restrict ourselves to adiabatic initial conditions, the most natural outcome of inflation. In our context this means that initially $\phi = \psi = \phi_0$, $\delta_\gamma = -8/3\phi_0$, and $v_\gamma = 0$. We have denoted ϕ_0 the initial amplitude of the potential fluctuations. We will take ϕ_0 to be a Gaussian random variable with power spectrum P_{ϕ_0}.

We have made enough approximations that the evaluation of the sources in the integral solution has become trivial. The solution for the density and velocity of the photon fluid at recombination are:

$$(\frac{\delta_\gamma}{4} + \psi)|_{\tau_R} \;\; = \;\; \frac{\phi_0}{3}(1+3R)\cos(kc_s\tau_R) - \phi_0 R$$

$$v_\gamma|_{\tau_R} \;\; = \;\; -\phi_0(1+3R)c_s\sin(kc_s\tau_R). \tag{19.3}$$

Equation 19.3 is the solution for a single Fourier mode. All quantities have an additional spatial dependence ($e^{ik\cdot x}$), which we have not included to make the notation more compact. With that additional term the solution of Equation 19.1 becomes

$$\Delta_T(\hat{n}) = e^{ikD_{LSS}\cos\theta}S$$

$$S = \phi_0\frac{(1+3R)}{3}[\cos(kc_s\tau_R)-\frac{3R}{(1+3R)}$$

$$-i\sqrt{\frac{3}{1+R}}\cos\theta\sin(kc_s\tau_R)], \tag{19.4}$$

where we have neglected the ψ on the left-hand side because it is a constant independent of $\hat{n}$. We have also ignored the ISW contribution. We have introduced $\cos\theta$, the cosine of the angle between the direction of observation and the wavevector k; for example, $k\cdot x = kD_{LSS}\cos\theta$. The term proportional to $\cos\theta$ is the Doppler contribution.

Once the temperature perturbation produced by one Fourier mode has been calculated, we need to expand it into spherical harmonics to calculate the $a_{lm} = \int d\Omega\, Y_{lm}^*(\hat{n})\, \Delta T(\hat{n})$. The power spectrum of temperature anisotropies is expressed in terms of the a_{lm} coefficients as $C_{Tl} = \sum_m |a_{lm}|^2$. The contribution to C_{Tl} from each Fourier mode is weighted by the amplitude of primordial fluctuations in this mode, characterized by the power spectrum of ϕ_0, $P_{\phi_0} = Ak^{n-4}$. We will take the power-law index to be $n = 1$ in our approximate formulas. In practice, fluctuations on angular scale l receive most of their contributions from wavevectors around $k^* = l/D_{LSS}$, so roughly the amplitude of the power spectrum at multipole l is given by the value of the sources in Equation 19.3 at k^*.

After summing the contributions from all modes, the power spectrum is roughly given by

$$l(l+1)C_{Tl} \approx A\{[\frac{(1+3R)}{3}\cos(k^*c_s\tau_R)-R]^2 +$$

$$\frac{(1+3R)^2}{3}c_s^2\sin^2(k^*c_s\tau_R)\}. \tag{19.5}$$

Equation 19.5 can be used to understand the basic features in the CMB power spectra shown in Figure 19.3. The baryon drag on the photon-baryon fluid reduces its sound speed below $1/3$ and makes the monopole contribution dominant [the one proportional to $\cos(k^*c_s\tau_R)$]. Thus, the C_{Tl} spectrum peaks where the monopole term peaks, $k^*c_s\tau_R = \pi, 2\pi, 3\pi, \cdots$, which correspond to $l_{peak} = n\pi D_{LSS}/c_S\tau_R$. More detailed discussions of the physics of the acoustic peaks can be found in reviews such as that by Hu & Dodelson (2002) or Hu (2003).

It is very important to understand the origin of the acoustic peaks. In this model the Universe is filled with standing waves; all modes of wavenumber k are in phase, which leads to the oscillatory terms. The sine and cosine in Equation 19.5 originate in the time dependence of the modes. Each mode l receives contributions preferentially from Fourier modes of a particular wavelength k^* (but pointing in all directions), so to obtain peaks in C_l, it is crucial that all modes of a given k are in phase. If this is not the case, the features in the C_{Tl} spectra will be blurred and can even disappear. This is what happens when one considers the spectra produced by topological defects (e.g., Pen, Seljak, & Turok 1997). The phase coherence of all modes of a given wavenumber can be traced to the fact that perturbations were produced very early on and had wavelengths larger than the horizon during many expansion times.

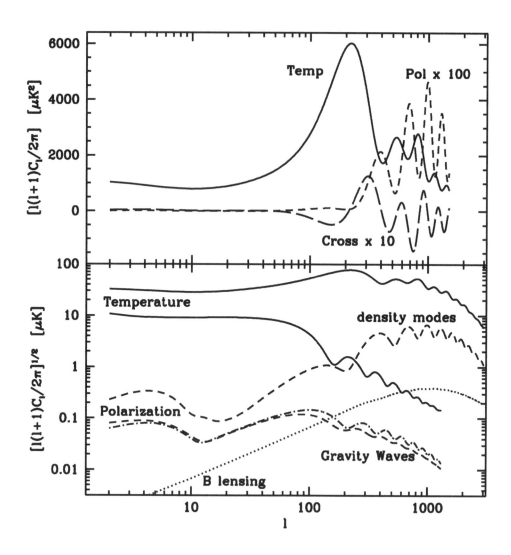

Fig. 19.3. Temperature, polarization, and cross-correlation power spectrum produced by density perturbations in the fiducial ΛCDM model ($\Omega_m = 0.3$, $\Omega_\Lambda = 0.7$, $h = 0.7$, $\Omega_b h^2 = 0.02$). For display purposes the polarization (cross-correlation) power spectrum was multiplied by a factor of 100 (10).

There are additional physical effects we have neglected. The Universe was radiation dominated early on, and modes of wavelength smaller and bigger than the horizon at matter-radiation equality behave differently. During the radiation era the perturbations in the photon-baryon fluid are the main source for the gravitational potentials which decay once a mode enters into the horizon. The gravitational potential decay acts as a driving force for the oscillator in Equation 19.2, so a feedback loop is established. As a result, the acoustic oscillations for modes that entered the horizon before matter-radiation equality have a higher amplitude. In the C_{Tl} spectrum the separation between modes that experience this feedback and those that do not occur at $l \sim D_{LSS}/\tau_{eq}$, where $\tau_{eq} \approx 16(\Omega_m h^2)^{-1}$ Mpc is the conformal time of matter–radiation equality. Larger l values receive their contributions from modes that entered the horizon before matter–radiation equality. Finally, when a mode is inside the horizon during the radiation era the gravitational potentials decay.

There is a competing effect, Silk damping (Silk 1968), that reduces the amplitude of the large-l modes. The photon–baryon fluid is not a perfect fluid. Photons have a finite mean free path and thus can random walk away from the peaks and valleys of the standing waves. Thus, perturbations of wavelength comparable to or smaller than the distance the photons can random walk get damped. This effect can be modeled by multiplying Equation 19.4 by $\exp(-k^2/k_s^2)$, with $k_s^{-1} \propto \tau_R^{1/2}(\Omega_b h^2)^{-1/2}$. Silk damping is important for multipoles of order $l_{Silk} \sim k_s D_{LSS}$.

Finally, the last-scattering surface has a finite width (Fig. 19.2). Perturbations with wavelength comparable to this width get smeared out due to cancellations along the line of sight. This effect introduces an additional damping with a characteristic scale $k_w^{-1} \propto \delta \tau_R$.

19.3 Polarization

In this section we will describe how polarization is characterized and how it is generated by density perturbations. We will stress the similarities and differences with temperature anisotropies. We will focus on polarization generated by density perturbations and leave the signal of GW for a later section. The physics of polarization has been previously reviewed in Hu & White (1997), Kosowsky (1999), and Hu (2003).

19.3.1 Characterizing the Radiation Field

The aim of this part is to summarize the mathematical tools needed to describe the CMB anisotropies. The anisotropy field is characterized by a 2×2 intensity tensor I_{ij}. For convenience, we normalize this tensor so it represents the fluctuations in units of the mean intensity ($I_{ij} = \delta I/I_0$). The intensity tensor is a function of direction on the sky, $\hat{n}$, and two directions perpendicular to $\hat{n}$ that are used to define its components ($\hat{e}_1, \hat{e}_2$). The Stokes parameters Q and U are defined as $Q = (I_{11} - I_{22})/4$ and $U = I_{12}/2$, while the temperature anisotropy is given by $T = (I_{11} + I_{22})/4$ (the factor of 4 relates fluctuations in the intensity with those in the temperature, $I \propto T^4$). When representing polarization using "rods" in a map, the magnitude is given by $P = \sqrt{Q^2 + U^2}$, and the orientation makes an angle $\alpha = \frac{1}{2}\arctan(U/Q)$ with $\hat{e}_1$. In principle the fourth Stokes parameter V that describes circular polarization is needed, but we ignore it because it cannot be generated through Thomson scattering, so the CMB is not expected to be circularly polarized. While the temperature is invariant under a right-handed rotation in the plane perpendicular to direction $\hat{n}$, Q and U transform under rotation by an angle ψ as

$$(Q \pm iU)'(\hat{n}) = e^{\mp 2i\psi}(Q \pm iU)(\hat{n}), \tag{19.6}$$

where $\hat{e}_1' = \cos\psi\, \hat{e}_1 + \sin\psi\, \hat{e}_2$ and $\hat{e}_2' = -\sin\psi\, \hat{e}_1 + \cos\psi\, \hat{e}_2$. The quantities $Q \pm iU$ are said to be spin 2.

We already mentioned that the statistical properties of the radiation field are usually described in terms the spherical harmonic decomposition of the maps. This basis, basically the Fourier basis, is very natural because the statistical properties of anisotropies are rotationally invariant. The standard spherical harmonics are not the appropriate basis for $Q \pm iU$ because they are spin-2 variables, but generalizations (called $_{\pm 2}Y_{lm}$) exist. We can expand

$$(Q \pm iU)(\hat{n}) = \sum_{lm} a_{\pm 2,lm} \; _{\pm 2}Y_{lm}(\hat{n}). \tag{19.7}$$

Q and U are defined at each direction $\hat{n}$ with respect to the spherical coordinate system $(\hat{\mathbf{e}}_\theta, \hat{\mathbf{e}}_\phi)$ (Kamionkowski, Kosowsky, & Stebbins, 1997b; Zaldarriaga & Seljak 1997). To ensure that Q and U are real, the expansion coefficients must satisfy $a^*_{-2,lm} = a_{2,l-m}$. The equivalent relation for the temperature coefficients is $a^*_{T,lm} = a_{T,l-m}$.

Instead of $a_{\pm 2,lm}$, it is convenient to introduce their linear combinations $a_{E,lm} = -(a_{2,lm} + a_{-2,lm})/2$ and $a_{B,lm} = i(a_{2,lm} - a_{-2,lm})/2$ (Newman & Penrose 1966). We define two quantities in real space, $E(\hat{n}) = \sum_{l,m} a_{E,lm} Y_{lm}(\hat{n})$ and $B(\hat{n}) = \sum_{l,m} a_{B,lm} Y_{lm}(\hat{n})$. E and B completely specify the linear polarization field.

The temperature is a scalar quantity under a rotation of the coordinate system, $T'(\hat{n}' = \mathcal{R}\hat{n}) = T(\hat{n})$, where $\mathcal{R}$ is the rotation matrix. We denote with a prime the quantities in the transformed coordinate system. While $Q \pm iU$ are spin 2, $E(\hat{n})$ and $B(\hat{n})$ are invariant under rotations. Under parity, however, E and B behave differently, E remains unchanged, while B changes sign.

As an illustration, in Figure 19.4 we show polarization patterns that have positive and negative E and B. It is clear that B patterns are "curl-like," having different properties than the E patterns under parity transformation. When reflected across a line going through the center, the E patterns remain unchanged, whereas the B patterns change from positive to negative. We want to stress at this point that whether a polarization field has an E or B component is a property of the pattern of polarization rods *around* a particular point, and not at the point itself. In this sense E and B are not local quantities.

There is a clear analogy between vector fields and polarization fields with regard to their geometrical properties. The polarization field is a spin-2 field, which means that if one rotates the coordinate system by 180° one goes back to the same components of the field, as opposed to 360° needed for vector fields. However, conceptually things are very similar, with E and B playing the roles of the gradient and curl parts. See Bunn et al. (2003) for a more detailed description of similarities and differences between vectors and polarization fields.

To characterize the statistics of the CMB perturbations, only four power spectra are needed, those for T, E, B and the cross correlation between T and E. The cross correlation between B and E or B and T vanishes if there are no parity-violating interactions because B has the opposite parity to T or E. Examples of models where this is not true were presented in Lue, Wang, & Kamionkowski (1999). The power spectra are defined as the rotationally invariant quantities $C_{Tl} = \frac{1}{2l+1} \sum_m \langle a^*_{T,lm} a_{T,lm} \rangle$, $C_{El} = \frac{1}{2l+1} \sum_m \langle a^*_{E,lm} a_{E,lm} \rangle$, $C_{Bl} = \frac{1}{2l+1} \sum_m \langle a^*_{B,lm} a_{B,lm} \rangle$, and $C_{Cl} = \frac{1}{2l+1} \sum_m \langle a^*_{T,lm} a_{E,lm} \rangle$. The brackets $\langle \cdots \rangle$ denote ensemble averages.

We will not discuss how to define correlation functions in real space for the polarization field. The reader is referred to the literature (e.g., Kamionkowski et al. 1997b; Zaldarriaga 1998; Tegmark & de Oliveira-Costa 2001). We only want to point out that the spin nature of the Stokes parameters needs to be considered to properly define correlation functions.

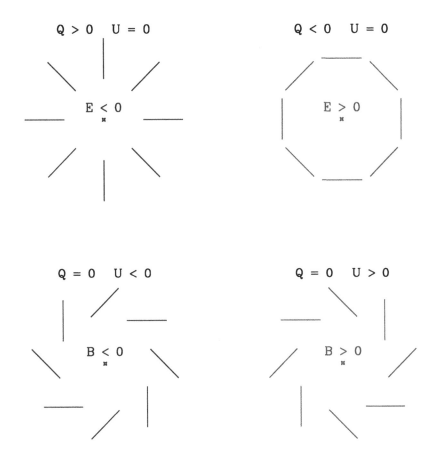

Fig. 19.4. Example of *E*-type and *B*-type patterns of polarization. Note that if reflected across a line going through the center the *E* patterns are unchanged, while the *B* patterns switch with one another.

19.3.2   The Physics of Polarization

Polarization is generated by Thomson scattering between photons and electrons, which means that polarization cannot be generated after recombination (except for reionization, which we will discuss later). But Thomson scattering is not enough. The radiation incident on the electrons must also be anisotropic. In fact, its intensity needs to have a quadrupole moment. This requirement of having both Thomson scattering and anisotropies is what makes polarization relatively small. After recombination, anisotropies grow by free streaming, but there are no scattering to generate polarization. Before recombination there were so many scatterings that they erased any anisotropy present in the photon-baryon fluid.

In the context of anisotropies induced by density perturbations, velocity gradients in the photon-baryon fluid are responsible for the quadrupole that generates polarization. Let us consider a scattering occurring at position x_0: the scattered photons came from a distance of order the mean free path (λ_T) away from this point. If we are considering photons traveling in direction $\hat{n}$, they roughly come from $x = x_0 + \lambda_T \hat{n}$. The photon-baryon fluid at that point was moving at velocity $v(x) \approx v(x_0) + \lambda_T \hat{n}_i \partial_i v(x_0)$. Due to the Doppler effect the temperature seen by the scatterer at x_0 is $\delta T(x_0, \hat{n}) = \hat{n} \cdot [v(x) - v(x_0)] \approx \lambda_T \hat{n}_i \hat{n}_j \partial_i v_j(x_0)$, which is quadratic in $\hat{n}$ (i.e., it has a quadrupole). *Velocity gradients in the photon-baryon fluid lead to a quadrupole component of the intensity distribution, which, through Thomson scattering, is converted into polarization.*

The polarization of the scattered radiation field, expressed in terms of the Stokes parameters Q and U, is given by $(Q + iU) \propto \sigma_T \int d\Omega' (\mathbf{m} \cdot \hat{\mathbf{n}}')^2 T(\hat{\mathbf{n}}') \propto \lambda_p m^i m^j \partial_i v_j|_{\tau_R}$, where σ_T is the Thomson scattering cross-section and we have written the scattering matrix as $P(\mathbf{m}, \hat{\mathbf{n}}') = -3/4 \sigma_T (\mathbf{m} \cdot \hat{\mathbf{n}}')^2$, with $\mathbf{m} = \hat{\mathbf{e}}_1 + i\hat{\mathbf{e}}_2$. In the last step, we integrated over all directions of the incident photons $\hat{\mathbf{n}}'$. As photons decouple from the baryons, their mean free path grows very rapidly, so a more careful analysis is needed to obtain the final polarization*(Zaldarriaga & Harari 1995):

$$(Q + iU)(\hat{\mathbf{n}}) \approx \epsilon \delta \tau_R m^i m^j \partial_i v_j|_{\tau_R}, \tag{19.8}$$

where $\delta \tau_R$ is the width of the last-scattering surface and is giving a measure of the distance photons travel between their last two scatterings, and ϵ is a numerical constant that depends on the shape of the visibility function. The appearance of $m^i m^j$ in Equation 19.8 assures that $(Q + iU)$ transforms correctly under rotations of $(\hat{\mathbf{e}}_1, \hat{\mathbf{e}}_2)$.

If we evaluate Equation 19.8 for each Fourier mode and combine them to obtain the total power, we get the equivalent of Equation 19.5,

$$l(l+1) C_{El} \approx A\epsilon^2 (1 + 3R)^2 (k^* \delta \tau_R)^2 \sin^2(k^* c_s \tau_R), \tag{19.9}$$

where we are assuming $n = 1$ and that l is large enough that factors like $(l+2)!/(l-2)! \approx l^4$. The extra k^* in Equation 19.9 originates in the gradient in Equation 19.8. As will be discussed later, density perturbations produce no B, so only three power spectra are needed to characterize the maps.

The curves in Figure 19.3 illustrate the differences between temperature and polarization power spectra. The large-angular scale polarization is greatly suppressed by the $k\delta\tau_R$ factor. Correlations over large angles can only be created by the long-wavelength perturbations, but these cannot produce a large polarization signal because of the tight coupling between photons and electrons prior to recombination. Multiple scatterings make the plasma very homogeneous; only wavelengths that are small enough to produce anisotropies over the mean free path of the photons will give rise to a significant quadrupole in the temperature distribution, and thus to polarization. Wavelengths much smaller than the mean free path decay due to photon diffusion (Silk damping) and so are unable to create a large quadrupole and polarization. As a result polarization peaks at the scale of the mean free path.

On sub-degree angular scales, temperature, polarization, and the cross-correlation power spectra show acoustic oscillations (Fig. 19.3). In the polarization and cross-correlation spectra the peaks are much sharper. The polarization is produced by velocity gradients of the photon-baryon fluid at the last-scattering surface (Equ. 19.8). The temperature receives contributions from density and velocity perturbations (Equ. 19.5), and the oscillations in each

---

* The velocity in this equation is in the conformal gauge.

partially cancel one another, making the features in the temperature spectrum less sharp. The dominant contribution to the temperature comes from the oscillations in the density (Equ. 19.3), which are out of phase with the velocity. This explains the difference in location between the temperature and polarization peaks. The extra gradient in the polarization signal, Equation 19.8, explains why its overall amplitude peaks at a smaller angular scale.

19.3.3 The Information Encoded in the Polarization Generated at the LSS

Now that we have reviewed the physics at recombination, we consider what can be learned from a measurement of polarization on sub-degree angular scales. The temperature and polarization anisotropies produced by density perturbations are characterized by three power spectra. Once polarization is measured, we have two extra sources of information from which to extract constraints on parameters. Although constraints should get tighter in principle, in practice because polarization anisotropies are significantly smaller than temperature ones, and thus more difficult to detect, parameter constraints do not improve that much from the measurement of the polarization. This is specially true if we restrict ourselves to the standard cosmological parameters such as those describing the matter content of the Universe, but not necessarily true for others or when some degeneracies are considered (Zaldarriaga et al. 1997; Eisenstein et al. 1999). This fact can be viewed in two different ways. One could say that a measurement of polarization in the acoustic peak region is not as relevant because it would not dramatically alter our constraints on parameters. Alternatively, one could say that it will provide a consistency check because, once the parameters of the model are determined from the temperature anisotropies, polarization is accurately predicted. We should keep in mind that, although everything seems to be falling in line with theoretical predictions, there are remarkably few real consistency checks, when the same quantity is measured accurately in two independent ways. We will mention three consistency checks that can be performed using polarization on sub-degree angular scales.

In what is quickly becoming the standard model, perturbations are generated during a period of inflation. The potential of CMB anisotropies to test inflationary models has long been recognized (e.g., Hu & White 1996; Hu, Spergel, & White 1997). One of the main predictions of inflation is that perturbations are correlated over scales larger than the horizon at recombination and that modes have phase coherence, which leads to the peaks in the power spectrum. Because the speed of the acoustic waves is smaller than the speed of light, one can produce the acoustic peaks in the temperature and polarization spectra without resorting to above-horizon perturbations (e.g., Turok 1996a, b). Moreover, because temperature can be produced after recombination (when the horizon is bigger) through the ISW effect, correlations on scales larger than two degrees (the scale of the recombination horizon) are not proof of above-horizon perturbations. Correlations of the polarization, on the other hand, provide a clearer test because there is no equivalent to the ISW effect (Spergel & Zaldarriaga 1997). This test has now been performed by the *WMAP* team, further strengthening the case for inflation (Pieris et al. 2003)

It is customary to assume that the power spectrum of primordial fluctuations is a simple power law, perhaps with a logarithmically varying spectral index. Although this is well motivated in the context of inflation, it should be checked. Moreover, preliminary determinations of the power spectrum, combining results from *WMAP* and large-scale structure, hint that something strange may be going on. Constraints on the power spectrum of primordial fluctuations come from comparing the amplitude of temperature perturbations at different angular scales. At least some of the changes produced by differences in the primordial spectrum can be mimicked by changes in the cosmological parameters. If both temperature and polarization are measured, there are three independent measures of the level of fluctuations on each scale. Moreover, temperature and polarization depend differently on cosmological parameters. Thus, the simultaneous measurement of temperature and polarization can be used to separate the early (inflation era) and later Universe physics affecting the CMB (e.g., Tegmark & Zaldarriaga 2002).

In the simplest inflationary models, perturbations only come in the adiabatic form. However, in more complicated inflationary models or in other classes of early-universe "scenarios," isocurvature perturbations can arise. In a model with photons, baryons, neutrinos and cold dark matter, there are actually four isocurvature modes in addition to the adiabatic one (e.g., Bucher, Moodley, & Turok 2000; Rebhan & Schwarz 1994; Challinor & Lasenby 1999). The simultaneous measurement of temperature and polarization will allow us to put constraints on small admixtures of these components that would be impossible otherwise (Bucher, Moodley, & Turok 2001).

Finally, because polarization perturbations are accurately predicted once temperature is measured, polarization can be used to make several consistency checks on the way recombination happened. In particular, changes in fundamental constants such as the fine structure constant, the gravitational constant, the speed at which the Universe was expanding during recombination, or the presence of ionizing photons on top of those from the CMB should be severely constrained once the data are in (Kaplinghat, Scherrer, & Turner 1999; Peebles, Seager, & Hu 2000; Hannestad & Scherrer 2001; Landau, Harari, & Zaldarriaga 2001; Uzan 2003; Zahn & Zaldarriaga 2003).

19.4   The $E-B$ Decomposition and the Imprint of Gravity Waves

The study of the anisotropies produced by GW has a long history (e.g., Polnarev 1985; Crittenden, Davis, & Steinhardt 1993; Coulson, Crittenden, & Turok 1994; Crittenden, Coulson, & Turok 1995). Perhaps the biggest driving force behind the push to measure polarization is the possibility of detecting a stochastic background of GW. The same mechanism that creates the seeds for structure formation, the stretching of vacuum fluctuations during inflation, is also expected to generate a stochastic background of GW. The amplitude of the GW in the background is directly proportional to the Hubble constant during inflation, or equivalently the square of energy scale of inflation.

As photons travel in the metric perturbed by a GW $[ds^2 = a^2(\tau) [-d\tau^2 + (\delta_{ij} + h_{ij}^T) dx^i dx^j]]$, they get redshifted or blueshifted depending on their direction of propagation relative to the direction of propagation of the GW and the polarization of the GW. For example, for a GW traveling along the z axis, the frequency shift is given by $\frac{1}{\nu}\frac{d\nu}{d\tau} = \frac{1}{2}\,\hat{n}^i\hat{n}^j\dot{h}_{ij}^{T(\pm)} = \frac{1}{2}\,(1 - \cos^2\theta)e^{\pm i2\phi}\,\dot{h}_t\,\exp(ik\cdot x)$, where (θ,ϕ) describe the direction of propagation of the photon, the $\pm$ correspond to the different polarizations of the GW, and h_t gives the time-dependent amplitude of the GW. During the matter-dominated era, for example, $h_t = 3j_1(k\tau)/k\tau$. This effect is analogous to the ISW effect: time changes in the metric lead to frequency shifts (or equivalently shifts in the temperature of the black body spectrum). Notice that the angular dependence of this frequency shift is quadrupolar in nature. As a result, the temperature fluctuations induced by this effect as photons travel between successive scatterings before recombination produce a quadrupole intensity distribution, which, through Thomson scattering, lead to polarization. *We want to stress that polarization is always generated by Thom-*

son scatterings, whether we talk about polarization generated by density perturbations or by GW. All that differs in these two cases is what is responsible for the quadrupole anisotropies.

Figure 19.3 shows the temperature and polarization anisotropy power spectrum produced by GW. We have arbitrarily normalized the GW contribution so that the ratio of $l = 2$ contributions from tensors and density perturbations $T/S \approx 0.1$, corresponding to an energy scale of inflation, $E_{\text{inf}} \approx 2 \times 10^{16}$ GeV. The temperature power spectrum, produced mainly after recombination, is roughly flat up to $l \approx 100$, and then rapidly falls off. By contrast, polarization is produced at recombination (except for the reionization bump at low values of l, to be discussed later). The power spectrum of polarization peaks at $l \approx 100$ because, just as for density perturbation, GW much larger than the mean free path of the photons at decoupling cannot generate an appreciable quadrupole. The spectrum falls on small scales as well because the amplitude of GW decays when they enter the horizon, as they redshift away. The decay occurs at horizon crossing, much before the Silk damping scale.

So far we have ignored the spin 2 nature of polarization. Both E and B power spectrum are generated by GW (Kamionkowski et al. 1997a; Seljak & Zaldarriaga 1997). The current push to improve polarization measurements follows from the fact that density perturbations, to linear order in perturbation theory, cannot create any B-type polarization. As a rough rule of thumb, the amplitude of the peak in the B-mode power spectrum for GW is $[l(l+1)C_{Bl}/2\pi]^{1/2} = 0.024(E_{\text{inf}}/10^{16}\text{GeV})^2 \mu$K.

One can understand why GW can produce B, while density perturbations cannot, by analyzing the symmetries of the problem. We recall that we are dealing with a linear problem, so one Fourier mode of density perturbations or a single GW needs to be analyzed. When calculating the temperature and polarization maps that would be observed in a universe with a single Fourier mode of the density, one realizes that the problem has two symmetries. There is symmetry of rotation around the k vector and symmetry of reflection along any plane that contains the k vector. The pattern of polarization produced by this mode has to respect these two symmetries, which in turn implies that there cannot be any B modes generated in this case. In Figure 19.4, if the cross represents the tip of the k vector, it is clear that only the E patterns respect the symmetries. For one polarization of the GW, these symmetries are not satisfied, as is obvious from the pattern of redshifts they produce, $\frac{1}{\nu}\frac{d\nu}{d\tau} \propto \frac{1}{2}(1-\cos^2\theta)e^{\pm i2\phi}$. Gravitational waves can produce a B pattern.

19.5 After Recombination

In this section we will discuss a few different effects that can change the polarization pattern after recombination.

19.5.1 Reionization

So far we have ignored the fact that hydrogen became ionized when the first sources of UV radiation started shining. The exact time and way this happened is not yet fully understood, the recent results from *WMAP* indicate that reionization started pretty early, with a prefer value of $z \approx 20$ (Bennett et al. 2003). There are also indications from studies of absorption toward high-redshift quasars that something happened around redshift $z \approx 6-8$ (Fan et al. 2002). It could well be that reionization is not a one-stage process. Depending on the details of the sources of radiation, a more complicated ionization history could have occurred (e.g., Wyithe & Loeb 2003).

In any event, at these redshifts the density of hydrogen was so much lower than at recombination that even though hydrogen is fully ionized the optical depth is still relatively low; somewhere around 17% is what *WMAP* favors. Even though only a fraction of the photons scatter after recombination, this still has a dramatic effect on polarization on large angular scales, of order 10 degrees. On these scales the polarization produced at recombination is minimal because these scales are much larger than the mean free path at recombination. As can be seen in Figure 19.3, polarization has a power-law decay toward large scales.

After recombination, photons were able to travel a significant distance over which the large-wavelength modes can induce a quadrupole. This quadrupole leads to a bump in the polarization power spectrum shown in Figure 19.3 that is orders of magnitude higher than what polarization would be if it were not for reionization (e.g., Zaldarriaga 1997). If there were no reionization the power law decay would continue up to $l = 2$.

A detection of this large-scale signal will allow a precise measurement of the total optical depth and provide some additional information on how and when reionization happened. The height of the reionization bump is proportional to the total optical depth, while the shape and location of the peak contains information about when reionization happened (Zaldarriaga 1997; Kaplinghat et al. 2003a). The total amount of information that can be extracted is, however, limited because at these large scales there are only a few multipoles that one can measure (i.e., cosmic variance is large; e.g., Hu & Holder 2003). However, a measurement of the optical depth is very important because it breaks degeneracies between determinations of several parameters (Zaldarriaga et al. 1997; Eisenstein et al. 1999). This excess of large-scale polarization is what *WMAP* used to determine a redshift of reionization around $z \approx 20$ (Spergel et al. 2003).

19.5.2   *Weak Lensing*

As photons propagate from the last-scattering surface they get gravitationally deflected by mass concentrations, the large-scale structure of the Universe. This gravitational lensing effect changes both temperature and polarization anisotropies (e.g., Seljak 1996; Bernardeau 1997, 1998; Hu 2000), but has rather profound consequences for the pattern of CMB polarization. Even a polarization pattern that did not have any *B* component at recombination will acquire some *B* due to gravitational lensing (Zaldarriaga & Seljak 1998). The effect is simple to understand. Because of lensing, a photon originally traveling in direction $\hat{n}$ will be observed toward direction $\hat{n}'$. We can use the analogy with a vector field and assume we start with the pattern that is a perfect gradient. As vectors are slightly shifted around due to lensing, the field will develop a curl component.

In Figure 19.3 we show the *B* component generated by lensing of the *E* mode. It is clear from the figure that if the level of the GW background is much lower than shown there, then the lensing signal would be larger at almost all scales. Note that the lensing *B* does not have the reionization bump at low l. This is so because the power is actually coming from "aliasing" of the small-scale polarization power rather than from a rearrangement of the original large-scale power.

The ultimate limitation for detecting the stochastic background of GW comes form the spurious *B* modes generated by lensing. The lensing distortions to the temperature and polarization maps make them non-Gaussian (e.g., Bernardeau 1997, 1998; Zaldarriaga 2000; Cooray & Hu 2001; Okamoto & Hu 2002). Methods have been developed to use this non-Gaussianity to measure the projected mass density (Seljak & Zaldarriaga 1999a; Zaldarriaga & Seljak 1999; Hu 2001a,b; Okamoto & Hu 2002), which can be used to clean, at least partially, this contamination. With the methods proposed so far, the lowest energy scale that seems measurable is $E_{inf} = 2 \times 10^{15}$ GeV (Kesden, Cooray, & Kamionkowski 2002; Knox & Song 2002). The energy scale of inflation is at present only very loosely constrained, so it is perfectly possible that the GW background is too small to be observed. However, if inflation is related to the Grand Unification scale around 10^{16} GeV, then there is a good chance that the GW background will be detected.

The lensing effect is not only a nuisance for detecting GW; it is interesting to constraint the large-scale structure that is doing the lensing. Because the last-scattering surface is at such high redshift, lensing of the CMB may eventually provide one of the deepest probes for large-scale structure. The level of structure on scales of order $2 - 1000$ Mpc at redshifts from $z \approx 10$ to $z \approx 0$ may eventually be constrained with this technique (e.g., Zaldarriaga & Seljak 1999). Lensing will not only allow a measurement of the level of fluctuations but may lead to actual reconstructed maps of the projected mass density that can be correlated with maps produced by CMB experiments and other probes (e.g., Goldeberg & Spergel 1999; Seljak & Zaldarriaga 1999b; Spergel & Goldberg 1999; Van Waerbeke, Bernardeau, & Benabed 2000; Benabed, Bernardeau, & van Waerbeke 2001; Song et al. 2003) and to strong constraints on parameters such as the mass of the neutrinos (Kaplinghat, Knox, & Song 2003b). On small scales, large mass concentrations such as cluster may leave a detectable signature (Seljak & Zaldarriaga 2000).

19.5.3 Foregrounds

As we have discussed, a significant amount of information is encoded in the large-scale polarization. The ability of polarization to determine the reionization history and with that help break some of the degeneracies between cosmological parameters relies on the large scales (Zaldarriaga et al. 1997).

Measuring polarization over the whole sky might be a problem when it comes to foregrounds. First, it means that one may have to use patches of the sky that are not that foreground free. Second, it appears that at least for the unpolarized component galactic foregrounds have rather red spectra, affecting large scales the most.

At microwave frequencies synchrotron, free-free, and dust emission are foreground contaminants. At present we have a fairly good understanding of the unpolarized component of the emission from our Milky Way and from distant galaxies (usually referred to as point sources).

The situation with regard to polarization is far worse. Both synchrotron and dust emission are expected to be polarized. In the case of synchrotron the theoretical maximum is 70%. Most of our knowledge about synchrotron polarization comes from surveys at relatively low frequencies. Extrapolation to frequencies where CMB experiments operate is severely hampered by Faraday rotation. Moreover, most modern surveys have concentrated on regions near the Galactic plane. How to extrapolate to higher latitudes remains unclear.

We will not dwell on foregrounds much longer because at this point there is not that much we can say. Time will tell if we are lucky again, as with the temperature anisotropies, or if foregrounds will spoil the potential fun of studying polarization. The interested reader is referred to the articles in the volume edited by de Oliveira-Costa & Tegmark (1999). More recent analysis of existing maps, some of which deal with the issue of whether foregrounds contaminate equally the E and B components can be found in a number of recent studies (Tucci et al. 2000, 2002; Baccigalupi et al. 2001; Bruscoli et al. 2002; Burigana & La

Porta 2002; Giardino et al. 2002). A recent summary including constraints coming from the PIQUE experiment is presented in de Oliveira-Costa et al. (2004). In the case of the DASI detection, the spectral index of the temperature anisotropies was very tightly constrained (ruling out any significant contamination there), and the temperature and polarization maps were correlated. This, together with the (rather weak) spectral index constraints from the polarization data, argued against any significant foreground contamination (Kovac et al. 2002).

19.5.4   *E − B Mixing: Systematic Effects*

There are a variety of systematic effects that lead to mixing between E and B modes. In finite patches of sky, the separation cannot be done perfectly (Lewis, Challinor, & Turok 2002; Bunn et al. 2003). This is analogous to trying to decompose a vector field into its gradient and curl parts when it is measured on a finite part of a plane. Vector fields that are gradients of a scalar with zero Laplacian will have both zero curl and divergence. In fact, in a finite patch one can construct a basis for the polarization field in which basis vectors can be split into three categories. There are pure E, pure B, and a third category of modes that is ambiguous, which receives contributions from both E and B. The number of ambiguous modes is proportional the number of pixels along the boundary.

Aliasing due to pixelization mixes E and B; the power that is aliased from sub-pixel scales leaks into both E and B. This is particularly important because E polarization is expected to be much larger than B modes, and because the E polarization power spectra is relatively very blue.

Finally, other effects such as common mode and differential gain fluctuations, line crosscoupling, pointing errors, and differential polarized beam effects will create a spurious B signal from temperature and/or E components (Hu, Hedman, & Zaldarriaga 2003).

19.6   Conclusions

We have summarized the physics behind the generation of a small degree of polarization in the CMB. Quadrupole anisotropies in the radiation intensity at the last-scattering surface through Thomson scattering lead to a small degree of linear polarization. These quadrupole anisotropies can be generated by both density perturbations (mainly through "free streaming" of the Doppler effect) and by GW.

The quadrupole generated by GW leads to a distinct pattern of polarization on the sky. Such a pattern has a curl component, and thus the CMB polarization can serve as an indirect GW detector. If inflation is the source of the density perturbations, it is also expected to generate a stochastic background of GW. Searching for this background through CMB polarization has become one of the driving forces for the field. The level of the B component produced by GW is expected to be quite small, so the first generation of polarization experiments should see B modes consistent with zero.

After recombination several processes can affect polarization. Gravitational lensing distorts the polarization patter generating a B component, even in the absence of GW. This source of noise could become the final limit to the detectability of the GW background.

The reionization of the Universe provides a new opportunity for the CMB photons to scatter. It leaves a signature in the large-scale polarization, a bump in the power spectrum. If detected, it would help constrain the epoch of reionization and break many of the degeneracies that occur in CMB fits of cosmological parameters.

The DASI experiment has recently detected polarization. It found a pattern of polarization consistent with having no *B* modes, just as expected. It also favors a spectrum that rises toward small scales, just as the theory predicts. Moreover, it provides a tentative detection of a cross correlation between temperature and polarization at the level predicted by the model. The *WMAP* satellite already released a high signal-to-noise ratio measurement of the cross correlation between temperature and polarization. These results lead to the conclusion that the Universe reionized at a surprisingly high redshift and provided further evidence in favor of inflation. Everything is looking good. Time will tell if more sensitive polarization experiments will eventually fulfill their promise and help us solve some of the remaining mysteries about our Universe.

References

Baccigalupi, C., Burigana, C., Perrotta, F., De Zotti, G., La Porta, L., Maino, D., Maris, M., & Paladini, R. 2001 A&A, 372, 8

Benabed, K., Bernardeau, F., & van Waerbeke, L. 2001, Phys. Rev. D, 63, 43501

Bennett, C. L., et al. 2003, ApJS, 148, 1

Bernardeau, F. 1997, A&A, 324, 15

——. 1998, A&A, 338, 767

Bruscoli, M., Tucci, M., Natale, V., Carretti, E., Fabbri, R., Sbarra, C., & Cortiglioni, S. 2002, NewA, 7, 171

Bucher, M., Moodley, K., & Turok, N. 2000, Phys. Rev. D, 62, 083508

——. 2001, Phys. Rev. Lett., 87, 191301

Bunn, E. F., Zaldarriaga, M., Tegmark, M., & de Oliveira-Costa, A. 2003, Phys. Rev. D, 67, 023501

Burigana, C., & La Porta, L. 2002, in Astrophysical Polarized Backgrounds, ed. S. Cecchini et al. (Melville, NY: AIP), 54

Challinor, A., & Lasenby, A. 1999, ApJ, 513, 1

Coulson, D., Crittenden, R. G., & Turok, N. G. 1994, Phys. Rev. Lett., 73, 2390

Cooray, A., & Hu, W. 2001, ApJ, 548, 7

Crittenden, R. G., Coulson, D., & Turok, N. G. 1995, Phys. Rev. D, 52, 5402

Crittenden, R., Davis, R. L., & Steinhardt, P. J. 1993, ApJ, 417, L13

de Oliveira-Costa, A. & Tegmark, M., ed., 1999, Microwave Foregrounds (San Francisco: ASP)

de Oliveira-Costa, A., Tegmark, M., Zaldarriaga, M., Barkats, D., Gundersen, J. O., Hedman, M. M., Staggs, S. T., & Winstein, B. 2003, Phys. Rev. D, 67, 023003

de Oliveira-Costa, A., Tegmark, M., O'Dell, C., Keating, B., Timbie, P., Efstathiou, G., & Smoot, G. 2004, Phys. Rev. D, submitted (astro-ph/0212419)

Dodelson, S., Kinney, W. H., & Kolb, E. W. 1997, Phys. Rev. D, 56, 3207

Eisenstein, D. J., Hu, W., & Tegmark, M. 1999, ApJ518, 2

Fan, X., et al. 2002, AJ, 123, 1247

Giardino, G., Banday, A. J., Górski, K. M., Bennett, K., Jonas, J. L., & Tauber, J. 2002, A&A, 387, 82

Goldberg, D. M., & Spergel, D. N. 1999, Phys. Rev. D, 59, 103002

Hannestad, S., & Scherrer, R. J. 2001, Phys. Rev. D, 63, 083001

Hedman, M. M., Barkats, D., Gundersen, J. O., McMahon, J. J., Staggs, S. T., & Winstein, B. 2002, ApJ, 573, L73

Hinshaw, G., et al. 2003, ApJS, 148, 135

Hu, W. 2000, Phys. Rev. D, 62, 43007

——. 2001a, Phys. Rev. D, 64, 83005

——. 2001b, ApJ, 557, L79

——. 2003, Annals Phys., 303, 203

Hu, W., & Dodelson, S. 2002, ARA&A, 40, 171

Hu, W., Hedman, M. M., & Zaldarriaga, M. 2003, Phys. Rev. D, 67, 043004

Hu, W., & Holder, G. P. 2003, Phys. Rev. D, 68, 023001

Hu, W., & Okamoto, T. 2002, ApJ, 574, 566

Hu, W., Seljak, U., White, M., & Zaldarriaga, M. 1998, Phys. Rev. D, 57, 3290

Hu, W., Spergel, D. N., & White, M. 1997, Phys. Rev. D, 55, 3288

Hu, W., & Sugiyama, N. 1995, Phys. Rev D, 51, 2599

Hu, W., & White, M. 1996, Phys. Rev. Lett., 77, 1687

——. 1997, NewA, 2, 323

Kamionkowski, M., & Kosowsky, A. 1999, Annu. Rev. Nucl. Part. Sci., 49, 77

Kamionkowski, M., Kosowsky, A., & Stebbins, A. 1997a, Phys. Rev. Lett., 78, 2058

——. 1997b, Phys. Rev. D, 55, 7368

Kaplinghat, M., Chu, M., Haiman, Z., Holder, G., Knox, L., & Skordis, C. 2003a, ApJ, 583, 24

Kaplinghat, M., Knox, L., & Song, Y.-S. 2003b, astro-ph/0303344

Kaplinghat, M., Scherrer, R. J., & Turner, M. S. 1999, Phys. Rev. D, 60, 023516

Keating, B. G., O'Dell, C. W., de Oliveira-Costa, A., Klawikowski, S., Stebor, N., Piccirillo, L., Tegmark, M., & Timbie, P. T. 2001, ApJ560, L1

Kesden, M., Cooray, A., & Kamionkowski, M. 2002, Phys. Rev. Lett., 89, 11304

Kinney, W. H. 1998, Phys. Rev. D, 58, 123506

Knox, L., & Song, Y.-S. 2002, Phys. Rev. Lett., 89, 11303

Kogut, A., et al. 2003, ApJS, 148, 161

Kovac, J. M., Leitch, E. M., Pryke, C., Carlstrom, J. E., Halverson, N. W., & Holzapfel, W. L. 2002, Nature, 420, 772

Kosowsky, A. 1999, NewAR, 43, 157

Landau, S. J., Harari, D. D., & Zaldarriaga, M. 2001, Phys. Rev. D, 63, 83505

Leitch, E. M., et al. 2002, Nature, 420, 763

Lewis, A., Challinor, A., & Turok, N. 2002, Phys. Rev. D, 65, 23505

Lue, A., Wang, L., & Kamionkowski, M. 1999, Phys. Rev. Lett., 83, 1506

Ma, C.-P., & Bertschinger, E. 1995, ApJ, 455, 7

Mather, J. C., et al. 1994, ApJ, 420, 439

Mukhanov, V. 2003, astro-ph/0303072

Newman, E., & Penrose, R. 1966, J. Math. Phys., 7, 863

Okamoto, T., & Hu, W. 2002, Phys. Rev. D, 66, 63008

Ostriker, J. P., & Vishniac, E. T. 1986, ApJ, 306, L51

Peebles, P. J. E., Seager, S., & Hu, W. 2000, ApJ, 539, L1

Pen, U.-L., Seljak, U., & Turok, N. 1997, Phys. Rev. Lett., 79, 1611

Penzias, A. A., & Wilson, R. W. 1965, ApJ, 142, 419

Pieris, H. V., et al. 2003, ApJS, 148, 213

Polnarev, A. G. 1985, Soviet Astron., 62, 1041

Rebhan, A. K., & Schwarz, D. J. 1994, Phys. Rev. D, 50, 2541

Rees, M. J. 1968, ApJ, 153, L1

Sachs, R. K., & Wolfe, A. M. 1967, ApJ, 147, 73

Seager, S., Sasselov, D. D., & Scott, D. 1999, ApJ, 523, L1

Seljak, U. 1994, ApJ, 435, L87

——. 1996, ApJ, 482, 6

Seljak U., & Zaldarriaga M. 1996, ApJ, 469, 7

——. 1997, Phys. Rev. Lett., 78, 2054

——. 1999a, Phys. Rev. Lett., 82, 2636

——. 1999b, Phys. Rev. D, 60, 43504

——. 2000, ApJ, 538, 57

Silk, J. 1968, ApJ, 151, 459

Smoot, G. F., et al. 1992, ApJ, 369, L1

Song, Y.-S., Cooray, A., Knox, L., & Zaldarriaga, M. 2003, ApJ, 590, 664

Spergel, D. N., et al. 2003, ApJS, 148, 175

Spergel, D. N., & Goldberg, D. M. 1999, Phys. Rev. D, 59, 103001

Spergel, D. N., & Zaldarriaga, M. 1997, Phys. Rev. Lett., 79, 2180

Sunyaev, R. A., & Zel'dovich, Y. B. 1972, CommAp, 4, 173

Tegmark, M., & de Oliveira-Costa, A. 2001, Phys. Rev. D, 64, 63001

Tegmark, M., & Zaldarriaga, M. 2002, Phys. Rev. D, 66, 103508

Tucci, M., Carretti, E., Cecchini, S., Fabbri, R., Orsini, M., & Pierpaoli, E. 2000, NewA, 5, 181

Tucci, M., Carretti, E., Cecchini, S., Nicastro, L., Fabbri, R., Gaensler, B. M., Dickey, J. M., & McClure-Griffiths, N. M. 2002, ApJ, 579, 607

Turok, N. 1996a, Phys. Rev. Lett., 77, 4138

——. 1996b, Phys. Rev. D, 54, 3686

Uzan, J.-P. 2003, Rev. Mod. Phys., 75, 403

Van Waerbeke, L., Bernardeau, F., & Benabed, K. 2000, ApJ, 540, 14

Weinberg, S. 2001a, Phys. Rev. D, 64, 123511
———. 2001b, Phys. Rev. D, 64, 123512
Wyithe, S., & Loeb, A. 2003, ApJ, 586, 693
Zahn, O., & Zaldarriaga, M. 2003, Phys. Rev. D, 67, 063002
Zaldarriaga, M. 1997, Phys. Rev. D, 55, 1822
———. 1998, ApJ, 503, 1
———. 2000, Phys. Rev. D, 62, 63510
Zaldarriaga, M., & Harari, D. 1995, Phys. Rev. D, 52, 3276
Zaldarriaga, M., & Seljak, U. 1997, Phys. Rev. D, 55, 1830
———. 1998, Phys. Rev. D, 58, 23003
———. 1999, Phys. Rev. D, 59, 123507
Zaldarriaga, M., Spergel, D. N., & Seljak, U. 1997, ApJ, 488, 1

20

The Wilkinson Microwave Anisotropy Probe

LYMAN A. PAGE
Department of Physics, Princeton University

Abstract

The *Wilkinson Microwave Anisotropy Probe (WMAP)* has mapped the full sky in five frequency bands between 23 and 94 GHz. The primary goal of the mission is to produce high-fidelity, all-sky, polarization-sensitive maps that can be used to study the cosmic microwave background. Systematic errors in the maps are constrained to new levels: using all-sky data, aspects of the anisotropy may confidently be probed to sub-μK levels. We give a brief description of the instrument and an overview of the first results from an analysis of maps made with one year of data. The highlights are (1) the flat ΛCDM model fits the data remarkably well, whereas an Einstein-de Sitter model does not; (2) we see evidence of the birth of the first generation of stars at $z_r \approx 20$; (3) when the *WMAP* data are combined and compared with other cosmological probes, a cosmic consistency emerges: multiple different lines of inquiry lead to the same results.

20.1   Introduction

We present the first results from the *Wilkinson Microwave Anisotropy Probe (WMAP)*, which completed its first year of observations on August 9, 2002. Dave Wilkinson, for whom the satellite was renamed, was a friend and colleague to many of the conference participants. He was a leader in the development of the cosmic microwave background (CMB) as a potent cosmological probe. He died on September 5, 2002, after battling cancer for 17 years, all the while advancing our understanding of the origin and evolution of the Universe. He saw the data in their full glory and was pleased with what I am able to report to you in the following.

At the time of the conference, the data were not ready to be released. The presentation focused on what *WMAP* would add to the considerable advances that had been made over the past few years. The emphasis was on the data; a companion talk by Ned Wright showed how the data are interpreted in the context of the currently favored models. Some effort was made to show what we knew firmly, like the position of the first acoustic peak in the CMB, as opposed to quantities that we knew better through the combination of multiple experiments, such as the characteristics of the second acoustic peak and the possible level of foreground contamination.

Since the conference, there have been 17 papers by the *WMAP* team[*] describing the instrument and data. This contribution draws heavily from those papers. Bennett et al. (2003a)

---

[*] Science team members are C. Barnes, C. Bennett (PI), M. Halpern, R. Hill, G. Hinshaw, N. Jarosik, A. Kogut, E. Komatsu, M. Limon, S. Meyer, N. Odegard, L. Page, H. Peiris, D. Spergel, G. Tucker, L. Verde, J. Weiland, E. Wollack, and E. Wright.

Table 20.1. *Characteristics of the Instrument*

Band	ν_{center} (GHz)	$\Delta\nu_{noise}$ (GHz)	N_{chan}	$\Delta T/pix$ (μK)	θ_{FWHM} (deg)
K	23	5	2	20	0.82
Ka	33	7	2	20	0.62
Q	41	7	4	21	0.49
V	61	11	4	24	0.33
W	94	17	8	23	0.21

The $\Delta T/pix$ is the Rayleigh-Jeans sensitivity per 3.2×10^{-5} sr pixel for the four-year mission for all channels of one frequency combined.

lay out the goals of the mission and give an overview of the instrument. More detailed descriptions of the radiometers, optics, and feed horns are found in Jarosik et al. (2003a), Page et al. (2003a), and Barnes et al. (2002). The inflight characterization of the receivers, optics, and sidelobes are described in Jarosik et al. (2003b), Page et al. (2003b), and Barnes et al. (2003). The data processing and checks for systematic errors are described by Hinshaw et al. (2003a), and Hinshaw et al. (2003b) present the angular power spectrum. Bennett et al. (2003c) present the analysis of the foreground emission. The temperature-polarization correlation is described in Kogut et al. (2003), and Komatsu et al. (2003) show that the fluctuations are Gaussian. Verde et al. (2003) present the details of the analysis of the angular power spectrum and show how the *WMAP* data can be combined with external data sets such as the 2dFGRS (Colless et al. 2001). Spergel et al. (2003) present the cosmological parameters derived from the *WMAP* alone, as well as the parameters derived from *WMAP* in combination with the external data sets. Peiris et al. (2003) confront the data with models of the early Universe, in particular inflation. The features in the power spectrum are interpreted in Page et al. (2003c). The scientific results are summarized in Bennett et al. (2003b).

20.2 Overview of the Mission and Instrument

The primary goal of *WMAP* is to produce high-fidelity, polarization-sensitive maps of the microwave sky that may be used for cosmological tests. *WMAP* was proposed in June 1995, at the height of the "faster, better, cheaper" era; some building began in June 1996, although the major push started in June 1997 after the confirmation review. *WMAP* was designed to be robust, thermally and mechanically stable, built of components with space heritage,* and relatively easy to integrate and test.

There is no *single* thing on *WMAP* that sets the mission apart from other experiments. Rather, it is a combination of attributes that are designed to work in concert to take full advantage of the space environment. The mission goal is to measure the microwave sky on angular scales of 180° as well as on scales of 0°.2 with micro-Kelvin *accuracy*. During the planning, building, and integration of the satellite, the emphasis was on minimizing

---

* Other than the NRAO HEMT-based (High Electron Mobility Transistors) amplifiers, which were custom designed, all components were "off-the-shelf."

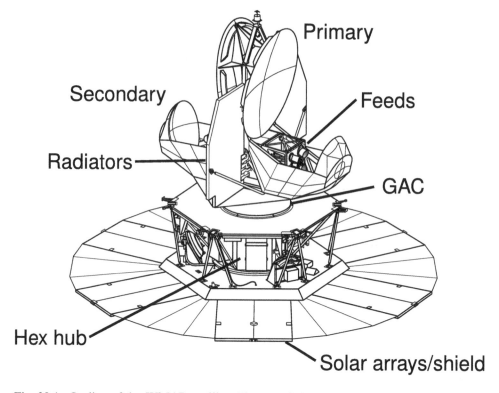

Fig. 20.1. Outline of the *WMAP* satellite. The overall height is 3.6 m, the mass is 830 kg, and the diameter of the large disk on the bottom is 5.1 m. Six solar arrays on the bottom of this disk supply the 400 W to power the spacecraft and instrument. Thermal blanketing between the hex hub and GAC, and between the GAC and radiators, shield the instrument from thermal radiation from the support electronics and attitude control systems.

systematic errors so that the data could be straightforwardly interpreted. There are many levels of redundancy that permit multiple cross-checks of the quality of the data.

WMAP was launched on June 30, 2001 into an orbit that took it to the second Earth-Sun Lagrange point, L_2, roughly 1.5×10^6 km from Earth. The L_2 orbit is oriented so that the instrument is always shielded from the Sun, Earth, and Moon by the solar panels and flexible aluminized mylar/kapton insulation, as shown in Figure 20.1. The satellite has one mode of operation while taking scientific data. It spins and precesses at constant insolation, continuously measuring the microwave sky.

The instrument measures only temperature differences from two regions of the sky separated by roughly 140°. The instrument is composed of 10 symmetric, passively cooled, dual polarization, differential, microwave receivers. As shown in Table 20.1, there are four receivers in W band, two in V band, two in Q band, one in Ka band, and one in K band. The receivers are fed by two back-to-back Gregorian telescopes.

The instrument is passively cooled. The optics, shielded from Sun and Earth, radiate to free space and cool to 70 K. Two large (5.6 m² net) and symmetric radiators passively cool the front-end microwave electronics to less than 90 K. One can just make out the heat straps that connect the base of the radiators to the microwave components housed below

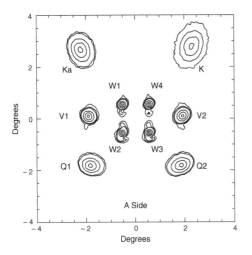 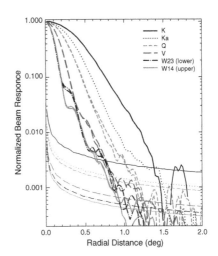

Fig. 20.2. *Left:* The A-side focal plane obtained from observations of Jupiter. Note that the beam widths are a function of frequency. The farther a beam is from the central focus, the more ellipsoidal it is. In the V and W bands, the beams have substructure. This is due to deformations in the surface of the primary and was anticipated. The scan pattern symmetrizes all the beams. The contour levels correspond to 0.9, 0.6, 0.3, 0.09, etc. of the peak value. *Right:* The normalized, symmetrized beam response. Except in the K band, the beams are mapped to better than −30 dB of their peak value. The lower set of lines shows the noise level in the maps. (From Page et al. 2003b.)

the primary reflectors in Figure 20.1. A hexagonal structure, "hex hub," between the solar panel array and the 1 m diameter gamma alumina cylinder (GAC) holds the power supplies, instrument electronics, and attitude control systems. The GAC supports a 190 K thermal gradient. There are no cryogens or mechanical refrigerators and thus no onboard source of thermal variations.

20.2.1 Optics

The optics comprise two back-to-back shaped Gregorian telescopes (Barnes et al. 2002; Page et al. 2003a). The primary mirrors are 1.4 m × 1.6 m. The secondaries are roughly a meter across, though most of the surface simply acts as a shield to prevent the feeds from directly viewing the Galaxy. The telescopes illuminate 10 scalar feeds on each side, a few of which are visible in Figure 20.1. The primary optical axes are separated by 141° to allow differential measurements over large angles on a fast time scale. The feed centers occupy a 18 cm × 20 cm region in the focal plane, corresponding to a 4° × 4.5° array on the sky, as shown in Figure 20.2

At the base of each feed is an orthomode transducer (OMT) that sends the two polarizations supported by the feed to separate receiver chains. The microwave plumbing is such that a single receiver chain (half of a "differencing assembly") differences electric fields with two nearly parallel linear polarization vectors, one from each telescope.

Precise knowledge of the beams is essential for accurately computing the CMB angular spectrum. The beams are mapped in flight with the spacecraft in the same observing mode as for CMB observations (Page et al. 2003b). Using Jupiter as a source, we measure the beam to less than −30 dB of its peak value. Because of the large focal plane the beams are not symmetric, as shown in Figure 20.2; neither are they Gaussian. In addition, as anticipated, cool-down distortions of the primary mirrors distort the W-band and V-band beam shapes. Fortunately, the scan strategy effectively symmetrizes the beam, greatly facilitating the analysis.

It is as important to understand the sidelobes, as it is to understand the main beams. We use three methods to assess them. (1) With physical optics codes* we compute the sidelobe pattern over the full sky and compute the current distributions on the optics. (2) We built a specialized test range to make sure that, by measurement, we could limit the Sun as a source of spurious signal to < 1 μK level in all bands. This requires knowing the beam profiles down to roughly −45 dBi (gain above isotropic) or −105 dB from the W-band peak. We find that over much of the sky, the measured profiles differ from the predictions at the −50 dB level due to scattering off of the feed horns and the structure that holds them. (d) During the early part of the mission, we used the Moon as a source to verify the ground-based measurements and models.

With a combination of the models and measurements, we assess both the polarized and unpolarized Galactic pickup in the sidelobes (Barnes et al. 2003). In Q, V, and W bands, the rms Galactic pickup through the sidelobes (with the main beams at $|b| > 15°$) is 2 μK, 0.3 μK, 0.5 μK, respectively, *before any modeling*. As this adds in quadrature to the CMB signal, the effect on the angular power spectrum is negligible. In K band, because of its low frequency and large sidelobes, the maps are corrected by 4% for the sidelobe contribution. The polarized pickup, which comes mostly from the passband mismatch, is ≪ 1 μK, except in K band where it is ∼ 5 μK.

20.2.2   Receivers

The *WMAP* mission was made possible by the HEMT-based amplifiers developed by Marian Pospieszalski (1992) at the National Radio Astronomy Observatory (NRAO). These amplifiers achieve noise temperatures of 25–100 K at 80 K physical temperature (Pospieszalski & Wollack 1998). Of equal importance is that the amplifiers can be phase matched over a 20% fractional bandwidth, as is required by the receiver design. *WMAP* uses a type of correlation receiver to measure the difference in power coming from the outputs of the OMTs at the base of the feeds (Jarosik et al. 2003a). All receiver were fully characterized before flight. The flight sensitivity was within 20% of expectations.

Figure 20.3 shows the signal path. From the output of the polarization selector (OMT), radiation from the two feeds is combined by a hybrid Tee into $(A+B)/\sqrt{2}$ and $(A-B)/\sqrt{2}$ signals, where A and B refer to the amplitudes of the electric fields from one linear polarization of each feed horn. In one arm of the receiver, both A and B signals are amplified first by cold (90 K) and then by warm (290 K) HEMT amplifiers. Noise power from the amplifiers, which far exceeds the input power, is added to each signal by the first amplifiers. Ignoring the phase switch for a moment, the two arms are then recombined in a second hybrid and both outputs of the hybrid are detected after a band-defining filter. Thus, for each differencing assembly, there are four detector outputs, two for each polarization. In a perfectly balanced system, one detector continuously measures the power in the A signal plus the average radiometer noise, while the other continuously measures the power in the B signal plus the average radiometer noise.

---

* We use a modified version of the DADRA code from YRS Associates (rahmat@ee.ucla.edu).

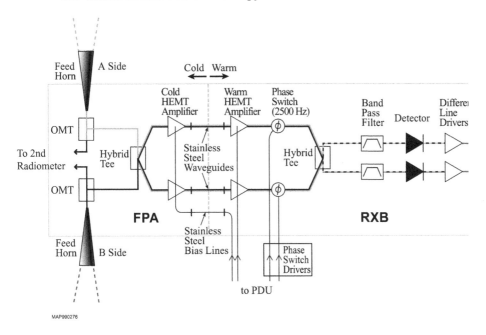

Fig. 20.3. One half of a differencing assembly (DA) for detecting one polarization component. A hybrid Tee ("magic tee") splits the inputs into two arms, where the signal is amplified and phase modulated before recombining. There are two stages of amplification, one at 90 K and one at 290 K. The recombined signal is then filtered and square law detected. The differential output that goes into the map-making algorithm is derived from the difference of the outputs of the two line drivers. (From Jarosik et al. 2003a.)

If the voltage amplification factors for the fields are G_1 and G_2 for the two arms, the difference in detector outputs is $G_1 G_2 (A^2 - B^2)$. Note that the average power signal present in both detectors has canceled so that small gain variations, which have a "$1/f$" spectrum†, act on the difference in powers from the two arms, which corresponds to 0–3 K, rather than on the total power, which corresponds to roughly 100 K in the W band. The system is stabilized further by toggling one of the phase switches at 2.5 kHz and coherently demodulating the detector outputs. (The phase switch in the other arm is required to preserve the phase match between arms and is jammed in one state.) The 2.5 kHz modulation places the desired signal at a frequency above the $1/f$ knee of the detectors and audio amplifiers, as well as rejecting any residual effects due to $1/f$ fluctuations in the gain of the HEMT amplifiers. The power output of each detector is averaged for between 51 and 128 ms and telemetered to the Earth. In total, there are 40 signals (two for each radiometer) plus instrument housekeeping data, resulting in a data rate of 110 MBy/day.

It is difficult to overemphasize the importance of understanding the radiometer noise and the stability of the satellite. One of the requirements for producing high-quality maps is stable receivers. One aims to measure the largest and smallest angular scales before the instrument can drift. The main challenge is to design a system that is stable over the spin rate, when the largest angular scales are measured. The drift away from stability is generically

† We use f for audio frequencies (< 20 kHz) and ν for microwave frequencies.

termed "$1/f$"; the effects are similar to $1/f$ noise in amplifiers. The *WMAP* noise power spectrum is nearly white between 0.008 Hz, the spin rate of the satellite, and 2.5 kHz; however, there is a slight amount of $1/f$. With the exception of the W4 DA, the $1/f$ knee is below 9 mHz. For W4 it is 26 and 47 mHz for the two polarizations. To account for this the data are prewhitened in the map-making process. This is the primary correction made to the data and affects only the maps and not the power spectra, as discussed below.

Although *WMAP*'s differential design was driven by the $1/f$ noise in the amplifiers, it is also very effective at reducing the effects of $1/f$ thermal fluctuations of the spacecraft itself. The thermal stability of deep space combined with the insensitivity to the spacecraft's slow temperature variations results in an extremely stable instrument. The measured peak-to-peak variation in the temperature of the cold end of the radiometers at the spin rate is $< 12 \ \mu K$. Using the measured radiometer gain susceptibilities, this translates to a < 20 nK radiometric signal. If one subtracts the sky signal from the data, the residual peak-to-peak radiometer output is measured to be $< 0.17 \ \mu K$ at the spin rate. Using the sky-subtracted signal, one can show that the receiver noise is Gaussian over 5 orders of magnitude (Jarosik et al. 2003). It is because of the stability, and our ability to verify it, that we can continuously average data together to probe cosmic signals at the sub-μK level.

20.2.3   Scan Strategy

High-quality maps are characterized by their fidelity to the true sky and their noise properties. Striping, the bane of the map-maker, results from scans of the sky in which the radiometer output departs from the average value for a significant part of the scan. Such departures, which can be produced by a number of mechanisms, are quantified as correlations between pixels.

A highly interlocking scan strategy is essential for producing a map with minimal striping. In any measurement, a baseline instrumental offset, along with its associated drift, must be subtracted. Without cross-hatched scans this subtraction can preferentially correlate pixels in large swaths, resulting in striped maps and substantially more involved analyses. *WMAP*'s noise matrix is nearly diagonal: the correlations between pixels are small. A typical off-diagonal element of the pixel-pixel covariance matrix is $< 0.1\%$ the diagonal value, except in W4, where it is $\sim 0.5\%$ at small lag (Hinshaw et al. 2003a).

To achieve the interlocking scan, *WMAP* spins around its axis with a period of 2 min and precesses around a $22°\!.5$ cone every hour so that the beams follow a spirograph pattern, as shown in Figure 20.4. Consequently, $\sim 30\%$ of the sky is covered in one hour, before the instrument temperature can change appreciably. The axis of this combined rotation/precession sweeps out approximately a great circle as the Earth orbits the Sun. In six months, the whole sky is mapped. The combination of *WMAP*'s four observing time scales (2.5 kHz, 2.1 min, 1 hour, 6 months) and the heavily interlocked pattern results in a strong spatial-temporal filter for any signal fixed in the sky.

Systematic effects at the spin period of the satellite are the most difficult to separate from true sky signal. Such effects, driven by the Sun, are minimized because the instrument is always in the shadow of the solar array and the precession axis is fixed with respect to the Sun-*WMAP* line. Spin-synchronous effects are negligible; there is no correction for them in the data processing.

With this scan, the instrument is continuously calibrated on the CMB dipole. The dominant signal in the timeline, the CMB dipole, averages to zero over the 1 hour precession

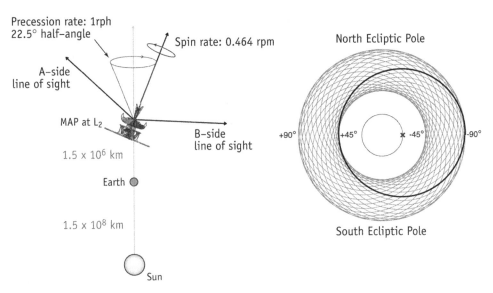

Fig. 20.4. *WMAP's* scan pattern from L_2. The dark circle on the left-hand drawing depicts the path covered by two beams for one rotation; the innermost circle is the path of the spin axis during one precession. The orbit follows a low-maintenance Lissajous pattern with the Sun, Earth, and Moon always behind. Corrections to the orbit are made roughly every three months. The scan motion is accomplished with three spinning momentum wheels so that the net angular momentum of the satellite is near zero.

period, enabling a clean separation between gain and baseline variations. The final absolute calibration is actually based on the component of the dipole due to the Earth's velocity around the Sun. As a result, *WMAP* is calibrated to 0.5% (which will improve).

20.3   Basic Results

The primary scientific result from *WMAP* is a set of maps of the microwave sky of unprecedented precision and accuracy. The maps have a well-defined systematic error budget (Hinshaw et al. 2003a) and may be used to address the most basic cosmological questions, as well as to understand Galactic and extragalactic emission (Bennett et al. 2003).

The process of going from the differential radiometer data to the maps is described in Hinshaw et al. (2003a) and Wright, Hinshaw, & Bennett (1996). Each step of the process is checked with simulations that account for all known effects in the data stream.

There are many components of the map-making process that must be included to produce maps with minimal systematic error. They include masking bright sources when they are in one beam, deleting data around the 21 glitches, and solving for the baseline drift and gain. These effects would be part of any pipeline. In all, 99% of the raw data goes directly into the maps.

There are three systematic errors, all of which were anticipated, for which corrections are made. They are (1) the 4% sidelobe correction in K band that is applied directly to the K-band map, (2) the $\sim$ 1% difference in loss between the A and B sides, and (3) the prewhitening described above. The last two are determined with flight data and are applied to the time stream. By any standards, this constitutes a remarkably little amount of correction. Artifacts from the gain, baseline, and pointing solutions are negligible.

Figures 20.5 and 20.6 show the maps in K through W bands. All maps have been smoothed to one degree. They are all in thermodynamic units relative to a 2.725 K black-body, so that the CMB anisotropy has the same contrast in all maps. Emission from the Galactic plane saturates the color scale at all frequencies. At the lowest frequency, Galactic emission extends quite far off the plane. Still, even at 23 GHz, the anisotropy is clearly evident at high Galactic latitude. As one moves to higher frequency, there is less Galactic emission. When averaged over regions of high Galactic latitude, foreground contamination is minimum in V band.

There are a number checks that can be made of the maps. The simplest is shown in Figure 20.7. The sum of two channels of the same frequency gives a robust signal whereas the difference gives almost no signal. With the manifold instrumental redundancy, a myriad of other internal consistency checks is possible. We can also compare to the *COBE*/DMR map and see that, to within the limits of the noise, *WMAP* and DMR are the same. This is a wonderful confirmation of both satellites: they observe from different places, use different techniques, and have different systematic errors.

20.3.1   Analysis of Maps

The first step in going from the maps to cosmology is to select a region of sky for analysis. We mask regions that are significantly contaminated by diffuse emission from our Galaxy and by point source emission. Our selection process is described in Bennett et al. (2003c).

The diffuse contribution comes from free-free emission, synchrotron emission, and dust emission. To find the most contaminated sections, one masks regions that are above a threshold in the K-band maps. The nominal cut excludes 15% of the sky in a contiguous region near the Galactic plane.

One of the surprises from *WMAP* was that a successful model of the foregrounds could be made that did not include a spinning dust component (Draine & Lazarian 1998). Rather, we find that there is a strong correlation between synchrotron emission and dust, and that the synchrotron spectral index varies significantly over the Galactic plane. Surprising to many, the synchrotron emission is better correlated with the Finkbeiner, Davis, & Schlegel (1999) dust map derived from the *COBE*/DIRBE and *IRAS* at $\nu > 200$ GHz than with the Haslam et al. (1982) synchrotron map at $\nu = 0.41$ GHz.

In addition to diffuse emission, we mask a $0°6$ circle around 700 bright sources that might contaminate our results. The sources are selected based on radio catalogs, as discussed in Bennett et al. (2003c). No correlation with infrared sources was found. Contamination by point sources is a potential problem because their angular power spectrum has constant C_l, and thus rises as l^2 on the standard plot of the anisotropy. Our approach to identifying sources is three-pronged: (1) we search the data with a matched filter for sources, (2) we marginalize over a point source contribution in the determination of the angular power spectrum, and (3) we use a version of the bispectrum tuned to find point sources to directly

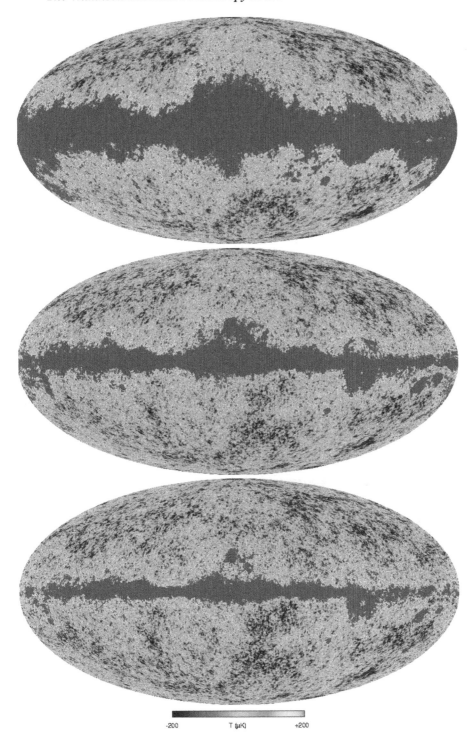

Fig. 20.5. K-band (top), Ka-band (middle), and Q-band (bottom) intensity maps in Galactic coordinates shown in a Mollweide projection. The Galactic Center is in the middle of each map. All maps are in CMB thermodynamic units. (Adapted from Bennett et al. 2003b.)

Fig. 20.6. V-band (top) and W-band (bottom) maps in the same coordinates and scale as the previous three maps. (Adapted from Bennett et al. 2003b.)

determine their contribution in a map. All three methods give consistent results when the sources are left in the map (Komatsu et al. 2003). All methods indicate that with our masking the sources do not contribute to the power spectrum.

After applying the diffuse and point source masks, the remaining parts of the maps are simultaneously fit to templates of the Finkbeiner et al. (1999) dust emission, an Hα map that has been correct for extinction that traces free-free emission (see Finkbeiner 2003, and references therein), and a synchrotron map extrapolated from the 408 MHz Haslam et al. map.

The template fit removes any remaining residual Galaxy emission, even though the fit coefficients do not correspond to the best Galactic model. Tests are made to ensure that the masking and fitting do not bias the results.

With the maps in hand, we show that the fluctuations in the CMB are Gaussian to the

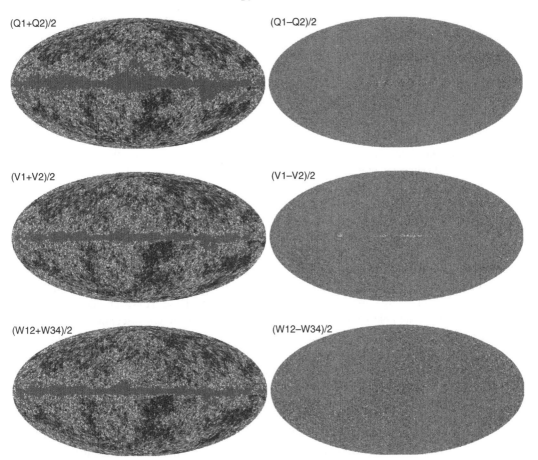

Fig. 20.7. This plot shows the sum (left) and difference (right) maps for Q, V, and W, respectively. A slight bit of contamination is evident in V1–V2. This arises because the finite bandwidth of the passbands. The maps are calibrated with the CMB dipole, which has a different spectrum than the Galactic emission. In the bottom right panel especially, one can see that the noise in the maps is larger in the ecliptic plane. This is because the scan pattern preferentially samples the ecliptic poles. (From Hinshaw et al. 2003a.)

level that can be probed with *WMAP* (Komatsu et al. 2003). This is a key step in the interpretation of the data. It means that all of the information in the CMB is contained in the angular power spectrum. In the parlance of CMB analyses, it means that if the temperature distribution is expanded in spherical harmonics as

$$T(\theta, \phi) = \sum_{lm} a_{lm} Y_{lm}(\theta, \phi), \tag{20.1}$$

that the real and the imaginary parts of the a_{lm} are normally distributed. In other words, the phases between all the harmonics are random. There are no features in the CMB like strings or textures that require some definite relation between the phases. The Gaussianity of the CMB is a triumph for models such as inflation. However, the degree of non-Gaussanity expected from these models is a factor of 1000 or more below where *WMAP* can probe (Maldacena 2002).

From the masked and cleaned maps, we produce the angular power spectrum, as detailed in Hinshaw et al. (2003). All known effects are taken into account in going from the maps to the spectrum. We account for the uncertainty in the beam profile, foreground masking, and uneven weighting of the sky. Monte Carlo simulations are used to check that the noise is handled correctly. The stability of the instrument permits us to determine the noise contribution to the percent level. In producing the power spectrum we use only the cross correlations between the eight individual Q, V, and W intensity maps that have had the two polarizations combined (Hinshaw et al. 2003). Because the noise in the maps is, for all intents and purposes, independent, there are no effects from prewhitening or from the peculiarities of any one DA in the power spectrum. Of the possible 36 separate measures of the power spectrum, we use 28. The results are shown in the top panel of Figure 20.8.

We also produce maps of the Stokes Q and U polarization components (Hinshaw et al. 2003a; Kogut et al. 2003). These maps are produced from the difference of the two differential measurements common to one pair of feeds and use lines of constant Galactic longitude as a reference direction. Though the polarization maps pass a series of null tests, there was not enough time to prepare them for the first release. However, the cross correlation between the polarization and temperature maps is much less susceptible to systematic error than the polarization maps alone and may be more easily determined with confidence. At $> 10\sigma$, we find a correlation between the Stokes I (temperature) and Q maps. This is due to the E mode of the polarization (Kamionkowski, Kosowsky, & Stebbins 1997; Zaldarriaga & Seljak 1997). In Q, V, and W bands, the TE (Stokes I cross Q) has a thermal frequency spectrum; it is not due to foregrounds. The correlation between Stokes I and U, which is significant for B modes, is consistent with zero.

The TE cross-power spectrum is shown in the bottom panel of Figure 20.8. We see that there is a model that describes the TT and TE data beautifully. As E modes at $1°$ scales are generated only during the transition from an ionized to a neutral Universe, the agreement indicates that we understand the physics of the decoupling process. Insofar as the $\ell > 10$ TE correlation is predicted from the TT data, it does not contribute to our understanding of the cosmological parameters.

The TE correlation for $\ell < 10$, however, considerably changes our understanding of cosmic history. It was a surprising discovery. The only way to polarize the CMB at these large angular scales is through reionization*. In our fiducial model, we assume that the first generation of stars completely reionized the Universe at a fixed redshift, z_r. The optical depth, τ, is then just the line integral out to z_r. We find that $\tau = 0.17 \pm 0.04$. Because quasar spectra show there is neutral hydrogen at $z = 6$ (Becker et al. 2001; Djorgovski et al. 2001; Fan et al. 2002), the ionization history may not be as simple as in our model. After accounting for uncertainties in our model, we find $z_r = 20^{+10}_{-9}$ (Kogut et al. 2003).

The large optical depth changes the way one interprets the TT power spectrum. It means that the intrinsic CMB fluctuations are 30% larger than the TT plot shows. This in turn affects the σ_8 (or overall normalization) one gets for the CMB data.

It is the maps and the power spectra that we derive from them that have the lasting value.

---

* Tensor modes can generate an $\ell < 10$ TE signal, but the required tensor amplitude is inconsistent with the TT data.

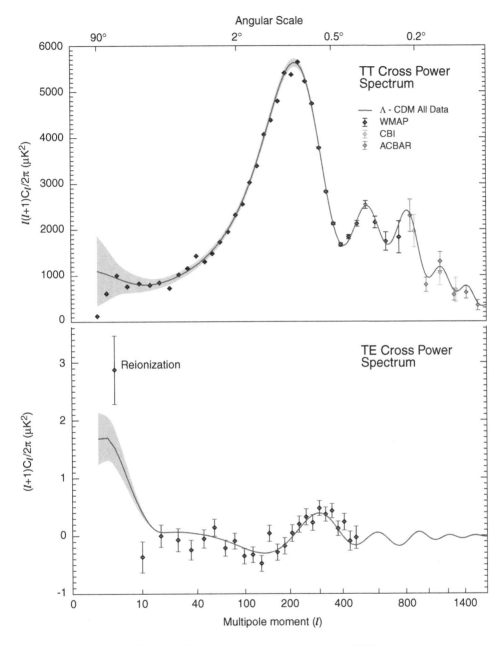

Fig. 20.8. The top panel shows the net temperature-temperature (TT) cross-power spectrum between all channels. The measurement error bars are shown; the grey band shows the cosmic variance limit. *WMAP* is cosmic variance dominated up to roughly $\ell = 350$. Below $\ell = 100$, the noise is completely cosmic variance dominated and so we use only the V and W bands. For $\ell > 700$ we augment *WMAP* with the data from CBI (Mason et al. 2003) and ACBAR (Kuo et al. 2004). The bottom panel shows the temperature-polarization (TE) cross-power spectrum. The point labeled "Reionization" is the weighted average of the lowest 8 multipoles. The red line is the best-fit model as discussed in §20.3.1. Note that the y-axis of the TE data is scaled to emphasize the low-ℓ region of the spectrum. (From Bennett et al. 2003b.)

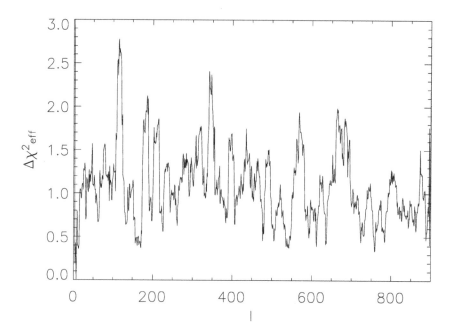

Fig. 20.9. The contribution to the χ^2 as a function of ℓ. The spikes in this plot may be identified with departures from the best-fit model in Figure 20.8. The binning here is per ℓ, whereas Figure 20.8 has values binned together. (From Spergel et al. 2003.)

Producing them is the primary goal of the mission. The maps do not depend on a particular cosmological model, the temperature power spectrum does not depend on a model, and the polarization does not depend on a model. We next focus more on to the cosmological interpretation of the data.

20.3.2 *Cosmic Parameters from Maps*

One can deduce the parameters of the cosmological model that fit the data. The procedure is conceptually straightforward, though the devil is in the details. Advances have been made since the first estimates. One now uses numerically friendly combinations of parameters (e.g., Kosowsky, Milosavljević, & Jimenez 2002), customized versions of CMBFAST-like programs (Seljak & Zaldarriaga 1996), and Markov Chain Monte Carlos (Christensen et al. 2001).

For any model fitting, one assumes some prior information. For example, one might limit oneself to only adiabatic ΛCDM models, or, more restrictively, to models with $h > 0.5$, or flat models, or models that agree with large-scale structure data. A goal is to say as much as possible with the minimum number of priors.

For the simplest models of the CMB, there is an intrinsic parameter degeneracy called the "geometric degeneracy" (Bond et al. 1994; Zaldarriaga, Spergel, & Seljak 1997). Using the TT CMB data alone, one cannot separately determine Ω_m, h, and Ω_Λ even with cosmic variance limited data out to $\ell = 2000$. One can play these parameters off one another to produce

Table 20.2. *Sample of Best-fit Cosmological Parameters*

Parameter	WMAP only, flat	WMAP+others	Non-CMB estimates
A	0.9 ± 0.1	$0.75^{+0.08}_{-0.07}$	...
τ	$0.166^{+0.076}_{-0.071}$	$0.117^{+0.057}_{-0.053}$	...
$\Omega_b h^2$	0.024 ± 0.001	0.0226 ± 0.0008	$0.021 - 0.025$
$\Omega_m h^2$	0.14 ± 0.02	0.133 ± 0.006	$0.062 - 0.18$
h	0.72 ± 0.05	0.72 ± 0.03	$h = 0.72 \pm 3 \pm 7$
n_s	0.99 ± 0.04	0.96 ± 0.02	...
χ^2_{eff}/ν	$1431/1342 = 1.066$	...	...

The best-fit parameters for the *WMAP* data for two of the models tested. The second column is for *WMAP* alone; the third is for a combination of *WMAP*+CBI+ACBAR+2dFGRS+Lyα. The χ^2 is not given because the Lyα data are correlated. The non-CMB estimates are from studies of the Lyα D/H and DLA systems, cluster number counts, galaxy velocities, and the *HST* Key Project. They come from many studies and are discussed in Spergel et al. (2003).

identical spectra. The degeneracy is broken by picking a value of the Hubble constant, assuming a flat geometry, or something similar. With the *WMAP* data, supernovae data (Riess et al. 1998; Perlmutter et al. 1999), and the *HST* Key Project value of $h = 0.72 \pm 3 \pm 7$ (Freedman et al. 2001), $\Omega_T = 1.02 \pm 0.02$. With the *WMAP* data and a prior of $h > 0.5$, $0.98 < \Omega_T < 1.08$ (95% confidence). While this does not *prove* that the Universe is geometrically flat, Occam's razor and the Dicke arguments that in part inspired inflation lead one to take a flat Universe as the basic model*.

Unless otherwise noted, all the parameters quoted have a prior that the Universe is geometrically flat. With this, the values in Table 20.2 are obtained. We give the values for just power-law models for the index. Values for a variety of models and parameter combinations are given in Spergel et al. (2003).

It is of interest that the probability of exceeding the reduced χ^2 for the best fit model is $\sim 5\%$. The reduced χ^2 is slightly large, but on its own does not signal that the model is wrong. In fact, no matter what model we try, χ^2/ν more or less stays the same. If isocurvature modes are added to fit the low-ℓ region of the spectrum, $\chi^2/\nu = 1468/1378 = 1.065$ (Peiris et al. 2003); if one adds tensor modes, allows the spectral index n_s to run, and the 2dFRGS data are added, $\chi^2/\nu = 1465/1379 = 1.062$; if a step is added to the power spectrum that attempts to fit bumps and wiggles in the *WMAP* angular power spectrum, $\chi^2/\nu = 1422/1339 = 1.062$ (Peiris et al. 2003). In other words, the admixture of additional model elements does not substantially improve the fit over the basic model. The contributions to χ^2/ν are shown in Figure 20.9. It is likely that, as the sky is covered more thoroughly, as we understand the beams better, and as we account for more detailed astrophysical effects (Spergel et al. 2003), χ^2 will decrease.

Though the simple model is a wonderful fit to the *WMAP* data, the small quadrupole and the related lack of correlation at large angular separations is striking. Though largely obscured by cosmic variance, there appears not even a hint of an upturn at low ℓ, as expected in ΛCDM models. (However, *COBE* also found no evidence for such an upturn). This

---

* If $\Omega_\Lambda = 0$, then just the position of the first peak shows that the Universe is flat (Kamionkowski, Spergel, & Sugiyama 1994).

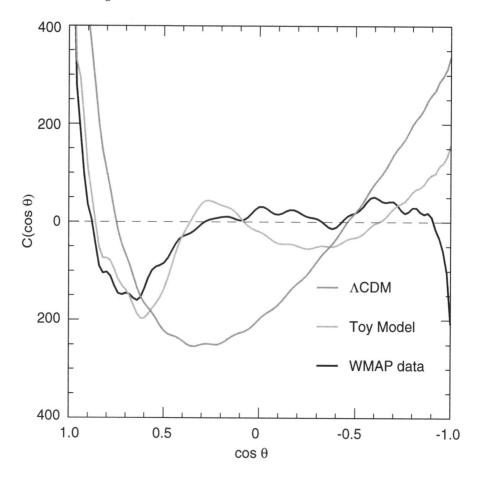

Fig. 20.10. Correlation function of the *WMAP* data. The green line corresponds to a ΛCDM model. The black line is the *WMAP* data. The "Toy" model, described in Spergel et al. (2003), has been tailored to fit the data by putting discrete modes in the matter power spectrum. (From Spergel et al. 2003.)

departure from the model constitutes a small (though possibly important) fraction of the total fluctuation power and does not cast doubt on the interpretation of the $\ell > 40$ spectrum. The correlation function is shown in Figure 20.10. For $\theta > 60°$ the probability that the *WMAP* data agree with the best-fit ΛCDM models *at these angular scales* is $\sim 1/500$. One must keep in mind that the statistic is *a posteriori*, but it is still unsettling*.

20.4   Summary and the Future

Before it was widely appreciated that so many of the cosmological parameters could be deduced directly from the CMB (e.g., Jungman et al. 1995), the motivation for studying the anisotropy was that it would tell us the initial conditions for the formation of

---

* David Spergel was overheard saying, "One should not obsess about the low quadrupole. One should not obsess about the low quadrupole. Really, one should not obsess about the low quadrupole."

cosmic structure and would tell us the cosmogony in which to interpret that process. In this vein, we summarize here some of the grander conclusions we have learned from *WMAP*.

(1) A flat Λ-dominated CDM model with just six parameters (A, τ, Ω_m, Ω_b, h, and n_s) gives an excellent description of the statistical properties of $> 10^6$ measurements of the anisotropy over 85% of the sky. This does not mean that a simple flat ΛCDM model is complete, but any other model must look very much like it.

(2) An Einstein-de Sitter model, which is flat with $\Omega_\Lambda = 0$, is ruled out at $> 5\sigma$. This statement is based on just the CMB, with no prior information (e.g., no prior on h). Stated another way, if $h > 0.5$ then *WMAP* requires $\Omega_\Lambda \neq 0$ in the angular-diameter distance to the decoupling surface.

(3) A closed model with $\Omega_m = 1.28$ and $\Omega_\Lambda = 0$ fits the data but requires $h = 0.33$, in conflict with *HST* observations. It also disagrees with cluster abundances, velocity flows, and other cosmic probes.

(4) The fluctuations in the metric are superhorizon (Peiris et al. 2003). Turok (1996) showed that in principle one can construct models based on subhorizon processes that can mimic the TT ΛCDM spectra. However, these models cannot mimic the observed TE anticorrelation at $50 < \ell < 150$ (Spergel & Zaldarriaga 1997).

By combining *WMAP* data with other probes we can constrain the mass of the neutrino and constrain the equation of state, w. As our knowledge of the large-scale structure matures, in particular of the Lyα forest (McDonald et al. 2000; Zaldarriaga, Hui, & Tegmark 2001; Croft et al. 2002; Gnedin & Hamilton 2002; Seljak, McDonald, & Markarov 2003), we can use the combination of these measurements to probe in detail the spectral index as a function of scale, thereby directly constraining inflation and other related models. No doubt correlations with other measurements will shed light on other cosmic phenomena. With *WMAP*, physical cosmology has moved from assuming a model and deducing cosmic parameters to detailed testing of a standard model.

The *WMAP* mission is currently funded to run through October 2005, though as of this writing there is nothing to limit the satellite from operating longer. The quality of the maps will continue to improve as the mission progresses. Not only will the noise integrate down, but our knowledge of the beams will continue to improve, and we will make corrections that were too small to consider for the year one analysis. For example, increased knowledge of the beams will reduce the uncertainties in the power spectrum at high ℓ and will permit an even cleaner separation of the CMB and foreground components near the Galactic plane. It is the set of high-fidelity maps of the microwave sky that will be *WMAP*'s most important legacy.

With four-year maps, the polarization at intermediate and large angular scale should be seen and the epoch of reionization will be much better constrained. It may even be possible to detect the lensing of the CMB by the intervening mass fluctuations.

On the horizon is the *Planck* satellite, which, with its higher sensitivity, greater resolution, and larger frequency range should, significantly expand upon the picture presented by *WMAP*. In addition, a host of other measurements that push to yet finer angular scales and a future satellite dedicated to measuring the polarization in the CMB, especially the B modes, promise a rich and exciting future for CMB studies.

More information about *WMAP* and the data are available through: http://lambda.gsfc.nasa.gov/

References

Barnes, C., et al. 2002, ApJS, 143, 567

——. 2003, ApJS, 148, 51

Becker, R. H., et al. 2001, AJ, 122, 2850

Bennett, C. L., et al. 2003a, ApJ, 583, 1

——. 2003b, ApJS, 148, 1

——. 2003c, ApJS, 148, 97

Bond, J. R., Crittenden, R., Davis, R. L., Efstathiou, G., & Steinhardt, P. J. 1994, Phys. Rev. Lett., 72, 13

Christensen, N., Meyer, R., Knox, L., & Luey, B. 2001, Classical and Quantum Gravity, 18, 2677

Colless, M., et al. 2001, MNRAS, 328, 1039

Croft, R. A. C., Weinberg, D. H., Bolte, M., Burles, S., Hernquist, L., Katz, N., Kirkman, D., & Tytler, D. 2002, ApJ, 581, 20

Djorgovski, S. G., Castro, S., Stern, D., & Mahabal, A. A. 2001, ApJ, 560, L5

Draine, B. T., & Lazarian, A. 1998, ApJ, 494, L19

Fan, X., et al. 2002, AJ, 123, 1247

Finkbeiner, D. P. 2003, ApJS, 146, 407

Finkbeiner, D. P., Davis, M., & Schlegel, D. J. 1999, ApJ, 524, 867

Freedman, W. L., et al. 2001, ApJ, 553, 47

Gnedin, N. Y., & Hamilton, A. J. S. 2002 MNRAS, 334, 107

Haslam, C. G. T., Stoffel, H., Salter, C. J., & Wilson, W. E. 1982, A&AS, 47, 1

Hinshaw, G., et al. 2003a, ApJS, 148, 63

——. 2003b, ApJS, 148, 135

Jarosik, N., et al. 2003a, ApJS, 145, 413

——. 2003b, ApJS, 148, 29

Jungman, G., Kamionkowski, M., Kosowsky, A., & Spergel, D. N. 1995, Phys. Rev. D., 54, 1332

Kamionkowski, M., Kosowsky, A., & Stebbins, A. 1997, Phys. Rev. D, 55, 7368

Kamionkowski, M., Spergel, D. N., & Sugiyama, N. 1994, ApJ, 426, L57

Kogut, A., et al. 2003, ApJS, 148, 161

Komatsu, E., et al. 2003, ApJS, 148, 119

Kosowsky, A., Milosavljević, M., & Jimenez, R. 2002, Phys. Rev. D, 66, 63007

Kuo, C. L., et al. 2004, ApJ, submitted (astro-ph/0212289)

Maldacena, J. 2002, preprint (astro-ph/0210603)

Mason, B. S., et al. 2003, ApJ, 591, 540

McDonald, P., Miralda-Escudé, J., Rauch, M., Sargent, W. L. W., Barlow, T. A., Cen, R., & Ostriker, J. P. 2000, ApJ, 543, 1

Page, L. A., et al. 2003a, ApJ, 585, 566

——. 2003b, ApJS, 148, 39

——. 2003c, ApJS, 148, 233

Peiris, H., et al. 2003, ApJS, 148, 213

Perlmutter, S., et al. 1999, ApJ, 517, 565

Pospieszalski, M. W. 1992, Proc. IEEE Microwave Theory Tech., MTT-S 1369

——. 1997, Microwave Background Anisotropies, ed. F. R. Bouchet (Gif-sur-Yvette: Editions Frontièrs), 23

Pospieszalski, M. W., & Wollack, E. W. 1998, IEEE Microwave Theory Tech., MTT-S Digest, Baltimore, MD 669

Riess, A. G., et al. 1998, AJ, 116, 1009

Seljak, U., McDonald, P., & Markarov, A. 2003, MNRAS, 342, L79

Seljak, U., & Zaldarriaga, M. 1996, ApJ, 469, 437

Spergel, D. N., et al. 2003, ApJS, 148, 175

Spergel, D. N., & Zaldarriaga, M. 1997, Phys. Rev. Lett., 79, 2180

Turok, N. 1996, Phys. Rev. D, 54, 3686

Verde, L., et al. 2003, ApJ, submitted (astro-ph/0302218)

Wright, E. L., Hinshaw, G., & Bennett, C. L. 1996, ApJ, 458, L53

Zaldarriaga, M., Hui, L., & Tegmark, M. 2001, ApJ, 557, 519

Zaldarriaga, M., & Seljak, U. 1997, Phys. Rev. D, 55, 1830

Zaldarriaga, M., Spergel, D. N., & Seljak, U. 1997, ApJ, 488, 1

21

Interferometric observations of the
cosmic microwave background radiation

ANTHONY C. S. READHEAD and TIMOTHY J. PEARSON
California Institute of Technology

Abstract
Radio interferometers are well suited to studies of both total intensity and polarized intensity fluctuations of the cosmic microwave background radiation, and they have been used successfully in measurements of both the primary and secondary anisotropy. Recent observations with the Cosmic Background Imager operating in the Chilean Andes, the Degree Angular Scale Interferometer operating at the South Pole, and the Very Small Array operating in Tenerife have probed the primary anisotropy over a wide range of angular scales. The advantages of interferometers for microwave background observations of both total intensity and polarized radiation are discussed, and the cosmological results from these three instruments are presented. The results show that, subject to a reasonable value for the Hubble constant, which is degenerate with the geometry in closed models, the geometry of the Universe is flat to high precision ($\sim 5\%$) and the primordial fluctuation spectrum is very close to the scale-invariant Harrison-Zel'dovich spectrum. Both of these findings are concordant with inflationary predictions. The results also show that the baryonic matter content is consistent with that found from primordial nucleosynthesis, while the cold dark matter component can account for no more than $\sim 40\%$ of the energy density of the Universe. It is a requirement of these observations, therefore, that $\sim 60\%$ of the energy content of the Universe is not related to matter, either baryonic or nonbaryonic. This *dark energy* component of the energy density is attributed to a nonzero cosmological constant.

21.1   Introduction

Interferometers are playing a key role in the determination of fundamental cosmological parameters through observations of the cosmic microwave background (CMB). Three such instruments have been constructed and deployed over the last few years—the Cosmic Background Imager (CBI), operating at 5080 m altitude in the Chilean Andes (Padin et al. 2001, 2002); the Degree Angular Scale Interferometer (DASI), operating at 2800 m altitude at the South Pole (Leitch et al. 2002a); and the Very Small Array (VSA) operating at an altitude of 2400 m in Tenerife (Watson et al. 2003), for which a prototype was the Cambridge Anisotropy Telescope (CAT) (Scott et al. 1996). In this paper we review the characteristics of interferometers that render them particularly well suited to CMB observations, and we then discuss the cosmological results from observations with these three instruments.

It is well known that bolometers have also played a crucial role in the development of this field, and the interested reader will find an account of the bolometer CMB results in

Fig. 21.1. The Cosmic Background Imager (CBI), located at 5080 m altitude in the Chilean Andes, was the first instrument to operate permanently from this, the future ALMA site. The CBI is a "stand-alone" operation with its own power plant and other observatory facilities. Oxygenated working and living quarters, pioneered here, have proven essential to the project.

the companion review in this volume by Lange. While we do not discuss the bolometer results here, they are in excellent agreement with the interferometer results described below (de Bernardis et al. 2000; Hanany et al. 2000; Lee et al. 2001; Netterfield et al. 2002), and all of these results are in excellent agreement with the recently released *WMAP* results (Bennett et al. 2003).

21.2   The CBI, the DASI, and the VSA

The CBI, shown in Figure 21.1, was the first of the new generation of CMB interferometers to be brought into operation (Padin et al. 2001), in January 2000. At that time the CBI began routine total intensity observations as well as preliminary polarization observations using a single cross-polarized antenna. We describe the CBI here in some detail since both the DASI and the VSA share most of these characteristics, and we point out the principal features in the VSA and the DASI in which they differ from the CBI. All three instruments operate in the Ka waveguide band (26 – 40 GHz) and use low-noise amplifiers (LNAs) based on high electron mobility transistors (HEMTs), which were designed by Marian Pospieszalski of the National Radio Astronomy Observatory (Pospieszalski et al. 1994, 1995). The specifications of the three instruments are given in Table 21.1.

As described by Padin et al. (2002), the CBI has 13 90-cm diameter antennas mounted

Table 21.1. *Specifications of CMB Interferometers*

Specification	CBI	DASI	VSA
antennas	13	13	14
center frequency (GHz)	31	31	31
channels	10	10	1(tunable)
bandwidth per channel (GHz)	1	1	1.5
system temperature	~ 30 K	~ 30 K	~ 30 K
l range	300–3500	150–850	150–1400
l resolution	~ 140	~ 80	~ 80
location	Atacama Desert	South Pole	Tenerife
altitude (m)	5080	2800	2400

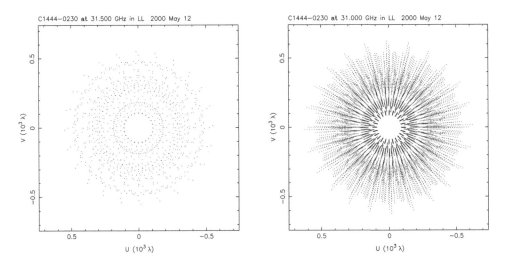

Fig. 21.2. The aperture-plane, (u,v), coverage of the CBI for a 6-hour observation. The left-hand panel shows the coverage for a single channel, and the right-hand panel shows the coverage for all 10 channels.

on a 6-m platform. In addition to the usual altazimuth rotation axes, the CBI platform can be rotated about the optical axis. This third axis of rotation enables the array to maintain a constant parallactic angle while tracking a celestial source; it also greatly facilitates polarization observations and calibration, and, in addition, it provides a powerful method of discriminating between sky signals and instrumental cross-talk. For each baseline both the real and imaginary channels are correlated in a complex correlator, and there are 10 1-GHz bandwidth frequency channels spanning the range 26–36 GHz. Thus, the CBI comprises a total of 780 complex interferometers. The fundamental data rate is 0.84 s. For typical total intensity observations the data are averaged over 10 samples, so that on each interferometer the complex visibility is measured every 8.4 s.

The aperture-plane, or (u,v), coverage of the CBI is shown in Figure 21.2. In this figure the (u,v) coverage corresponding to a single baseline should be the convolution of two apertures centered on the (u,v) point corresponding to the baseline length and orientation. For clarity we have shrunk the aperture convolution to a single point to make the distribution of the coverage clearer. There is, of course, much overlap in the coverage when the full diameters are used, since the antenna diameter of 90 cm is close to the size of the shortest baselines (100 cm). Mosaic observations of contiguous fields improve the resolution in multipole space. For example, in the CBI, mosaic observations have been used to improve the multipole resolution from $\delta l \approx 500$ to $\delta \approx 140$. CBI mosaic observations of three fields are shown in Figure 21.3.

The DASI is shown in Figure 21.4. The DASI was designed to complement the CBI in multipole coverage, and many of the detailed CBI designs, including those for the correlator, the receiver control cards, and the channelizer, as well as the telescope control and data acquisition software, were shared with the DASI team, who duplicated these components in the DASI. For these reasons, the DASI is in many respects a copy of the CBI, but ~ 4 times smaller. The DASI achromatic polarizers were duplicated by the DASI team for use on the CBI in its upgraded phase for full polarization observations, which commenced in October 2002. The DASI observed in total intensity mode during the austral winter of 2000, and was then outfitted with achromatic polarizers and both a ground screen and a sunscreen, shown in Figure 21.4, for the austral winter of 2001 when it made polarization observations of two fields, as described by Zaldarriaga (2004).

The VSA (Figure 21.5) and the DASI are similar in size, and originally covered similar multipole ranges (and hence very similar angular scales), extending from $l \approx 150$ to $l \approx 900$ with a resolution in multipoles of $\delta l \approx 80$. In 2001 the original VSA horns were replaced with a larger set that extended the VSA range up to $l \approx 1400$.

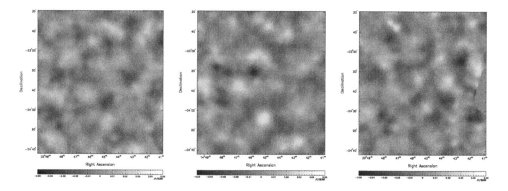

Fig. 21.3. The CBI has made mosaic observations of three regions separated by 6 hr in Right Ascension. Mosaic images from the first year of observations, which cover an area of $\sim 2° \times 2°$, are shown above. In these images the seeds that gave rise to clusters of galaxies are seen for the first time.

Fig. 21.4. The Degree Angular Scale Interferometer (DASI), located at 2800 m at the South Pole, was outfitted with a sunscreen for the polarization observations of 2001.

Fig. 21.5. The Very Small Array (VSA) is sited at 2400 m in Tenerife.

The CBI covers the range $l \approx 300$ to $l \approx 3500$ with a multipole resolution in mosaicing mode of $\delta l \approx 140$. Apart from this primary difference in multipole range and resolution, the other major difference is in the correlators—the CBI and the DASI each have 10 1-GHz channels, operating between 26 GHz and 36 GHz, while the VSA has a single tunable 1.5-GHz channel, which operates over the same frequency range.

21.3 The Sky Brightness Distribution of the CMB

We denote the CMB temperature in direction $\hat{\mathbf{n}}$ (angular coordinates θ, ϕ) by $T(\hat{\mathbf{n}})$, and hence the variation in temperature about the mean $\langle T(\hat{\mathbf{n}}) \rangle$ by

$$\Delta(\hat{\mathbf{n}}) = \frac{T(\hat{\mathbf{n}})}{\langle T(\hat{\mathbf{n}}) \rangle} - 1, \tag{21.1}$$

which may be expanded into spherical harmonics

$$\Delta(\hat{\mathbf{n}}) = \sum_{l=0}^{\infty} \sum_{m=-l}^{l} a_{lm} Y_{lm}(\hat{\mathbf{n}}), \tag{21.2}$$

where a_{lm} are the multipole moments

$$a_{lm} = \int Y_{lm}^*(\hat{\mathbf{n}}) \Delta(\hat{\mathbf{n}}) d\Omega \quad . \tag{21.3}$$

If the CMB is isotropic it must be rotationally invariant so that the fluctuations can be expressed in terms of a one-dimensional angular spectrum, $C_l \equiv \langle a_{lm}^2 \rangle$, and the variance of $\Delta(\hat{\mathbf{n}})$ can be expressed in terms of this angular power spectrum:

$$\langle \Delta(\hat{\mathbf{n}})^2 \rangle = \sum_{l=0}^{\infty} \frac{2l+1}{4\pi} C_l \quad . \tag{21.4}$$

For large l,

$$\langle \Delta(\hat{\mathbf{n}})^2 \rangle \approx \int \frac{l(2l+1)}{4\pi} C_l \, d\ln l, \tag{21.5}$$

showing that the contribution to the variance from a logarithmic interval of l is proportional to $l(2l+1)C_l/4\pi$. It is therefore often convenient to plot this quantity versus $\log l$, so this convention is adopted by many authors in displaying angular spectra of the CMB. It should be borne in mind, however, that the signal is C_l, so that, in plotting $l(2l+1)C_l/4\pi$ we are artificially boosting the apparent variance at high l by a factor $\propto l^2$. *Observations at high l therefore require far greater sensitivity than is immediately apparent from this conventional way of plotting the angular power spectrum.*

The angular correlation function is defined as

$$C(\theta) = C(\hat{\mathbf{n}}_1, \hat{\mathbf{n}}_2) = \langle \Delta(\hat{\mathbf{n}}_1) \Delta(\hat{\mathbf{n}}_2) \rangle, \tag{21.6}$$

where by isotropy and homogeneity C is a function only of θ, and $\cos\theta = \hat{\mathbf{n}}_1 \cdot \hat{\mathbf{n}}_2$. It is easy to show that

$$C(\theta) = \frac{1}{4\pi} \sum_{l=0}^{\infty} (2l+1) C_l P_l(\cos\theta), \tag{21.7}$$

where $P_l(\cos\theta)$ is the Legendre polynomial. The inverse result is

$$C_l = 2\pi \int_{-1}^{1} C(\theta)P_l(\cos\theta)d\cos\theta \quad . \tag{21.8}$$

These two results are the analogs on the sphere of the theorem relating the covariance function to the Fourier transform of the power spectrum.

In interferometric observations of the CMB the fields of view, even for mosaic observations, are small enough that the small-angle approximation may be used. That is, we can assume that

$$\theta \ll 1 \text{ and } l \gg 1. \tag{21.9}$$

In this case

$$P_l(\cos\theta) \approx J_0\{(l+1/2)\,\theta\}, \tag{21.10}$$

so that

$$
\begin{aligned}
C_l &\approx 2\pi \int_{-1}^{1} C(\theta)J_0\{(l+1/2)\,\theta\}d\cos\theta \\
&\approx 2\pi \int_{0}^{\infty} C(\theta)J_0\{(l+1/2)\,\theta\}\theta\,d\theta \quad .
\end{aligned}
\tag{21.11}
$$

If we set $l+1/2 = 2\pi\nu$ then this is a Hankel transform, which is simply the two-dimensional Fourier transform for a circularly symmetric function; i.e., as expected, the angular correlation function and the power spectrum form a Fourier transform pair in the small-angle approximation.

In practice we observe the sky with instruments of finite size, so that what we observe is the convolution of the sky brightness distribution with the instrument response function, corresponding to a product in the Fourier transform (spectral) domain:

$$C(\theta) = \frac{1}{4\pi}\sum_{l=0}^{\infty}(2l+1)C_l W_l(\cos\theta), \tag{21.12}$$

where $W_l(\cos\theta)$ is called the window function of the instrument.

21.4 Radio Interferometric Observations of the CMB

We do not review here the basic theory of radio interferometry, but refer the reader to the comprehensive text of Thompson, Moran, & Swenson (2001). Useful discussions of the application of standard radio interferometric techniques to observations of the CMB have been given by Hobson, Lasenby, & Jones (1995), White et al. (1999), Hobson & Maisinger (2002), and Myers et al. (2003).

21.5 Interferometer Characteristics

The following properties of interferometers make them particularly well suited to observations of the CMB:

- Automatic subtraction of the mean signal (to high precision)
- Precise knowledge of the beamshape is easy to obtain (and is not as important as it is in single-dish observations)
- Direct measurements of visibilities (which are very nearly the desired Fourier components of the sky brightness distribution)

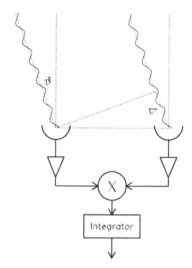

Fig. 21.6. The Multiplying Interferometer subtracts off the constant term in the field under observation, as was first demonstrated by Martin Ryle in 1952 (see text).

- Precision radiometry (through observations of planets, supernova remnants, quasars, and radio galaxies)
- Precision polarimetry (and, in the case of the CBI, making use of the third axis of rotation about the optical axis)
- Repeated baselines enable a wide variety of instrumental crosschecks

We discuss each of these interferometer properties separately below, illustrating some of them with examples from the CBI. It should be clear that similar advantages apply, in most cases, to the DASI and the VSA.

21.5.1 Automatic Mean Subtraction

A multiplying interferometer (Fig. 21.6) has the advantage over adding interferometers and other total-power detection systems in that the mean signal is subtracted automatically to high precision (Ryle 1952). In the adding interferometer the voltages from the two antennas are added and then squared in a square-law detector, so that the power output from the square-law detector is proportional to $(V_1 + V_2)^2$. Ryle introduced the phase-switched interferometer in which the voltages are alternately summed and differenced by introducing a $\pi/2$ phase offset in one of the signals, and then detected by a phase-synchronous detector, which is in phase with the phase switch. In this system the output power is the time average of $(V_1 + V_2)^2 - (V_1 - V_2)^2$, and so it is proportional the the product of the voltages from the two antennas. The power output in this system is thus independent of the mean level of the signal, $V_1^2 + V_2^2$, and it measures the correlation between the two signals, $\langle V_1 V_2 \rangle$. Modern interferometers, such as those discussed here, accomplish the same effect by using a correlator. Since, in a multiplying or phase-switching interferometer, the mean signal does not appear, and only the spatially varying signal appears, this eliminates many sources of spurious systematic errors. This approach is particularly advantageous in CMB observations, in

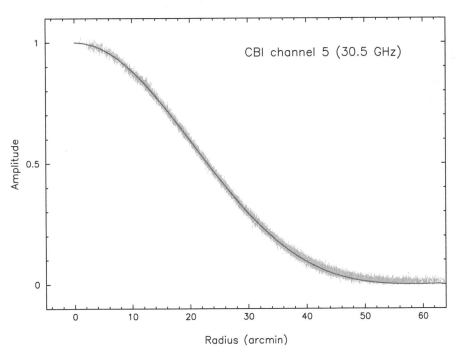

Fig. 21.7. Radial profile of the CBI primary beam in one of the 10 frequency channels; data from all 78 baselines are superimposed. The observations were of Tau A and are shown by the error bars. The curve shows the profile computed by taking the square of the Fourier transform of the aperture illumination pattern, assumed to be circularly symmetric. These observations show that the 13 antennas have very similar beams (see Pearson et al. 2003).

which the spatial fluctuations in temperature are over 10^5 times smaller than the mean signal of 2.725 K (Mather et al. 1999). For this reason a number of potential sources of systematic error are reduced to negligible levels by interferometry.

21.5.2 *Precise Knowledge of Beamshape*

The resolution of an interferometer is set by the baseline length between the antenna pair, rather than by the primary beam of the individual antennas. Thus, in interferometric observations precise knowledge of the primary beamshape is not required in order to measure the power spectrum on small angular scales. Of course, precise knowledge of the primary beamshape is needed in computing the covariance function of the instrument, but this can be determined to the required accuracy of a few percent by measurements of the beamshape on bright unresolved radio sources. The profile of the mean primary beamshape for the CBI is shown in Figure 21.7.

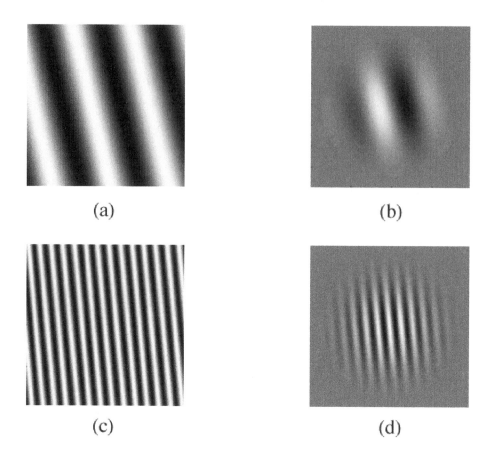

Fig. 21.8. Panels (a) and (c) illustrate two multipole components of the sky brightness distribution over a $1.5° \times 1.5°$ field of view, typical for the CBI. An interferometer measures directly these components multiplied by the primary beam, which is fixed by the antenna size and illumination. The product of the primary beam with the multipole component measured by an interferometer, which is set by the baseline length, is shown for the two multipole components above in panels (b) and (d). In the case of the CBI, (a) and (b) represent the one-meter baselines, while (c) and (d) represent the five-meter baselines.

21.5.3   *Direct Measurements of Visibilities*

An interferometer measures visibility, which is the Fourier transform of the product of the primary antenna beam, $A(\mathbf{x})$, with the sky brightness, $I(\mathbf{x})$. The visibilities, $V(\mathbf{u}) = V(u,v)$, measured by an interferometer are related to the sky brightness by

$$V(\mathbf{u}) = \int_{-\infty}^{\infty} \int_{-\infty}^{\infty} A(\mathbf{x})I(\mathbf{x})e^{-2\pi i \mathbf{u}\cdot\mathbf{x}}d\mathbf{x} \quad . \tag{21.13}$$

The visibility is thus the convolution of the Fourier transform of the sky brightness, $\tilde{I}(\mathbf{u})$ (i.e., the angular spectrum of the sky brightness distribution), and of the primary beam, $\tilde{A}(\mathbf{u})$,

$$V(\mathbf{u}) = \tilde{I}(\mathbf{u}) * \tilde{A}(\mathbf{u}) \quad . \tag{21.14}$$

An interferometer therefore measures directly the angular spectrum of the sky brightness distribution, which is the desired result, convolved with the Fourier transform of the primary beam. An example of a single Fourier component on the sky and of this component multiplied by the primary beam is shown in Figure 21.8.

The matrix of the covariance between all the visibility measurements is the observed *covariance matrix*, $C_{ij} = \langle V_i V_j^* \rangle$, and this is made up of two parts,

$$C = M + N, \tag{21.15}$$

where M is the sky covariance matrix and N is the noise covariance matrix. Hobson et al. (1995) have shown that the sky covariance matrix is

$$
\begin{aligned}
M_{jk} &= \langle V(\mathbf{u}_j, \nu_j) V^*(\mathbf{u}_k, \nu_k) \rangle \\
&= \int \int d^2\mathbf{v} \tilde{A}(\mathbf{u}_j - \mathbf{v}, \nu_j) \tilde{A}(\mathbf{u}_k - \mathbf{v}, \nu_k) S(\mathbf{v}, \nu_j, \nu_k),
\end{aligned}
\tag{21.16}
$$

where $S(\mathbf{v}, \nu_j, \nu_k)$ is a generalized power spectrum of the intensity fluctuations. Hence, converting to temperature fluctuations, we find that for CMB fluctuations,

$$
\begin{aligned}
M_{jk} &= \langle V(\mathbf{u}_j, \nu_j) V^*(\mathbf{u}_k, \nu_k) \rangle \\
&= \left(\frac{2\nu^2 k_B T_0 g(\nu)}{c^2} \right)^2 \int \int d^2\mathbf{v} \tilde{A}(\mathbf{u}_j - \mathbf{v}, \nu_j) \tilde{A}(\mathbf{u}_k - \mathbf{v}, \nu_k) C(\nu) \\
&= \left(\frac{2\nu^2 k_B T_0 g(\nu)}{c^2} \right)^2 \int_0^\infty W_{jk}(\nu) C(\nu) \nu d\nu,
\end{aligned}
\tag{21.17}
$$

where $g(\nu) = x^2 e^x / (e^x - 1)^2$, $x = h\nu / k_B T_0$, and $C(\nu) = C_l$ for $\sqrt{l(l+1)} \approx l + \frac{1}{2} = 2\pi\nu$, and

$$W_{jk}(\nu) = \int_0^{2\pi} \tilde{A}(\mathbf{u}_j - \mathbf{v}, \nu_j) \tilde{A}(\mathbf{u}_k - \mathbf{v}, \nu_k) d\theta_\nu \tag{21.18}$$

is the *window function*.

21.5.4 Precision Radiometry

A great advantage of interferometric measurements is precision radiometry. Provided that there are sufficiently bright unresolved and nonvarying radio sources, the calibration of interferometers is straightforward. In the case of the CBI, for example, the primary calibrators were Jupiter and Tau A, and secondary calibrators were Mars, Saturn, and 3C 274. Of these only Tau A is significantly resolved on the CBI, but it is easily modeled with an elliptical Gaussian brightness profile. The internal calibration consistency on the CBI is considerably better than 1%, and the flux density scale, originally set by our own absolute calibrations at the Owens Valley Radio Observatory with an uncertainty of 3.3% (Mason et al. 1999), has now been improved by comparison with the *WMAP* temperature measurement of Jupiter (Page et al. 2003) to an uncertainty of 1.3% (Readhead et al. 2004).

21.5.5 Precision Polarimetry

The instrumental polarization of radio interferometers is much easier to calibrate than single-dish instruments, because there is an elegant way of distinguishing between the polarization of the instrument and that of the source (Conway & Kronberg 1969). The instrumental polarization can be expressed in terms of *leakage factors*. In the CBI each antenna observes either left- or right-circular polarization. The voltage output from a left-circularly polarized antenna, p, may be written

$$V_L(p) = E_L(p) + E_R(p)\epsilon_p e^{j\phi_p}, \tag{21.19}$$

where $E_L(p)$ is the electric field that we wish to measure, and $\epsilon_p e^{j\phi_p}$ represents the instrumental leakage of the unwanted right-circular polarization electric field, $E_R(p)$, on antenna p. Similarly, the voltage output of a right-circular polarized antenna, q, may be written

$$V_R(q) = E_R(q) + E_L(q)\epsilon_q e^{j\phi_q}, \tag{21.20}$$

where $\epsilon_q e^{j\phi_q}$ represents the instrumental leakage of left-circular polarization on antenna q.

For a point source with zero circular polarization the correlator output on the LR cross-polarized baselinem which combines antenna p with antenna q, can be written

$$
\begin{aligned}
V_L(p)V_R^*(q) &= \frac{E_L(p)E_R^*(q)e^{2j\eta} + I(\epsilon_p e^{j\phi_p} + \epsilon_q e^{-j\phi_q})}{(1+\epsilon_p^2)^{1/2}(1+\epsilon_q^2)^{1/2}} \\
&= \frac{mIe^{2j\chi}e^{2j\eta} + I(\epsilon_p e^{j\phi_1} + \epsilon_q e^{-j\phi_2})}{(1+\epsilon_p^2)^{1/2}(1+\epsilon_q^2)^{1/2}},
\end{aligned} \tag{21.21}
$$

where m is the linear polarization, I is the total intensity, χ is the polarization angle on the sky, and we have neglected terms of order ϵ^2 and higher. The angle η is the azimuthal angle of the antennas about the line of sight to the source. In the CBI and the DASI, the antennas are mounted on a rotatable deck, so that, in addition to the usual two (alt-az) axes, the instruments can be rotated about the optical axis, thereby varying η for all antennas simultaneously. The expression given above applies for instruments such as the CBI and the DASI in which a single rotation by η of the deck upon which the antennas are all mounted leads to a phase advance of η in one polarization and a phase retardation of η in the other polarization—thereby giving rise to a relative phase shift of 2η on cross-polarized baselines.

By means of rotation of the deck, it is easy to measure the instrumental polarization on an instrument such as the CBI or the DASI: we simply need to observe a bright unresolved polarized source, for which m and χ are known, such as 3C 279, and measure $V_L(p)V_R^*(q)$ while varying η by rotating the deck. The complex number $V_L(p)V_R^*(q)$ then traces out an ellipse that is closely approximated by a circle of radius

$$mI/(1+\epsilon_p^2)^{1/2}(1+\epsilon_q^2)^{1/2} \tag{21.22}$$

centered on the point

$$\frac{I(\epsilon_p e^{j\phi_p} + \epsilon_q e^{-j\phi_q})}{(1+\epsilon_p^2)^{1/2}(1+\epsilon_q^2)^{1/2}}. \tag{21.23}$$

By observing a bright source of known polarization on all baselines, it is then possible to solve for the individual antenna leakage factors to high precision. Both the CBI and the DASI use achromatic polarizers with leakage factors $< 3\%$ designed by John Kovac (Kovac et al. 2002), so that, after correcting for instrumental polarization, uncertainties in polarization due to instrumental effects are $\ll 1\%$.

The polarization of the CMB can be expressed in terms of a curl-free mode and a curl mode. By analogy with electromagnetic theory these are designated as the E-mode and the B-mode. The strongest polarized signal in the CMB is due to Thomson scattering by electrons of the anisotropic radiation field due to quadrupole velocity anisotropy, and this is an

E-mode component. The much weaker *B*-mode components would be caused by gravitational radiation or gravitational lensing (see, e.g., Hu 2003).

It can be shown that for the case of infinitely small antennas the sum of the LR+RL visibilities yields the Fourier transform of the *E*-mode, while the difference LR−RL yields the Fourier transform of the *B*-mode (Zaldarriaga 2001). For the realistic case of finite antennas there is also a small amount of mixing of these modes due to the finite aperture. This is a direct analog of the arguments given above showing that the visibility in total intensity is the convolution of the angular spectrum with the Fourier transform of the primary beam.

21.5.6 Instrumental Crosschecks

On the CBI there are many duplicated baselines in any given orientation of the deck, and further baseline duplication can be accomplished by deck rotation. This is very useful in tracking down and eliminating various sources of systematic errors (e.g., cross-talk between electronic components). It also provides for crosschecks of visibilities on calibrators measured on different baselines, etc.

21.6 CMB Spectra from Interferometry Observations

The CBI began CMB observations at the Chilean site in January 2000, and the DASI and the VSA began their observations shortly thereafter. In this section we discuss the details of the spectra revealed by these interferometric observations, and in the following section we discuss the constraints that these place on key cosmological parameters.

21.6.1 The CBI CMB Spectrum

The first results from the new generation of interferometers were published by the CBI group in January 2001. These were the first observations of the CMB with both the sensitivity and the resolution to make images of the mass fluctuations on scales corresponding to clusters of galaxies (Padin et al. 2001; Mason et al. 2003; Pearson et al. 2003), and the images show clearly, and for the first time, the seeds that gave rise to galaxy clusters (Fig. 21.3). The straight line in Figure 21.9 shows the mass within a spherical region at the epoch of last scattering as a function of the angular diameter of the sphere, assuming a ΛCDM cosmology. Also shown are the angular scales and the corresponding mass scales that have been probed by key CMB experiments over the last three years. In addition, these were the first CMB observations to show the reduction in spectral power at high multipole numbers due to the damping tail at small scales caused by photon viscosity and the finite thickness of the last-scattering region. Thus, this major pillar of the standard ΛCDM cosmology has been confirmed by the CBI observations. The CBI spectrum for observations from January 2000 through November 2001 is shown in Figure 21.10.

An intriguing feature of the CBI spectrum is the excess of power seen at $l \approx 2500$ (Mason et al. 2003). The reality of this feature and its possible significance are discussed below.

At the frequency of operation of the CBI (26 – 36 GHz), and at multipoles above $l \approx 500$, there is significant foreground contamination due to radio galaxies and quasars. These are dealt with in the CBI data by a constraint matrix approach, such as was first used on interferometry data by the DASI (Leitch et al. 2002a), and which effectively "projects out" the point sources. The application to the CBI is described in detail by Mason et al. (2003). Thus far it has been necessary to use the 1.4 GHz NRAO VLA Sky Survey (NVSS) (Condon et al.

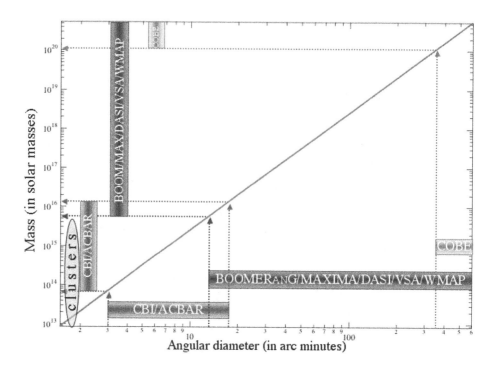

Fig. 21.9. The mass of material within a sphere at the last-scattering surface is shown as a function of the angular diameter of the sphere. Here we have assumed a ΛCDM model in which $\Omega_{matter} h^2 = 0.3$ and $\Omega_{total} = 1$. The mass scales accessed by various CMB observations are shown. Only the CBI and ACBAR cover masses of galaxy clusters.

1998) for identifying possible contaminating sources, because there is no higher-frequency radio survey that covers the CBI fields. Only $\sim 20\%$ of the NVSS sources are flat-spectrum objects that are bright enough at the CBI frequencies to cause detectable contamination, and a survey of NVSS sources in the CBI fields at ~ 30 GHz would enable us to identify these sources, thus reducing by a factor of 5 the number of sources that must be projected out of the CBI data. This would significantly increase the amount of CBI data that is retained, and it would substantially reduce the uncertainties in the CBI spectrum at multipoles higher than $l \approx 1000$. It is possible to measure the flux densities of the NVSS sources in the CBI frequency range with either the Bonn 100-m telescope or with the Green Bank Telescope.

In 2002 the CBI was upgraded with the DASI-style achromatic polarizers, and the antennas were moved into a close-packed configuration that concentrates the observations into the multipole range $500 < l < 2000$. Since October 2002 it has been observing in this configuration.

21.6.2 *The DASI CMB Spectrum and the Detection of Polarized CMB*

The first DASI results were released in April 2001 (Halverson et al. 2002). The DASI spectrum is shown in Figure 21.11. We see here a clear detection of the first and second acoustic peaks, and a hint of the third peak. These DASI results were the first in-

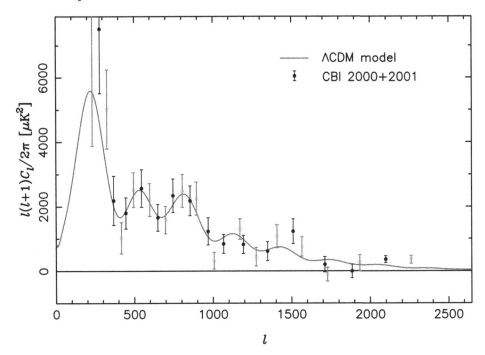

Fig. 21.10. The CBI spectrum shown here was obtained with a combination of deep observations of single fields and mosaic observations of $2° \times 4°$ fields. The results for two alternate binning schemes are shown by the solid error bars and closed circles, and the dashed error bars and open circles. These are not independent points, but there is more information in the data than can be displayed in a single spectrum, and this also shows that there is nothing peculiar about the particular binning scheme being used. The spectrum shows four important features for the first time: (1) anisotropy on multipoles above $l \approx 1000$, corresponding to mass scales typical of clusters of galaxies; (2) the damping tail due to photon viscosity in the scattering region and the finite thickness of this region; (3) the second, third, and fourth acoustic peaks, with strong hints of the first and fifth peaks; and (4) an excess of power in the multipole range $2000 < l < 3500$, ascribed to the Sunyaev-Zel'dovich effect in either galaxy clusters or the ionized regions produced by Population III stars (see text).

terferometry results to show the first acoustic peak, which had first been seen clearly in the TOCO observations and then with very high signal-to-noise ratio in the BOOMERanG results released in April 2000 (de Bernardis et al. 2000).

In the second year of operation (2001) the DASI was outfitted with achromatic polarizers and carried out polarization observations that yielded a detection of polarized emission attributed to the CMB (Kovac et al. 2002; Leitch et al. 2002b). Thus, another key pillar of the standard cosmological model has been confirmed by an interferometer. The DASI polarization results are discussed in the article by Zaldarriaga (2004) in this volume.

21.6.3 *The VSA CMB Spectrum*

The VSA was deployed in 2000 with a compact array (Scott et al. 2003; Watson et al. 2003), and the following year the horns were replaced with larger ones and the array was extended to longer baselines, thereby permitting the VSA to make observations of the

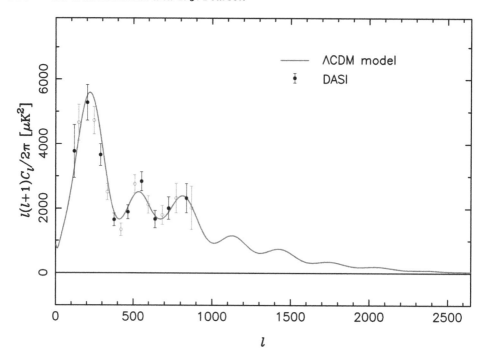

Fig. 21.11. The DASI spectrum. The symbols are as for the caption to Figure 21.10.

CMB spectrum up to multipoles $l \approx 1400$. The VSA CMB spectrum from the combination of observations with the original and extended arrays is shown in Figure 21.12 (Grainge et al. 2003; Scott et al. 2003). Here we can see clearly the first, second and third acoustic peaks.

21.7   Cosmological Results

The CBI, DASI, and VSA CMB spectra are shown in Figure 21.13 (Halverson et al. 2002; Grainge et al. 2003; Pearson et al. 2003; Readhead et al. 2004). Here, in order to avoid confusion, we have plotted only a single binning of the data from each instrument, but the reader should be aware that there is considerably more information than can be displayed in this one-dimensional plot. There is excellent agreement between these three independent experiments in the regions of overlap. Some of the key cosmological results from these three interferometers are given in Table 21.2. The "weak prior" assumptions used here are that the age of the Universe is greater than 10 billion years, $45 \, \text{km s}^{-1} \, \text{Mpc}^{-1} < H_0 < 90 \, \text{km s}^{-1} \, \text{Mpc}^{-1}$, and $\Omega_{\text{matter}} > 0.1$. It can be seen that all three interferometers provide strong evidence for a flat geometry ($\Omega_k \approx 0$) and a scale-invariant spectrum ($n_s \approx 1$), both of which are expected for inflationary universes. Given the strong evidence here for a flat geometry, we have also determined key parameters for the case of a flat Universe, with the results shown in Table 21.3, where it can again be seen that the results are in excellent agreement, and it is clear that the matter density is only about one-third of the critical density, with the predominant matter component being nonbaryonic. Thus, the remainder of the energy density must be made up of something other than matter, and is here attributed to the

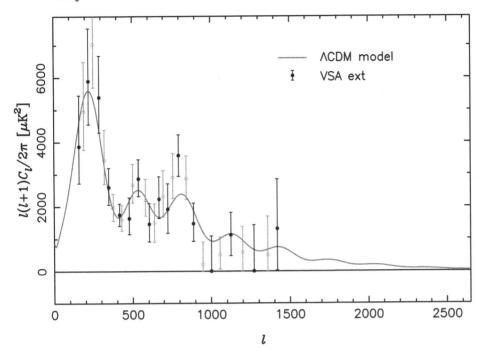

Fig. 21.12. The VSA spectrum shown above was obtained by combining results from the first and second sets of horns (see text). The symbols are as for the caption to Figure 21.10.

cosmological constant. It is interesting to compare the CBI results, which depend almost entirely on high multipoles ($l > 600$), and are therefore independent of the first peak, with the DASI and VSA results, which depend entirely on low multipoles, especially on the first peak. We see that the cosmological parameter values derived from both the DASI and the VSA agree very well with those derived from the CBI. This provides strong justification for the assumption of a featureless, primordial density fluctuation spectrum.

This is important because it is conceivable that the structures observed in these angular spectra are not simply acoustic peaks, but contain significant features that are either present in the primordial spectrum or are produced by other, as yet undetermined, physical effects. If this were the case there could be significant errors in the derived cosmological parameters. However, it is almost inconceivable that we would derive the *same* incorrect cosmological parameters from two different parts of the angular spectrum. This conclusion has been strengthened by the recent ACBAR results (Goldstein et al. 2004), which cover much of the same range of multipoles as the CBI.

We have carried out an analysis of the CBI+*WMAP* observations in order to determine the additional constraints placed on cosmological parameters by the extension of the spectrum beyond the range covered by *WMAP* (Bennett et al. 2003; Readhead et al. 2004). The results are shown in Table 21.4. Comparison of this table with Tables 21.2 and 21.3 shows clearly the much tighter cosmological constraints from *WMAP*. It can be seen here, however, that the extension of the spectral coverage to high l by the CBI yields a significant reduction in many of the parameter uncertainties.

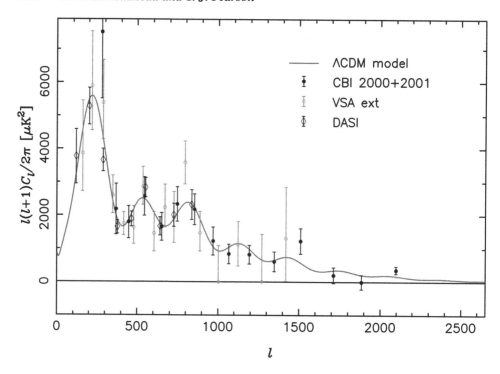

Fig. 21.13. The combined CBI, DASI, and VSA spectra show the excellent agreement between these three experiments in the region of overlap. The combination of these three
spectra provides stringent constraints on key cosmological parameters and shows a number
of the key spectral features predicted by theory: a large first acoustic peak followed by successively smaller harmonics of this peak, with a strong damping tail. The excess detected
by the CBI at $l \approx 2500$ was not predicted and must be confirmed (see text).

21.7.1   *The Spectrum above $l \approx 2000$*

There is excellent agreement between all of the CMB observations in the regions
of overlap, and these are well fitted by a ΛCDM model. At higher multipoles, however,
the CBI results are not consistent with the predicted levels of the anisotropies, but show an
excess that is significant at the $\sim 3\sigma$ level (Mason et al. 2003; Bond et al. 2004).

In the 2001 season the CBI concentrated on mosaic observations in order to increase the
resolution in l, whereas deep observations of a small number of fields give the best sensitivity
to observations at high l. Nevertheless, the 2001 observations do contribute to this region.
The observations from 2000 alone yield a value of $508^{+116}_{-149}\,\mu\text{K}^2$, which is an excess of 3.1
σ over the predicted ΛCDM model in the multipole range $2000 < l < 3500$; whereas the
observations from 2000 and 2001 combined yield $360^{+102}_{-95}\,\mu\text{K}^2$, which is significantly lower
in absolute terms. The resulting significance relative to the ΛCDM model has therefore
dropped to $2.3\,\sigma$. The tantalizing detection of the excess persists, therefore, albeit at lower
significance, in the combined 2000+2001 CBI data set (see Readhead et al. 2004), and more
observations are required to confirm or disprove this excess.

If real, the excess at high multipoles detected by the CBI could have significant consequences, as has been spelled out in a number of papers (e.g., Oh, Cooray, & Kamionkowski

Table 21.2. *Cosmological Constraints from CMB Interferometers*[†]

Parameter	CBI	DASI	VSA	CBI+DASI+VSA
Ω_k	$-0.10^{+0.007}_{-0.008}$	$-0.06^{+0.05}_{-0.05}$	$-0.07^{+0.08}_{-0.08}$	$-0.06^{+0.04}_{-0.04}$
n_s	$1.03^{+0.10}_{-0.07}$	$1.03^{+0.12}_{-0.07}$	$1.07^{+0.10}_{-0.08}$	$0.98^{+0.08}_{-0.05}$
$\Omega_{cdm} h^2$	$0.11^{+0.05}_{-0.03}$	$0.12^{+0.04}_{-0.03}$	$0.17^{+0.06}_{-0.05}$	$0.12^{+0.03}_{-0.02}$
$\Omega_b h^2$	$0.040^{+0.013}_{-0.014}$	$0.023^{+0.004}_{-0.004}$	$0.034^{+0.007}_{-0.007}$	$0.024^{0.004}_{-0.003}$
Ω_Λ	$0.62^{+0.15}_{-0.23}$	$0.58^{+0.16}_{-0.22}$	$0.43^{+0.22}_{-0.25}$	$0.58^{+0.14}_{-0.17}$

[†] Weak prior.

2003; Bond et al. 2004). Both of these papers ascribe the excess to the Sunyaev-Zel'dovich effect. The paper by Bond et al. attributes the excess to the Sunyaev-Zel'dovich effect in clusters of galaxies, while that of Oh et al. attributes it to the Sunyaev-Zel'dovich effect in hot gas resulting from supernova explosions in the first generation of stars (Population III). The *WMAP* results (Bennett et al. 2003) have shown that reionization started early, at around a redshift of 20, so it may be that both the CBI and *WMAP* results provide evidence for Population III stars.

21.8 Conclusions

The results from the three CMB interferometers show excellent agreement and provide compelling evidence for an approximately flat Universe with an approximately scale-invariant spectrum. The derived baryonic matter density is consistent with big bang nucle-

Table 21.3. *Cosmological Constraints from CMB Interferometers*[†]

Parameter	CBI	DASI	VSA	CBI+DASI+VSA
n_s	$0.99^{+0.08}_{-0.06}$	$1.00^{+0.08}_{-0.06}$	$1.05^{+0.11}_{-0.07}$	$0.96^{+0.05}_{-0.04}$
$\Omega_{cdm} h^2$	$0.15^{+0.04}_{-0.04}$	$0.14^{+0.03}_{-0.03}$	$0.19^{+0.05}_{-0.05}$	$0.13^{+0.03}_{-0.02}$
$\Omega_b h^2$	$0.028^{+0.010}_{-0.009}$	$0.022^{+0.004}_{-0.003}$	$0.031^{+0.006}_{-0.006}$	$0.023^{+0.003}_{-0.003}$
Ω_Λ	$0.57^{+0.18}_{-0.29}$	$0.59^{+0.16}_{-0.26}$	$0.46^{+0.22}_{-0.28}$	$0.66^{+0.11}_{-0.20}$

[†] Flat + weak priors.

Table 21.4. *Cosmological Constraints from WMAP and from WMAP+CBI*

Parameter	WMAP	WMAP+CBI
Ω_k	$-0.063^{+0.050}_{-0.028}$	$-0.071^{+0.064}_{-0.023}$
n_s	$0.975^{+0.032}_{-0.020}$	$0.962^{+0.022}_{-0.013}$
$\Omega_{cdm}\,h^2$	$0.125^{+0.015}_{-0.0092}$	$0.120^{+0.0072}_{-0.0092}$
$\Omega_b\,h^2$	$0.0234^{+0.0012}_{-0.0008}$	$0.0231^{+0.0010}_{-0.0005}$
Ω_Λ	$0.437^{+0.243}_{-0.075}$	$0.446^{+0.289}_{-0.059}$

osynthesis calculations, and the total matter content is only ∼40% of the critical density required for a flat Universe. Therefore, it is clear that the major fraction of the energy density of the Universe is provided by something other than matter. This is generally assumed to be a nonzero cosmological constant.

The high-*l* excess detected by the CBI, if real, is the most interesting result of the CMB interferometry experimental results. It is unexpected, and would likely be due to secondary anisotropy. Both the suggested explanations, that it is due to the Sunyaev-Zel'dovich effect in clusters of galaxies (Bond et al. 2004) or in the aftermath of Population III stars (Oh et al. 2003), would provide an important new window on the processes of structure formation.

References

Bennett, C. L., et al. 2003, ApJS, 148, 1
Bond, J. R., et al. 2004, ApJ, submitted (astro-ph/0205386)
Condon, J. J., Cotton, W. D., Greisen, E. W., Yin, Q. F., Perley, R. A., Taylor, G. B., & Broderick, J. J. 1998, AJ, 115, 1693
Conway, R. G., & Kronberg, P. P. 1969, MNRAS, 142, 11
de Bernardis, P., et al. 2000, Nature, 404, 955
Goldstein, J. H., et al. 2004, ApJ, submitted (astro-ph/0212517)
Grainge, K., et al. 2003, MNRAS, 341, L23
Halverson, N., et al. 2002, ApJ, 568, 38
Hanany, S., et al. 2000, ApJ, 545, L5
Hobson, M. P., Lasenby, A. N., & Jones, M. 1995, MNRAS, 275, 863
Hobson, M. P., & Maisinger, K. 2002, MNRAS, 334, 569
Hu, W. 2003, Annals of Physics, 303, 203
Kovac, J., Leitch, E. M., Pryke, C., Carlstrom, J. E., Halverson, N. W. & Holzapfel, W. L. 2002, Nature, 420, 720
Lee, A., et al. 2001, ApJ, 561, L1
Leitch, E. M., et al. 2002a, ApJ, 568, 28
——. 2002b, Nature, 420, 763
Mason, B. S., et al. 2003, ApJ, 591, 540
Mason, B. S., Leitch, E. M., Myers, S. T., Cartwright, J. K., & Readhead, A. C. S. 1999, AJ, 118, 2908
Mather, J. C., Fixsen, D. J., Shafer, R. A., Mosier, C., & Wilkinson, D. T. 1999, ApJ, 512, 511
Myers, S. T., et al. 2003, ApJ, 591, 575
Netterfield, B., et al. 2002, ApJ, 571, 604

Oh, S. P., Cooray, A., & Kamionkowski, M. 2003, MNRAS, 342, L20

Padin, S., et al. 2001, ApJ, 549, L1

——. 2002, PASP, 114, 83

Page, L., et al. 2003, ApJS, 148, 39

Pearson, T. J., et al. 2003, ApJ, 591, 556

Pospieszalski, M. W., et al. 1995, IEEE MTT-S International Symp. Digest, 95.3, 1121

Pospieszalski, M. W., Nguyen, L.D., Lui, T., Thompson, M. A., & Delaney, M. J. 1994, IEEE MTT-S
    International Symp. Digest, 94.3, 1345

Readhead, A. C. S., et al. 2004, in preparation

Ryle, M. 1952, Proc. Roy. Soc., 211, 351

Scott, P. F., et al. 2003, MNRAS, 341, 1076

Scott, P. F., Saunders, R., Pooley, G., O'Sullivan, C., Lasenby, A. N., Jones, M., Hobson, M. P., Duffet-Smith, P.
    J., & Baker, J. 1996, ApJ, 461, L1

Thompson, A. R., Moran, J. M. & Swenson G. W. 2001, Interferometry and Synthesis in Radio Astronomy, 2nd
    ed. (New York: Wiley)

Watson, R. A., et al. 2003, MNRAS, 341, 1057

White, M., Carlstrom, J. E., Dragovan, M., & Holzapfel, W. L. 1999, ApJ, 514, 12

Zaldarriaga, M. 2001, Phys. Rev. D, 64, 103001

——. 2004, in Carnegie Observatories Astrophysics Series, Vol. 2: Measuring and Modeling the Universe, ed. W.
    L. Freedman (Cambridge: Cambridge Univ. Press), in press

22

Conference summary: observational cosmology

SANDRA M. FABER

UCO/Lick Observatory, University of California, Santa Cruz

Abstract

I review the major concordances and controversies of the meeting concerning observations. Cosmologists are nearing agreement on the global cosmological parameters of the Universe. A few parameters still have large error bars, notably Ω_m, but most of these windows will close soon with upcoming accurate observations of the cosmic microwave background. The major era of chasing the major cosmological parameters is now closing. Understanding galaxy formation in all its messy detail will continue to occupy cosmologists for some time to come. Development of highly accurate standard candles, such as inspiraling black holes, holds out hope for understanding Λ, whether it is truly constant or evolving. This is the major prospective contribution from cosmology to fundamental physics in the next generation. I close with a discussion of anthropic cosmology. Anthropic reasoning has been shown to be correct at least three times in the history of cosmology. Applying it now leads us to take seriously the prospect of other universes, a notion that should be pursued seriously by theoreticians.

22.1   Introduction

If the birth of cosmology can be reckoned from the first in-depth survey of the Universe beyond our Galaxy, that birth dates to the beginning of the 20th century, when the first comprehensive photographic survey of external galaxies was compiled by Keeler at Lick. Dated thus, cosmology has reached its centennial milestone, coinciding providentially with the centennial milestone of the Carnegie Institution of Washington. At this conference we therefore celebrate not one but two birthdays, and it is appropriate to appraise the progress in our field after a century of effort.

The efficiency of the human scientific endeavor is such that one hundred years of work in a field typically yield enormous fruit, and cosmology is no exception. Given the remoteness of the object of our study in both distance and time, what we have learned in 100 years is nothing short of stupendous. The basic outlines of the subject are now known, the fundamental questions have been framed, and many have even been answered. We now have a theory for the global geometry of spacetime, we have a general description of the time evolution of the system from inflation to now, and we are closing in on the values of the dozen or so global parameters that are needed to characterize the observable Universe. We have mapped the heavens and seen structure, and we have a rough theory for why that structure exists and how it developed. In short, the state of our field can accurately be characterized as "mature."

In this ocean of knowledge, three islands of ignorance still stand out. The first is the messy baryonic physics of galaxy formation, which is challenging but not in any sense fundamental. It will keep us busy for another couple of decades but will ultimately yield to the heavy artillery of future high-powered computing. There is no fundamentally new physics to be discovered here.

The second island is the domain of the question, what came "before" the Big Bang, and indeed, whether we can presently frame any meaningful question at all along these lines. The "cause" of the Big Bang takes us to the very depths of epistemology and natural philosophy. I shall have a few words to say about this at the end of this talk, but the basic conclusion is that, as scientists, we do not appear ready to grapple meaningfully with this mother of all questions at the present time.

The third island, by virtue of its in-between status, is the most exciting for exploration now. This is the realm of the *dark energy*. Dark energy is not so far removed from established principles of particle physics that it is unapproachable by known methods, and it is also accessible to measurement via observations of the distant Universe. Its pursuit promises to enlarge our basic concept of "thing" by providing an entirely new thing to consider in addition to radiation and matter, and in the process it may cause us to examine what other new "things" can in principle exist. It also compels us to question the future evolution of our Universe and whether new physics might appear in the future that will govern its ultimate fate (I am here invoking an analogy to early inflation, whose physics was the mother of our present-day Universe). We are very lucky, I think, to have discovered this third land, as cosmology would otherwise be confined today to the challenging yet basically trivial pursuit of galaxy formation, versus the fundamentally impenetrable and inapproachable realm of the Big Bang. Dark energy provides something meaty yet doable in between and promises to keep cosmology intellectually vibrant for some time.

I was amused at this conference at how some of us evidently feel discomfited by recent success. We are closing in not only on broad concepts but also on specific details, yet some attendees felt (almost reflexively?) compelled to express doubt. Consider, for example, Virginia Trimble, who said pessimistically, "I wouldn't bet the window on Ω_m to be any smaller than 0.1 to 0.5 with 95% confidence." Or Malcolm Longair, who anguished whether "we might be extrapolating too far...," "there might be surprises..," or "fundamental misconceptions" in the present picture. Or Bernard Sadoulet, who wondered whether the discovery of Λ might even signal "the first hint of a failure of gravity." I began to wonder myself whether WIMPs had actually been detected at this conference for the first time!

But MACHOS were also here in quantity. Mike Turner exuded typical confidence with statements like: "There are no current controversies," and "Now that we're closing in." Andreas Albrecht advised us to "stop whining [over small discrepancies] and get to work!" I personally identify more with the MACHOs—the data are fitting so well together (by virtue of nonzero Λ) that it seems highly unlikely that the whole edifice could be seriously undermined at this point. Musings about "missing something big" strike me as wishful thinking that the game of chasing cosmological parameters might continue indefinitely, when in fact a major era is closing.

Tactics in this end game are shifting, reflecting the new situation. When a science is young, a single observer with a simple apparatus can make a paradigm-shifting discovery. Thus, we cosmologists used to go to the telescope with the goal of measuring a single number, like H_0 or q_0. No more. With the possible exception of H_0, hardly any experiments measure a single number any more, but rather increasingly complicated functions of the form $F(h, \Omega_{tot}, \Omega_m, \Omega_b, \Omega_\Lambda, \sigma_8, b,...)$. Where these surfaces intersect is the sweet spot of the Universe. Each experiment in this new era has only a piece of the truth, and this has led to a new era of coordination, cooperation, and intercomparison, but the old satisfaction of pioneering on alone is somewhat diminished.

22.2   Current Status and Prognostications

That said, recent progress has been outstanding, and I list here five major experiments that all of us would agree have provided big breakthroughs.

- *The H_0 Key Project* value 72 ± 8 (Freedman, Jensen) came just in time to anchor measurements of other cosmic parameters from cosmic microwave background (CMB) experiments and redshift surveys.
- *CMB surveys* (BOOMERANG, MAXIMA, DASI, CBI; Wright) confirmed the flat universe predicted by inflation, and the rest of the CMB spectrum tightly constrained other combinations of important cosmic parameters.
- *Type Ia supernovae* (Filippenko, Perlmutter) directly detected $\Omega_\Lambda \approx 0.7$, exactly the right value to fill the empty gap between $\Omega_{tot} = 1$ from the CMB and $\Omega_m \approx 0.3$ from large-scale structure (LSS) and dynamics. Cosmologists might have wrangled endlessly about the reality of an Ω_Λ deduced arithmetically from Ω_{tot} and Ω_m, but the direct detection from Type Ia's has virtually put that controversy to rest.
- *Big Bang nucleosynthesis* (Tytler, Steigman) yielded the first model-independent estimate of Ω_b, from primordial deuterium.
- *Massive nearby redshift surveys* (2dF, Colless; SDSS, Bernardi) tightened the noose on large-scale structure, constrained σ_8 and bias, b, and demonstrated the power of huge samples for studying structure formation. Bringing the fire-power of these surveys to bear on galaxy formation will be the logical next step.

From this cascade of recent data, certain fundamental conclusions have emerged:

- The Universe is essentially flat: $\Omega_{tot} = 1$, as predicted by inflation.
- Dark matter exists and is mainly nonbaryonic. Taking $\Omega_b = 0.04h_{70}^{-2}$ and $\Omega_m = 0.15$–0.35 (see below), we find Ω_{tot}/Ω_b in the range 4–8, with the value of 1 totally excluded.
- Ω_Λ is nonzero and in the range 0.65–0.85. This is implied indirectly by the previous two bullets but also comes directly, as I have noted, from Type Ia supernovae.
- Primordial fluctuations created in an early epoch of inflation planted the seeds for later formation of galaxies and large-scale structure via gravitational instability.

In contrast, discussion at this conference shows that several important parameters still remain insecure:

(1) The total window on Ω_m quoted at this meeting was 0.15–0.35, which is still quite large. Neta Bahcall marshaled an impressive case for 0.2, but many of the methods she quoted are fairly model dependent. Values of Ω_m from peculiar motions, large-scale structure, and the CMB in contrast tended to hover near or above 0.3 (Dekel, Colless, Bernardi). The best strategy is to wait, as the most accurate value will come from the next round of CMB measurements (*WMAP*).

(2) Similar controversy swirled around σ_8, which is expected because most methods measure the product $\sigma_8 \Omega_m^{0.5-0.6}$, so when one goes up the other goes down. Quoted values of σ_8 ranged from 0.7 to 1.0, with Bahcall favoring the higher range and Colless the lower. New data from the Cosmic Background Imager (Readhead) on short wavelengths also favor $\sigma_8 \approx 1$. In retrospect, the σ_8 scale is a bad choice for normalizing the power spectrum because it sits between Lyα and large-scale structure surveys, so that quoted values from these data sets are often entangled with the assumed spectral index, n. The final value of σ_8 will require a joint analysis of the entire body of fluctuation data (CMB, LSS, Lyα), which again awaits the next round of CMB data.

(3) We still do not understand the density run of matter in galaxies, and uncertainty here generated a modest rear-guard attack on the Key Project value of H_0. Broadly, dark matter halo models (e.g., Navarro, Frenk, & White 1997) predict a fairly shallow dark matter mass profile on the scale of galaxy-galaxy lensing, $\rho \propto r^{-\eta}$, where η is about 2. This is consistent with gravitational lensing time delays only if $H_0 = 50$ (Kochanek). For power-law mass models, $H_0 \propto (\eta - 1)$; hence, to permit the Key Project value of $H_0 = 72$ would require $\eta \approx 2.5$, which is already that of the light alone; adding dark matter would only make this shallower, reducing H_0 below 72. At smaller radii the picture is also confused. Here, adiabatic contraction of dark matter by baryonic infall should produce rather steep central slopes and densities (Blumenthal et al. 1987), yet a variety of information points to the contrary, including inner galaxy rotation curves (e.g., de Blok & Bosma 2002; Swaters et al. 2003), lensing around central cluster galaxies (Ellis), rotation curve amplitudes (Alam, Bullock, & Weinberg 2002), the amplitude of central bars (Weiner, poster paper), and the scarcity of dwarf galaxies (Dekel).

The disconnect between theory and observations for $\rho(r)$ in halos has led some to posit warm or self-interacting dark matter (reviewed by Silk). Alternatively, halo models may need revision to include stirring of dark matter cusps via bars or mergers (Katz), strong feedback and consequent ejection of baryons (Silk, Dekel), photoionization to retard baryonic infall (Katz), and stripping (Katz). A third possibility, not much discussed here, is a tilt to the fluctuation spectrum, n, to reduce the strength of fluctuations on galactic scales, though that might imply late structure formation and consequently galaxy formation too late to match the number counts of Lyman-break galaxies and early QSOs (Somerville, Primack, & Faber 2001). My guess is that the $\rho(r)$ discrepancy will be resolved by a combination of several small things, such as small changes to the lensing time delays (which are accurate to only $\sim 15\%$; Chartas, Treu), improvements to galaxy collapse models (as noted above), a small reduction in small-scale fluctuation amplitudes via spectral tilt, and a small downward adjustment of H_0 to the mid-60's, still within the current Key Project window. Should future data confirm the high value of $H_0 = 72$ from the Key Project, this will exacerbate the tension between model slopes and density data, motivating further scrutiny of the collapse models.

(4) The above discussion makes clear that H_0 is not quite as rock-solid as we would wish, and so it is worth reviewing the values of H_0 discussed at this conference. In addition to the revised Key Project value of 72 ± 8 (Freedman, Jensen), four other independent values were mentioned. The combination of the CMB plus nearby large-scale structure is consistent with $H_0 \approx 70$ (Colless). The CMB itself implies that $H_0 \approx 70$ if the Universe is perfectly flat. The maser galaxy NGC 4258 implies an upward correction of 10% to the Key Project value to 80 or so (Freedman), but the photometry is not *HST*'s best and should be redone. Finally, there are gravitational lenses, which, if η really is 2.5, imply a downward correction to $H_0 = 50$ (Kochanek). Several speakers lamented the shaky distance to the Large Magellanic Cloud, which underpins the Key Project value. Some also mentioned a downward correction as large as 10% to the Key Project value owing to our location in a local low-density bubble. My own guess, $H_0 = 65$, is a compromise reached by stretching the error bars of all methods to achieve overlap. Finally, if cosmologists

are willing simply to *posit* that the Universe is exactly flat, CMB data will determine the value of H_0 to exquisite accuracy, and that might be in fact be our final adopted value.

(5) David Tytler's review of Ω_b from Big Bang deuterium was illuminating yet a bit scary. The number of QSOs that has been analyzed is only five, of which only two are really firm. Steigman stressed that there are residual discrepancies for both He and Li obtained when using the Ω_b measured from deuterium. Given the scatter and small numbers, I am concerned that the data presently do not merit an error bar as small as the 10% that is often quoted. Again, this question will be resolved soon when the second peak in the CMB spectrum is accurately measured.

The following table is a playful attempt to predict how some of these issues will be resolved in future:

2006 $H_0 = 65 \pm 3$ is determined by using a combination of the CMB and LSS.

2008 Dark matter is detected in the lab as the lightest supersymmetric particle; somebody (Bernard Sadoulet?) wins the Nobel prize.

2008 *GAIA* claims to measure $H_0 = 75 \pm 3$ by providing accurate parallaxes for a few Cepheids but cosmologists take no notice; the distance-ladder approach is by then deemed dead.

2010 Moore's law rescues "gastrophysics." The cusp/concentration problem with baryons and dark matter in galaxies is resolved via a bit of everything: fewer baryons falling into galaxies, stirring of central cusps by bars and mergers, expulsion of gas (especially in dwarfs) by winds, etc.

2012 *LISA* measures $w=-1.01\pm0.01$, $w' = 0.03\pm0.05$ using the newly developed "ultimate" standard candle, inspiraling central black holes (Phinney). Confronted finally by what appears to be a genuine cosmological constant, string theorists retire in droves.

2015 The fourth and final reanalysis of *Planck* data teases out the tensor B modes (Cooray, Zaldarriaga)...but barely at the $2\text{-}\sigma$ level of significance. Undeterred, NASA trumpets to Congress, "Now we've *really* seen the face of God."

2030 Neal Katz retires, having finally removed all but one free parameter from galaxy formation models. The one remaining remaining free parameter is, of course, the star formation rate.

The fifth entry concerning inspiraling black holes is particularly interesting. If these objects can be made into few-percent standard candles, as Sterl Phinney outlined, they hold the prospect for actually measuring w', and thus testing whether Λ is indeed constant or evolving.

22.3 Anthropic Cosmology

I close this summary by coming out of the closet as a believer in anthropic cosmology. Much has been written on this topic, both pro and con, some of it needlessly complicated, and I'd like to take this opportunity to state my view, since I envision anthropic reasoning to play a greater role in cosmological discussions in the future. Anthropic arguments are a kind of data, though not of the conventional kind.

To illustrate, consider the plight of an intelligent cosmologist back in the geocentric Aristotelian era. In the then-current world model, the radius of the Earth would have looked like a fundamental constant of the Universe, analogous to the radius of curvature of today's Universe. Thinking like a modern cosmologist, our Greek would have felt compelled to understand where this value, the radius of the Earth, came from. There would have been two choices: search for an argument rooted in physical principles that shows why this radius could have one and only one conceivable value, namely that observed. Or, posit the existence of an infinite (or at least very large) ensemble of spherical bodies with a spectrum of radii, and then argue that the actual Earth must occupy a narrow window within that spectrum that is picked out *a posteriori* by the existence of life as we know it on this Earth (the precise location of the radius within this window would be random.) The first route might be termed the "physics approach." I argue that the situation of the Greek is not fundamentally different than the situation today with modern cosmology, and that the physics route, if followed 2000 years ago, would have been demonstrably sterile. The correct approach for the Greek would in fact be the anthropic approach.

So far, this is familiar, but I now make three points that have not been stressed previously. First, cosmology has been faced with many explanatory challenges in the past similar to the ones we now face, and in all cases the correct approach was in fact anthropic. Classic examples are the nature of the Sun, the nature of the Solar System, and the nature of the Galaxy. We now see that there is nothing unique about any of these objects and that their properties were in fact picked out of a much larger ensemble by anthropic requirements. Thus, anthropic arguments are not speculative, they have been proven correct several times over.

The second point is that an anthropic argument makes sense only if you accept the actual existence of the larger ensemble—*even if you have not yet observed it.* The larger ensemble is not merely hypothetical, it is really out there! This is the real power of an anthropic argument—*not* to explain a particular cosmological parameter but to alert us in to the existence of a much larger (though possibly still unseen) ensemble. Again, this is borne out by history. In the above cases, astronomers actually went on to discover semi-infinite ensembles of suns and galaxies, and we are now (2000 years later) well on our way to verifying a semi-infinite ensemble of planets. In all cases, the singular object that we were once so fixated on was revealed to be merely a member of a much larger sample.

The third point is that, once the anthropic approach is invoked, the thrust of the science shifts from trying to explain the properties of the singular object to understanding the *properties of the ensemble.* Not, why is our Sun the way it is, but rather, what does the total ensemble of star-like objects look like? What are the physics of stars in general, how do they form, what is the range of properties spanned by the class as a whole, and how does one characterize a given star within the class? Again, this has been the standard route of astronomical inquiry, which has borne tremendous fruit.

In our present situation as cosmologists, I argue that taking an anthropic approach to explaining the properties of our Universe is a rational strategy based on historical success. This presumably includes not only the dozen or so macroscopic cosmological parameters cataloged by Freedman in her review (this conference), but also all the messy 40-odd parameters of the particle physics Standard Model (however they might ultimately be revised). Together these parameters are simply the suite of numbers needed to characterize our Universe. Anthropic reasoning then leads us to accept, or at least hypothesize, the existence of the larger ensemble, namely *other* universes. It would be reassuring as we do so to have at least a glimmer of a physical process that might have created that ensemble. Fortunately, the speculative fringes of particle physics and cosmology have come up with a few ideas revolving around chaotic inflation and multiple dimensions. Unfortunately, there does not seem to be the prospect of directly observing these other universes any time soon. Nevertheless, that should not stop us from trying to deduce their properties, any more than the Greek cosmologist should have been deterred from thinking about other planets. His attempts might have failed for lack of proper tools and understanding at the time, but they could not in any sense have been termed "unscientific."

My last point is a slight stepping back from a purely anthropic approach and is occasioned by the fact that at least one cosmological parameter, Ω_{tot}, *has* been explained by appeal to fundamental physical principles. The fundamental process generating Ω_{tot} is inflation, which in its simplest version produces a very flat universe. Inflation, in turn, is believed to occur under a wide range of conditions, and is thus regarded by particle physicists as at least generic, if not ubiquitous. The case of Ω_{tot} thus warns us that perhaps not all current cosmological parameters are equal, in the sense that some may one day be derived from others via generic physical arguments. To be more precise, I am reasoning here that the capacity of our Universe to support inflation (and hence have $\Omega_{tot} \approx 1$) was determined by a combination of other, more fundamental cosmological parameters. If this reasoning is true, it suggests that a continuing job of cosmologists will be to discover hidden relationships among the current parameters based on physical principles, and in the process to reduce the total number of independent parameters. This activity will resemble conventional physics, i.e., explaining one thing by another. However, the history of astronomy and cosmology strongly suggests that even diligent application of this method will leave some number of cosmological parameters ultimately unaccounted for. The leftover ones will be the truly fundamental ones and thus, I argue, will have to be accounted for anthropically. The frontier will shift to discovering the properties of the larger ensemble spanned by these remaining fundamental parameters. Not, why is our Universe the way it is, but rather, what does the greater ensemble of universes, the *Meta-universe*, look like?

In closing, I would be the first to admit that the anthropic explanation for our Universe does not provide the ultimate explanation for the Universe that we are looking for. It does not address the origin of the Meta-universe, and hence postpones by only one step the reckoning of ultimate causes—why something exists rather than nothing, and indeed whether that question has meaning. Perhaps the Theory of Everything, if it lives up to its name, will explain the existence and properties of the Meta-universe from first principles. Until then, I find it thrilling to contemplate the possibility of multiple universes in parallel with our own, and count that awareness a major step forward in the growing cosmological sophistication of our species.

There is a convention in astronomy (not strictly observed but still useful) that capitalizes the name of an object when it refers to our local example, as distinct from an object in the larger ensemble. Thus, "earth" becomes "Earth," "sun becomes "Sun," and "galaxy" becomes the "Galaxy." In keeping with this tradition, I suggest that anthropic cosmologists might capitalize the word "Universe" when referring to our own, to express explicitly our willingness to contemplate the existence of the larger ensemble.

References

Alam, S. M. K., Bullock, J. S., & Weinberg, D. H. 2002, ApJ, 572, 34
Blumenthal, G. R., Faber, S. M., Flores, R., & Primack, J. R. 1986, ApJ, 301, 27
de Blok, W. J. G., & Bosma, A., 2002, A&A, 385, 816
Navarro, J. F., Frenk. C. S., & White, S. D .M. 1997, ApJ, 490, 493
Somerville, R. S., Primack, J. R., & Faber, S. M. 2001, MNRAS, 320, 504
Swaters, R. A., Madore, B. F., van den Bosch, F. C., & Balcells, M. 2003, ApJ, 583, 732

23

Measuring and modeling the Universe: a theoretical perspective

ROGER D. BLANDFORD
Kavli Institute for Particle Astrophysics and Cosmology, Stanford University

Abstract

A brief summary of a symposium discussing the measuring and modeling of the Universe is presented from a theoretical perspective. This whole field has been transformed by the cumulative success of microwave background observations and the large-scale galaxy surveys. After a long history of confusion and contradiction, the standard model of the Universe that has emerged is internally consistent with prior theoretical expectation, save for the now dominant feature of dark energy that accounts for over 70 percent of the energy density of the Universe. The dark matter that is required to account for local dynamics within galaxies and clusters is found to be present on the cosmological scale and has a quite plausible identity as a stable supersymmetric particle that has been invoked within particle physics. The prospects for significantly improving the precision of these measurements over the next decade and for identifying a dark matter particle are good. Considerable theoretical challenges also lie ahead. An immediate task is to reexamine the physics that fixes the two numbers that most strongly characterize our Universe, the entropy per baryon and the amplitude of the fluctuation spectrum. Another goal is to probe the very early Universe and to provide further corroboration of the principle that the Universe must have undergone a period of inflationary expansion. The hardest challenge of all is to account for the apparent acceleration of the Universe and its interpretation in terms of a dark energy density some 120 orders of magnitude smaller than the natural, Planckian value.

23.1   Introduction

Scepticism is the chastity of the intellect and it is shameful to surrender it too soon or to the first comer. G. Santayana

It is both an honor and a pleasure to participate in the celebration of the Centennial of the Carnegie Observatories. Collectively, the world of astronomy, and more recently that of physics, owes an immense debt to the Carnegie Institution and its distinguished astronomical faculty for discovering an impressive fraction of what we now take for granted about the Universe and its major constituents. There is also a personal debt that I would like to acknowledge. As a child growing up in the north Birmingham (England) suburb of Erdington, I obtained my very first books about astronomy and physics from the local public library which had been endowed with a grant from Andrew Carnegie.*

The title of this symposium highlights two important and, I believe, distinguishable fea-

---

* Malcolm Longair tells a similar story in his introductory talk.

tures of the recent explosive progress in cosmology. The first is that we are now making pretty secure and reproducible kinematic measurements. This is an activity that is sometimes called "cosmography" and it has long been associated with the Carnegie Observatories, especially through the work of Allan Sandage and Wendy Freedman. Cosmography is basically surveying and is not especially concerned with the dynamics of the Universe and the social histories of its inhabitants. The second feature is encapsulated in the somewhat pejorative term "modeling." This is what we do when we are ignorant of the physics, either because it is too complicated to consider from first principles, as is true of galaxy evolution, or because it is too mysterious, which is true of the phenomenon described by the term "dark energy" or "quintessence." That we are now measuring with such confidence is the giant step forward; that we are once again reduced to modeling much of the Universe is a prudent step backward. Put another way, when we are confident of the physics as, happens with the microwave background, we can construct elaborate deductive models and then test them observationally. When we do not have this confidence, it is better to restrict our science to simple inferences from direct observations.†

When I was asked to give this talk, I was suggested that I try to take a critical and provocative look at recent developments from the perspective of one who has worked more on the fringes of cosmology than at its core. This means that I will inevitably raise some concerns that are apparently not shared by other theorists. In order to assist myself in this unpopular task, I will appeal to a few apposite quotations from one of the more pungent academic commentators, George Santayana.

23.2   The Standard Model

The world is a perpetual caricature of itself; at every moment it is the mockery and the contradiction of what it is pretending to be. G. Santayana

There is some similarity between the development of the standard model of particle physics and what is becoming the standard model of cosmology, which, for the purposes of this talk, I will define as ΛCDM and a scale-invariant scalar fluctuation spectrum. Both represent a stage of organization and measurement allowing predictions to be made and tested, although cosmology is trailing particle physics by roughly 20 years. Both standard models are refutable and are characterized by a small set of parameters that can be measured. In particle physics, none of the experimental challenges have stood up although there is evidence of and hopes for new physics beyond the standard model, specifically neutrino mass and supersymmetry, respectively. In cosmology, there is impressive progress toward a concordance, and none of the serious observational challenges so far have been upheld. In addition, in both fields there are fundamental questions, for example the origin of mass in particle physics and the value of the scale of the density fluctuation spectrum in cosmology, that cannot be understood at present and may turn out to be metaphysical. Indeed most of the exciting theoretical (as opposed to phenomenological) research today in cosmology is addressing issues beyond the standard model, and the goals of the next generation of accelerators and space missions are to discover and characterize supersymmetry and see direct evidence for tensor modes from inflation.

As brought out in the excellent historical talks by Longair, Christiansen, and Trimble, it

---

† It is an ironic commentary on recent progress that we are more confident in the application of physics to the Universe at large before reionization than after.

is surprising how deep are the standard model's intellectual roots. The cosmological constant derives from the first relativistic cosmology, the Einstein static universe, and figured prominently in world models introduced over the following decade, most notably the de Sitter model. McCrea in 1951 understood the point that a substance with $P = -\rho$ would have similar dynamical consequences. This offered a more physically palatable cosmology than Hoyle's C-field. Bondi, in the first edition of his cosmology textbook, published in 1952, presents the analytical solution for the scale factor of the ΛCDM world model:

$$a(t) = \left(\frac{\Omega_0}{1-\Omega_0}\right)^{1/3} \sinh^{2/3}\left[\frac{3(1-\Omega_0)H_0 t}{2}\right], \tag{23.1}$$

where H_0 is the contemporary Hubble constant and Ω_0 is the current matter density parameter. He also understood the distinction between hot and cold matter.

However, all of this was comfortably theoretical. It took the discovery of the microwave background in 1965 to give a firm observational basis to physical cosmology. Within a few years, the theory of helium formation, recombination, the transfer of microwave background radiation, and gravitational instability were all given a secure foundation by Zel'dovich, Sunyaev, Peebles, and many others. This foundation that has been built upon over the last 35 years by successive generations of theorists is a remarkable, collective achievement, to provide the robust framework that can be used to interpret the equally impressive observations of the microwave background fluctuation and galaxy power spectra, most recently from *WMAP* and SDSS. The result, which was celebrated at this meeting and in papers submitted soon afterwards, is that, out of the suite of possible cosmological models, the one that describes the contemporary Universe is as generally anticipated (flat, dark matter-dominated, older than its constituents, endowed with scale-free perturbations). However, it is accelerating, which was not expected. As discussed here by Filippenko, Perlmutter, and Phillips, prior credit for this discovery should go to the Type Ia supernova teams who demonstrated that these explosions were far more standardized than it seemed reasonable to hope and that, at the 10 percent level, their luminosity distances increased slower with redshift than in an Einstein-de Sitter model. (I must remark that X-ray astronomers also deserve credit for anticipating this discovery with their steadfast claim that, granted the theory of nucleosynthesis, the ratio of the dark to the luminous mass in rich clusters of galaxies is much smaller than needed to close the Universe. The intercept with the flat Universe solution agrees with the supernova intercept to within the errors.)

I will come to the implications of this discovery later, but, for the moment I would like to consider a more limited approach to discussing the observations. This is based upon admitting that we do not understand the dynamics at all and so should just work in terms of kinematics. As we are claiming that the Universe was decelerating in the past and is now accelerating, it seems natural to introduce the third derivative of $a(t)$. Engineers often call this quantity the jerk, and following cosmological tradition it is natural to introduce a jerk parameter

$$j(t) = \frac{a''' a^2}{a'^3} \quad . \tag{23.2}$$

One reason why this is a nice thing to do, at least for observational cosmologists, is that the standard model—ΛCDM–has the curious property that $j = 1$.

We can use this as a basis for a purely kinematic description of the expansion of the Universe. The standard, dynamical first-order differential equation for the scale factor is

$$a' = H_0 [\Sigma_i \Omega_i(a)]^{1/2}, \tag{23.3}$$

where the functions $\Omega_i(a)$ measure the various contributions to the density. In the standard model, these are $\Omega_0 a^{-1}$ and $(1 - \Omega_0)a^2$. If we adopt the convention that $a(t_0) = 1$ as a boundary condition, then we measure the Hubble constant to give a scale to size and age of the Universe and fit a single parameter Ω_0 to give the shape of the expansion curve. (Note that the deceleration parameter is $q_0 = 1.5\Omega_0 - 1$.) If we go beyond the standard model and allow the dark energy to have a nontrivial equation state, then we must fit additional parameters.

In the kinematic approach, we replace this equation with the third-order differential equation

$$a^2 a''' = j a'^3, \tag{23.4}$$

with a constant parameter j. Now, we need three boundary conditions and choose $a(0) = 0$, $a(t_0) = 1$ and $a'(t_0) = H_0$. It is straightforward to express t_0 in terms of q_0 and to regard q_0 as a parameter. We can then compute a three-parameter family of world models $a(t; H_0, q_0, j)$, similar to the three-parameter family $a(t; H_0, \Omega_0, w)$ used in the currently standard formalism.

Now, most astronomical measurements produce values for the comoving distance

$$d \equiv a d_L \equiv d_A/a = \int_t^{t_0} dt'/a \tag{23.5}$$

as a function of $a \equiv 1/(1+z)$. We can then use the measured values of $d(a)$, for example from a sample of Type Ia supernovae (Filippenko, Perlmutter), to fit for j. Just as it is possible to introduce a second parameter w' to characterize the dark energy, it is possible to add additional parameters such as j' to the kinematic description and to test the hypothesis that $j = 1, j' = 0$. The family of world models described by this parametrization seems no less well-motivated than that described by the $w - w'$ formalism.

23.3   Standard Parameters

Science is nothing but developed perception, interpreted intent, common sense, rounded out and minutely articulated. G. Santayana

Arguably the most important recent deduction that has been made from microwave background observations is that the geometry of the Universe is spatially flat to an accuracy of about 2 percent. Not only will this make life easier for future generations of astronomy graduate students but it provides an affirmation of a principle that had been expected, although not universally, as a consequence of the theory of inflation, which offers a mechanism for achieving this flatness. The way it comes about is more complicated than generally appreciated, as it depends upon knowing the size of the measuring rod at the last scattering surface (which comes from the theory of acoustic oscillations) as well as the angular diameter distance of this surface. Its measurement is likely to get much more accurate, especially after the *Planck* mission, and I suppose that there is a strong prediction that the spatial curvature of the Universe, insofar as it can be isolated from the physics of recombination, should remain unmeasurable.

As I am not an angel, let me tread fearlessly into a Carnegie speciality, the measurement of the Hubble constant. All the measurements reported here are consistent with the

WMAP value of $h = 0.72$ (with 4 to 7 percent errors, depending upon whether or not non-*WMAP* information is included) (Page). However, the full quoted errors from some of the non-*WMAP* methods standing alone is still disappointingly large (Filippenko, Perlmutter, Phillips, Hamuy, Jensen, Reese). The problem is that all of these approaches are subject to systematic error that is as difficult to bound as to reduce. It is possible that theorists will be able to contribute more to this measurement through making better Cepheid models and providing comprehensive models of rich clusters of galaxies that address all of the puzzles that they now present when viewed throughout the electromagnetic spectrum.

Gravitational lens determinations are subject to similar problems (Kochanek). All approaches rely on measuring time delays, and the accuracy of these measurements in the best-studied of these systems is no longer limiting. The major problem is with the modeling. This is one of the few techniques that still has a chance of being competitive with the microwave background determination. Here, I think there are two approaches. The first relies on *a priori* models of the lensing galaxies. These are generally pretty simple. There has been much progress in understanding the sensitivity of the result to the details, and understanding the radial profile of the mass distribution is crucial. (Measuring the galaxy velocity dispersion and its radial variation can help a lot.) The image flux ratios are proving to be interesting as they can be diagnostic of the amount of substructure in the lens potential. These small-scale clumps are generally expected on the basis of numerical simulations, but they are under-represented in visible light. The best way to quantify the modeling uncertainty is to study enough lensing systems in detail, and the large CLASS and SDSS surveys (Bernardi, Colless) are proving invaluable.

The second approach to the modeling of lenses is nonparametric. Basically, when one resolve extended sources, whose unresolved nuclei vary, one can plot out a pattern of multiply imaged isophotes around the Einstein ring. As surface brightness is conserved, these isophotes map onto each other, and so the only freedom is to slide along the isophotes. However, it turns out that if, as is true for a single lens, the deflection field is the gradient of a scalar potential, then this freedom is restricted and subject to certain conditions, and it is possible to predict the ratio of time delays up to a constant inversely proportional to h. Perfect data should furnish a perfect answer. Of course, in the real world the method is compromised by finite pixelation, telescope aberration, reddening, difficulty in distinguishing the source from the image, and so on, and it is difficult to determine how useful this approach will be in the foreseeable future, using a combination of VLBI and ACS imaging on sources like B1608+656. However, it does come with enough internal consistency checks that it will be possible to assess this using relatively few examples.

The major source of systematic error will ultimately be the so-called mass-sheet degeneracy. This relies on the observation that a large sheet of matter in front of the source acts like a converging lens, which will appear to shrink the Universe. This mass can be contributed by matter along the line of sight, for example in an intervening, rich cluster of galaxies, just as much as by matter in the lens plane. This is very similar to the old uncertainty that existed when we did not know which world model to use, which afflicts all determinations when the uncertainty associated with the inhomogeneity of the Universe becomes important. It can be removed either statistically or by tracing the mass distribution out to large radius,

The *WMAP*-inferred age of the Universe, 13.7 Gyr, is closely linked to the Hubble constant, and it is a minor coincidence that it is so close to being its reciprocal. This is another pleasing consistency check, as it is now concordant with the best direct chronometers, a far cry from the days when the Einstein-de Sitter model held sway when the age was 30 percent smaller. As this is now more reliable than the direct measurements, it will be very useful in understanding the history of our Galaxy and its oldest stars.

The mean baryonic density was first supplied by the theory of nucleosynthesis when combined with the measurement of intergalactic deuterium (Steigman), as well as consistent determinations of the mean density of other light elements. This approach has received remarkable vindication from the microwave background fluctuation analysis, which is now the method of choice and gives $\Omega_b = 0.044$. It is also consistent with the direct measurement of the column density of intergalactic gas in quasar absorption spectra when the Universe was ~ 4 Gyr old, provided one trusts the estimate of the intergalactic ultraviolet radiation density at this epoch, which is necessary to fix the ratio of ionized to neutral hydrogen. There was no appreciable amount of nonbaryonic dark matter. We can now turn the argument on its head and accept the microwave background determination and very simple classical physics to infer that nuclear physics cross sections were the same when the Universe was roughly a minute old as they are today.

The *WMAP* determination of the total matter density is close to $\Omega_0 = 0.27$, with and without priors supplied by other approaches. This is quite consistent with the X-ray measurements in the most relaxed clusters of galaxies (Silk). However, it is not consistent with local determinations of the intergalactic hydrogen column density using quasar absorption lines. We can see no more than a quarter of now. This is not a mystery because numerical simulations show that the gas outside clusters collides and will shock to temperatures $\sim 10^6$ K, such that the sound speed is comparable with the virial velocities (~ 200 km s^{-1}) present in the nonlinearly clustered dark matter distribution. Gas at this temperature is rather hard to detect, being too high for ultraviolet detectors like *FUSE* and too soft for hard X-ray detectors. There are clear indications that it is present, especially from *XMM-Newton* observations, but its detailed study will, sadly, have to wait for the next generation of X-ray detectors like *Constellation-X* and *XEUS*.

There are still some puzzles associated with the behavior of this gas and, although it is not the subject of this conference, I would like to mention, in passing, one feature of the astrophysics that has not been widely appreciated. We know from X-ray and radio observations of supernova remnants that, when astrophysical plasma passes through a high-Mach number, collisionless shock front, cosmic rays are also accelerated. We may not understand the details of how this occurs, but it does seem that a partial pressure of perhaps 0.3 is transmitted in the form of mildly relativistic GeV particles. The slope of the momentum space cosmic ray distribution function power law, $-q$, is typically ~ -4.2, so that the effective specific heat ratio of this phase is $\gamma_{CR} = q/3 \approx 1.4$, significantly less than that of the thermal plasma. Now, intergalactic shock fronts should surely behave the same, and I see no escape from the conclusion that the intergalactic gas has an appreciable cosmic ray component. As the gas expands in the post-shock flow, the dynamical importance of these cosmic rays will diminish. However, when the gas recompresses as it falls into dark matter potential wells that are eventually associated with galaxies, groups, and clusters, it will compress to a density that is ultimately greater than that with which it started. Furthermore, the gas will start to cool, thereby increasing the importance of the cosmic rays that have no where to escape. This provides a different approach to understanding the characteristic of intergalactic gas that has been called an entropy floor.

23.4   Cold Dark Matter

For an idea to be fashionable is ominous, since it must afterward be always old-fashioned. G. Santayana

The idea of dark matter can be recognized in a Newtonian context in papers going back to the nineteenth century. However, it is Zwicky who is generally credited with making the quantitative case for its presence through measuring the mass-to-light ratio of the galaxies in the Coma cluster (Silk). Nowadays, we distinguish baryonic from nonbaryonic dark matter.

The astronomical options for dark matter are now mostly ruled out, leaving particle physics as the most likely place to turn (Silk). The most popular candidate is now surely an irreducible supersymmetric particle—perhaps, though not necessarily a neutralino. The mass of this particle is generally argued on physical and astrophysical grounds to lie in the range 30–1000 times the mass of the proton. In the solar neighborhood, these particles are expected to be roughly 3–10 cm apart on average. This is timely as the supersymmetric sector will be accessible to experiment using the Large Hadron Collider. Supersymmetric dark matter is also already accessible to terrestrial detectors that hope to measure the recoils associated with scattering. Furthermore, direct detection of dark matter particles is possible by measuring the recoil of atomic nuclei (Sadoulet). A somewhat longer shot is that annihilations of these particles will be seen by the *GLAST* gamma ray observatory due to be launched in 2006. This will be sensitive to gamma rays with GeV energies and will be looking for narrow lines from clusters of galaxies.

Although rather overshadowed by the evidence for dark energy in recent years, it is surely the quest to understand dark matter and the exciting prospect of being able to study the detailed properties of a new class of elementary particles that one must be more optimistic about. Convincing evidence for the direct detection of cosmological dark matter particles could come any day, while the accelerator position should become clear within five years. For astronomers, the challenge is to try to find new ways to probe its particulate nature that go beyond studying its overall distribution.

An alternative possibility for the dark matter is the axion, which was invoked theoretically in an extension to the standard model of particle physics to solve the strong CP problem. Axions are sought experimentally using resonant microwave cavities. Here the parameter space that must be searched is even larger, but upper limits are already being reported.

23.5   Large-scale Structure

Chaos is a name for any order that produces confusion in our minds. G. Santayana

It is remarkable that we can connect the ancient and the modern Universe in a rational and consistent fashion using an essentially Newtonian theory for the growth of perturbations as the Universe ages from four hundred thousand years to fourteen billion years. This is a remarkable consistency check and an affirmation of our tacit assumption that the evolution of the Universe between recombination and the present day followed the laws of relativistic dynamics. Small-scale perturbations actually grow according to the laws of Newtonian dynamics. However, perturbations entering the horizon at early times need relativity for their prescription. This overlap between cosmic microwave background and galaxy survey power spectra is possible over comoving length scales of $\sim 30 - 600$ Mpc. The total fluctuation spectrum is impressively consistent with the Harrison-Zel'dovich scale-invariant form at the

5 percent level over 4 decades in wavenumber if one accepts the assumptions that go into including Lyman α measurements of the fluctuation spectra.

One of the worries that I have about this procedure, though, is that the bias—the relationship between the clustering of the luminous galaxies and the underlying dark matter—could still be a free parameter and used to adjust the two spectra. The very fact that elliptical galaxies and spiral galaxies cluster differently and have different luminosity functions assures us that the considerations that led to the introduction of bias in the first place cannot easily be dismissed.

For this reason, I see the future of "precision" large-scale structure investigations as lying with weak lensing studies, which study the mass directly rather than the light. Remarkable advances have been made using this technique, and it looks ready for application to enormous data sets garnered from special purpose telescopes in space and on the ground, the former for high spatial frequencies, the later for low frequencies.

23.6   Galaxies

There is no cure for birth and death save to enjoy the interval. G. Santayana

For the purposes of this meeting, galaxies are mostly elementary particles (and hosts of supernovae). However, they do have their own peculiarities. Ellipticals are more gregarious than spirals, and these differences will have to be included in more detailed comparisons of the microwave background fluctuation spectra with the galaxy correlation function.

There are two puzzles that have come from the otherwise extremely well-interpreted microwave background observations. They may be linked. The first of these is that the optical depth to Thomson scattering from the cross correlation of the polarization and the density spectrum is larger than expected, implying that reionization occurred when the Universe was a hundred million years young (Page). The other is that the high-angular frequency perturbations are stronger than predicted. This may be due to integrated Sunyaev-Zel'dovich fluctuations along the line of sight, but, if so, the fluctuation level must be significantly larger than measured by other means ($\sigma_8 \approx 1$). A common explanation of both of these results is that we are seeing the consequences of the formation of the first, clustered superstars (Readhead). Future observations will surely clarify this matter.

Looking at the other end of the galaxy life cycle, it is natural to ask if there is a significant population of dead galaxies that only formed high-mass stars and burned out. The evidence from gravitational lensing is negative, but this might be worth considering more carefully.

23.7   Inflation

Fanaticism consists in redoubling your efforts when you have forgotten your aim. G. Santayana

The microwave background observations are widely taken as a corroboration of the principle of the theory of inflation. This is because the Universe is flat and, as noted above, because a scale-invariant spectrum is measured. However, these observations do not furnish a proof of inflation. I would argue that the strongest reasons, and they are very strong, for believing that the very early Universe passed through a phase of exponential expansion remain theoretical. Even the prediction of flatness was not universal. When confronted with the evidence that $\Omega_0 \approx 0.3$ and presented with the choice of a negatively curved, dark matter-dominated Universe and an exquisitely tuned dark energy-dominated Universe, many theorists took the first option. I do not personally see any compelling reason to regard tiny curvature as a much less likely outcome than a tiny dark energy density when we do not know when inflation occurred, have no accepted theory to describe the inflationary potential and the strength of the perturbations that it leaves us *ab initio*, and no good understanding of the Universe's exit strategy from exponential expansion.

Furthermore, the idea of a scale-invariant spectrum preceded the theory of inflation by several years and is far more general. By itself, it does not seem to be the smoking gun for the theory, even if the theory provides a good mechanism for producing it. This is also true for the tensor spectrum, which may one day be measured using the *B*-modes of the microwave background polarization (Zaldarriaga). A radio map of an extragalactic radio source produces *B*-modes in abundance. There is nothing very special about them, and I suspect that alternative cosmological sources could be found by devil's advocates. (The prospects for direct detection by gravitational radiation observatories seem particularly bleak [Flanagan, Lazzarini].) Not until we have a quantitative understanding of the physics, and this will require some understanding the form of the inflationary potential from first principles, as well as being able to compute the fate of quantum fluctuations, can we say that inflation has been observed. Perhaps the best hope for providing a good "proof" of inflation lies in general quantitative relationships between the fluctuation spectra that exist under restrictive assumptions.

23.8   Second Inflation

Those who cannot remember the past are condemned to repeat it. G. Santayana

Although theoretical physicists dislike this connection, it does seem as though we are entering upon a second epoch of inflation. (Perhaps this has happened more than twice, although there are clear constraints on our past lightcone against this having taken place after nucleosynthesis.) As we look to the future, we can survey a bleak and depressing eschatology of dilution and decay. Alternatively, we can take the optimistic viewpoint that we have got out of this mess once in the past and if we try hard enough, we can do it again!

More seriously, and for reasons that I still do not understand, the theoretical description of dark energy has followed a different path from the description of inflation. Now, the simplest and most natural possibility is that we are dealing with a cosmological constant, which we would interpret nowadays as a form of vacuum energy. The stress energy tensor for dark energy is simply proportional to the metric tensor and is the same in all frames and at all times. Dark energy is a simple fact of nature, like the existence of electrons. (This is not quite as mysterious as one might think when one considers a uniform magnetic field. It, too, has a net stress along the field that is the negative of the energy density. Of course, it is not isotropic.) This is the basis of the ΛCDM model, and there is no good evidence that it is wrong.* Now, as with general relativity itself, which is the simplest and least fussy theory of gravity, we need to provide a framework for checking if the simple view is correct and, if it is not, for providing a framework for relating it to the expansion of the Universe.

Most general descriptions of dark energy, for which most observing proposals are written, treat it as an isotropic fluid endowed with a pressure and a density endowed with an equation of state

---

* The recent detection of the integrated Sachs-Wolfe effect provides a very neat affirmation that it must be approximately correct.

$$P = w\rho \quad . \tag{23.6}$$

(This, surely, is the equation of state, *not* the parameter w, a usage favored by many authors.) What this equation is proposing is a modified Boyle's law

$$PV^{1+w} = \text{negative constant}, \tag{23.7}$$

where $w = -1$ corresponds to the cosmological constant, and we already know that $w < -0.8$. However, if we interpret w as a pure number, there is a problem. If we consider small perturbations $\propto \exp[i(\vec{k} \cdot \vec{x} - \omega t)]$, then the dispersion relation is $\omega^2 = wk^2$ and there will be growth on all scales as long as $w < 0$.

Of course, this is not what was intended. This prescription is supposed to represent the dynamics of a scalar field ϕ with a potential $V(\phi)$. We use Noether's theorem to derive the stress energy tensor from the Lagrangian density, from which the density and pressure can be read off:

$$\rho = \phi'^2/2 + (\nabla\phi)^2/2 + V(\phi) \tag{23.8}$$

and

$$P = \phi'^2/2 - (\nabla\phi)^2/6 - V(\phi) \quad . \tag{23.9}$$

In addition, by setting the divergence of the stress energy tensor to zero, we can obtain the equation of motion for the field:

$$\phi'' + 3H\phi' - \nabla^2\phi + dV/d\phi = 0 \quad . \tag{23.10}$$

What is usually done is to ignore spatial gradients and the second derivative in the equation of motion. In this case, it is possible to relate w to V. This may be permissible on the scale of the horizon. However, the danger of instability remains, and, so far as I can see, the stability of the field and its impact on the growth of matter perturbations are not routinely considered. All of this must be computed directly from the field theory, and it does not seem to me that a fluid model of dark energy is of any more utility than a fluid model of the electromagnetic field has proven to be, despite the fact that it can be written down. Furthermore, the prescribed solutions with fixed w, say, may be quite different from those that could follow from a real field theory.

It turns out that, under reasonable assumptions, a knowledge of $a(t)$ can be used to infer $V(\phi)$ directly. It therefore seems much simpler to bypass the fluid treatment and either concentrate on kinematics, as I discussed above, or to introduce a simple set of parameterized potentials, as was done in nuclear physics. Observations can then be used to fit for these parameters, and theorists should then explore systematically the other consequences of these dynamical models.

There is no guarantee that second inflation is driven by a scalar field, and it is worth recalling that we have not detected any examples of fundamental scalar fields to compare with yet. To make this point, let us suppose that a blind astrophysicist was present during the radiation-dominated era and perceived the mean separation of electrons and protons increasing as the square root of time, and she tried to understand what was going on. She could find a solution to the scalar field equations that fit all the known facts—specifically it has $V(\phi)$ be a decaying exponential—and she would be completely wrong! Are we so sure that the only approach to explaining dark energy has to be cast in the language of a scalar field?

23.9   The String Connection

The cry was for vacant freedom and indeterminate progress. G. Santayana

The deductive approach to cosmology is surely from a fundamental theory of quantum gravity of which the overwhelmingly best-developed candidate is string theory (Schwartz). The challenge used to be understand why the cosmological constant vanishes. It has now become more interesting! Remarkably, there has been recent progress in understanding new classes of string vacua, some of which have quite unexpected classical behavior.

23.10   Anthropic Reasoning

Reason is not come to repeat the universe but to fulfil it. G. Santayana

There has recently been an upsurge in the use of anthropic arguments (Livio) and a serious reaction, particularly from theoretical particle physicists, who see it as dangerously close to religion and therefore a bad thing. I suspect that anthropic arguments do have a religious component but in a rather different sense from that attributed to them by the critics. Like many religions, anthropism provides a measure of solace in hard times and is set aside when things are going well. It is really hard to imagine that an anthropic explanation would be preferred to a rigorous physics calculation or that physicists would stop trying to make such calculations when told that the reason why things were as they were is because we are as we are. Nonetheless, anthropic arguments frequently illustrate beautifully the fundamentals of physics and may have made teleology respectable.

23.11   The Theoretical Challenge

Fashion is something barbarous for it produces innovation without reason and imitation without benefit. G. Santayana

At several times in the history of cosmology, there have been bold steps taken, for example by de Sitter, Friedman, Gamow, Guth, and Linde. Each of these advances was incomplete and had to be supplemented with corrections and amendments. However, they each had the characteristic that they were motivated by deep and fundamental physical questions that were real, and they faced up to these questions in ways that were constrained by what was then observed. Today, cosmology is a much more crowded and competitive field. It is also a more accomplished field, and much contemporary research is technically outstanding and reactive to detailed observational studies, sometimes retrofitting existing models, as in semi-analytical galaxy formation theory, occasionally making clean predictions and interpretations, as in the discussion of microwave background fluctuations.

However, the questions that cosmologists face are as deep as any in the history of the subject, and they are now quite crisply stated. At present it looks like the standard model is correct, and, if we set aside anthropic arguments and adopt physics as we find it locally, we are required to "explain" just two numbers, the entropy per baryon in units of k and the level of the fluctuation amplitude (in round numbers 10^{10} and 10^{-5}). The first of these requires a theory of baryogenesis, the second a theory that requires us to understand the connection to some form of dissipation at very early time. The standard ways of approaching these problems are to posit phase transitions at either the strong or the electroweak scale and to invoke a designer inflaton potential. I have no objections to either of these approaches and certainly have nothing special to offer in their place. However, I wonder if there are more possibilities (like, for example, the brane cosmologies) that promise quite different explanations, and which could impact physics on a scale of interest to laboratory physicists.

As we go beyond the standard model and anticipate even more accurate microwave background and large-scale structure observations, there is the prospect of measuring a slope parameter n different from 1 and detecting the tensor fluctuations. These would enable us to estimate the extent of the fluctuation spectrum and presumably the duration of exponential expansion under the inflationary paradigm. Features may appear in the primordial fluctuation spectrum, which would be incredibly prescriptive of new physics.

I am probably most optimistic about the prospects for studying supersymmetry using data acquired below, on, and above ground. Of course there are no guarantees and nothing is proven yet, but so many independent approaches point in this direction. If this happens, it will be a tremendous shot in the arm for particle physicists who, having developed one standard model, have to start all over again. It is nice to think of cosmologists repaying some of the debt that they owe particle physicists in this enterprise.

However, the biggest challenge of all is to understand why the Universe appears to be accelerating. For reasons to which I have alluded, I personally find the alternatives to pure vacuum energy unappealing. If it turns out that $w = -1$ (or $j = 1$), then we have a third single number to explain, the ratio of the dark energy density to the Planck density, which is 10^{-120} also in round numbers. This is surely an even greater challenge to fundamental thinking.

I started this review with a statement of my charge to be critical and provocative. I think I have largely failed. Perhaps it is the vicarious pleasure in seeing a theoretical (and observational) job well done (at least so far) and the clarity of the questions that have been uncovered in the process. Perhaps it is the realization that we have not reached the end of the road in terms of achievable observations and affordable facilities and that the next decade will surely provide several of the cosmological measurements that we will need. Perhaps I have been swept along by the hyperbolic public statements to which cosmologists are prey. However, the time does seem ripe for a new radical idea that might take our understanding of the Universe to a higher level and remove the need, which we currently have, to model as well as to measure.

Acknowledgements. I am indebted to Wendy Freedman and Luis Ho for their encouragement and patience and many of my theoretical colleagues for their instruction. I acknowledge support under NSF grant AST-0206286.

Credits

The following figures in this volume were reproduced with permission from the original author and publisher.

Figure 3.3: Bennett, C. L., et al. 2003, ApJS, 148, 1, "First-Year Wilkinson Microwave Anisotropy Probe (WMAP) Observations: Preliminary Maps and Basic Results." Reproduced with permission from the American Astronomical Society.

Figure 6.5: Tegmark, M., et al. 2004, ApJ, submitted (astro-ph/0310725), "The 3D Power Spectrum of Galaxies from the SDSS." Reproduced with permission from the American Astronomical Society.

Figure 8.2: Fassnacht, C. D., et al. 2002, ApJ, 581, 823, "A Determination of H_0 with the CLASS Gravitational Lens B1608+656. III. A Significant Improvement in the Precision of the Time Delay Measurements." Reproduced with permission from the American Astronomical Society.

Figure 12.4: Norberg, P., et al. 2002, MNRAS, 332, 827, "The 2dF Galaxy Redshift Survey: The Dependence of Galaxy Clustering on Luminosity and Spectral Type." Reproduced with permission from Blackwell Publishing.

Figure 13.1: Connolly, A. J., et al. 2002, ApJ, 579, 42, "The Angular Correlation Function of Galaxies from Early SDSS Data." Reproduced with permission from the American Astronomical Society.

Figure 13.2: Tegmark, M., et al. 2002, ApJ, 571, 191, "The Angular Power Spectrum of Galaxies from Early SDSS Data." Reproduced with permission from the American Astronomical Society.

Figure 13.4–13.5: Zehavi, I., et al. 2002, ApJ, 571, 172, "Galaxy Clustering in Early SDSS Data." Reproduced with permission from the American Astronomical Society.

Figure 17.3: Riess, A. G. et al. 1998, AJ, 116, 1009, "Observational Evidence from Supernovae for an Accelerating Universe and a Cosmological Constant." Garnavich, P. M., et al. 1998, ApJ, 509, 74, "Supernova Limits on the Cosmic Equation of State." Reproduced with permission from the American Astronomical Society.

Figure 17.5: Riess, A. G. et al. 1998, AJ, 116, 1009, "Observational Evidence from Supernovae for an Accelerating Universe and a Cosmological Constant." Coil, A. L., et al.

Lightning Source UK Ltd.
Milton Keynes UK
UKOW07n1928040516

273551UK00001B/3/P